T. Matsuzawa, M. Tomonaga, M. Tanaka (Eds.)
Cognitive Development in Chimpanzees

T. Matsuzawa, M. Tomonaga, M. Tanaka (Eds.)

Cognitive Development in Chimpanzees

With 237 Figures, Including 26 in Color

Tetsuro Matsuzawa, Ph.D.
Professor, Primate Research Institute, Kyoto University
41 Kanrin, Inuyama, Aichi 484-8506, Japan

Masaki Tomonaga, Ph.D.
Associate Professor, Primate Research Institute, Kyoto University
41 Kanrin, Inuyama, Aichi 484-8506, Japan

Masayuki Tanaka, Ph.D.
Assistant Professor, Primate Research Institute, Kyoto University
41 Kanrin, Inuyama, Aichi 484-8506, Japan

Spine: Ayumu, Ai's son, 1 year, 3 months. Photo by Akihiro Hirata
Front cover:
Upper left: Ayumu stacks blocks in front of Tetsuro Matsuzawa. Photo by Etsuko Nogami
Upper right: Sharing food. Ayumu watches as his mother, Ai, peels an orange. Photo by Etsuko Nogami
Center: Ai embraces her son, Ayumu. Photo by Tokufumi Inagaki (AERA, Asahi Newspaper Co.)
Back cover, from the top:
#1. Two infants in the Bossou community. Photo by Rikako Tonooka-Tomonaga
#2. A chimpanzee family in Bossou: Jire (mother), Jeje (5½-year-old male), and Jimato (2½-month-old male who died in the 2003 epidemic) in January 2003. Photo by Tetsuro Matsuzawa
#3. A juvenile chimpanzee in Bossou eating the fruit of a huge fig tree. Photo by Tetsuro Matsuzawa
#4. Ayumu (*left*, 4 years, 11 months) and Pal (*right*, 4 years, 8 months) on top of the climbing frames, April 9, 2005. Photo by Tomomi Ochiai
#5. In a computer-controlled finger drawing task, Ai learned to trace symbols used for writing (collaborative work by Iversen and Matsuzawa). Photo by Tetsuro Matsuzawa
#6. Pal examines the remnants of macadamia nuts that her mother, Pan, has cracked with stones. Photo by Etsuko Nogami
#7. Ayumu wears a big smile while playing. Photo by Akihiro Hirata (Mainichi Newspaper Co.)

This book was financially supported by the Japan Society for the Promotion of Science (Grant-in-Aid for Publication of Scientific Research result: Grant No. 175332).

Library of Congress Control Number: 2006920239

ISBN-10 4-431-30246-8 Springer-Verlag Tokyo Berlin Heidelberg New York
ISBN-13 978-4-431-30246-9 Springer-Verlag Tokyo Berlin Heidelberg New York

Springer is a part of Springer Science+Business Media
springer.com

Typesetting: SNP Best-set Typesetter Ltd., Hong Kong
Printing and binding: Shinano Inc., Japan

Printed on acid-free paper

Foreword by Jane Goodall

(*Photo*: Jane and Ayumu, November 2002)

This book describes, firstly, the unfolding of an imaginative and brilliantly conceived research project; secondly, the development of the fascinating relationship between the principal collaborators, Tetsuro Matsuzawa and Ai—a human being and a chimpanzee (subsequently a third major player was recruited: Ayumu, Ai's son, born in 2000); and thirdly, the journey of Matsuzawa the laboratory scientist into the African wilderness, where he adapted the lab researchers' methodology to a completely new environment: in the wild it is only the *methods* of data collection that can be controlled; the *movements and behavior* of the subjects cannot.

I first met Matsuzawa in 1986 at a conference "Understanding chimpanzees" in Chicago, where I was most impressed with the paper he presented on the accomplishments of a chimpanzee named Ai. The following year I met Ai herself. When I first saw her she was in her enclosure with other chimpanzees. We made eye contact, and I gave the soft panting grunts that chimpanzees utter when they greet each other. She did not reply. An hour later I was crouched, looking through a small pane of glass, so that I could watch her at her computer. Matsuzawa warned me: "She hates to make a mistake, and especially if a stranger is watching. She will bristle up and charge towards you and hit the glass window. But don't worry — it's bulletproof glass!"

For some time Ai made one correct response after another. Then came the first mistake. Sure enough, Ai glared at me, bristled, and charged towards me. But at the last moment she stopped, her hair sleeked, and, looking at me intently, she pressed her lips to the glass. I kissed in return. She made a total of three mistakes, and each time the same result—a mutual kiss through glass. Matsuzawa told me that had never happened before, and I don't think it has happened since! On my next visit I was allowed to spend an hour sitting with Ai in her room, just the two of us. It was a memorable experience, looking into her wise eyes, grooming her a little, playing for a while.

I first saw Ayumu when he was a small infant. When he was two and a half years old I was allowed to go into his room but was warned that he was quite rough with strangers. As I am familiar with boisterous chimpanzee youngsters in our African sanctuaries (orphans whose mothers have been shot by hunters), I was prepared to have my hair pulled and my hands and arms bitten through my clothing. But it wasn't like that at all. Ayumu was so gentle, grooming me, laughing softly as I tickled him, climbing all over me, and repeatedly kissing me. It reminded me of my first contact with his mother. Afterwards I watched Ayumu working with his computer. Like Ai, he seemed to have great concentration and loved to press the right panels for a small reward.

This book reveals sometimes surprising aspects of chimpanzee intelligence, discovered I believe because Matsuzawa's sensitivity for his primate collaborator allowed him to ask the right questions in the right way. His intuitive and sensitive reading of chimpanzee nature, coupled with the meticulous methodology and quick intellect of the true scientist, made him the perfect partner for Ai—herself a remarkably intelligent chimpanzee. One of the reasons she is able to master very complex tasks is because she has an incredible power of concentration, and because she truly wants to succeed. Indeed, if she gets a bad score after one 20-minute test session, she may actually ask for another session so that she can try to do better.

One incident illustrates how these qualities enhance her success. Ai was working at a difficult task that involved memorizing a sequence of numbers on one computer screen so that she could replicate it on a second screen. A film crew was present, as well as myself. Ai, who is used to peace and quiet whilst she works, began to lose her concentration as first one and then another member of the team moved to get a better view, often bumping into the cage. She began to make mistakes—and after a few minutes her hair began to bristle. I was sure she was about to vent her frustration in a stamping display. Instead, she suddenly stopped working altogether, her hair sleeked, she sat very still and seemed to be staring at a point midway between the two screens. For at least 30 seconds, and maybe longer, she remained motionless. Then she started to work again. For the remainder of the session she paid no further attention to her noisy human observers. It was exactly as though she had decided that she must either give up or else pull herself together and get on with the job! At any rate, whatever that pause meant, she made no further mistakes!

This book represents an important contribution to our overall understanding of the intelligence of our closest living relatives, and the development of a variety of mental and social skills, in some remarkable individuals. And it provides a valuable bridge between research conducted in the controlled conditions of the laboratory and that which takes place in the natural environment. Finally, it will, I believe, help many to understand the role that can be played by empathy between scientist and subject, especially when there is such a close evolutionary relationship between the two.

Preface

Humans and chimpanzees last shared a common ancestor about 5 to 6 million years ago. Single-nucleotide substitutions occur at a mean rate of 1.23% between copies of the human and chimpanzee genome. Chimpanzees are our closest living evolutionary relatives, and therefore it is reasonable to suggest that understanding them may provide us with clues to understanding human nature. The present book focuses on cognition and its development in chimpanzees, providing a window through which the reader can glimpse the cognitive world of another species. Across the 28 chapters, a large variety of topics are covered, including perception, cognition, emotion, memory, face recognition, folk biology, categorization, concept formation, object manipulation, tool manufacture and use, decision making, learning, possible instances of teaching, education by master–apprenticeship, communication, origin of speech, gaze following, mutual gaze, smiling, social referencing, food sharing, neophobia, mother–infant bonds, parental care, self-awareness, intentionality, imitative processes, understanding others' minds, cooperation, deception, altruism, reciprocity, personality, social networks, culture, social learning, and ecological constraints.

The approach advocated in this book clearly sets our work apart from previous studies. The so-called ape-language paradigm has focused on single home-raised subjects: isolated apes were taught human-like skills and were forced to adapt to human ways of communication and the human environment. In contrast, we study chimpanzees in a captive, yet much more natural, setting. Infants are reared by their own mothers, and live within a community comprising three generations of chimpanzees in an enriched outdoor environment. In our "participation observation" method, the infants take part in tests run by human experimenters with the assistance of chimpanzee mothers. These studies in the laboratory have already illuminated developmental changes in chimpanzee cognition over the first 5 years of life through intensive observation and many controlled experiments.

In addition, in parallel with our laboratory work we also conduct field studies in Africa with the aim of learning more about chimpanzees in their natural habitat. Long-term research at Bossou, in Guinea, West Africa, has just entered its fourth decade. Field experiments on tool use have created a logical bridge between field observations and laboratory experiments—drawing together evidence from these distinct sources, we aim to elucidate chimpanzee cognition as a whole.

Each human community has its own unique cultural traditions. According to the Christian calendar, I am writing this in 2006, but in Japan the year is 2666, which, incidentally, is also the year of the dog. The Japanese calendar begins with the first emperor, Jimmu (660 B.C.), whose 125th successor rules today. Westerners use 26 letters of the alphabet (plus some additional variants) to express ideas and communicate with one another in written form. In contrast, Japanese primary schoolchildren have to learn the basic 45 letters of the *kana* script to express information phonetically, as well as 1006 *kanji* (Japanese and Chinese) characters, each of which has a unique shape and meaning. The Japanese use a pair of chopsticks to eat *sashimi*, but it does not follow that all humans use a pair of sticks as a pinching device to eat raw fish. Each culture is unique, different, but at the same time each shares some characteristics common to all human cultures. Chimpanzees also have a rudimentary form of culture. Researchers now recognize that, similarly to the human examples just given, each community of wild chimpanzees has its unique set of behavioral traditions. For example, chimpanzees in Gombe are well known to fish for termites, while those in Bossou do not perform this behavior. Bossou chimpanzees eat termites that emerge from their mound, but seldom use fishing tools to reach those that cannot be seen from the outside. They do, however, have ways of obtaining other hidden foods: Bossou chimpanzees use a pair of stones as hammer and anvil to crack open hard-shelled oil-palm nuts. Gombe chimpanzees do not use stone tools for nut cracking even though both oil-palm trees and stones are readily available. They instead eat only the outer red, soft tissue of oil-palm nuts, leaving behind the kernel concealed inside the hard shell. Each community of chimpanzees has its own unique culture, not only in terms of tool use but also for such matters as greetings and possibly dialects.

In many living organisms, genetic channels are important for the transmission of information from one generation to the next. However, in both humans and chimpanzees, the cognitive-learning channel also plays an essential role in the cross-generational transfer of knowledge, skills, and values. Chimpanzees survive 40 to 50 years, or even longer. Infants continue to suckle for the first 4 to 5 years of life. Details of how, when, what, where, and from whom to whom information is transmitted between generations is a fundamental issue for understanding chimpanzee behavior. Chapters in this book focus on various aspects of chimpanzee cognition and the developmental changes associated with them. The core part of the book is the collaborative product of the three editors, their postdoctoral assistants, graduate students, and Japanese and foreign collaborators, working at the Primate Research Institute of Kyoto University (KUPRI), Japan. The research project we refer to as "Cognitive Development in Chimpanzees" (CDC) began in 2000, with the birth of three chimpanzee infants in the KUPRI community.

This book has a sister volume entitled *Primate Origins of Human Cognition and Behavior*, published in 2001 by Springer. It covers various topics related to comparative cognitive science in chimpanzees and more than 90 species of non-human primates, in an attempt to synthesize fieldwork and laboratory work. I encourage readers to consult the earlier book for a broader perspective and historical background.

The following Internet sites provide useful information about our ongoing projects:

Laboratory work: http://www.pri.kyoto-u.ac.jp/ai/index-E.htm
Fieldwork: http://www.pri.kyoto-u.ac.jp/chimp/index.html
References: http://www.pri.kyoto-u.ac.jp/koudou-shinkei/shikou/index.html
Green Corridor Project: http://www.phytoculture.co.jp/greenbelt-top-E.html
HOPE International Collaboration: http://www.pri.kyoto-u.ac.jp/hope/index.html
SAGA for conservation and welfare: http://www.saga-jp.org/

I deeply thank all the contributors for their efforts to make this book possible. Each of the authors may have his or her own long list of acknowledgments, as all research requires help from many people. The three editors wish especially to express their thanks to Kiyoko Murofushi, who began the chimpanzee project (now known as the "Ai Project") at the Primate Research Institute in 1978. We also thank Toshio Asano, Shozo Kojima, Kisou Kubota, Tetsuya Kojima, Kazuo Fujita, Nobuo Masataka, Shinichi Yoshikubo, Jyunichi Yamamoto, Takao Fushimi, Shoji Itakura, Koji Hikami, Rikako Tonooka-Tomonaga, Noriko Inoue-Nakamura, Kazuhide Hashiya, So Kanazawa, Akira Satoh, Shuji Suzuki, and Akihiro Izumi for their collaboration in running the laboratory. Sumiharu Nagumo developed computer programs to assist our cognitive research. Masuhiro Suzuki tended the plants in the compounds and orchards for the chimpanzees. Special thanks are due to the veterinarians and caretakers: Kiyoaki Matsubayashi, Shunji Gotoh, Satoru Oda, Junzo Inagaki, Juri Suzuki, Yoshikazu Ueno, Norikatsu Miwa, Nobuko Matsubayashi, Masamitu Abe, Yoshiro Kamanaka, Mayumi Morimoto, Chihiro Katsuta-Hashimoto, Akino Kato, Akihisa Kaneko, Kiyonori Kumazaki, Norihiko Maeda, Yoshitaka Fukiura, Shino Yamauchi, Shohei Watanabe, and Takashi Kageyama. We also thank Michiko Sakai for many years of secretarial work. We are grateful to the group of Gifu University veterinary students who looked after the chimpanzees every Sunday for the past 4 years: Akihisa Kaneko, Masato Kobayashi, Atsushi Kodama, Nami Nakayama, Shino Tanaka, Tomoya Kaneko, and Mami Kondo.

The research project CDC 2000 was originally set up by the three editors of this book together with two postdoctoral researchers, Masako Myowa-Yamakoshi (Shiga Prefectural University at present) and Satoshi Hirata (Hayashibara GARI), and help from the following people: Shozo Kojima, Akihiro Izumi, Noriko Inoue-Nakamura, Tomomi Ochiai-Ohira, Chisato Douke-Inoue, Cláudia Sousa, Maura Celli, Dora Biro, Ari Ueno, Yuu Mizuno, Makiko Uchikoshi, Gaku Ohashi, Sanae Okamoto-Barth, Noe Nakashima, Tomoko Imura, Midori Uozumi, Toyomi Matsuno, and Misato Hayashi. Many other collaborators have made unique contributions to the project, and without their efforts and dedication we would not have been able to put together such a comprehensive picture of chimpanzee cognition. In particular, we appreciate the cooperation of our collaborators from other disciplines, such as morphology, neuroscience, and genomics: Osamu Takenaka, Akichika Mikami, Yuzuru Hamada, Takeshi Nishimura, Kaoru Chatani, Hirohisa Hirai, Motoharu Hayashi, Keiko Shimizu, and Miho Inoue-Murayama. The following researchers outside KUPRI are involved in collaborative studies with us: Kazuo Fujita, Shoji Itakura, Kazuhide Hashiya, Nobuyuki Kawai, Hideko Takeshita, Yukuo Konishi, Gentaro Taga, Rieko Takaya, Tatsushi Tachibana, Satoru Ishikawa, Daisuke Kosugi, Yuko Kuwahata, Chizuko Murai, So Kanazawa, Masami Yamaguchi, Naruki Morimura, Sumirena Sekine, Toshiko Uei-Igarashi, Naoki Horimoto, Seiichi Morokuma, Shohei Takeda, Orie Nakagawa, Reiko Oeda, Kikuko Tsutsui, Yusuke Moriguchi, Masako Matsuzawa, and Aya Saitoh. Our collaborators abroad are also numerous: Iver Iversen, Joel Fagot, James Anderson, Dorothy Fragaszy, Kim Bard, Elisabetta Visalberghi, Celine Devos, Amelie Dreiss, Nadege Bacon, Carol Betsch, Dina Stolpen, Pan Jing, Sanha Kim, Mariko Yamaguchi, Patrizia Pozi, and Jessica Crast. The ongoing project continues to be assisted by Sana Inoue, Tomoko Takashima, Etsuko Nogami, Suzuka Hori, Shinya Yamamoto, Yoshiaki Sato, and Laura Martinez.

Fieldwork at Bossou over the past three decades has been carried out in collaboration with Institut de Recherche Environnementale de Bossou (IREB) and Direction Nationale de la Recherche Scientifique et Technique (DNRST) of the Republic of Guinea. We thank Kabine Kante, Fode Soumah, Coullibaly Bakary, Jeremie Koman, Tamba Tagbino, Momoudou Diakite, and Makan Kourouma, as well as other staff from these organizations. We also thank our local assistants and the villagers at Bossou who have given us their support over the many years since the project began in 1976. Special thanks are due to the two oldest guides who have been working with us since the 1970s: Guano Goumy and Tino Zogbila. The Bossou-Nimba project is currently run by the following members of the KUPRI international team: Yukimaru Sugiyama, Tetsuro Matsuzawa, Gen Yamakoshi, Hiroyuki Takemoto, Shiho Fujita, Satoshi Hirata, Gaku Ohashi, Misato Hayashi, Kazunari Ushida, Shinya Yamamoto, Asami Kabasawa, Gentaro Uenishi, Ryutaro Goto, Ryo Hasegawa, Tatyana Humle, Dora Biro, Cláudia Sousa, Kathelijne Koops, Kim Hockings, Nicolas Granier, and Susana Carvalho.

Our research activity both in the wild and in the laboratory has been filmed by ANC (Miho Nakamura and Tamotsu Aso), NHK, and the Mainichi Newspaper Company (Daisuke Yamada and Akihiro Hirata). Chukyo TV (Michiyo Owaki) has also kept a long-term film record, through periodic visits to our laboratory. We thank these people for their collaboration in documenting our research. The color photos reproduced in the book were provided by Akihiro Hirata (1c, 2ab, 3b–f, 4ab), Hiroki Sameshima (3a), Tomomi Ochiai (5ab), Gaku Ohashi (7a–e,g), Tatyana Humle(6b, 7f, 8b), and Tetsuro Matsuzawa (1ab, 6a, 8a).

Our studies in the laboratory and the field were financially supported by grants from MEXT (12002009, 16002001 to Tetsuro Matsuzawa; 11710035, 13610086, 16300084 to Masaki Tomonaga; and 12710037, 15730334 to Masayuki Tanaka) and from JSPS (21COE program for biodiversity A14, and 21COE program for psychological studies D10). Additional support came from the cooperative research program of KUPRI. The project HOPE (an anagram of "Primate Origins of Human Evolution"), a core-to-core program financially supported by JSPS, began on February 1, 2004, and encouraged us to publish our studies in English. HOPE aims to build an international network of collaborations to promote the study of nonhuman primates with the aim of understanding human nature. I thank the following scholars for their leadership and role in setting up the collaboration: Michael Tomasello, Josep Call, Svante Pääbo, Christophe Boesch, Jean-Jacques Hublin, Richard Wrangham, Marc Hauser, Frans de Waal, William McGrew, and Elisabetta Visalberghi. Conservation in Africa has been supported by the following agencies and organizations: Japanese Embassy in Guinea, Japan Fund for Global Environment, U.S. Fish and Wildlife Services, Conservation International (Primate Action Fund), Houston Zoo (USA), CCCC-Japan, SAGA, GRASP-Japan, Phytoculture Control Co., Nippon Keidanren, and Toyota.

In closing, I would like to express our gratitude to our publisher, Springer. I thank Tatiana and Dieter Czeschlik for their friendship and continued support. Thanks are also due to Aiko Hiraguchi, Akemi Tanaka, and Motoko Takeda for their editorial work, and to Susan Kreml and Winston Priest for the copyediting. Because this book is the result of a collaboration among such a large and varied group of people, I hope that it will provide stimulating reading to all those interested in human origins and our evolutionary neighbors.

Tetsuro Matsuzawa

Contents

Part 5 Conceptual Cognition

Part 6 Tools and Culture

List of Authors

ADDESSI, ELSA (CHAPTER 16)
Unit of Cognitive Primatology, Institute of Cognitive Sciences and Technologies, CNR, Via Ulisse Aldrovandi 16/b, 00197 Rome, Italy

ANDERSON, JAMES R. (CHAPTER 15)
Department of Psychology, University of Stirling, Stirling FK9 4LA, Scotland, UK

BIRO, DORA (CHAPTER 28)
Department of Zoology, University of Oxford, South Parks Road, Oxford OX1 3PS, UK

HAMADA, YUZURU (CHAPTER 6)
Primate Research Institute, Kyoto University, 41 Kanrin, Inuyama, Aichi 484-8506, Japan

HAYASAKA, IKUO (CHAPTER 7)
Kumamoto Primate Park, Sanwa Kagaku Kenkyusho, 990 Ohtao, Misumi-cho, Uki, Kumamoto 869-3201, Japan

HAYASHI, MISATO (CHAPTER 24)
Primate Research Institute, Kyoto University, 41 Kanrin, Inuyama, Aichi 484-8506, Japan

HAYASHI, MOTOHARU (CHAPTER 4)
Primate Research Institute, Kyoto University, 41 Kanrin, Inuyama, Aichi 484-8506, Japan

HIBINO, EMI (CHAPTER 7)
Faculty of Applied Biological Sciences, Gifu University, 1-1 Yanagito, Gifu 501-1193, Japan

HIRATA, SATOSHI (CHAPTERS 2, 7, 13, 17)
Great Ape Research Institute, Hayashibara Biochemical Laboratories, 952-2 Nu, Tamano, Okayama 706-0316, Japan

HUMLE, TATYANA (CHAPTER 27)
Department of Psychology, University of Wisconsin, Madison, 250 N. Mills St., Madison, WI 53706, USA

Imura, Tomoko (Chapter 19)
Department of Integrated Psychological Science, Kwansei Gakuin University, 1-1-155 Uegahara, Nishinomiya, Hyogo 662-8501, Japan

Inoue-Murayama, Miho (Chapter 7)
Faculty of Applied Biological Sciences, Gifu University, 1-1 Yanagito, Gifu 501-1193, Japan

Ito, Shin'ichi (Chapter 7)
Faculty of Applied Biological Sciences, Gifu University, 1-1 Yanagito, Gifu 501-1193, Japan

Izumi, Akihiro (Chapter 21)
National Institute of Neuroscience, National Center of Neurology and Psychiatry, 4-1-1 Ogawa-Higashi, Kodaira, Tokyo 187-8502, Japan

Kawai, Nobuyuki (Chapters 3, 20)
Graduate School of Information Science, Nagoya University, Furo-cho, Chikusa-ku, Nagoya 464-8601, Japan

Matsuno, Toyomi (Chapter 20)
Primate Research Institute, Kyoto University, 41 Kanrin, Inuyama, Aichi 484-8506, Japan

Matsuzawa, Tetsuro (Chapters 1, 7, 8, 15, 20, 24, 25, 28)
Primate Research Institute, Kyoto University, 41 Kanrin, Inuyama, Aichi 484-8506, Japan

Morimura, Naruki (Chapter 23)
Great Ape Research Institute, Hayashibara Biomedical Laboratories, 952-2 Nu, Tamano, Okayama 706-0316, Japan

Murai, Chizuko (Chapter 18)
Brain Science Research Center, Tamagawa University Research Institute, 6-1-1 Tamagawa Gakuen, Machida, Tokyo 194-8610, Japan

Murayama, Yuichi (Chapter 7)
National Institute of Animal Health, Tsukuba, Ibaraki 305-0856, Japan

Myowa-Yamakoshi, Masako (Chapters 2, 9, 14)
School of Human Cultures, The University of Shiga Prefecture, 2500 Hassaka-cho, Hikone, Shiga 522-8533, Japan

Nishimura, Takeshi (Chapter 5)
Laboratory of Physical Anthropology, Department of Zoology, Graduate School of Science, Kyoto University, Kitashirakawa Oiwake-cho, Sakyo-ku, Kyoto 606-8502, Japan

Ohashi, Gaku (Chapter 26)
Primate Research Institute, Kyoto University, 41 Kanrin, Inuyama, Aichi 484-8506, Japan

Okamoto-Barth, Sanae (Chapter 10)
Department of Cognitive Neuroscience, Faculty of Psychology, Maastricht University, P.O. Box 616, 6200 MD Maastricht, The Netherlands

Sousa, Cláudia (Chapters 25, 28)
Department of Anthropology, Faculty of Social and Human Sciences, New University of Lisbon, Avenida de Berna, 26–c, 1069-061 Lisbon, Portugal

Takenaka, Osamu (Chapter 7)
Primate Research Institute, Kyoto University, 41 Kanrin, Inuyama, Aichi 484-8506, Japan

Takeshita, Hideko (Chapters 2, 24)
School of Human Cultures, The University of Shiga Prefecture, 2500 Hassaka-cho, Hikone, Shiga 522-8533, Japan

Tanaka, Masayuki (Chapter 22)
Primate Research Institute, Kyoto University, 41 Kanrin, Inuyama, Aichi 484-8506, Japan

Tomonaga, Masaki (Chapters 10, 12, 19)
Primate Research Institute, Kyoto University, 41 Kanrin, Inuyama, Aichi 484-8506, Japan

Udono, Toshifumi (Chapter 6)
Kumamoto Primate Park, Sanwa Kagaku Kenkyusho, 990 Ohtao, Misumi-cho, Uki, Kumamoto 869-3201, Japan

Ueno, Ari (Chapter 11)
Center for Evolutionary and Cognitive Sciences, Graduate School of Arts and Sciences, The University of Tokyo, 3-8-1 Komaba, Meguro-ku, Tokyo 153-8902, Japan

Visalberghi, Elisabetta (Chapter 16)
Unit of Cognitive Primatology, Institute of Cognitive Sciences and Technologies, CNR, Via Ulisse Aldrovandi 16/b, 00197 Rome, Italy

Yagi, Akihiro (Chapter 19)
Department of Integrated Psychological Science, Kwansei Gakuin University, 1-1-155 Uegahara, Nishinomiya, Hyogo 662-8501, Japan

1a. (*Upper left*): Pal (*left*) and Ayumu (*right*) on October 23, 2001

1b. (*Upper right*): Cleo (*left*) and Ayumu (*right*) on October 23, 2001

1c. (*Lower*): The Ayumu-Cleo-Pal trio on February 1, 2002

2a. First calligraphy of the New Year by the chimpanzee Ai. January 1, 2000

2b. Participation observation based on the triadic relationship of the chimpanzee mother, the infant, and the tester. Ai's son, Ayumu, shown here at the age of 7 months, clings to Tetsuro Matsuzawa. December 4, 2000

3a. Ai and her son, Ayumu, with T. Matsuzawa. January 17, 2002

3b. Chloe and her daughter, Cleo, with Masaki Tomonaga. February 20, 2002

3c. Pan and her daughter, Pal, with Masayuki Tanaka. May 16, 2002

3d. Testing a human child, Mimori, with her mother

3e. Pan and Pal, cracking macadamia nuts

3f. Ayumu at the age of 2 years and 3 months already has started honey fishing

4a. Ayumu watching Ai carrying out a matching-to-sample task at the computer. Ai knows the names of 11 colors. As the lexigram (visual symbol) meaning "blue" is presented in the bottom center of the screen, she chooses the corresponding *kanji* (Japanese-Chinese) character

4b. At the age of 10 months Ayumu touches the image of an apple on the touch-sensitive monitor

5a. Outdoor compound for the KUPRI community in April 2005. During the daytime, chimpanzees spend about 80% of their time on the 15-m-high "triple towers"

5b. Pal at the age of 1 year and 8 months on the rope with an old female, Reiko. May 7, 2002

6a. Chimpanzees cracking oil-palm nuts at Bossou, Guinea. The infant closely watches her mother cracking nuts

6b. The chimpanzee Jire (estimated age, 47 years) and her children, Jeje (7-year-old male born in December 1997) and Joya (1-year-old female born in September 2004)

7a. Kai, an old woman (estimated age, 52 years), died in 2003

7b. Tua, the former alpha male of the Bossou community

7c. Poni, an adolescent male, taking a nap on the ground

7d. Yolo, the current alpha male of the Bossou community eating *Myrianthus arboreus*

7e. Yo, the mother of Yolo

7f. Jire with Joya, her 1-year-old daughter

7g. Veve (*left*) and Jimato (*right*), both infants, died in the flu-like epidemic of 2003

8a. Veve died on December 31, 2003, watched over by her mother, Vuavua, and grandmother, Velu. The mother continued to carry Veve's body even after it mummified

8b. The Nimba Mountains and savanna, located to the southeast of Bossou. To connect the Bossou community with groups of chimpanzees in the Nimba Mountains, the Green Corridor project was launched in 1997

Part 1
Introduction to Cognitive Development in Chimpanzees

1
Sociocognitive Development in Chimpanzees: A Synthesis of Laboratory Work and Fieldwork

Tetsuro Matsuzawa

1 Comparative Cognitive Science

The human mind is a product of evolution, as are the body and society. What is human nature? What is uniquely human? Where did we come from? To answer these questions, I have been studying chimpanzees. The chimpanzee is the closest living species to the human. More than 98.7% of the DNA sequence is the same between the two species (The Chimpanzee Sequencing and Analysis Consortium 2005).

The brain is the center of human cognition and behavior. Many disciplines, such as neuroscience, physiology, anatomy, and molecular biology, are trying to understand the structure and function of the brain. Other disciplines, such as computer science, artificial intelligence, and robotics, are trying to create systems functioning similarly to the human brain. Although these efforts are all important, there might remain one important question: namely, why and how the human brain evolved. This question is about the evolutionary basis of the human mind, in other words, its history.

The brain is a soft tissue, so that it cannot remain in fossils as do bones and teeth. Neanderthal man (*Homo neanderthalensis*) lived up to about 30,000 years ago, and Beijing man (*Homo erectus*) lived several hundred thousand years ago. If we could investigate these hominids, that would provide much information about human nature and evolutionary changes. We can know about them to some extent through the fossils and some other nearby remains. However, you cannot easily access the mind from these relicts. You have to look at and compare the living organisms to understand the unique character of the human species (Fig. 1).

The human (*Homo sapiens*) is close to the two living species of the genus *Pan*, chimpanzees (*Pan troglodytes*) and bonobos (*Pan paniscus*). The common characters shared by *Homo* and *Pan* should have come from their ancestors. DNA data and fossil records both suggest that there was a common ancestor of humans and chimpanzees (including bonobos) about 5 to 6 million years ago. The characters not found in chimpanzees but found in humans may be unique human characters acquired in the process of hominization, the evolution of

Primate Research Institute, Kyoto University, 41 Kanrin, Inuyama, Aichi 484-8506, Japan

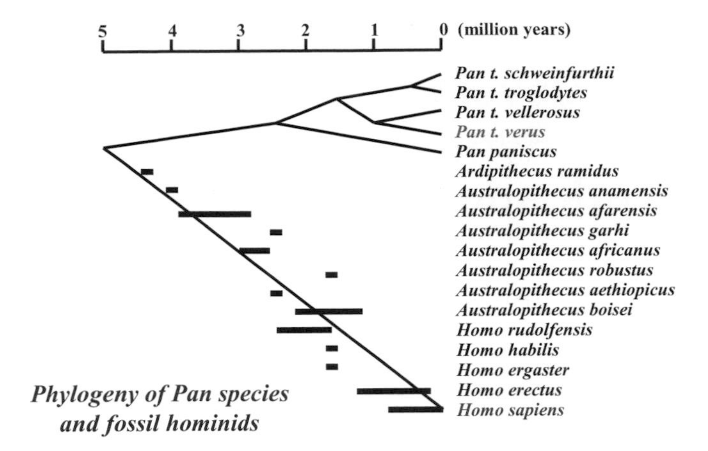

Fig. 1. Phylogeny of chimpanzees and fossil hominids. Data from the DNA analysis of humans and chimpanzees (Morin et al. 1994; Gonder et al. 1997) are juxtaposed on the estimated time range of the hominid fossil record (Asfaw et al. 1999; Aiello et al. 1999)

hominids departing from the human–chimpanzee ancestor. Applying the same logic, the characters found in chimpanzees but not found in humans may be unique chimpanzee characters developed during chimpanzee evolution. Of course, it must be noted that theoretically there is a possibility of not unique acquisition but rather unique loss in the species.

Humans to gibbons (in other words, humans and all the apes) comprise one group of living primates, called hominoids. Hominoid means a human-like creature, in contrast to monkeys. Monkeys have tails but hominoids have no tail at all. The chimpanzee has no tail, and the human also has no tail. Thus, the chimpanzee is close to humans, not close to monkeys. Comparison of humans and chimpanzees shows us the similarities and the differences. Then, we may ask about their evolutionary origins in a broader context. Suppose that humans and chimpanzees share a mental character. Where did it come from? To answer the question, we have to know more about gorillas (*Gorilla* spp.) living in Africa, the genus close to the common ancestor of humans and chimpanzees. Humans, chimpanzees, and gorillas all originated from Africa. The common characteristics among the three African hominoids, *Homo, Pan,* and *Gorilla,* should be compared with orangutans (*Pongo pygmaeus*), the only Asian hominoid, which lives in Borneo and Sumatra. This kind of comparative framework may be extended to gibbons (smaller apes), macaques (Old World monkeys), capuchins (New World monkeys), lemurs (prosimians), and so on.

Comparative cognitive science (Matsuzawa 2001) is a discipline that is looking for the evolutionary origins of human cognition and behavior by comparing living species. To understand human origins, you need to know about chimpanzees, their mind, body, society, and genome. You also need to know humans and chimpanzees among about 200 living species of primates, and primates should be put in the context of about 4,500 living species of mammals (see

Chapter 8 by Matsuzawa). For example, humans and chimpanzees, and most of the primates, have color vision, the trait developed in primates that is lacking in other mammals. We have hands and digits to manipulate objects. That is also a character unique among primates (see Chapter 24 by Hayashi et al.). Originating from the nocturnal and terrestrial common ancestors of mammals, the primate ancestors turned to being diurnal to see color and to being arboreal to grasp branches. On the other hand, we still keep the common trait that has descended from the mammalian ancestor. We raise our offspring by giving milk, which provides the fundamental basis of the mother–infant relationship and is rooted in our mammalian ancestors. You can compare a wide range of species such as mammals, birds, reptiles, amphibians, fishes, insects, bacteria, plants, and so on, and thus you will know more about the origin of the human mind, which is a product of the divergence and convergence of evolutionary processes.

The chimpanzee is a bridge to other living organisms who are sharing the earth with us. By understanding chimpanzees, you can understand the unique position of humans and also their responsibility. The human is just one species among the millions or tens of millions of species living on the earth. This biodiversity is very important, indeed, essential, for all the ecosystems of our earth, and it is threatened by human activity. The study of chimpanzees must come from intrinsic motivation to know this interesting creature, but it should also be a good way to know more about ourselves and to reflect on the relationship of the human species and other creatures.

This book aims to be the first one to describe the cognition and behavior of chimpanzees in sociodevelopmental perspectives. This chapter provides the background information for understanding the other chapters. The book has three unique features.

The first feature is **the developmental and sociocognitive perspective.** We have addressed the popular questions such as these: "Do chimpanzees have abilities such as language capability?" "A 4-year-old human child can do it. How about chimpanzees?" These questions clearly and naively neglect an important aspect: Just like humans, chimpanzees have developmental changes. When you mention the chimpanzees, you should not forget whether you are talking about neonates, infants, juveniles, adults, or elders. Moreover, you should not forget their rearing condition, cultural background, living situation, etc., just as with us. In contrast to previous studies, the chimpanzee infants discussed in this book were reared by their biological mothers. The mother–infant pairs are living in a community of chimpanzees of three generations. An ape that was isolated from its conspecifics and raised by humans cannot be a real ape.

The second concept is **the synthesis of laboratory work and fieldwork.** The book contains chapters that concern the wild chimpanzees in Africa (see Chapter 26 by Ohashi, Chapter 27 by Humle, and Chapter 28 by Biro et al., this volume). I believe that it is essential to combine laboratory work and fieldwork to understand the chimpanzees. That study should not be a simple parallel effort but should be a synthesis of the two approaches. We have introduced a "field experiment," an experimental manipulation in their natural habitat. We also introduced a method of research called "participation observation," an engagement or active enrollment with the captive community of chimpanzees. Putting these ideas together, the unique approach of synthesizing field and laboratory studies tried to understand the cognitive development of chimpanzees in its social context.

The third approach is **the broader perspective of looking at the mind from the physical and biological basis.** Beginning from the fetus, we tried to look at the developmental changes (see Chapter 2 by Takeshita et al. and Chapter 3 by Kawai). Spindle cells provide an important basis for the cognitive development of both humans and chimpanzees (see Chapter 4 by Hayashi). We found neonatal smiling in chimpanzees (see Chapter 8 by Matsuzawa), and we also found the descent of the larynx in chimpanzees (see Chapter 5 by Nishimura); we had previously believed that both these features are uniquely human. The origins of early communication through smiling and early speech correlated with the descent of the larynx should be further explored in the future. This is just one example. We should try to understand the chimpanzee mind from a broader perspective, in which we should relate the cognitive development of chimpanzees to its physical, neural, and genetic basis in addition to social context. Such a broader perspective will provide us with understanding of chimpanzees as a whole.

2 Chimpanzees: Their Life History

This section provides basic information on the life history of chimpanzees (Goodall 1986; Heltne and Marquardt 1989; McGrew et al. 1996; Wrangham et al. 1994). Chimpanzees live in the tropical rain forests and the surrounding savannas in Africa. Chimpanzee populations probably once spanned most of equatorial Africa, including at least 25 countries. They probably numbered more than a million just 100 years ago. Today they occur in 22 countries, and an estimate from the World Conservation Union (IUCN) in 2003 put their numbers in Africa between 172,700 and 299,700 (Kormos et al. 2003). This sudden decrease is linked to various human activities, such as deforestation, poaching, and trading in bush-meat, as well as the transmission of diseases.

Chimpanzees are usually classified into four subspecies (Gonder et al. 1997). From east to west, there are the east chimpanzee (*Pan troglodytes schweinfurthii*), central chimpanzee (*Pan troglodytes troglodytes*), Nigerian chimpanzee (*Pan troglodytes vellurosus*, which has only a small population), and west chimpanzees (*Pan troglodytes verus*). East chimpanzees are relatively smaller, and the face looks pale. West chimpanzees are relatively larger, and the face looks masked because there are black portions surrounding the eyes, just like wearing a mask. However, there is a huge difference of physical appearance among individuals, so that even the experts cannot easily determine the subspecies. Thanks to genetic analysis of blood, hairs, and feces, you can now easily identify the subspecies. For example, there are 349 chimpanzees in 57 zoos and facilities in Japan at present (in 2005). About two-thirds are west chimpanzees, which are the ones born in West Africa and shipped from places along the Guinean Gulf many years ago, and their descendants. One-third of the chimpanzees in Japan are hybrids of the subspecies of all possible combinations of the three major subspecies born in the human environment.

Chimpanzees (*Pan troglodytes*) have a living different species, called bonobo (*Pan paniscus*). Some hundreds of thousands of years ago, there were two species of hominids existing at the same time, *Homo sapiens* and *Homo erectus*. A recent study (Brown et al. 2004) reported that there was a third species, *Homo floresiensis*. This hominid was discovered from the Late Pleistocene of Flores, Indonesia. It was an adult hominin with stature and endocranial volume approximating 1 m and 380 cm^3, respectively, equal to the smallest known australopithecines and the living chimpanzees. Just as in the coexistence of different hominin species, there are two *Pan* species living at present in Africa. However, there is no overlap of their habitats. Bonobos are living only on the left bank of Zaire River, the Republic of Congo, Central Africa, where there are no chimpanzees or gorillas. Bonobos live in dense tropical forests. Once you recognize bonobos, it is easy to discriminate them from chimpanzees: bonobos are slender and their face is flatter. Their voice is tonal, rich in frequency modulation, in contrast to that of chimpanzees. The two species share a common ancestor, about 2 million years ago, and are equally close to humans.

There are currently six major research sites of chimpanzees in which long-lasting observation has been continuing for more than three or four decades, or even longer: Gombe (Goodall 1986) and Mahale (Nishida 1990) in Tanzania, Budongo (Reynolds 2005) and Kibale (Kanyawara community: Wrangham et al. 1996, and Ngogo community: Watts and Mitani 2000) in Uganda, Tai (Boesch and Boesch-Aschermann 2002) in Cote d'Ivoire, and Bossou (Matsuzawa 1994; Sugiyama 2004) in Guinea. All the information on the ecology, their life history, and their social life comes from the collective efforts of people who dedicated their lives to understanding the chimpanzees in their natural habitats.

Chimpanzees are living in the forests and the surrounding areas. They live in primary forests, secondary forests, gallery forests, and even in savanna. The population density is one to three individuals per square kilometer on average. According to a record in Bossou (Sugiyama and Koman 1992), the chimpanzees live on about 200 plant species among about 600 species available in the forests. They eat various parts of the plants such as fruit, leaves, flower, bark, stem, roots, and gum. They also eat insects, eggs, birds, mammals, etc. Bossou chimpanzees live on figs and other fruits. However, the food repertoire has huge differences among the communities. For example, the Bossou chimpanzees seldom eat meat. They seem to be like vegetarians. However, the chimpanzees in Tai, Kibale, and Mahale often engage in hunting for meat. Behavioral assessment based on time sampling revealed that Bossou chimpanzees use tools for getting food during 16% of their feeding time (Yamakoshi 1998), which is a much higher percentage than people had expected. The chimpanzees really need these tools for their survival.

Chimpanzees live in a group called a community or unit group. Each community consists of multiple males and multiple females, about 20 to 100 individuals. There are infants (0 to less than 4 years old), juveniles (4 to less than 8 years old), adolescents (8 to less than 12 years old), adults (12 to less than 36 years old), and olds (elders, more than 36 years old). The age category may be slightly different between the sexes and among the communities. The interbirth interval of chimpanzees is about 5 years, meaning that a mother gives birth to a single baby every 5 years on average. Twins are rare in comparison to humans. Weaning occurs at about 4 years of age. The infants actually suckle the nipples for such a long period. After weaning, the females restart the sexual menstrual cycle (about 35 days per cycle, slightly longer than that of humans). By the way, the menstrual cycle varies among mammals, for example, 21 days for horses and 4 days for mice; this means that the menstrual cycle neither follows the lunar calendar nor depends on species body size. The gestation period is about 235 days in chimpanzees, in contrast to 280 days in humans. The chimpanzee baby is born weighing a little less than 2 kg, whereas a human baby is about 3 kg at birth. During the first 5 years, by which time the younger brother or sister will be born, the infant chimpanzee is fully taken care of by the mother (Fig. 2).

Toward the age of 5, sex differences of behavior become apparent. Female juveniles have a tendency to continue to stay with the mother and take care of the younger siblings. Male juveniles have a tendency to follow males older than themselves. Males often patrol the periphery of the territory and also follow the estrous females, which have a huge pink swelling on their rump.

Chimpanzees have a patrilineal society, which means family lines from grandfather, to father, to the son, and so on (Fig. 3). Patrilineal is pitted against matrilineal, that is, family lines from grandmother, to mother, and the daughter, and so on. Most mammal societies are matrilineal; females remain in the natal com-

Fig. 2. Chimpanzee mother and her infant in the Bossou community. An infant chimpanzee is taken care of by the mother in the first 5 years of life. (Photograph by T. Humle)

Fig. 3. Chimpanzees have a patrilineal society. Three adult males of the Bossou community are participating in grooming in tandem fashion. Grooming is an important aspect of their social life. (Photograph by T. Humle)

munity while males move out. The Japanese monkey is one of the first primate species whose society has been fully analyzed by fieldwork since 1948 (Hirata et al. 2001a; Watanabe 2001). Solitary monkeys move independently from the group of monkeys, called a troop. They are all males, who are in the process of immigrating from the natal troop to the new one. In the case of chimpanzees, males stay in the natal community while females immigrate around the age of puberty. In the 30-year record of the Bossou community, all females born in the community left before giving birth for the first time or after the first birth. There are cases of females not immigrating, or who immigrated but returned, and so on. However, in general, chimpanzees have a strong bias toward a patrilineal society. In contrast, most mammalian species are matrilineal, which means that males immigrate from the natal community to avoid incest. Most primate species are also matrilineal, as are Japanese monkeys. You can find the patrilineal society in chimpanzees and bonobos as well as a few other species, such as the hamadryas baboon (*Papio hamadryas*), red colubus (*Colobus badius*), spider monkeys (*Ateles* spp.), woolly monkey (*Lagothrix lagotricha*), and Muriqui (woolly-spider monkey, *Brachiteles aracnoides*).

An estrous female chimpanzee gets widespread attention from males. All juvenile, adolescent, and adult males can access to her. Chimpanzee males start performing the penile thrust to the female vagina from a very early age, such as 2 years old. The testis of chimpanzees is quite large in comparison to that of evolutionarily close species, that is, humans, gorillas, and orangutans. It is believed that not competition of individual chimpanzees but rather sperm competition is going on.

In places where humans and chimpanzees are coexisting, chimpanzees often eat human crops so that they are called pest animals. I have been studying chimpanzees at Bossou, Guinea, West Africa, since 1986. For 20 years, I have been there once a year for 1 to 3 months. During the past two decades, the perception of chimpanzees by the local people was stable. The chimpanzees at Bossou are a totem for the villagers, which is the reason why the small community of about 20 chimpanzees has continued to survive in the forests next to a village crowded with about 1,500 people. Since 1989, because of the civil war in Liberia, only 10 km away from Bossou, the population of Bossou has doubled, as many refugees entered the area. The conflict between humans and chimpanzees is getting worse there. People need more land for survival. Contacts of chimpanzees with people, including researchers, tourists, and the villagers, became more frequent. For example, the Bossou community lost 5 of its 19 members to a contagious respiratory disease at the end of 2003. Similar stories are taking place all over Africa. Truly intensive efforts are necessary on our part to prevent the extinction of the cultural variation among chimpanzee communities that we have so recently begun to uncover.

In the last part of this section, I want to add some information about the physical aspect of chimpanzees. Chimpanzees are quadripedal, so that height is not an adequate measure. However, when they happen to take the upright posture, their height is about 120 cm in an adult male and about 110 cm in an adult female. Body weight is about 40 to 60 kg in males and 30 to 50 kg in females. The chimpanzee has the A blood type (most common) and the O type, following the ABO type classification in humans. We know that gorillas have type B only, but the reason is still unknown (Saitou et al. 1997).

Humans, the apes, and Old World monkeys share the same dental eruption, 20 deciduous teeth and 32 permanent teeth. The dental formula is expressed as "iicmm" in infants and "IICPPMMM" in adults, respectively. This shared character cannot be extended to the New World monkeys, which have 36 permanent teeth in the "IICPPPMMM" dental format (they have an extra premolar). Thus, dental eruption can be a good measure to compare physical development or maturation among the Old World monkeys such as Japanese monkeys, the apes, including chimpanzees, and humans. Based on the unitary scale, you can match the development and maturation of different species (Table 1).

3 Laboratory Work

3.1 The KUPRI Community in Japan

The Section of Language and Intelligence at Kyoto University's Primate Research Institute (KUPRI) has been focusing on the study of chimpanzee cognition both in the laboratory and in the wild. Our laboratory studies have mainly concentrated on the cognitive behavior of a group of 15 chimpanzees living at KUPRI (Table 2). In parallel, we have also been conducting a field study on wild chimpanzees at Bossou-Nimba and neighboring areas in Guinea, West Africa. The core of this book consists of compiling all our progress in chimpanzee research

Table 1. Comparison of life history of humans, chimpanzees, and Japanese monkeys based on dental eruption

Eruption of the teeth	Monkeys[a]	Chimpanzees	Humans
First deciduous tooth (i1)	1 week	2 months	8 months
Last deciduous tooth (m2)	6 months	1 year	2.5 years
First permanent tooth (M1)	1.5 years	3 years	6 years
Second molar (M2)	3.5 years	6.5 years	13 years
Third molar (M3)	6 years	11 years	20 years
Life span (years)	20–25	40–60	60–90
Body weight at birth (g)	500	1800	3000
Gestation period (days)	168	235	280

The dental formula is common to the three species: iicmm for the deciduous teeth and IICPPMMM for the permanent teeth; thus, you can obtain the comparable age among the species by matching the dental eruption

There is a tendency for early maturation in an earlier phase, as follows: monkeys > chimpanzees > humans

[a]Japanese monkeys (*Macaca fuscata*)

Table 2. Chimpanzees in the Kyoto University's Primate Research Institute (KUPRI) community

Name	Sex	Birth date[a]	Age[b]	Mother	Father	Born	Arrival year	Age
Reiko	f	*1966*	*40*	n.a.	n.a.	Africa	1968	2
Gon	m	*1966*	*40*	n.a.	n.a.	Africa	1979	13
Puchi	f	*1966*	*40*	n.a.	n.a	Africa	1979	13
Akira	m	*1976-6*	*29*	n.a.	n.a	Africa	1978	2
Mari	f	*1976-6*	*29*	n.a.	n.a.	Africa	1978	2
Ai	f	*1976-10*	*29*	n.a.	n.a.	Africa	1977	1
Pendesa	f	1977-2-2	28	Fujiko	Kenchi	JMC	1979	2
Chloe	f	1980-12-12	25	Charlotte	n.a.	Paris	1985	4
Popo	f	1982-3-7	23	Puchi	Gon	KUPRI	1982	0[c]
Reo	m	1982-5-18	23	Reiko	Gon	KUPRI	1982	0
Pan	f	1983-12-7	22	Puchi	Gon	KUPRI	1983	0[c]
Ayumu	m	2000-4-24	5	Ai	Akira	KUPRI	2000	0
Cleo	f	2000-6-19	5	Chloe	Reo	KUPRI	2000	0
Pal	f	2000-8-9	5	Pan	Akira	KUPRI	2000	0
Pico	f	2003-5-14	2	Puchi	Reo	KUPRI	2003	0

DNA analysis revealed that all chimpanzees are *Pan troglodytes verus* except Pendesa, a hybrid of *verus* and *troglodytes*

[a]Birth date given as year-month-day

[b]Age is calculated in January 2006; italic type shows that age is estimated

[c]Popo and Pan were human-reared because the mother rejected them at the time of delivery. Pico was also abandoned by the mother, but she was returned to the mother after 10 days of human rearing. Unfortunately, she died in June 9, 2005 at the age of 2 years because of an inherited malformation of the bones of her back

in recent years. Ongoing research covers various topics from perception, cognition, and memory to developmental and social aspects of the chimpanzee mind. Readers can explore the full diversity of research underway in both Japan and Africa.

The present book has a sister volume titled *Primate Origins of Human Cognition and Behavior*, published in 2001 by Springer (Matsuzawa 2001). This earlier volume, *Primate Origins of Human Cognition and Behavior*, dealt with a wide range of species; actually, more than 90 species of primates were the target of the chapters even though the main focus was on chimpanzees. The current book focuses on chimpanzees, especially their cognitive development.

Laboratory study is based on the KUPRI community. It started in 1968, when one infant female named Reiko was brought in for the study of bipedal locomotion. She was alone for 10 years; then, on November 30, 1977, a chimpanzee named Ai arrived at KUPRI at the age of about 1 year. She was expected to become the principal subject in a project that originally had the aim of developing into the first ape-language study in Japan. The project was led by Dr. Kiyoko Murofushi, an associate professor of the Section of Psychology at the time. She was flanked by three young assistant professors, Toshio Asano, Shozo Kojima, and myself. On April 15, 1978, Ai participated in a computer task inside an experimental booth for the very first time. This event marked the beginning of a long line of research, which has over the years produced a burgeoning list of publications thanks to many collaborative researchers and chimpanzee participants. The project is now known as the Ai project and still continues 28 years later (Matsuzawa 2003).

The KUPRI community of chimpanzees has been growing steadily (Fig. 4). In 2005, it comprised a community of 15 chimpanzees of three generations, ranging in age from 2 to 40 years old. There are three mother–infant pairs, with all three babies born in the year 2000 (Fig. 5). Many chapters in the present book cover

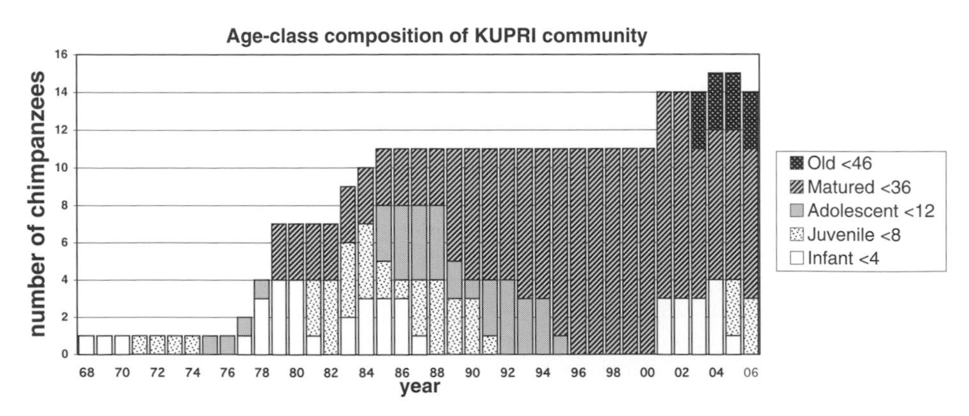

Fig. 4. Demography of chimpanzees in the Kyoto University's Primate Research Institute (KUPRI) community. The number of chimpanzees in each age category is plotted against years since 1968 when the community started

Fig. 5. Chimpanzee Ai gave birth to her son, named Ayumu, on April 24, 2000. This photo was taken 9 h after the delivery. You can still see the placenta and the cord. (Photograph by T. Matsuzawa)

topics related to cognitive development in these infants. Our studies of cognitive development in chimpanzees began in 2000 with the collaboration of three professors in the section of Language and Intelligence (which was created from the section of Psychology in 1993): Tetsuro Matsuzawa, Masaki Tomonaga, and Masayuki Tanaka, the editors of this book. All three have been helped greatly through the involvement of postdoctoral investigators, graduate students, visiting scientists from other countries, and other supporting staff.

3.2 Participation Observation

Bearing in mind the lessons learned from both the laboratory and fieldwork, we devised a new paradigm to study cognitive development in chimpanzees (see Chapter 12 by Tomonaga; Tomonaga et al. 2004). The research method can be likened to a form of "participation observation" (Fig. 6). The researchers are closely involved in the daily lives of the chimpanzees by interacting with them directly in their own space. The new paradigm is based on a triadic relationship among a mother chimpanzee, an infant chimpanzee, and a human tester. Thanks to a long-term relationship between the mother chimpanzee and the tester, the latter is able to test the infant chimpanzee in a face-to-face situation, much like a human infant cared for by its mother. We have thus been able to closely examine the cognitive development in infant chimpanzees and make comparisons with humans.

The project shows clear contrast to previous studies of chimpanzee cognition. There are many publications on chimpanzee cognition, most of which have focused in human-reared chimpanzees in an isolated situation (Fouts 1997;

Fig. 6. A scene of participation observation. A human tester can test an infant chimpanzee who is reared by the biological mother. The mother can be an assistant to the human tester because of their long-term friendship. The photo shows the mirror self-recognition test in an infant chimpanzee. (Photograph by A. Hirata)

Gardner and Gardner 1969; Kellog and Kellog, 1933; Ladygina-Kohts 2002; Povinelli 2003; Premack 1971; Premack and Woodruff 1978; Rumbaugh 1977; Rumbaugh et al. 1973; Savage-Rumbaugh et al. 1993; Terrace 1979; Yerkes and Yerkes 1929). Most of the "ape-language" studies are of a single subject, isolated from their conspecifics and tested for human-like skills. These papers reported, in other words, how the apes can adapt to the human environment because of their intellectual flexibility. They did not illuminate cognition in the social context as much. More recent studies have started to explore the social aspects of chimpanzee cognition (Hare et al. 2000; Tomasello et al. 1993; de Waal 2005; Whiten 2005; Whiten et al. 2005).

The main focus of the project since 2000 has been to illuminate cognitive development as well as the cultural transmission of chimpanzee knowledge and skills from one generation to the next. When, what, from whom to whom, and how are knowledge and skills passed on? (Fig. 7). The Ai project has now entered its third decade and has progressed from the study of a single individual to a simulation of the chimpanzee community as a whole.

As I have described, the core part of the present book is the study of cognitive development of the chimpanzees born in 2000. You can see here chapters covering various topics such as associative learning in the fetus (Chapter 3 by Kawai), neonatal imitation (Chapter 14 by Myowa-Yamakoshi), neonatal smiling (Chapter 8 by Matsuzawa; Mizuno et al. 2006), eye-to-eye contact, gaze detection, understanding of pointing, joint attention (Chapter 10 by Okamoto-Barth and Tomonaga), social referencing (Chapter 11 by Ueno), face recognition (Chapter 9 by Myowa-Yamakoshi), contagious yawning (Chapter 15 by

Fig. 7. Stone-tool use by a chimpanzee in the KUPRI community. The mother, named Chloe, is performing macadamia nut-cracking with stones, simulating oil-palm nut-cracking in the wild. The daughter, named Cleo, is observing the performance. (Photograph by A. Hirata)

Anderson and Matsuzawa), food sharing (Chapter 11 by Ueno), spontaneous categorization (Chapter 18 by Murai), processing of shadow information (Chapter 19 by Imura et al.), object manipulation (Chapter 24 by Hayashi et al.), tool use (Chapter 13 by Hirata), token use (Chapter 25 by Sousa), observation learning and the mothers' influence (Chapter 13 by Hirata), and lack of triadic relation (Chapter 12 by Tomonaga). Please enjoy reading each of the chapters to learn the new findings obtained in the project of cognitive development since 2000.

The three infants are around 5.5 years old at present (January 2006). They still suckle sometimes and sleep by their mother's side at night, but during the day they spend their time with other members of the community. They still come to the laboratory booth together with their mothers, but now have their own cognitive task to work on, presented on a separate computer screen, or allow a human tester testing the infant in a face-to-face situation. Our new "twin booth" testing paradigm (Fig. 8), where mother and infant chimpanzees can be separated in two adjacent booths, will illuminate new aspects of the chimpanzee mind.

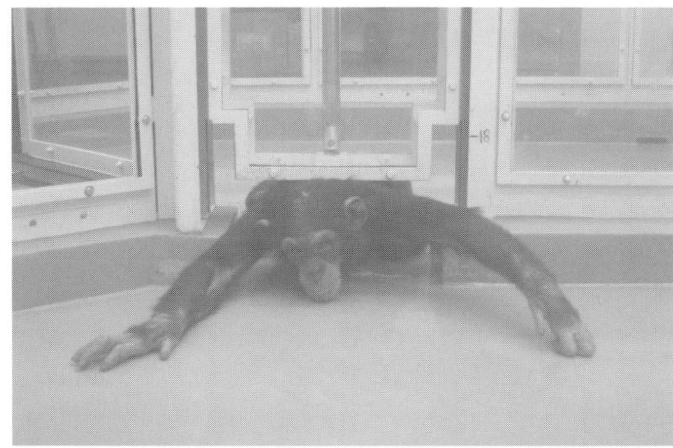

a

b

Fig. 8. **a** An infant chimpanzee named Pal is going through the door separating one booth from another. The twin booth allowed separating the infant from the mother based on their free will. (Photograph by E. Nogami). **b** The twin booth allowed a human tester to give a cognitive test to an infant chimpanzee in a face-to-face test situation. The mother is in the adjacent booth. Without interference by the mother, you can test the infant independently. (Photograph by T. Matsuzawa)

3.3 Animal Welfare and Environmental Enrichment

Laboratory work should be founded on efforts toward environmental enrichment for captive chimpanzees. We have been persistently striving to modify the physical environment of our captive chimpanzees in terms of animal welfare and environmental enrichment. There have been many publications and suggestions on environmental enrichment in chimpanzees (Brent 2001; Novak and Petto 1991). Here I would like to focus on four unique aspects of our efforts. First, we planted about 500 trees belonging to 60 species in our outdoor compound, which measures about 700 square meters. Chimpanzees in the wild show strong feeding

Fig. 9. Environmental enrichment of physical aspects in the KUPRI community. We built 15-m-high climbing frames in the outdoor compound. The chimpanzees spend about 80% of their time during the day on the frames and 20% of their time on the ground; this allows a rich vegetation. You can see the "outdoor booth" in front. (Photograph by A. Hirata)

selectivity. For example, chimpanzees at Bossou, Guinea, feed on about 200 species of plants and trees among the 600 or so species available in their habitat (Sugiyama and Koman 1992). Chimpanzees at KUPRI also show high selectivity when feeding on trees. For example, they eat the leaves of cherry trees but not those of plum. They feed on almost all deciduous trees, but only some of the evergreens and none of the conifers. At present, our outdoor compound remains covered by leafy green trees and grass all year round, even in the absence of further planting efforts (Ochiai and Matsuzawa, unpublished data).

Second, we built 15-m-high climbing frames, our so-called "triple towers" (Fig. 9). The initial tower, built in 1995, was 8 m tall; to this we added further structures in 1998 to produce our current 15-m frames. Ten years on, there are nine facilities including zoos and research institutes equipped with climbing frames more than 10 m tall in Japan. The high climbing frames become standard equipment in renovated facilities of the great apes. Chimpanzees in nature are half-arboreal and half-terrestrial. Members of the KUPRI community make extensive use of the climbing frames: according to our assessment (Ochiai and Matsuzawa, unpublished data), they spend about 80% of the daytime in the frames and only 20% on the ground. Thus, climbing frames keep the chimpanzees off the ground and in turn contribute to the protection and growth of the surrounding vegetation.

Fig. 10. An outdoor booth was placed in the chimpanzee compound. Through the underground tunnel, you can access the inside of the booth. The traditional idea of experiments was reversed; the tester and equipment are kept inside the booth while the subject chimpanzees are free outside of it. The photo shows that an old female named Reiko is fishing honey from the bottle. (Photograph provided by T. Matsuzawa)

Third, we created a small stream in the outdoor compound. Water is recycled by a pump, and the small stream provides a habitat for fish, amphibians, and small insects, and has also attracted birds. Although chimpanzees in general seldom play with water, the stream has enriched their environment by creating a small biosphere within the compound.

Fourth, we built a small outdoor booth connected to our main building through an underground tunnel (Fig. 10). In this way, we have reversed the traditional idea of the experimental room. Here, the experimenter and the apparatus are kept inside the booth while the subjects are free to roam outside (Celli et al. 2004; Tonooka et al. 1997). The booth has provided a unique opportunity for studying chimpanzees within their daily environment and as a social group.

In addition to enriching the physical environment, we have also made efforts toward feeding enrichment. We have been providing various fruits and vegetables, aiming to present the chimpanzees with a total of more than 100 species per year. In general, chimpanzees in captivity do not spend much time feeding (just under 5% of the daytime in our assessment), whereas chimpanzees in the wild customarily spend about 30% to 40% of the daytime in feeding. In a sense, the goal of feeding enrichment is to lengthen feeding time in captive chimpanzees. Thanks to our success at keeping the outdoor compound covered by natural vegetation, the chimpanzees of KUPRI are free to eat various parts of trees and other plants whenever they wish (Fig. 11).

Enrichment of the social environment has been achieved by keeping chimpanzees in a diverse community. KUPRI is currently home to three generations of chimpanzees, including mother–infant pairs and siblings. Two of our adult males, Akira and Reo, are highly competitive; as a result we now separate the

Fig. 11. Environmental enrichment of the feeding aspect in the KUPRI community. Chimpanzees are free to eat leaves and grass whenever they want. This increases the time allocation of feeding in their daytime activity budget. (Photograph by A. Hirata)

other members into two groups, Akira's and Reo's. The two groups represent something similar to two "parties" in a regional community in the wild. Females can move from one party to the other, so that party composition undergoes frequent changes and simulates the fission–fusion society in the natural habitat.

In addition to these three kinds of environmental enrichment, physical, feeding, and social, we have also attempted to enrich our chimpanzees' "cognitive environment." Chimpanzees in the wild live in a very demanding environment: they must possess various sorts of knowledge to survive tough conditions. They need a cognitive map to locate fruiting trees. They need to know when the fruits are available. They need to know which part of a plant can be eaten and which cannot. They need to be able to find detours to reach various goals, for example, ripe fig fruits on a branch that is just out of reach. Chimpanzees in the wild continually participate in such cognitively challenging tasks to obtain food. The tasks draw on capabilities such as cognitive mapping, route finding, memory search, decision making, and so forth.

Experimental tasks using computer systems in the laboratory are, of course, primarily employed in the analysis of cognitive performance by chimpanzees. However, I believe that the cognitive tasks themselves are a form of cognitive enrichment for captive chimpanzees, in fact, the most important and most neglected part of environmental enrichment (see Chapter 23 by Morimura). Chimpanzees in captivity have no freedom to access food, the availability of which is usually controlled by caretakers. For example, in many facilities, they may receive food three times a day, in amounts sufficient to maintain a physically healthy body. However, the most important point is freedom: free access to food.

The freedom to eat or not to eat should be one of the basic rights of captive chimpanzees. The chimpanzees in the KUPRI laboratory are free: it is completely up to each subject whether he or she will come to the booth to participate in a cognitive task or not. If they prefer not to participate, they may stay outside. It must be noted that choosing to participate in an experiment does not affect the total amount of food given to a chimpanzee per day: daily rations are fixed. For example, an apple given to a subject in the setting of a cognitive task is then deducted from the remainder of the daily ration. Suppose that a chimpanzee does decide to come to the booth. Again, it is up to them whether to start the first trial of the test session or not. The subject can begin the trial by touching the start key on the monitor. This means that nothing happens before the subject touches the key of his or her own will. When the chimpanzee makes a correct choice in the task, a small amount of food is delivered as a reward: half a raisin, or a tiny piece of apple. The size of the food is not a problem, and as already mentioned, the apple is taken from the daily ration and cut into pieces, such that the total amount of food given is kept constant within a day. However, the response-contingent delivery of food has a special value for the chimpanzee. Based on their free will, they work on a cognitively challenging task and as a result they are rewarded. This approach really does simulate situations in the wild, where chimpanzees search for food in the forest: they must climb trees, look for the best route, and finally obtain a piece of fruit.

I hope that our laboratory continues to succeed in illuminating the cognitive capabilities of chimpanzees. I also hope that our efforts at environmental enrichment will work to improve the lives of captive chimpanzees. The community of 15 individuals at KUPRI should provide an excellent model for the 349 chimpanzees living in Japan, about 2,800 chimpanzees in North America [only 299 in American Zoo and Aquarium Association (AZA)-registered zoos, and many in roadside zoos, many as pets in private homes, and about 1,700 in biomedical facilities], and others who are forced to live, for one reason or another, in places other than the forests of Africa.

4 Fieldwork

4.1 Bossou-Nimba Community in Guinea

After working for several years with my chimpanzee partners in the laboratory, the accumulation of knowledge about chimpanzee cognition elicited a naïve question in my mind. I had learned that the chimpanzee is astonishingly intelligent in laboratory tests of cognition. But how was such intelligence actually utilized in the natural habitat? In 1986, as I was taking a 2-year sabbatical leave in David Premack's laboratory at the University of Pennsylvania, I decided to go to Africa. I visited my senior colleague from KUPRI, Dr. Yukimaru Sugiyama, who was at the time working in the tropical forests of Africa, exploring the ecology and behavior of wild chimpanzees.

The majority of field studies reported in this volume were carried out in Bossou, Guinea, West Africa (Fig. 12). Dr. Sugiyama first settled in Bossou in 1976 after the pioneering efforts by Dutch scientists (Kortlandt 1986; Kortlandt and Holzhaus 1987). There has been a group of about 20 individuals for years (Fig. 13). Bossou has since become known as one of a handful of long-running field research sites of wild chimpanzees in Africa: others include Gombe and Mahale in Tanzania, Kibale and Budongo in Uganda, and Tai in Côte d'Ivoire. I joined

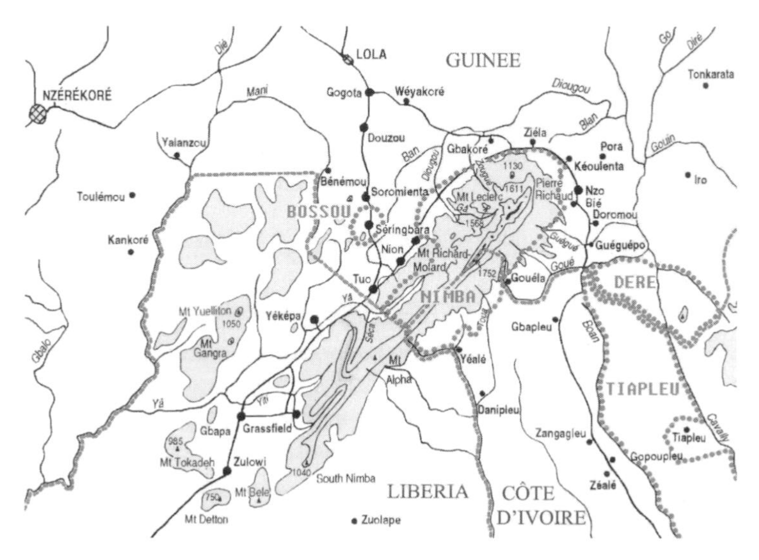

Fig. 12. A map of the field study site, Bossou and Nimba mountains in the Republic of Guinea, West Africa. (Map drawn by T. Humle, N. Granier, and L. Martinez)

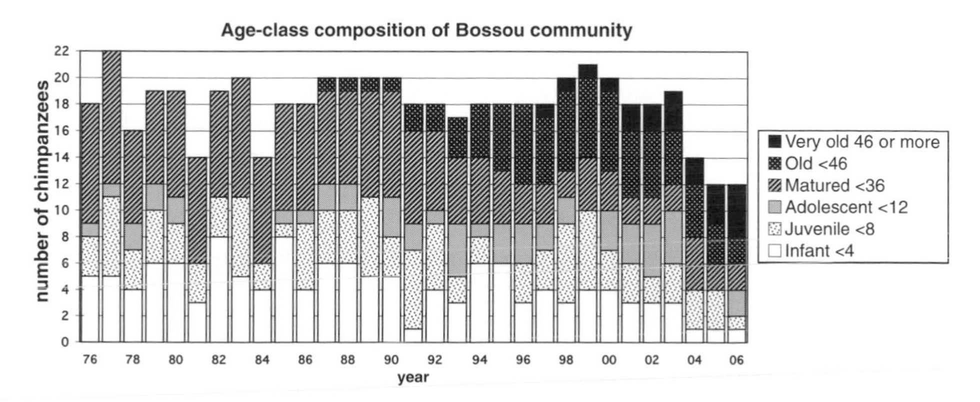

Fig. 13. Demography of chimpanzees in the Bossou community. The number of chimpanzees in each age category was plotted against years since 1976 when Japanese scientists started long-term observation

Fig. 14. Stone-tool use by the wild chimpanzees in the Bossou community, Guinea. The chimpanzees use a pair of stones as hammer and anvil to crack open oil-palm nuts. This practice is a unique cultural tradition of this community. (Photograph by T. Matsuzawa)

the Bossou site as the second researcher, and later became project leader, after Dr. Sugiyama's retirement in 1999.

My colleagues and I have been focusing on tool use and cultural behavior in chimpanzees (see Chapter 28 by Biro et al.). Chimpanzees at Bossou are well known to use a pair of stones as hammer and anvil to crack open the hard shells of oil palm nuts to obtain the edible kernel (Fig. 14). In addition to stone-tool use, the Bossou chimpanzees' repertoire of tool manufacture and use includes a variety of unique examples such as pestle-pounding, algae-scooping, hyrax-toying (Fig. 15), the use of folding leaves for drinking water, the use of leaves as cushions, and so on (see Chapter 26 by Ohashi). Up to the present, the tool-use behavior just described is limited to the Bossou community. It is well documented that each community of chimpanzees has developed its own cultural tradition (Boesch and Tomasello 1998; McGrew 1992, 2004; Whiten et al. 1999; Wrangham et al. 1994; Yamakoshi 2001). We need further examination of the cultural behavior (Hirata et al. 2001b). For example, ant-dipping is known in several communities across Africa. However, if you look at the target ant species, the material of the tool, the length of the tool, and the technique of using the tool, you will find a unique aspect to each community (see Chapter 27 by Humle; Humle and Matsuzawa 2002).

4.2 Field Experiments

Thanks to the continued efforts of many young colleagues, we have continued to find unique aspects of material and social culture in the Bossou community.

Fig. 15. An adolescent female named Vuavua (8 years old) was observed toying with a western tree hyrax (*Dendrohyrax dorsalis*, order Hyracoidea) at Bossou, Guinea, on January 18, 2000. The Bossou chimpanzee may capture animals but seldom eat the meat, except that of pangolins. Young adolescent females have been observed to practice mothering behavior with a hyrax or a log doll. See the details reported by Hirata et al. (2001b). (Photograph by S. Hirata)

In addition to the traditional method of fieldwork, we have also established a unique way of studying chimpanzee cognition in the wild. We refer to this approach as "field experiments" for tool use (Matsuzawa 1994; Biro et al. 2003; see Chapter 28 by Biro et al.). We have created an open-air "laboratory" for analyzing many aspects of stone-tool use in close detail. In the core part of the chimpanzees' ranging area, on top of a hill named Gban, we set up a laboratory site by laying out stones and nuts in a clearing. We then simply waited for chimpanzees to pass by and use the objects provided. Here, stone-tool use could be directly observed and video-recorded from behind a grass screen located about 15 m away from the chimpanzees.

This field experiment has brought to light many novel findings, such as the chimpanzees' perfect hand preference in hammering. Right-handed individuals always use their right hand for hammering, while left-handers use only the left. This was the first example where perfect hand preference had been shown in primates other than humans. There appears to be a critical age for acquiring the skill: chimpanzees learn to nut-crack by the age of 4 to 5 (minimum 3, maximum 7) years, while beyond this period they seem to have difficulty doing so (Matsuzawa 1994).

In addition to hand preference, we have made several other interesting findings. For example, each individual develops a clear preference of stone shape and size. They transport both stones and nuts. Young chimpanzees carefully observe the behavior of elder individuals, but not vice versa. Adult chimpanzees are more conservative, reluctant to extend their feeding repertoire to newly introduced species of nuts.

Our long-term observation has revealed a unique form of observational learning, which we have called "education by master-apprenticeship" (Matsuzawa

2001), somehow close to the idea of BIOL (Bonding- and Identification-based Observational Learning) by de Waal (de Waal 2001). This concept is characterized by infants' prolonged exposure to a model from birth, no active teaching, intrinsic motivation to create a copy of the model, and the mothers' high level of tolerance toward their infants. On the basis of these results, we look forward to many more exciting findings in the future.

Since 1997, we have also been recording intensively the use of leaves for drinking water at the same field experimental site. We drilled a hole in the trunk of a large tree and filled the hollow with fresh water (see Chapter 28 by Biro et al.). This setup created a unique opportunity to compare two different kinds of tool use at the same site at the same time. As a result, it has become clear that the use of leaves for drinking water is acquired at the age of about 2 years, much earlier than stone-tool use. This difference may result from the complexity of the tools involved: nut-cracking requires a set of tools, hammer and anvil, in comparison to the single tool (leaves) needed for drinking. Nevertheless, there are also characteristics that are common to the acquisition of both types of tool-use. Just as do the chimpanzees in our laboratory simulation, infants like to carefully observe the actions of their mothers and other older members of the community, especially just before they begin using tools or just after they fail to use them. Infants also have a strong tendency to use "leftover" tools, a set of stones or a drinking tool that was once used and then discarded by elder community members (see Chapter 13 by Hirata).

In addition to the scientific study of the chimpanzees at Bossou, particularly their tool-use behavior, I have been working toward two additional developments: gradually extending our research area and organizing a growing international research team.

First, my colleagues and I began to explore neighboring forests such as those of Nimba, Diecke, and Ziama (Biro et al. 2003; Humle and Matsuzawa 2001). Nimba, especially, has become an important research site since 1993 when Gen Yamakoshi and I first visited Yeale, a small village on the Côte d'Ivoire side (Matsuzawa and Yamakoshi 1996). Here we discovered ground-nests and evidence that showed that these chimpanzees cracked coula nuts instead of oil-palm nuts. The lives of Nimba chimpanzees thus appeared to be quite different from those at Bossou (Koops and Humle, personal communication). This observation convinced us that unique cultural traditions existed within each community of chimpanzees. The studies that followed, a combination of extensive surveys and intensive field experiments, made clear that the propagation of cultural traditions in tool use occurred both across generations and between communities.

Second, I tried to open up the long-running research at Bossou to scientists from other countries. The Bossou site is currently maintained by a truly international team. Past and present team members originate from countries as diverse as France, the U.K., the United States, the Netherlands, Portugal, Brazil, and Austria, in addition to Japan. The Guinean government has also created a new institute for environmental research at Bossou (Institut de Recherche

Fig. 16. The building at the research site of Bossou, Guinea. It was originally built in 1995 in the grass-roots aid program by the Japanese embassy in Guinea and the Japanese Ministry of Foreign Affairs (*to the left*). Then, the annex (*to the right*) was built in 2000 by the Guinean government to start a new institute called IREB (Institut Recherche Environmentale de Bossou) for environmental research in Bossou and Nimba, the only world heritage site of UNESCO (United Nations Educational, Scientific, and Cultural Organization) in Guinea. (Photograph by T. Matsuzawa)

Environnementale de Bossou, IREB). IREB was founded in October 2001, and since that time Guinean researchers and students have also began to take an important role in the study of chimpanzees in the Bossou-Nimba region (Fig. 16).

4.3 Conservation: Green Corridor Project

If the foundations of laboratory work lie in environmental enrichment, then fieldwork should have wildlife conservation at its core. As part of our continuing work at Bossou, we have made considerable effort to protect the chimpanzees and their forests.

Chimpanzees are endangered. This decrease is connected to various human activities such as deforestation, poaching or bush-meat trading, and contagious diseases. Bossou is no exception. The same story applies everywhere in Africa. Long-term, concentrated efforts are necessary to prevent the extinction of chimpanzee cultural variation.

Although legislation does not carry much weight in many African countries, Bossou has been designated as a reserve by the Guinean government for many years. However, the formally declared reserve area is only about 320 ha (3.2 km^2). In contrast, the core part of the ranging area of Bossou chimpanzees is about 5 to 6 km^2. This means that the chimpanzees frequently intrude into human areas and feed on many crops available in fields cultivated by humans (Fig. 17). They eat rice, manioc, corn, banana, papaya, cacao, mango, sugar cane, pineapple, orange, grapefruit, and so on, although they will not eat avocado, coffee, pepper, and okra, for example. From the human point of view, Bossou chimpanzees are pests raiding crops. There have been many instances of human–chimpanzee conflict, as well as conflict between people who wish to protect the chimpanzees and those who bear only hatred toward them.

Fig. 17. A Bossou chimpanzee is passing in front of a house in the village. Bossou chimpanzees coexist with humans. They often cross the roads in their home range, get into the cultivated fields, and eat the crops. The local people, the Manon, respect the chimpanzees as their totem

Despite these problems, chimpanzees continue to survive at Bossou and in the nearby small hills. However, there are no chimpanzees at all in the forests around neighboring villages even though the natural landscape is more or less the same—why? The reason is that chimpanzees are considered totems by the Manon people of Bossou. The founder families and their descendants have maintained respect toward the chimpanzees.

The Bossou community is at present isolated from neighboring groups. In 1982, Sugiyama noted the arrival of a single immigrant male to Bossou, yet no comparable additions to the group have been observed since then. The number of chimpanzees in Bossou has been stable, about 20, for decades (Sugiyama 2004). However, at the end of 2003 it suddenly decreased from 19 to 14, mainly through the death of five community members brought on by a respiratory disease that first appeared in November 2003. We can therefore infer the high likelihood of genetic problems arising in the near future. To ensure the continued survival of the Bossou community, we realized that we needed to launch a reforestation program, creating corridors to connect Bossou with neighboring groups of chimpanzees.

The nearest community lives in the Nimba mountains, at a distance of approximately 10 km from the center of Bossou. Savanna vegetation extends along a radius of at least 4 km between Bossou and Seringbara, a small village in Nimba. In the forest around Seringbara, we have identified a community of more than 30 chimpanzees (Koops and Humle, personal communication).

In January 1997, Japanese researchers in collaboration with local villagers launched a project aimed at creating a "Green Passage" (Green Corridor Project or Projet Corridor Vert) by planting trees in the savanna along a 300-m wide, 4-km-long stretch of land extending between Bossou and Nimba (Fig. 18). The

Fig. 18. Satellite image of the "green corridor" connecting the Bossou hills to the Nimba mountains. In 1997, we started a tree-planting project to facilitate gene exchange of the adjacent chimpanzee communities. (Image provided by the satellite of Digital Globe)

Fig. 19. The tree nursery for the green corridor. (Photograph by T. Humle)

project was funded by the Japanese embassy in Guinea and then by other agencies. IREB is now highly involved in the green corridor project. The corridor will make migration between the two groups, Bossou and Seringbara, possible.

The project devised a unique method for planting trees. Local guides collected chimpanzee feces, which were mixed with soil, then placed into a plastic sack each seed they contained from which a young tree would develop (Fig. 19). This method has two clear advantages. First, these trees will bear fruits that are consumed by the chimpanzees. Second, seeds that have passed through the chimpanzee digestive system actually have improved chances of germination.

Our initial efforts in 1997 consisted of creating a small botanical garden (Projet Petite Jardin) as a pilot attempt for the Green Passage. The garden was constructed on 0.36 ha (about 60 m × 60 m) in the peripheral savanna area of the chimpanzee habitat. Several local assistants cleared the land and planted nursery

Fig. 20. The regeneration of the forest in the green corridor. The highest tree reached 9 m high in January 2005 after the initial effort 7 years ago. (Photograph by T. Matsuzawa)

trees from 28 species. The total number of trees planted was 250. One and a half years later (July 1998), the trees in the garden were inspected by Hirata and Morimura (Hirata et al. 1998). The number of living trees had dropped to 125 (50.0% of those planted).

Eight years later, in January 2005, a second inspection was carried out. During these 8 years, no further attempts at planting trees had been made in the area marked. This assessment showed that 62 planted trees (24.8%) from 9 different species survived in the savanna. Among them, the following 4 species were notable: *Uapaca heudelotii, Parkia bicolor, Craterispermun codatum* (or *Craterispermun laurinum*), and *Albizia zygia*. The tallest of the planted trees reached 9 m in height (Fig. 20). In addition to our trees, we also found 386 young trees of 30 species not planted by us. These trees had appeared and grown naturally: their seeds were brought in by the wind, by birds, by other animals. This observation means that a great variety of species of vegetation can coexist even in such a small area. Among the newcomers, the following 3 species dominated: *Harungana madagascariensis, Nauclea latiforia, and Dychrostachys glomerata*.

Based on our efforts over the past 8 years, we can draw the following conclusions. First, some of the trees we planted were able to survive savanna conditions. We need to select those tree species for planting that can be utilized by the chimpanzees and which can survive in the savanna. Second, many species of savanna-growing trees appeared naturally, subsequent to our planting program and follow-up efforts of guarding the area from fire. Actually, more than 86% of the trees found in the garden arrived there naturally, by the natural regeneration of the forest. This result means that planting itself does not constitute our main contribution, but that ensuring protection of the area (from fire and other destructive forces such as young trees being eaten by sheep and goats) is a more important factor in the success of the project. With the help of these lessons learned, we may someday be able to transform parts of the savanna into forest through carefully designed reforestation programs.

5 Looking Toward the Future

Five years have passed since the three infant chimpanzees were born at KUPRI in 2000. Five years ago, I had no clear understanding or impressions about cognitive development in infant chimpanzees. Now, 5 years later, I believe that I have many clear memories of interesting scenes involving chimpanzee mothers and their infants. These images also remind me of scenes I have witnessed in the wild. Such facts and episodes may help us to paint a fuller picture of chimpanzee cognitive development. This book as a whole may provide a basis for any reader to construct his or her own story about chimpanzee cognition and its development.

Chimpanzees survive for 40 or 50 years, perhaps even longer. A 5-year-old chimpanzee corresponds to a human child approximately 7.5 years old. The three infant chimpanzees at KUPRI are now ready to graduate to a new stage of life: childhood. In the wild, chimpanzees 5 years of age gradually become independent from their mothers and may soon have younger siblings. Eight-year-old chimpanzees correspond to 12-year-old humans, which roughly marks the onset of puberty in both species. Therefore, the next stage in chimpanzees, childhood, spans the age from 5 to 8 years. We will continue to observe ontogenetic changes in the three young members of the KUPRI community.

Ayumu, Ai's son, began to learn Arabic numerals a year ago, at the age of 4 (Inoue and Matsuzawa, unpublished data). Before starting to practice, he had been observing his mother's performance on the computer since his birth. When his turn came, he first began by touching the numeral 1 followed by 2; this happened in April 2004. Later, he learned to touch 1-2-3, then 1-2-3–4, and so on. Gradually, he succeeded in touching all numerals from 1 through 9 in an ascending order just like his mother. Ayumu then proceeded to the next stage: memorizing the numerals. Imagine that five numerals appear on the monitor. When Ayumu touches the first numeral, the other numerals turn into white rectangles (Kawai and Matsuzawa 2000), yet he is able to touch the rectangles in the correct order. For this task, Ayumu has to memorize the numerals and their respective positions before he makes his first touch. Ayumu's performance in memorizing five numerals at a glance now exceeds that of his mother, and also that of human adults (Fig. 21a,b).

We may still have underestimated the cognitive capabilities of chimpanzees. We do not yet have a full picture of their cognitive development. Ai is still only 29 years old and has nearly half her life ahead of her. My hope is that my colleagues and I will, through our parallel efforts in the laboratory and in the wild, continue to contribute to our understanding of chimpanzee life as a whole.

a

b

Fig. 21a,b. A test of numeric memory span (Kawai and Matsuzawa 2000) was applied to an infant chimpanzee named Ayumu at the age of 5. He can memorize the numerals, that is, which numeral appeared in which position, at a glance. He first touched the smallest numeral "1". Then, he was trying to touch the white rectangle that was "3". He can reconstruct the asending order in his memory. (Photograph by T. Matsuzawa)

Acknowledgments

The present study was financially supported by grants from MEXT (#07102010, #12002009, #16002001 to T. Matsuzawa; #11710035, #13610086, #16300084 to M. Tomonaga; and #15730334 to M. Tanaka), as well as by the following grants: 21st Century COE program for biodiversity (A14), JSPS core-to-core program HOPE, and the cooperative research program of KUPRI. I would like to thank my colleagues, students, and the administration staff at the Primate Research Institute of Kyoto University. I want to give special thanks to the veterinary staff and caretakers of the chimpanzees: Kiyoaki Matsubayashi, Juri Suzuki, Shunji Gotoh, Yoshikazu Ueno, Akino Kato, Akihisa Kaneko, Kiyonori Kumazaki, Norihiko Maeda, Shino Yamauchi, Shohei Watanabe, and others. Thanks are also due to the ex-students such as Chisato Douke-Inoue, Noe Nakashima, Yuu Mizuno, Ari Ueno, Midori Uozumi, and Sanae Okamoto-Barth for their participation in caring for the chimpanzees. Without their efforts, we could not continue our study of chimpanzees in KUPRI. I also want to thank our field assistants in Bossou, Seringbara, and Yeale in Africa for their help. Thanks are also due to the government of the Republic of Guinea, especially DNRST (Direction Nationale de la Recherche Scientifique et Technique) and IREB (Institut Recherche Environmentale de Bossou), and the government of Côte d'Ivoire. I also appreciate Miho Nakamura and Tamotsu Aso of ANC Corporation, Michiyo Owaki of Chukyo TV Company, and Akihiro Hirata and Daisuke Yamada of Maiinichi News Paper for recording the development of the infant chimpanzees in photos and films. Finally, I also thank Sana Inoue and Laura Martinez for their drawing of figures, Dora Biro for English corrections, Michiko Sakai for secretarial work, and Sumiharu Nagumo for his computer programming.

References

Aiello LC, Wood B, Key C, Lewis M (1999) Morphological and taxonomic affinities of the Olduvai ulna (OH36). Am J Phys Anthropol 109:89–110

Asfaw B, White T, Lovejoy O, Latimer B, Simpson S, Suwa G (1999) *Australopithecus garhi*: a new species of early hominid from Ethiopia. Science 284:629–635

Biro D, Inoue-Nakamura N, Tonooka R, Yamakoshi G, Sousa C, Matuzawa T (2003) Cultural innovation and transmission of tool use in wild chimpanzees: evidence from field experiments. Anim Cogn 6:213–223

Boesch C, Boesch-Achermann H (2000) The chimpanzees of the Tai Forest: behavioural ecology and evolution. Oxford University Press, New York

Boesch C, Tomasello M (1998) Chimpanzee and human cultures. Curr Anthropol 39:591–614

Brent L (2001) The care and the management of captive chimpanzees. American Society of Primatologists, San Antonio

Brown P, Sutikna T, Morwood MJ, Soejono RP, Jatmiko W, Saptomo E, Rokus Awe Due (2004) A new small-bodied hominin from the Late Pleistocene of Flores, Indonesia. Nature (Lond) 431:1055–1061

Celli ML, Hirata S, Tomonaga M (2004) Socioecological influences on tool use in captive chimpanzees. Int J Primatol 25:1267–1281

de Waal F (2001) The ape and the Sushi master: cultural reflections by a primatologist. Basic Books, New York

de Waal F (2005) A century of getting to know the chimpanzees. Nature (Lond) 437:56–59

Fouts R (1997) Next of kin: What chimpanzees have taught me about who we are. William Morrow and Company, New York

Gardner RA, Gardner BT (1969) Teaching sign language to a chimpanzee. Science 165:664–672

Gonder MK, Oates JF, Disotell TR, Forstner MRJ, Morales JC, Melnick DJ (1997) A new West African chimpanzee subspecies? Nature (Lond) 388:337

Goodall J (1986) The chimpanzees of Gombe: patterns of behavior. Harvard University Press, Cambridge

Hare B, Call J, Agnetta B, Tomasello M (2000) Chimpanzees know what conspecifics do and do not see. Animal Behaviour 59:771–786

Heltne P, Marquardt L (1989) Understanding chimpanzees. Harvard University Press, Cambridge

Hirata S, Morimura N, Matsuzawa T (1998) Green passage plan (tree-planting project) and environmental education using documentary videos at Bossou: a progress report. Pan African News 5:18–20

Hirata S, Watanabe K, Kawai M (2001a) "Sweet-potato washing" revisited. In: Matsuzawa T (ed) Primate origins of human cognition and behavior. Springer, Tokyo, pp 487–508

Hirata S, Yamakoshi G, Fujita S, Ohashi G, Matsuzawa T (2001b) Capturing and toying with hyraxes (*Dendrohyrax dorsalis*) by wild chimpanzees (*Pan troglodytes*) at Bossou, Guinea. Am J Primatol 53:93–97

Humle T, Matsuzawa T (2001) Behavioural diversity among wild chimpanzee populations of Bossou and neighboring areas, Guinea and Côte d'Ivoire, West Africa. Folia Primatol 72:57–68

Humle T, Matsuzawa T (2002) Ant-dipping among the chimpanzees of Bossou, Guinea, and some comparisons with other sites. Am J Primatol 58:133–148

Kawai N, Matsuzawa T (2000) Numerical memory span in a chimpanzee. Nature (Lond) 403:39–40

Kellog W, Kellog L (1933) The ape and the child. McGraw-Hill, New York (Revised edition, Hafner, New York, 1967)

Kortlandt A (1986) The use of stone tools by wild-living chimpanzees and earliest hominids. J Hum Evol 15:77–132
Kortlandt A, Holzhaus E (1987) New data on the use of stone tools by chimpanzees in Guinea and Liberia. Primates 28:473–496
Kormos R, Boesch C, Bakarr M, Butynski T (2003) West African chimpanzees: status survey and conservation action plan. IUCN/SSC Primate Specialist Group, Cambridge
Ladygina-Kohts N (2002) Infant chimpanzee and human child. Oxford University Press, New York (first published in 1935 in Russian)
Matsuzawa T (1994) Field experiments on the use of stone tools by chimpanzees in the wild. In: Wrangham R, de Waal F, Heltne P (eds) Chimpanzee cultures. Cambridge University Press, Cambridge, pp 196–209
Matsuzawa T (ed) (2001) Primate origins of human cognition and behavior. Springer, Tokyo
Matsuzawa T (2003) The Ai project: historical and ecological contexts. Anim Cogn 6:199–211
Matsuzawa T, Yamakoshi G (1996) Comparison of chimpanzee material culture between Bossou and Nimba, West Africa. In: Russon A, Bard K, Parker ST (eds) Reaching into thought. Cambridge University Press, Cambridge, pp 211–232
McGrew WC (1992) Chimpanzee material culture. Cambridge University Press, Cambridge
McGrew WC (2004) The cultured chimpanzee: reflections on cultural primatology. Cambridge University Press, Cambridge
McGrew WC, Marchat LF, Nishida T (1996) Great ape societies. Cambridge University Press, Cambridge
Mizuno Y, Takeshita H, Matsuzawa T (2006) Behavior of infant chimpanzees during the night in the first four months of life: smiling and suckling in relation to arousal levels. Infancy (in press)
Morin PA, Moor JJ, Chakraborty R, Jin L, Goodall J, Woodruff DS (1994) Kin selection, social structure, gene flow, and the evolution of chimpanzees. Science 265:1193–1201
Nishida T (1990) The chimpanzees of the Mahale mountains. University of Tokyo Press, Tokyo
Novak M, Petto A (1991) Through the looking glass: issues of psychological well-being in captive nonhuman primates. American Psychological Association, Washington, DC
Povinelli D (2003) Folk physics for apes: the chimpanzee's theory of how the world works. Oxford University Press, New York
Premack (1971) Language in chimpanzee? Science 172:808–822
Premack D, Woodruff G (1978) Does the chimpanzee have a theory of mind? Behav Brain Sci 4:515–526
Reynolds V (2005) The chimpanzees of the Budongo forest: ecology, behaviour, and conservation. Oxford University Press, New York
Rumbaugh DM (1977) Language learning by a chimpanzee: the Lana Project. Academic Press, New York
Rumbaugh DM, Gill TV, von Glasersfeld EC (1973) Reading and sentence completion by a chimpanzee (*Pan*). Science 182:731–733
Saitou N, Yamamoto F (1997) Evolution of primate ABO blood group genes and their homologous genes. Mol Biol Evol. 14:399–411
Savage-Rumbaugh S, Murphy J, Sevcik RA, Brakke KE, Williams SL, Rumbaugh D (1993) Language comprehension in ape and child. Monogr Soc Res Child Dev (Ser 233) 58(3–4):1–254
Sugiyama Y (2004) Demographic parameters and life history of chimpanzees at Bossou, Guinea. Am J Phys Anthropol 124:154–165
Sugiyama Y, Koman J (1992) The flora of Bossou: its utilization by list of chimpanzees and humans. Afr Study Monogr 13:127–169
Terrace H (1979) Nim. Knopf, New York
The Chimpanzee Sequencing and Analysis Consortium (2005) Initial sequence of the chimpanzee genome and comparison with the human genome. Nature (Lond) 437:69–87
Tomasello M, Kruger AC, Ratner HH (1993) Cultural learning. Behav Brain Sci 16:495–552
Tomonaga M, Tanaka M, Matsuzawa T, Myowa-Yamakoshi M, Kosugi D, Mizuno Y, Okamoto S, Yamaguchi M, Bard K (2004) Development of social cognition in infant chimpanzees (*Pan troglodytes*): face recognition, smiling, gaze, and the lack of triadic interactions. Jpn Psychol Res 46:227–235

Tonooka R, Tomonaga M, Matsuzawa T (1997) Acquisition and transmission of tool making and use for drinking juice in a group of captive chimpanzees (*Pan troglodytes*) Jpn Psychol Res 39:253–265
Watanabe K (2001) A review of 50 years of research on the Japanese monkeys of Koshima: status and dominance. In: Matsuzawa T (ed) Primate origins of human cognition and behavior. Springer, Tokyo, pp 405–417
Watts DP, Mitani J (2000) Infanticide and cannibalism by male chimpanzees at Ngogo, Kibale National Park, Uganda. Primates 41:357–365
Whiten A (2005) The second inheritance system of chimpanzees and humans. Nature (Lond) 437:52–55
Whiten A, Goodall J, McGrew W, Nishida T, Reynolds V, Sugiyama Y, Tutin C, Wrangham R, Boesch C (1999) Cultures in chimpanzees. Nature (Lond) 399:682–685
Whiten A, Horner V, de Waal FBM (2005) Conformity to cultural norms of tool use in chimpanzees. Nature (Lond) 437:737–740
Wrangham R, McGrew W, de Waal F, Heltne P (1994) Chimpanzee cultures. Harvard University Press, Cambridge
Wrangham R, Chapman C, Clark-Arcadi A, Isabirye-Basuta I (1996) Social ecology of Kanyawara chimpanzees: implications for understanding the costs of great ape groups. In: McGrew W, Marchant L, Nishida T (eds) Great ape societies. Cambridge University Press, Cambridge, pp 45–57
Yamakoshi G (1998) Dietary responses to fruit scarcity of wild chimpanzees at Bossou, Guinea: possible implications for ecological importance of tool use. Am J Phys Anthropol 106: 283–295
Yamakoshi G (2001) Ecology of tool use in wild chimpanzees: toward reconstruction of early hominid evolution. In: Matsuzawa T (ed) Primate origins of human cognition and behavior. Springer, Tokyo, pp 537–556
Yerkes RM, Yerkes A (1929) The great apes. Yale University Press, New Haven

Part 2
Behavioral and Physical Foundation

2
A New Comparative Perspective on Prenatal Motor Behaviors: Preliminary Research with Four-Dimensional Ultrasonography

Hideko Takeshita[1], Masako Myowa-Yamakoshi[1], and Satoshi Hirata[2]

1 A Unique Characteristic in the Development of the Basic Orientation System in Primate Neonates

Organisms adopt various postures and actions while adjusting their bodies to the environment. Orientation to gravity, the surfaces, and media of the environment is the most fundamental prerequisite for organisms to perform any functional activities such as foraging and reproduction (Reed 1996). It is noteworthy that in primates such a basic orientation system develops through mother–infant interactions immediately after birth. For a primate neonate, the mother's body functions as an environmental substrate. The neonates sense the speed and direction of actions through the movements of their mothers, to whom they cling and by whom they are carried. The mothers support their neonates to maintain physical contact with them, and the neonates explore and learn how to coordinate their own postures and actions with those of their mothers while sensing any other maternal stimuli, such as warmth, taste, and softness of the skin. The dynamic organization of actions and perceptions that emerges from the mother–infant interactions underlies the early development of motor behaviors in primates.

2 Immaturity in Postural Control and Its By-Products in Human Neonates

When born, humans are immature in terms of locomotion, and they exhibit incompetence when required to support the weight of their own bodies during the first few months after birth. Although closely related primate species share this developmental characteristic, it is most conspicuous in human neonates, as is evident in the longitudinal comparative studies on postural reactions. Previous

[1]School of Human Cultures, The University of Shiga Prefecture, 2500 Hassaka-cho, Hikone, Shiga 522-8533, Japan
[2]Great Ape Research Institute, Hayashibara Biochemical Laboratories, 952-2 Nu, Tamano, Okayama 706-0316, Japan

studies found that similar postural reactions are induced when primate neonates are compelled to remain in unstable postures while separated from their mothers (Takeshita et al. 1989, 2002; Matsuzawa 2001). These reactions develop in at least three developmental stages that are common among primate species, as follows: (1) in the first stage, both forelimbs and hindlimbs are flexed and do not function to support the body; (2) in the second stage, the forelimbs are extended such that they support the body; and (3) in the third stage, both forelimbs and hindlimbs are extended such that they support the body together. In humans, reactions develop in the second stage after 4 months of age. Of the ten primate species studied, inclusive of apes, macaques, and capuchins, the relative length of the sum of the first and second stages is the longest in humans.

Such immaturity in postural control necessitates that human mothers use both hands to cradle the neonates and provide attentive care for longer periods than the mothers of any other primate species do. However, human neonates are not completely passive during interactions with their mothers, but they play an active role to change their postures. For example, when separated from their mothers, they cry to be cradled, and when their mothers do not cradle them in the proper manner, they fret until they are repositioned. Neonatal vocalization serves to evoke immediate maternal care and is incorporated into the early development of the basic orientation system in humans. Facial expressions such as neonatal smiling and neonatal imitation have a similar function (Meltzoff and Moore 1977, 1983; Myowa-Yamakoshi et al. 2004; Tomonaga et al. 2004; see also chapters by Matsuzawa, by Myowa-Yamakoshi, and by Tomonaga, this volume).

The emergence of general movements (GMs) and other specific motor behaviors such as hand–mouth contacts (HMCs), hand–hand contacts (HHCs), or foot–foot contacts (FFCs) in the supine position also appear to be remarkable by-products of immaturity in postural control in human neonates. GMs are complex movements involving the head, trunk, arms, and legs in the supine position in the absence of any stimulus; these movements emerge during early fetal life and disappear approximately 5 months postterm. The duration of these movements varies from a few seconds to some minutes (Prechtl and Hopkins 1986; Taga et al. 1999). HMCs are also known to originally emerge in fetuses from the 12th week of postmenstrual age (de Vries et al. 1982). These movements are also observed in preterm neonates; the frequency of HMCs and the complexity of GMs decrease in the 2nd month of age. These U-shaped developmental changes in the GMs and HMCs suggest the reorganization of the neural functions (Taga et al. 1999; Takaya et al. 2003).

GMs and other specific motor behaviors involving the limbs, such as HMCs, emerge when neonates are in the supine position, that is, when they are not cradled by their mothers. While experiencing these motor behaviors, human neonates explore and learn how to remain in the supine position through their own postural control, independent of their mothers. The neonates manifest a variety of limb movements to explore the immediate environment and their own bodies, long before acquiring voluntary control of locomotion; this is a unique characteristic of human motor development (Takeshita 1999). The proprioceptive, tactile, and visual experiences that they gain through these activities might promote a rudimentary perception of the "ecological" self (Neisser 1991, 1995; Rochat 2001).

3 Coordinated Self-Oriented Actions by Human Fetuses

Limb movements such as GMs and HMCs emerge during fetal life. Exploratory behaviors such as grabbing the umbilical cord and pushing the uterine wall have also been reported in human fetuses (Sparling et al. 1999). In the womb, the fetuses do not cling to their mothers. These prenatal limb movements could be the basis of those observed when the neonates lie in the supine position after birth.

The recently introduced technique of four-dimensional (4-D) ultrasonography has enabled the continuous monitoring of fetal faces and other surface features of the fetus, such as fetal limbs (Kurjak et al. 2003, 2004, 2005) (Fig. 1). Our recent study has demonstrated that human fetuses are already capable of manifesting coordinated behaviors such as HMCs such that the mouth is open before the hand makes contact with it (Myowa-Yamakoshi and Takeshita, in preparation).

Twenty pregnant Japanese women with singleton fetuses and gestational age of 19 to 35 weeks participated in the study. Using the 4-D ultrasound system Accuvix XQ (Medison, Seoul, Korea) with a 4–7 MHz transabdominal transducer, 4-D images of the fetuses were displayed on the screen and videotaped during the observational period. The videotapes were reviewed at 0.5-s intervals from the time point identified as the moment when an HMC emerged, wherein the fetal hand movements resulted in the contact of the thumb or fingers with the mouth and lips, that is, the oral region (Kurjak et al. 2003). For the purpose of analysis, we defined the HMCs as satisfying the following two criteria: (1) the mouth was closed before the movement began, and (2) the upper limbs were above the waist until the hand came into contact with the mouth.

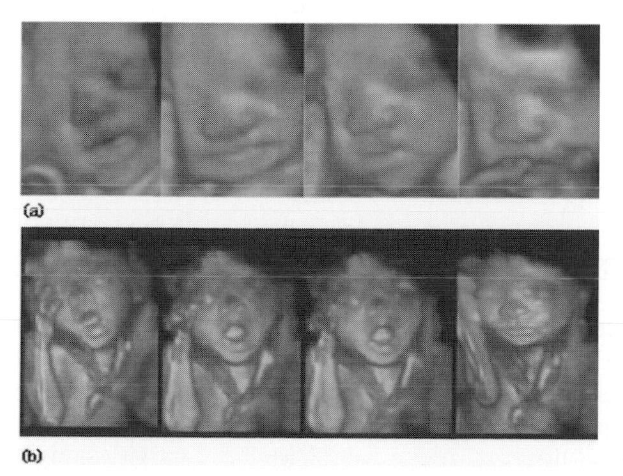

Fig. 1. Facial expressions of a fetus at 32 weeks of gestational age (**a**) and yawning by a fetus at 25 weeks of gestational age (**b**: *left* to *right*)

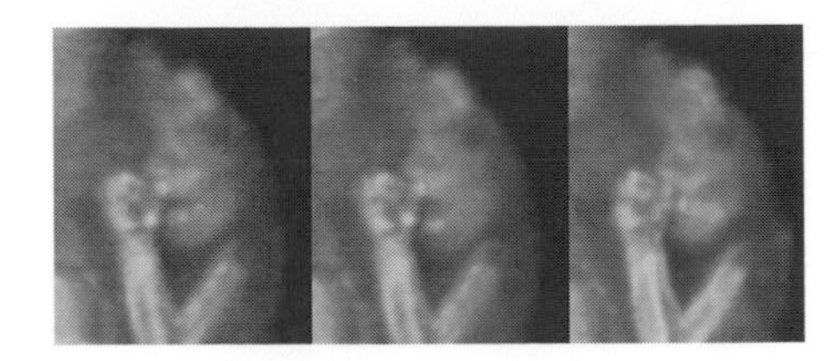

Fig. 2. A human fetus at 25 weeks of gestational age. The mouth of the fetus opens before contact with the hand and the hand makes contact with the mouth

We found 26 HMCs in 9 fetuses. Of these, 17 HMCs (65.4%, 8 fetuses) comprised those in which the mouth was opened wide before the hand came into contact with it (Fig. 2), while this was not the case in the other 9 HMCs (34.6%, 1 fetus). The following four patterns of behavioral sequences were identified before the HMCs began: (1) mouth opening (MO)–hand approaching the mouth (HA)–hand–mouth contact (HMC), (2) MO–head approaching the hand (HE)–HA–HMC, (3) HA–MO–HMC, and (4) HE–HA–MO–HMC. The most characteristic sequence was the MO–HA–HMC pattern; it accounted for 10 of the 17 HMCs (58.8%). All the patterns contained elements of both MO and HA. We investigated these patterns to determine which element began earlier. Of the 17 HMCs, 13 (76.5%) were those in which the mouth was opened before the approach of the hand.

To the best of our knowledge, this is the first study that noted several fetal HMCs in which the mouth was opened wide before the hand came into contact with it. Interestingly, such types of HMCs have been reported in human neonates (Butterworth and Hopkins 1988; Rochat 2001). These HMCs may be referred as an "anticipating" or "expecting" behavior based on the proprioceptive calibration of the body. It is possible that human fetuses have a prior perception of how they should move their hands to make contact with their mouths. It is also possible that the fetuses might be just beginning to perceive their own bodies through the experience gained from well-coordinated MO and HA movements. In any case, we may observe a developmental continuity of hand–mouth coordination from the prenatal to neonatal period.

Another important finding is the repetition of fetal HMCs. After the first HMC, the fetus was often observed to repeat it 2.4 times on average, within 5-s intervals. The most frequently observed case of repetition was that in which fetuses repeated the HMCs 6 times. It is possible that such circular fetal behavior is used by fetuses to explore the intersensorimotor relations of their bodies and to enhance their learning of their "ecological" selves (Neisser 1991, 1995; Rochat 2001).

4 Comparative Data from Chimpanzees

As the closest related species, it is likely that chimpanzees share these fundamental characteristics of early development of motor behaviors with humans. Chimpanzee mothers often cradle their neonates with both hands and manifest the "tripedal" walk while supporting the neonates with one of their forelimbs.

The chimpanzee neonates also fuss to be cradled better when they are no longer in a comfortable position. On most such occasions, the chimpanzee mothers change their manner of cradling (Mizuno et al. 2004).

During a few months in the corresponding stage of postural reactions, that is, in the first stage, when they are separated from their mothers, the chimpanzee neonates manifest motor behaviors such as GMs and hand–foot contacts (HFCs) that are similar to those observed among human neonates in the supine position. However, differences have also been reported. In chimpanzees, HHCs, which involve the firm clasping of one hand in the other, are sometimes observed immediately after birth; these appear related to the grasping reflex. However, a more-elaborate pattern of HHCs in the supine position, involving complex movements that require the entwining of the fingers of one hand around those of the other, are scarcely observed before 5 months after birth (Takeshita 1999). On the other hand, in humans, this pattern of HHCs is frequently observed 3 to 4 months after birth. With regard to GMs, Takaya et al. (2002) found that in humans, the complexity of forelimb movements is significantly greater than that of hindlimb movements, whereas no significant difference was observed in the complexity of the forelimb and hindlimb movements in chimpanzees, who manifest GMs a few months after birth.

On considering the similarities and differences reported between both species after birth, our research concern is to examine how motor behaviors develop in the womb in chimpanzees. Because it is difficult to secure the participation of pregnant chimpanzee subjects in the study without administering anesthesia, fetal behavioral data for chimpanzees have rarely been obtained thus far, although a few studies on other aspects of fetal development have been conducted (Hayashi et al. 2001; Kawai et al. 2004; see also chapters by Hayashi and by Kawai, this volume). We recently conducted another study that corresponded to our research interest and enabled us to overcome this difficulty (Myowa-Yamakoshi et al. 2005).

Tsubaki, a 9-year-old female chimpanzee who belongs to the Great Ape Research Institute, Hayashibara Biochemical Laboratories, 952-2 Nu, Tamano, Okayama 706-0316, Okayama, Japan, participated in the study. We introduced the same 4-D ultrasound system to observe her fetus during 22 to 32 weeks of gestational age (Fig. 3).

Before beginning the first test session, it was necessary to familiarize the chimpanzee mother with the experimental settings in which she was required to be in contact with the gel on the probe. This situation was made possible because of the close relationship that the chimpanzee shared with one of the keepers, who operated the probe, and several training sessions conducted over more than 2 months. The training sessions had the following three stages: (1) the gel was applied on the belly, (2) the probe with the gel was placed in contact with the belly, and (3) the probe with the gel was then moved over the belly. Each test session lasted for 6 to 20 min (mean, 10.5 min) and was repeated two or three times a week. Recordings were conducted for a total of 367 min (35 times).

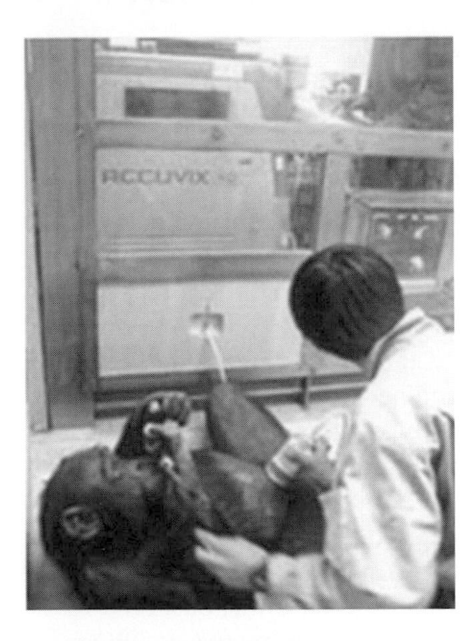

Fig. 3. A chimpanzee mother, Tsubaki, participated in the study with four-dimensional (4-D) ultrasonography

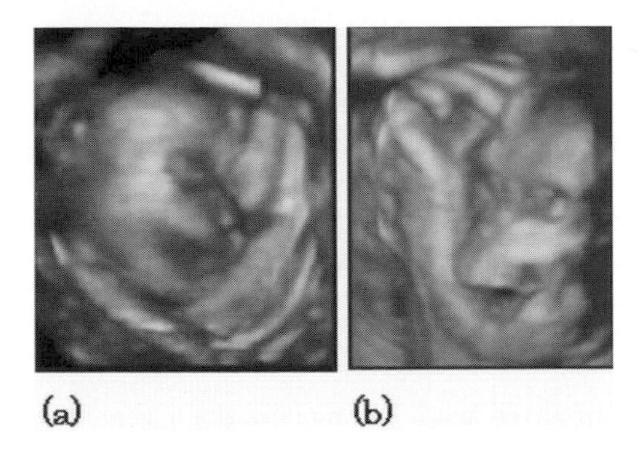

Fig. 4. A chimpanzee fetus observed at 23 weeks of gestational age (**a**) and at 25 weeks of gestational age (**b**)

We found a distinct difference in the pattern of forelimb movements in the chimpanzee fetus as compared with the data obtained from human fetuses (Kurjak et al. 2004, 2005). The chimpanzee fetus showed frequent forelimb contacts with the head (Fig. 4), whereas human fetuses showed relatively more frequent forelimb contacts with other parts of the face, including the eyes, nose, and mouth. HHCs were not observed in the fetal chimpanzee subject but were observed in humans (Fig. 5).

We are unable to generalize the results based on a single subject. The study, however, provided an important implication for developmental comparisons between both species; that is, both fetal somatic and environmental constraints might influence the feasibility of the body movements. The relative size of the fetal forelimbs to the upper body in chimpanzees is considerably larger than that in humans (Fig. 6). In contrast, in chimpanzees, the relative size of the womb in

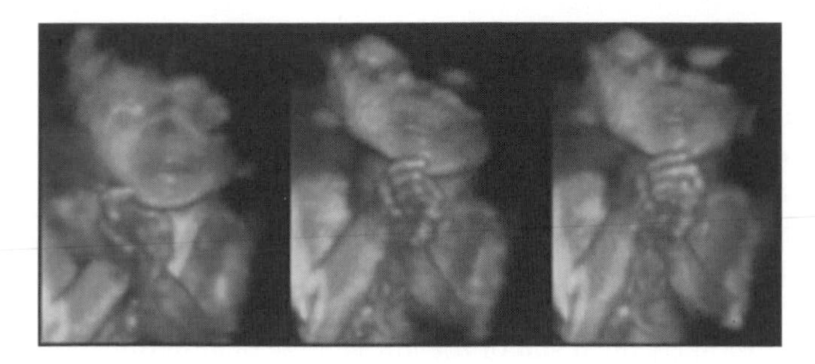

Fig. 5. Finger movements during a hand–hand contact (HHC) made by a human fetus at 25 weeks of gestational age. The change in the movements occurred during every second (*left* to *right*)

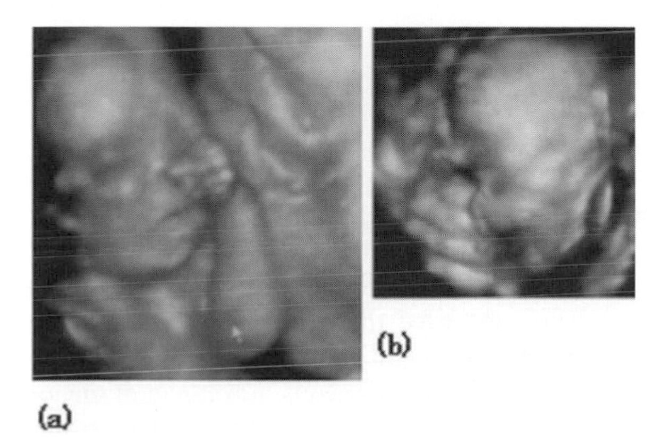

Fig. 6. A human fetus at 21 weeks of gestational age (**a**) and a chimpanzee fetus at 27 weeks of gestational age (**b**)

which the limbs could move appears to be smaller than that in humans. Two-dimensional (2-D) images that were obtained by using 4-D ultrasonography suggested that chimpanzees might have less amniotic fluid; one of its presumed functions is to facilitate the generation of fetal motor behaviors.

5 Limb Movements for Exploring the Entities of the Environment and Self

As already mentioned, human neonates are born with the ability to develop close interactions with their mothers; this compensates for their immature postural control at birth (Parker and McKinney 1999; Takeshita 1999; Falk 2004; see also chapters by Matsuzawa, by Myowa-Yamakoshi, and by Tomonaga, this volume). Facial expressions such as neonatal smiling (which is exclusively observed in the state of active sleep) and neonatal imitation (facial responses that match the facial stimuli provided by the adult demonstrator) appear to draw and hold the attention of the mothers and serve to promote the early development of the basic orientation system. Although we do not yet know the precise mechanism that enables such "social" interactions in the neonates immediately after birth, facial expressions that are a part of these activities are observed to emerge and continue during fetal life (Kurjak et al. 2003, 2004, 2005). Recent studies revealed both neonatal smiling and neonatal imitation in chimpanzees (Mizuno et al. 2006; Myowa 1996; Myowa-Yamakoshi et al. 2004). Chimpanzees also manifest immature postural control at birth. We were unable to obtain clear video images of the movements in the oral regions because of the low quality of the 4-D ultrasonography, which was most likely the result of the low volume of amniotic fluid in chimpanzees. However, facial movements corresponding to those of neonatal smiling and neonatal imitation in human fetuses are likely to be found in chimpanzee fetuses as well. In that case, human neonatal facial expressions that could contribute to the development of mother–infant bonds might be considered to have their evolutionary origin in immature birth, which is observed in both species.

Another aspect that draws our attention is the variety of fetal limb movements in humans. Human fetuses manifest varied limb movements in the uterine environment. They often touch their own body parts with their hands; such double-touch stimulus produced through their activities might develop their perception of their "ecological" self (Rochat 2001). Rochat and Hespos (1997) demonstrated that within 24 h of birth neonates are able to discriminate between self-produced tactile stimulation (self-stimulation) and tactile stimulation from a non-self or external origin (allostimulation). They observed the rooting responses of the neonates following stimulation of either their right or left cheeks by either the experimenter's finger (allostimulation) or the spontaneous movement of one of their hands to their faces (self-stimulation). The neonates displayed a greater tendency to turn their heads and root toward the experimenter's finger than toward their own hands. This finding suggests that the ability to discriminate their bodies from other entities is acquired in the womb. Human fetuses make contact with the various entities in the uterine environment (e.g., the umbilical cord, uterine wall, and amniotic fluid as well as their own body) by using their hands. Active exploration through limb movements and interactions with the entities in the environment begins in the womb itself. Repeated exploration would lead to fetal learning of the "ecological" self in the course of fetal neural development.

Interestingly, the relative number of limb movements observed in a chimpanzee fetus subject was very low. The differences in fetal limb movements between humans and chimpanzees might be attributed to the different courses of postnatal development, especially the development of self-perception, which is substantially important in human cognitive development.

Among the forelimb movements, HMCs and HHCs attract the most attention. Both the mouth and hands are the primary sensorimotor organs used to explore one's own body as well as the external environment during the prenatal and postnatal periods. Combination of the activities of the hand and mouth (HMCs) and those of both hands (HHCs) would promote synergy between the roles of an active subject and a passive object in the activities involving an organ, the mouth, and the hand. Thus, the most complex versions of double-touch activities and their routines would be obtained with HMCs or HHCs, which might promote the rudimentary perception of the "ecological" self (Neisser 1991, 1995, Rochat 2001).

Recently, Itakura et al. (2002) observed the same ability in a chimpanzee neonate as that observed in humans by using the double-touch stimulus paradigm (Rochat and Hespos 1997). It is necessary to accumulate much more precise and detailed data of fetal motor behaviors in both species to acquire a plausible scenario integrating the likely principal factors discussed here. We believe that further comparative research of fetal motor behaviors will elucidate the evolutionary and developmental origin of coordinated motor behavior, higher social cognition, and complex composition of self-recognition among humans.

Acknowledgments

We thank all the participants and collaborators for their understanding and cooperation for studying prenatal behaviors. K. Fuwa, K. Sugama-Seki, T. Takechi, A. Kaide, and K. Yuri extended useful technical support throughout the project. We gratefully acknowledge the helpful comments provided by T. Sato, Y. Kuniyoshi, G. Idani, Y. Konishi, G. Taga, R. Takaya, S. Itakura, M. Tanaka, M. Tomonaga, and T. Matsuzawa. The studies were supported by Grants-in-Aid for Scientific Research (A) to H. Takeshita (#16203034), Young Scientists (A) to M. Myowa-Yamakoshi (#16683003), Creative Scientific Research, "Synthetic Study of Imitation in Humans and Robots" to T. Sato, and the Core-to-Core Program HOPE by the Japan Society for the Promotion of Science (JSPS) and by Grants-in-Aid for Special Promotion Research (#12002009, #16002001) to T. Matsuzawa and the 21st Century COE Programs (A14 to Kyoto University) funded by the Ministry of Education, Culture, Sports, Science and Technology (MEXT) and The Cooperation Research Program of the Primate Research Institute, Kyoto University.

References

Butterworth G, Hopkins B (1988) Hand-mouth coordination in the new-born baby. Br J Dev Psychol 6:303–314

de Vries JIP, Visser GHA, Prechtl HFR (1982) The emergence of fetal behaviour: quantitative aspects. Early Hum Dev 7:301–322

Falk D (2004) Prelinguistic evolution in early hominins: whence motherese? Behav Barin Sci 27:491–541

Hayashi M, Ito M, Shimizu K (2001) The spindle neurons are present in the cingulate cortex of chimpanzee fetus. Neurosci Lett 309:97–100

Itakura S, Izumi A, Myowa M, Tomonaga M, Tanaka M, Matsuzawa T (2002) Sense of self in baby chimpanzees. Presented at 2nd International Workshop of Epigenetic Robotics, Edinburgh

Kawai N, Morokuma S, Tomonaga M, Horimoto N, Tanaka M (2004) Associative learning and memory in a chimpanzee fetus: learning and long-lasting memory before birth. Dev Psychobiol 44:116–122
Kurjak A, Azumendi G, Veček N, Kupešic S, Solak M, Varga D, Chervenak F (2003) Fetal hand movements and facial expression in normal pregnancy studied by four-dimensional sonography. J Perinat Med 31:496–508
Kurjak A, Stanojevic M, Andonotopo W, Salihagic-Kadic A, Carrera JM, Azumendi G (2004) Behavioral pattern continuity from prenatal to postnatal life: a study by four-dimensional (4D) ultrasonography. J Perinat Med 32:346–353
Kurjak A, Stanojevic M, Andonotopo W, Scazzocchio-Duenas E, Azumendi G, Carrera JM (2005) Fetal behavior assessed in all three trimesters of normal pregnancy by four-dimensional ultrasonography. Croat Med J 46:772–780
Matsuzawa T (2001) Primate foundations of human intelligence: a view of tool use in non-human primates and fossil hominids. In: Matsuzawa T (ed) Primate origins of human cognition and behavior. Springer, Tokyo, pp 3–25
Meltzoff AN, Moore MK (1977) Imitation of facial and manual gestures by human neonates. Science 198:75–78
Meltzoff AN, Moore MK (1983) Newborn infants imitate adult facial gestures. Child Dev 54:702–709
Mizuno Y, Tomonaga M, Takeshita H (2004) Crying in infants and mother-infant interactions in chimpanzees. Folia Primatol 75(S1):304
Mizuno Y, Takeshita H, Matsuzawa T (2006) Behavior of infant chimpanzees during the night in the first 4 months of life: smiling and suckling in relation to behavioral state. Infancy 9:215–234
Myowa M (1996) Imitation of facial gestures by an infant chimpanzee. Primates 37:207–213
Myowa-Yamakoshi M, Tomonaga M, Tanaka M, Matsuzawa T (2004) Imitation in neonatal chimpanzees (*Pan troglodytes*). Dev Sci 7:437–442
Myowa-Yamakoshi M, Hirata S, Fuwa K, Sugama-Seki K, Takeshita H (2005) Development of motor behaviors in a chimpanzee fetus: using 4D ultrasound imaging. Primate Res 21:s19–s20 (in Japanese)
Neisser U (1991) Two perceptually given aspects of the self and their development. Dev Rev 11:197–209
Neisser U (1995) Criteria for an ecological self. In: Rochat P (ed) The self in infancy: theory and research. Advances in psychology, vol 112. Elsevier Science, Amsterdam, pp 17–34
Parker ST, McKinney ML (1999) Origins of intelligence: the evolution of cognitive development in monkeys, apes, and humans. Johns Hopkins University Press, Baltimore
Prechtl HFR, Hopkins B (1986) Developmental transformations of spontaneous movements in early infancy. Early Hum Dev 4:233–238
Reed E (1996) Encountering the world. Oxford University Press, Oxford
Rochat P (2001) The infant's world. Harvard University Press, Cambridge
Rochat P, Hespos SJ (1997) Differential rooting response by neonates: evidence for an early sense of self. Early Dev Parenting 6:105–112
Sparling JW, Van Tol J, Chescheir NC (1999) Fetal and neonatal hand movement. Phys Therapy 79:24–39
Taga G, Takaya R, Konishi Y (1999) Analysis of general movements of infants towards understanding of developmental principle for motor control. Proc IEEE Int Conf Syst Man Cybern V:678–683
Takaya R, Taga G, Konishi I, Takeshita H, Mizuno Y, Itakura S, Tomonaga M, Matsuzawa T (2002) Comparative study of spontaneous movement of human and chimpanzee infants in the first few months of life. 13th Int Conf Infant Stud (Toront)
Takaya R, Konishi Y, Bos AF, Einspieler C (2003) Preterm to early postterm changes in the development of hand-mouth contact and other motor patterns. Early Hum Dev 75: S193–S202
Takeshita H (1999) Early development of human mind and language: comparative studies of behavioral development in primates. University of Tokyo Press, Tokyo (in Japanese)

Takeshita H, Tanaka M, Matsuzawa T (1989) Development of postural reactions and object manipulation in primate infants. Primate Res 5:111–120 (in Japanese with English summary)

Takeshita H, Mizuno Y, Matsuzawa T (2002) Development of postural reaction. In: Tomonaga M, Tanaka M, Matsuzawa T (eds) Cognitive and behavioral development in chimpanzees: a comparative approach. Kyoto University Press, Kyoto, pp 292–295 (in Japanese)

Tomonaga M, Tanaka M, Matsuzawa T, Myowa-Yamakoshi M, Kosugi D, Mizuno Y, Okamoto S, Yamaguchi MK, Bard KA (2004) Development of social cognition in infant chimpanzees (*Pan troglodytes*): Face recognition, smiling, gaze, and the lack of triadic interactions. J Psychol Res 38:163–173

3
Cognitive Abilities Before Birth: Learning and Long-Lasting Memory in a Chimpanzee Fetus

Nobuyuki Kawai

1 Introduction: The Dawn of Research on Prenatal Cognition

Everyone has a naive question about ontogenesis ("the ontogenetic origin") of our intelligence. When do we start to learn about events and memorize them? To address this question, two species have been intensively investigated: rats (*Rattus norvegicus*) and humans (*Homo sapiens*). In this chapter, I briefly review the literature of prenatal and postnatal learning of the two species. Then, I discuss our recent research on learning and memory by a chimpanzee fetus.

Once not only laymen but also researchers believed that neonates did not have most of the cognitive abilities of adults (Douglas 1975) because the immature brains of altricial infants are under development for some years or even until adolescent years (Paus et al. 1999). This notion corresponded with general behavioral development, especially in humans. It takes almost 12 months before a human infant begins bipedal locomotion and 6 months or more before an infant utters a meaningful word. Therefore, it is not surprising that it was long believed that a neonate is not yet prepared for many physical and cognitive abilities.

In the late 1970s, however, the journal *Science* reported unexpected behavioral and cognitive capabilities of rat and human newborns in succession: (1) infant imitation (Melzoff and Moore 1977), (2) instrumental conditioning in 1-day-old rats (Johanson and Hall 1979), (3) instrumental conditioning in human newborns (DeCasper and Fifer 1980), and (4) instrumental conditioning and its memory in human newborns (Rovee-Collier et al. 1980). A curtain was opened for research on cognition during infancy by these studies.

Graduate School of Information Science, Nagoya University, Furo-cho, Chikusa-ku, Nagoya 464-8601, Japan

2 Learning Abilities in Prenatal and Postnatal Animals

2.1 Learning in Postnatal Animals

Although Pavlovian conditioning in newborn rats had been demonstrated in the 1960s (Caldwell and Werboff 1962), the obtained levels of performance were weak. Caldwell and Werboff (1962) trained 1-day-old rat pups by giving a pair of vibrotactile stimuli to the rat's chest as the conditioned stimulus (CS) and an electric shock (the unconditioned stimulus; US) to the forelimb 80 times. The highest level of conditioned response (leg flexion by the CS) attained was only 32% for the two best groups. This level of performance was significantly below that traditionally reported in the literature of Pavlovian conditioning of adult rats (Hilgard and Marquis 1961).

Later, it found that newborn rats are ready to associate events when olfactory and/or gustatory stimuli are employed. For instance, Rudy and Cheatle (1977) found that 2-day-old rat pups exposed to a single paired presentation of lemon scent and nausea induced by lithium chloride (LiCl) displayed a reduced preference for the lemon scent after 6 days of testing. Subsequent studies demonstrated that even 1-day-old pups could readily establish conditioned odor aversions (Cheatle and Rudy 1978). Even more, 1-day-old rat pups learned to probe upward into a puddle when they were rewarded with small infusions of milk into their mouths, namely, instrumental learning (Johanson and Hall 1979). These results strongly suggest that newborn rats are born equipped with the ability to recognize events and learn from them.

2.2 Prenatal Learning in Animals

There is no reason to distinguish cognitive ability between before and after birth. It should be plausible to assume that the ability is available before birth. Smotherman (1982) revealed that fetal rats are capable of rapidly acquiring olfactory aversions. In that study, the flavor of apple juice introduced into the amniotic fluid was paired with injections of LiCl into fetuses 2 days before the normal end of gestation. When these infant rats were tested 10 days after birth, they demonstrated marked aversion to the odor of apple juice (Smotherman and Robinson 1991). Appetitive conditioning has also been reported in the same species (Robinson et al. 1993). Robinson et al. (1993) demonstrated that rat fetuses exposed to chemosensory stimuli are capable of retaining associations after birth. These studies indicate that the rat fetus seems well prepared to process chemical stimuli (e.g., amniotic fluid and milk) that are critical to its survival.

2.3 Some Restrictions to Prenatal and Postnatal Associative Learning in Rats: Limited Abilities in Early Visual and Auditory Senses

Nevertheless, there seem to be some constraints in learning even during infancy. In the early period of development, not all the stimulations establish learning. Although significant odor aversion was produced in 2-day-old rats by both LiCl and intraperitoneal shock US, foot shock was not effective until pups were 14 days old (Haroutunian and Campbell 1979).

Rats are born with immature sensory organs. Through quick development, weaning in rats occurs at about postnatal day 21. The rat's auditory and visual systems, however, start to function after about postnatal day 14. The external meatus opens the ear canal to sound at about 13 days of age (Kelly et al. 1987), and the eyes open at 15 days (Spear and Rudy 1991). These completions of organic maturation do not mean that rats are ready to learn by these sensory systems. Learning by these systems is delayed a few days after functioning of these sensory systems has begun. Hyson and Rudy (1984) report that it was not until the rat pups were 14 days old that they were conditioned to that tone, whereas pups 12 days old were able to detect the 2,000-Hz tone used as the CS. Moye and Rudy (1985) found that even though the 15-day-old rat pups could detect the flashing light CS, the light paired with shock did not elicit a conditioned freezing response until the pups were 17 days old. This finding cannot be attributed to the ineffectiveness of the shock in reinforcing the conditioning because 15-day-old rats were conditioned to both auditory and olfactory stimuli paired with the same shock. Therefore, the learning ability of perinatal rats relies on the chemical (olfactory and/or gustatory) stimulation for which the mammalian fetus appears to be prepared. This result is not surprising, because the fetus and newborn need to process chemical stimuli (amniotic fluid and milk, etc.) in their intrauterine and perinatal life (Papini 2002) (Table 1).

3 Cognitive Abilities in Human Fetuses

3.1 The Human Newborn Is Sensitive to the Mother's Voice and Learns by Hearing the Voice

The human fetus and newborn are unique because they are sensitive to visual and auditory stimuli from just after birth. Among all, newborns selectively respond to the stimuli produced by humans (Melzoff and Moore 1977). Not only are they sensitive, but also they can change their own behavior to listen to the voice, namely, instrumental learning. Infants younger than 3 days can rapidly be conditioned using sucking as the instrumental response and tape recordings of the mother's voice as the reinforcement (DeCasper and Fifer 1980). Not only do very young infants have the ability to learn, but they are also able to distinguish the sound of a human voice from other kinds of sounds, and they seem to prefer

Table 1. The onset of sensory modalities and its availability in learning in rats

		Age in days				
Sensation	Cognitive function	Perinatal fetus	Birth	12 days	14 days	17 days
Olfaction/ gustation	Perception	Yes		Yes	Yes	Yes
	Learning	Yes		Yes	Yes	Yes
Audition	Perception	No		Yes	Yes	Yes
	Learning	No		No	Yes	Yes
Vision	Perception	No		No	Yes	Yes
	Learning	No		No	No	Yes

this sound (Butterfield and Siperstein 1974). Human neonates readily display a preference for their mother's voice as opposed to an unfamiliar female voice, whereas they show no significant preference for their fathers' voice (DeCasper and Prescott 1984). These results suggest that the prenatal auditory experience of the human infant influences postnatal auditory preferences.

3.2 The Human Fetus Shows Habituation to Acoustic Stimuli

In fact, human newborns prefer the sound of a passage recited over the last 6 weeks of gestation to the sound of a passage from a novel (DeCasper and Spence 1986). Prenatal auditory experience exerts a change in fetal response per se (Birnholz and Benacerraf 1983; Murphy and Smyth 1962). Typically, human fetuses decrease their response when a sound or vibration is repeatedly presented to them (Lecanuet et al. 1986). Such a decrement in response has been interpreted to reflect habituation rather than receptor fatigue (Madison et al. 1986) and implies that the human fetus also has a simple form of learning ability.

3.3 Can the Human Fetus Learn Within the Uterus?

This result does not imply, however, that human fetuses are capable of more complex learning, such as associative and moreover discriminative learning. Associative learning can be distinguished from habituation, because habituation is a behavioral and/or attentional change to a single stimulus, whereas associative learning requires the ability to associate more than two events that are characterized in terms of a relationship between two or more environmental events. In addition, habituation does not persist for a long period. For instance, a recent study (van Heteren et al. 2000) reported that human fetuses demonstrated habituation to vibroacoustic stimulation (VAS) in the uterus, but it was only maintained for 24h. Further, although the auditory system of human fetuses already functions so that they can distinguish a slight difference between syllables (Lecanuet et al. 1992), hearing ability does necessarily mean that the fetus is ready to form auditory associations.

As already mentioned, although rat pups that were 10 and 12 days old could detect a 2,000-Hz tone that served as the CS, they did not become conditioned to that tone until they were 14 days old (Rudy and Hyson 1984). During the development of a particular sensory system, there seems to be a period when the system can detect and respond reflexively to a relevant stimulus source and yet be unable to mediate associative learning involving that stimulus (Moye and Rudy 1985; Rudy and Hyson 1984). Therefore, it is still unclear whether a human fetus can form an association between events originating from outside the uterus.

3.4 Does the High Sensitivity to Acoustic Stimuli by Human Fetuses Evolve in the Human Lineage?

As mentioned, the human fetus responds to various sounds from outside the uterus. Behavioral (Ramus et al. 2000) and neurophysiological (Peña et al. 2003) studies report that newborns already distinguish their own language from unfamiliar ones. Is this advanced auditory sensitivity related to our greater vocal communication after birth and only limited to the human fetus?

Other than humans, evidence that prenatal auditory experience can exert a heavy influence on postnatal behavior has been limited to precocial mammals (Vince 1979) and birds (Gottlieb 1976). So far, we have no information on whether this well-developed auditory sensitivity is shared with other fetal primates. If a chimpanzee fetus can establish associative learning mediated by its auditory system, then we can infer that our closest relative, the chimpanzee, shares the superior auditory sensitivity of the human fetus. In other words, we can infer that the advanced auditory sensitivity of human fetuses and newborns has not evolved in the *Homo* lineage for our rich vocal communication.

3.5 Toward Decisive Evidence of Prenatal Learning in a Natural Situation

To our knowledge, there has been no decisive evidence demonstrating fetal associative learning in primates, including humans. Although substantial evidence on associative learning capacity has been provided by studies on rat fetuses, these involved directly stimulating the rat fetus via an incision in the maternal abdomen. Of special interest here is whether a fetus that remains untouched within the uterus can form an association between stimuli presented from an extrauterine environment. To address this, we employed the fetus within a captive chimpanzee as the subject. These particular primates not only afford many opportunities for daily experiments but allow a close comparison with humans.

The primary purpose of the present study was to assess whether a chimpanzee fetus could undergo associative learning. Because chimpanzees have long life spans and it is not easy to increase their numbers in captivity, for practical reasons our treatment had to be limited to a single fetal subject. For this limited number of subjects, we employed differential conditioning as the control procedure. We compared responsiveness to two tones, one of which was paired with an unconditioned stimulus of VAS, which produces exaggerated responses in fetuses, whereas the other tone was never paired with VAS. We hypothesized that if an association were formed between a tone and the VAS, the subject would demonstrate active movement to the tone, because fetal responses to the VAS are essentially startled. Two other chimpanzee infants served as the control subjects to assess their unconditioned potential to respond to the tones employed in the conditioning.

4 Associative Learning and Long-Lasting Memory Before Birth in a Chimpanzee

4.1 Conditioning with a Chimpanzee Fetus

We investigated whether a chimpanzee fetus can form associations between external stimuli by using Pavlovian conditioning (Kawai et al. 2004). The conditioning was initiated at 201 days gestational age (GA). An experimenter (MT) well known to the chimpanzee mother (Pan) came into the same booth (cf. Kawai and Matsuzawa 2000), and, after calming the pregnant chimpanzee, was able to position the equipment on her lower abdomen. Before each conditioning, another experimenter (NK) outside the booth monitored the fetus ultrasonically through the mother's abdominal wall and confirmed the fetus was behaviorally active. Activity was defined as any substantial movement of the arms, legs, or whole body for 1 min before each trial. If the fetus was not active, conditioning was postponed until activity resumed. Once activity was confirmed, the speaker and stimulator were placed on the lower maternal abdomen, then differential conditioning was applied to the fetus (Fig. 1). Two pure 1-s tones were employed as conditioned stimuli (110 db), with one tone (500 Hz; CS+) always followed by a VAS of 80 Hz (110 Gal) applied near the fetus, while another (1,000 Hz; CS–) was never followed by the VAS. The conditioning was conducted for 156 trials in total until labor at 233 days GA.

The conditioned fetus was born as a result of natural delivery and reared by her own mother. The tests were done on the 33rd and 58th days after birth along with various other kinds of behavioral, cognitive, and developmental experiments and observations (see other chapters in this volume). In the test session, the conditioned infant (Pal) was taken by anesthetizing her mother and was placed supine on a wide white bed in another room. She was then presented with

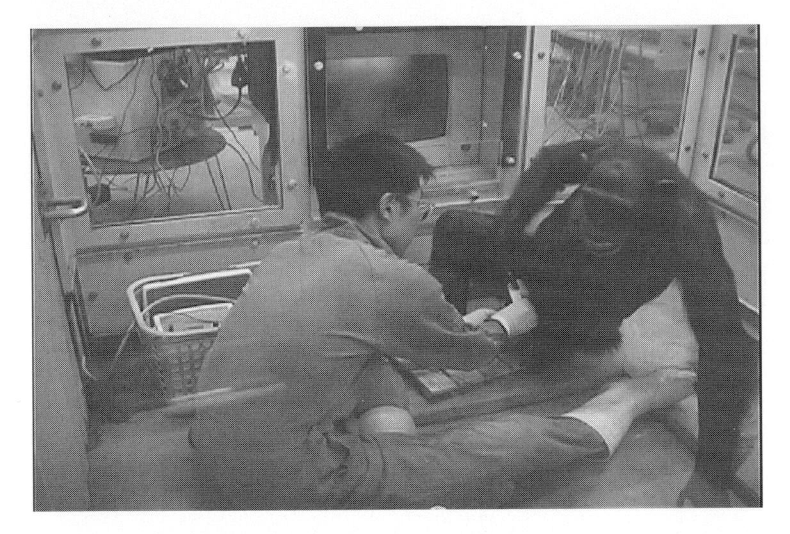

Fig. 1. The pregnant chimpanzee and an experimenter to whom she was well habituated during the conditioning treatment with speaker and stimulator placed on her lower abdomen

the CS from 10 cm above her head six times (–, +, +, –, +, –) with a 1-min intertrial interval (Fig. 2). The two other infant chimpanzees underwent this test in exactly the same way. Pico was tested when 34 and 57 days old and Cleo was tested when 121 days old. Behavior and vocalizations were recorded with a digital video camera suspended 1 m above the subjects.

4.2 Data Analysis for the Test

4.2.1 Behavioral Measures: Body Movements

The degree of body movement was calculated graphically by subtracting adjacent frames at 100-ms intervals for the first 1 s after each stimulus was presented (Fig. 2). The original images were captured on digital video and saved in 256-step gray-scale mode. We arbitrarily established a rectangular region of interest (see Fig. 2) to cover the whole body of the infant for each session. We calculated the absolute difference in brightness for corresponding pixels between adjacent frames. If this value exceeded a predetermined threshold of 20, a black dot was placed on the white background to establish an image of subtracted brightness. The body movement index for each of those brightness-subtracted images was the proportion of black dots to the total region of interest. The mean body movement index for each trial was the average of ten brightness-subtracted images.

4.2.2 Behavioral Measures: Observer Rating

In addition, the video tape recording was edited into silent video clips of the first 5 s after the CS so that observers could evaluate the activity of subjects. Five

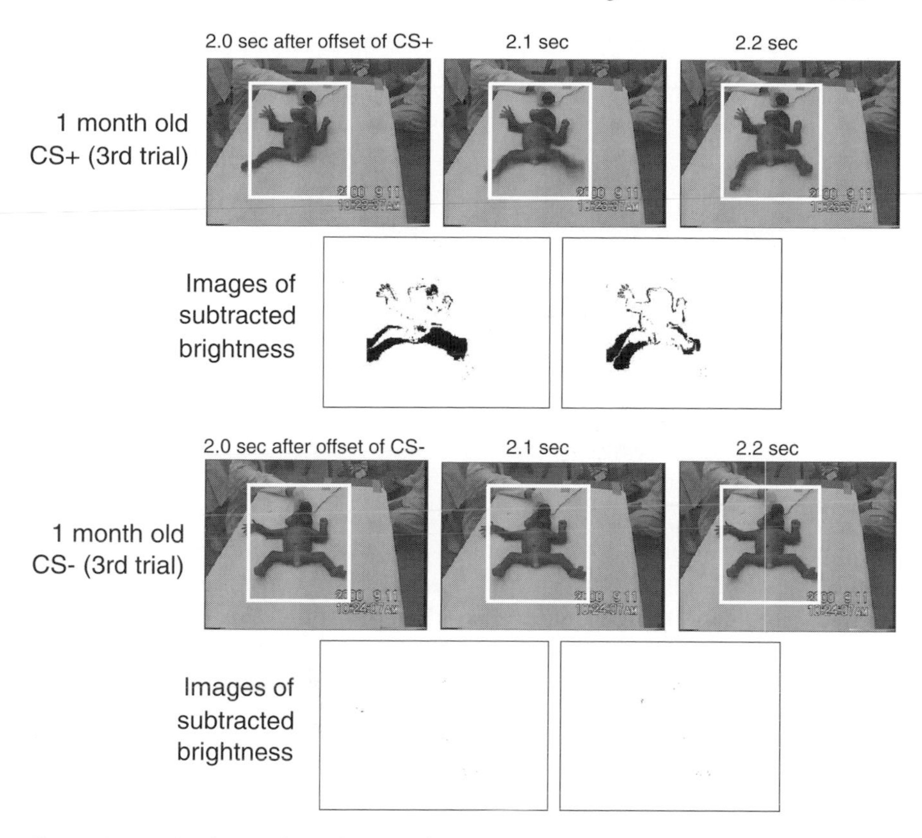

Fig. 2. Successive frames from digital video recordings and the graphically subtracted images

experimentally blind observers rated the subjects' activity according to a five-point Likert scale (1 being "completely inactive" and 5 "very active"). These ratings for behavioral activity were averaged. The observers were unfamiliar with the chimpanzees and could not distinguish among the three subjects. The clips were presented in random order with 7-s intervals.

4.3 Evidence of Fetal Learning by the Chimpanzee

Pal (the conditioned subject) was activated by the CS+, but not by the CS–, in which the conditioned infant went through frenzied movements and cries (i.e., surprised) for CS+ but not for CS– presentations. This behavior was observed in both tests at 33 and 58 days old. However, Pico and Cleo (the two chimpanzee infants), who experienced no conditioning, did not show any response to either CS.

These differences are evident in the two indices. Figure 3, which shows the mean body-movement indices calculated graphically, shows that Pal (the conditioned subject) demonstrated greater activity after the CS+ than the CS–,

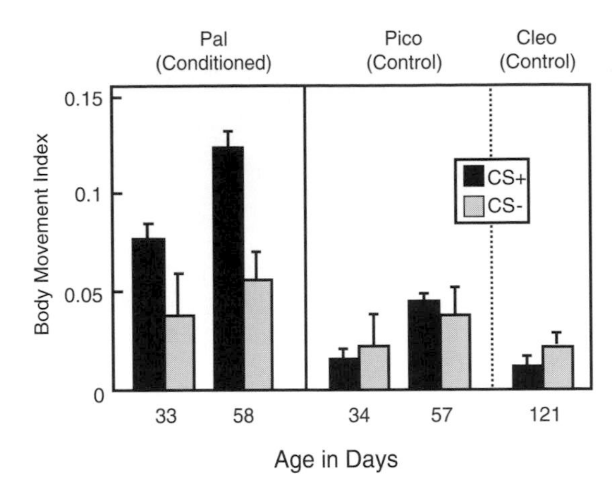

Fig. 3. Mean body movement of the conditioned (*left panel*) and control (*middle and right panels*) infants calculated graphically by subtraction between adjacent pictures for each 100 ms of the first second after stimulus presentations

whereas the body movements of the control subjects were limited and indistinguishable between the two trials. Statistical analysis on body-movement indices revealed that the conditioned infant reacted significantly more after CS+ than after CS– presentations [F (1, 2) = 75.24, $P < 0.05$). The effects of age [F (1, 2) = 1.62, n.s.] and interaction [F (1, 2) = 1.71, n.s.] were, however, not significant. The activity of Pico (the control infant of the same age) was lower and indistinguishable between CSs [F (1, 2) < 1, n.s.] The effects of age [F (1, 2) = 2.54, n.s.) and interaction [F (1, 2) < 1, n.s.) were also not significant. The activity level of Cleo (the unconditioned 121-day-old chimpanzee infant) was also lower and indistinguishable between CSs [$t(4) = 0.12$].

Exactly the same picture can be drawn by the behavior scores rated by experimentally naive observers. Scores on the behavior of Pal (the conditioned subject) after the CS+ period were evaluated as being more active than those after the CS–, but Pico's and Cleo's behaviors (the control subjects) were rated as less active and indistinguishable (Table 2). Pal's mean CS+ and CS– scores were 3.8 and 2.1, respectively, at 33 days, and 4.6 and 3.5 at 58 days. Those of Pico (the same age control infant) were 2.2 and 2.3 at 34 days, for the CS+ and CS–, respectively, and 2.9 and 2.7 at 57 days. Those of Cleo (the 121-day-old infant) were 1.2 and 1.1 for the CS+ and CS–, respectively. Statistical analysis confirms that only Pal's behavior after the CS+ was active.

These results can be summarized as follows: (a) a chimpanzee fetus is capable of associative learning mediated by its auditory system, and (b) its memory persists for at least 2 months. This learning was assessed by comparing responses to two CS. The conditioned chimpanzee demonstrated greater response to the 500-Hz tone, which had been paired with VAS during the prenatal period, than to the 1,000-Hz tone that had never been paired with VAS. These differential responses to the two tones suggest that the chimpanzee fetus distinguished them in utero and was already capable to inhibit responding to the CS–.

One may argue that the fetus could only detect the 500-Hz tone. Although the present study cannot exclude this possibility, intrauterine recordings indicate

Table 2. Behavior scores as rated by experimentally blind observers

Pal (conditioned subject)				Pico				Cleo	
33 days old		58 days old		34 days old		57 days old		121 days old	
CS+	CS–	CS+	CS–	CS+	CS–	CS+	CS–	CS+	CS–
3.8 (0.20)	2.1 (0.12)	4.6 (0.20)	3.5 (0.15)	2.2 (0.13)	2.3 (0.15)	2.9 (0.22)	2.7 (0.20)	1.2 (0.06)	1.1 (0.12)

Note: Behavioral activity rated by experimentally blind observer (1 being "completely inactive", 5 "very active")
Values within parentheses represent the standard error of the means

that sound frequencies below 1,000 Hz, which hardly attenuate acoustic energy, are detected by the fetus (Armitage et al. 1980; Querler and Renard 1981). Furthermore, some studies have reported that human fetuses respond to a 2,000-Hz tone at 100 dB (Dwornicka et al. 1964) and 110 dB (Gelman et al. 1982). Regardless of the stimulus control tone of 1,000 Hz, it is certain that the 500-Hz tone activated the conditioned infant. The exaggerated responses to the 500-Hz tone by the conditioned infant cannot be attributed to any unconditioned potential because it did not elicit any response in the unconditioned control infants. So far, there have been no reports on fetal associative learning mediated by the auditory system. The present demonstration of associative learning by a chimpanzee fetus suggests that its auditory system is already functioning and ready for associative learning. We expect that a near-term human fetus, which is sensitive to auditory stimulation, would also demonstrate the ability to form associations if the same procedure as used in the present study were followed.

4.4 Long-Lasting Memory in the Chimpanzee Infant

Interestingly, the associative learning demonstrated in our experiment (Kawai et al. 2004) remained for at least 58 days. Such retention seems surprising, given that studies with human infants report less-persistent memory during infancy (Fagen and Rovee-Collier 1983; Rovee-Collier et al. 1980). In the case of the human fetus, the longest retention interval of habituation is just 24 h (van Heteren et al. 2000)! Nevertheless, the relatively long-term memory noted in our study may be the result of the biological significance of reinforcement. Reinforcement used in studies on human infant memory seems to be biologically less significant than that used in associative learning by animals. In most studies on human infant memory (cf. Grosset al. 2002), visual stimuli used as reinforcers did not elicit unconditional responses. In contrast, the VAS we used in our study usually evoked greater activity in the fetus within the uterus. Other studies on nonhuman infant learning have also employed biologically significant stimuli as reinforcers, such as milk (Johanson and Hall 1979, 1982), LiCl (Rudy and Cheatle 1977), and electric shock (Caldwell and Werboff 1962). Smotherman (1982) reports that infant rats tested postnatally at 10 days maintained an association acquired 2 days before the end of gestation. Consequently, it appears that the more biologically significant the reinforcement, the longer the retention span.

It is worth to mention that Pal might retain the conditioning for a longer time. We conducted extinction training at the age of 8 months. Pal came into the experimental booth along with her mother, Pan, for the behavioral and cognitive tests. The CS were presented from outside the booth. In the beginning, Pal rushed back to her mother when the CS+ was presented. This result does not mean that Pal was merely surprised by the loud sound because this kind of behavior was not observed when the CS– (the same intensity) was presented. In the latter trials, Pal paid attention to the CS without overt behavior. Then, she came to ignore the CS. Although the purpose of the extinction was to extinguish the potential association and we did not expect her memory, behavioral changes by the CS+ presentation suggest she retained the memory for 8 months.

4.5 Long-Term Effects of Early Experience in Other Animals

So far, long-term effects of early experience have been documented for insects (Alloway 1972) and amphibians (Hepper and Waldman 1992). For instance, injecting orange extract into the eggs of frog embryos resulted in a preference for locations doused with that odor in both tadpoles and adult frogs (Hepper and Waldman 1992). More-complex learning can be trained in larvae of the crested newt in a visual discrimination task by food reward (Hershkowitz and Samuel 1973). The newt as adults responded to the visual stimulus previously followed by food (which were now non-rewarded). Similarly, Miller and Berk (1977) trained both tadpoles and frogs in a one-way avoidance situation. After completion of the learning, a retention interval of 35 days was inserted, in which period the tadpoles became adult. The learning was preserved regardless of the biological status at the time of acquisition. These results suggest behavioral habits induced by conditioning in immature amphibians can exert an effect on behavior of adults, even after metamorphosis.

4.6 Infantile Amnesia and Brain Development: Limits of Long-Term Effects of Early Memory

Nevertheless, we can hardly remember our own experiences in early life. This phenomenon is known as "infantile amnesia." Comparative-developmental studies suggest this phenomenon is caused by the immature brain. Campbell et al. (1974) trained rats, an altricial species, and guinea pigs, a precorcial species, in an escape-from-shock situation. In the case of rats ranging from 15 to 35 days of age, the performance was the same as the original training after a 1-day retention interval. However, with a 14-day retention interval, performance was degraded substantially for the young unweaned infants (less than 20 days of age). In contrast, 5-day-old and 100-day-old guinea pigs learned rapidly, and behavioral levels after a 75-day-long retention interval were well preserved when original training had many trials and were indistinguishable between the two age groups.

Infantile amnesia, however, does not necessarily mean that the forgotten information is irreparably lost. It is possible, within certain limits, to reactivate early memories otherwise forgotten (cf. Springer and Miller 1972). For instance, human infants at the age of 3 months can learn to activate an overhead crib mobile by operant foot kicking; however, forgetting had occurred after 1 week. Nonreinforced exposure to a visual reminder of the event that existed in the training session (i.e., mobile, etc.) reactivates their memory (Fagen and Rovee-Collier 1983).

This recovery by reactivation treatments suggests that forgetting in the early period of life may not necessarily involve the destruction of relevant memories. Rather, it may be caused by retrieval failure. The neural network responsible for retrieving (and/or storing) memories may be overlaid by networks that develop subsequently.

4.7 Brain Development and Cognitive Abilities in the Fetus

Brain development correlates not only with memory but also with learning ability in humans. Our recent study showed that not the chronological age of human fetuses but the development of the central nervous system (CNS) determines the habituation ability to the VAS presented from a maternal abdomen. Morokuma et al. (2004) divided 26 fetuses at 32 to 37 weeks of gestation into three groups using combined criteria of gestational age (GA) and behavioral indicators. First, based on GA at the time of testing, the fetuses were divided into two groups, 32 to 34 weeks and 35 to 37 weeks of GA, because it is known that the neurologically normal fetus shows RRM (repetitive mouthing movement) at or above 35 weeks of gestation. Then, the fetuses less than 34 weeks of GA were divided in terms of the positive or negative of three behavioral indicators: EM/NEM (alternation of the eye movement/no eye movement periods), REM/SEM (rapid and slow eye movement patterns), and RMM. Group II (younger age) showed RMM (positive), but Group I in the same age group did not show RMM. This difference between the same-age groups was regarded as reflecting a difference in CNS development. As results, fetuses showed habituation from at least 32 weeks of gestation. Nevertheless, fetuses less developed (Group I) from the behavioral standpoint took significantly more trials to achieve habituation than developed fetuses even in the same gestational age. These findings strongly suggest a relationship between brain development measured by behavior scores and habituation ability.

5 Conclusion: Toward Comparative and Physiological Analysis of Cognitive Ability by Fetuses

The prenatal learning capacity of mammals has been documented, particularly in rats and human. Our study adds a decisive evidence of prenatal learning in the closest primate to humans, namely, the chimpanzee. The fact that the chimpanzee fetus can distinguish the two tones provided from the maternal abdomen and form an association within uterus indicates that the advanced auditory sense in the human fetus has not evolved in human lineage. This result, however, does not always mean early acoustic experience is not a prerequisite to human rich vocal communication. Early experience (even a mere exposure) exerts an obvious effect on adult behaviors, even when the brain is undergoing substantial changes in maturation during early development. The best evidence for this is provided by human language acquisition.

The question is no longer whether fetal learning occurs, but how complex prenatal learning abilities are and how much early experiences influence adult behavior. Furthermore, we have to assess the time at which a fetus begins to acquire information. Our treatment of conditioning was initiated at 201 days GA and continued until labor at 233 days GA. It is not certain that the fetus at 201 days GA was mature enough to establish learning or whether it was formed in the latter prenatal age. Obviously, more detailed information ranging from rodents to primates including human fetuses is needed to address the question of the ontogenetic and phylogenetic origins of our cognition.

Acknowledgment

I am very grateful to T. Matsuzawa, M. Tomonaga, N. Horimoto, and M. Tanaka for giving me the opportunity to conduct this research. A part of this study has been published in *Developmental Psychobiology*, in collaboration with S. Morokuma, M. Tomonaga, N. Horimoto, and M. Tanaka. I wish to express my appreciation to them.

References

Alloway TM (1972) Retention of learning through metamorphosis in the grain beetle (*Tenebrio molitor*). Am Zool 12:471–477

Armitage SE, Baldwin BA, Vince MA (1980) The fetal sound environment of sheep. Science 208:1173–1174

Birnholz JC, Benacerraf BR (1983) The development of human fetal hearing. Science 222:516–518

Butterfield EC, Siperstein GN (1974) Influences of contingent auditory stimulation upon non-nutritional sucking. In: Proceedings of third symposium on oral sensation and perception: the mouth of the infant. Thomas, Springfield

Caldwell DF, Werboff J (1962) Classical conditioning in newborn rats. Science 136:1118–1119

Campbell BA, Misanin JR, White BC, Lytle LD (1974) Species differences in ontogeny of memory: indirect support for neural maturation as a determinant of forgetting. J Comp Physiol Psychol 87:193–202

Cheatle MD, Rudy JW (1978) Analysis of second-order odor-aversion conditioning in perinatal rats: implications for Kamin's blocking effect. J Exp Psychol Anim Behav Proc 4:237–249

DeCasper AJ, Fifer WR (1980) Of human bonding: newborns prefer their mothers' voice. Science 208:1174–1176

DeCasper AJ, Prescott P (1984) Human newborn's perception of male voices: preference, discrimination, and reinforcing value. Dev Psychobiol 17:481–491

DeCasper AJ, Spence MJ (1986) Prenatal maternal speech influences newborns perception of speech sounds. Infant Behavior & Development 9:133–150

Douglas RJ (1975) The development of hippocampal function: implications for theory and for therapy. In: Isaacson RL, Pribram KH (eds) The hippocampus: a comprehensive treatise. vol 2. Plenum Press, New York, pp 327–361

Dwornicka B, Jasienska A, Smolarz W, Wawryk R (1964) Attempt of determining the fetal reaction to acoustic stimulation. Acta Otolaryngol 57:571–574

Fagen J, Rovee-Collier CW (1983) Memory retrieval: a time-locked process in infancy. Science 222:1349–1351
Gelman SR, Wood S, Spellacy WN, Abrams RM (1982) Fetal movements in response to sound stimulation. Am J Obstet Gynecol 143:484–485
Gottlieb G (1976) Conceptions of prenatal development. Psychol Rev 83:215–234
Gross J, Hayne H, Herbert J, Sowerby P (2002) Measuring infant memory: does the ruler matter? Dev Psychobiol 40:183–192
Haroutunian V, Campbell BA (1979) Emergence of interoceptive and exteroceptive control of behavior in rats. Science 205:927–928
Hepper PG, Waldman B (1992) Embryonic olfactory learning in frogs. Q J Exp Psychol 44B:179–197
Hershkowitz M, Samuel D (1973) The retention of learning during metamorphosis of the crested newt (*Triturus cristatus*). Anim Behav 21:83–85
Hilgard ER, Marquis DG (1961) Conditioning and learning, 2nd edn. Appleton-Century-Crofts, New York
Hyson RL, Rudy JW (1984) Ontogeny of learning: II. Variation in the rat's unlearned and learned response to acoustic stimulation. Dev Psychobiol 17:263–283
Johanson IB, Hall WG (1979) Appetitive learning in 1-day old rat pups. Science 205:419–421
Johanson IB, Hall WG (1982) Appetitive conditioning in neonatal rats: conditioned orientation to a novel odor. Dev Psychobiol 15:379–397
Kawai N, Matsuzawa T (2000) Numerical memory span in a chimpanzee. Nature (Lond) 403:39–40
Kawai N, Morokuma S, Tomonaga M, Horimoto N, Tanaka M (2004) Associative learning and memory in a chimpanzee fetus: learning and long lasting memory before birth. Dev Psychobiol 44:116–122
Kelly JB, Judge PW, Fraser IH (1987) Development of the auditory orientation response in the albino rat (*Rattus norvegicus*). J Comp Psychol 101:60–66
Lecanuet JP, Graier-Deferre C, Cohen H., Le Houezec R, Busnel MC (1986) Fetal responses to acoustic stimulation depend on heart rate variability pattern, stimulus intensity, and repetition. Early Hum Dev 13:269–283
Lecanuet JP, Graier-Deferre C, Jacquet AL, Busnel MC (1992) Decelerative cardiac responsiveness to acoustical stimulation in the near term fetus. Q J Exp Psychol 44B:279–303
Madison LS, Adubato SA, Madison JK, Nelson RW, Anderson JC, Erickson J, Kuss L, Goodlin RC (1986) Fetal response decrement: true habituation? Dev Behav Pediatr 7:14–20
Melzoff AN, Moore MK (1977) Imitation of facial and manual gestures by human neonates. Science 198:75–78
Miller RR, Berk AM (1977) Retention over metamorphosis in the African claw-toed frog. J Exp Psychol Anim Behav Proc 3: 343–356
Morokuma S, Fukushima K, Kawai N, Tomonaga M, Satoh S, Nakano H (2004) Fetal habituation correlates with functional brain development. Behav Brain Res 153:459–463
Moye TB, Rudy JW (1985) Ontogenesis of learning: VI. Learned and unlearned responses to visual stimulation in the infant hooded rat. Dev Psychobiol 18:395–409
Murphy KP, Smyth CN (1962) Response of foetus to auditory stimulation. Lancet 11:972–973
Papini MR (2002) Comparative psychology: evolution and development of behavior. Prentice Hall, Englewood Cliffs
Paus T, Zijdenbos A, Worsley K, Collins DL, Blumenthal J, Giedd JN, Rapoport JL, Evans AC (1999) Structural maturation of neural pathways in children and adolescents: in vivo study. Science 283:1908–1911
Peña M, Maki A, Kovačić D, Dehaene-Lamberts G, Koizumi H, Bouquet F, Mehler J (2003) Sounds and silence: an optical topography study of language recognition at birth. Proc Natl Acad Sci U S A 100:11702–11705
Querler D, Renard K (1981) Les perceptions auditives du foetus humain. Med Hyg 39:2102–2110
Ramus F, Hauser MD, Miller C, Morris D, Mehler J (2000) Language discrimination by human newborns and by cotton-top tamarin monkeys. Science 288:349–351

Robinson SR, Arnold HM, Spear NE, Smotherman WP (1993) Experience with milk and an artificial nipple promotes conditioned opioid activity in the rat fetus. Dev Psychobiol 26:375–387

Rovee-Collier C, Sullivan MW, Enright M, Lucas D, Fagen JW (1980) Reactivation of infant memory. Science 208:1159–1161

Rudy JW, Cheatle MD (1977) Odor-aversion learning by neonatal rats. Science 198:845–846

Rudy JW, Hyson RL (1984) Ontogenesis of learning: III. Variation in the rat's differential reflexive and learned response to sound frequencies. Dev Psychobiol 17:285–300

Smotherman WP (1982) Odor aversion learning by the rat fetus. Physiol Behav 29:769–771

Smotherman WP, Robinson SR (1991) Conditioned activation of fetal behavior. Physiol Behav 50:73–77

Spear NE, Rudy JW (1991) Tests of the ontogeny of learning and memory: issues, methods, and results. In: Shair HN, Barr GA, Hofer MA (eds) Developmental psychobiology: new methods and changing concepts. Oxford University Press, New York, pp 84–113

Springer AD, Miller RR (1972) Retrieval failure induced by electroconvulsive shock: reversal with dissimilar training and recovery agents. Science 177:628–630

van Heteren CF, Boekkooi PF, Jongsma HW, Nijhuis JG (2000) Fetal learning and memory. Lancet 356:1169–1170

Vince MA (1979) Postnatal effects of prenatal sound stimulation in the guinea pig. Anim Behav 27:908–918

4
Spindle Neurons in the Anterior Cingulate Cortex of Humans and Great Apes

Motoharu Hayashi

1 Introduction

The anterior cingulate cortex (Brodmann's areas 24 and 25), which is a part of the limbic system, lies ventral and rostral to the corpus callosum. In addition to regulating autonomic and endocrine functions, the area has been shown to be involved in emotional learning, attention, error recognition, and pain. Furthermore, it is involved in vocalization, singing, and word processing, suggesting that the area is of importance to higher brain functions such as communication and language (for review, see Bush et al. 2000; Devinsky et al. 1995; Paus 2001; Posner and Rothbart 1998; Vogt et al. 1992).

Recent anatomical studies have indicated that an unusual type of neuron (spindle neuron) is present in layer Vb of the anterior cingulate cortex (ACC) (Nimchinsky et al. 1995, 1999). The spindle neurons are characterized by large vertical fusiform morphology and a type of projection neuron. These neurons have been observed only in humans and great apes such as bonobos, common chimpanzees, gorillas, and orangutans, whereas they are absent in gibbons as well as in New World and Old World monkeys. Furthermore, the density of the spindle neurons in layer V and the volume of the cell body have been found to vary as a function of relative brain size (encephalization) across humans and great apes.

In this chapter, I review the structures and functions of the ACC of humans and primates and discuss the significance of the presence of the spindle neurons in the ACC of a chimpanzee fetus. I also discuss the relationship between spindle neurons and brain-derived neurotrophic factor (BDNF), which is one of the neurotrophic factors in the vertebrate central nervous system (CNS).

2 Structures of the ACC

The cingulate cortex is a cerebral neocortical region located between the cingulate sulcus and the parieto-occipital sulcus (Fig. 1). Cytoarchitecturally, the cingulate cortex has two different regions, the ACC (Brodmann's areas 24 and 25)

Primate Research Institute, Kyoto University, 41 Kanrin, Inuyama, Aichi 484-8506, Japan

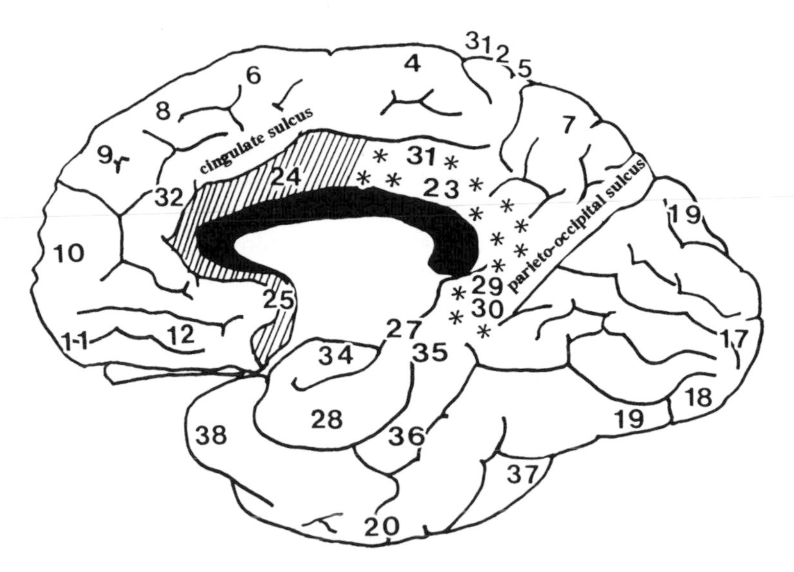

Fig. 1. Location of the anterior cingulate cortex (ACC) (areas *24* and *25*) and the posterior cingulate cortex (areas *23*, *29*, *30*, and *31*) in human cerebral cortex based on Brodmann's (1909) map. *Black area* shows the position of the corpus callosum

and the posterior cingulate cortex (Brodmann's areas 23, 29, 30, and 31). The ACC is distinct from the posterior cingulate cortex in that it lacks layer IV. Studies on cortical afferents and efferents in rhesus monkeys (Vogt and Pandya 1987; Vogt et al. 1987) have indicated that most cortical input to the ACC originates in the prefrontal cortex, temporal cortex, cingulate cortex, insula, amygdala, hippocampus, and thalamus. Most corticocortical connections arise from layers III and V, and projections primarily terminate in layers I to III of the ACC.

One of the most interesting features of the connections is the strong structural connection with the lateral prefrontal cortex. In fact, recent functional magnetic resonance imaging (fMRI) studies have indicated coactivations of the ACC and the lateral prefrontal cortex during the performance of a variety of tasks, including the Stroop task (i.e., naming the colors of words printed in nonmatching colored ink) (MacDonald et al. 2000; Pardo et al. 1990).

3 Functions of the ACC

With respect to the possible functions of the ACC (Devinsky et al. 1995; Paus 2001), changes in blood pressure as well as heart and respiratory rates have been observed by electrical stimulation in the ACC, indicating that the ACC plays a role in autonomic regulation. Furthermore, this brain region is of importance in the regulation of endocrine functions: control of the secretion of gonadal hormones, erection of the penis, and aggressive behavior.

Positron emission tomography (PET) studies have shown a significant change in pain-evoked activity within the human ACC (Rainville et al. 1997). Hutchison et al. (1999) have found that single neurons in the human ACC respond selectively to painful thermal and mechanical stimulation. Moreover, feelings of sadness or happiness change elevations in blood flow within the human ACC (Barrett et al. 2004; George et al. 1995). Metabolic activity is significantly reduced in the ACC of depressed patients (Rogers et al. 2004). These results indicate that this brain area participates in the coding of emotion. As already mentioned, the ACC is activated during the performance of various tasks, including the Stroop task. Interestingly, significant increases in cerebral blood flow in the ACC are more likely to occur during the performance of difficult tasks (Paus et al. 1998), suggesting that the ACC is of importance in error recognition, correction, and problem solving. Sanders et al. (2002) have recently proposed that cognitive deficits such as disturbances in attention, working memory, and verbal production in schizophrenia may be linked to the dysfunction of the ACC in these patients.

One of the most interesting functions of the ACC seems to be its involvement in vocalization, speech, and communication. For example, bilateral lesions of the human ACC have been found to result in akinetic mutism (Barris and Schuman 1953). In the squirrel monkey, cackling and growling calls are induced by electrical stimulation in the ACC (Jurgens and Ploog 1970). In addition, during single-word processing (Petersen et al. 1988) and singing (Perry et al. 1999), strong activations of the human ACC have been observed. Interestingly, single photon emission computed tomography shows an asymmetrical blood flow in the ACC of stutterers (Pool et al. 1991).

Concerning neuroactive molecules in the ACC, the densest innervation of dopaminergic fibers in the human ACC (Gaspar et al. 1989) suggests that dopamine may be an important neurotransmitter for the functions of this brain area. In fact, Ross and Stewart (1981) have reported that a patient with akinetic mutism responded to treatment with dopamine receptor. The activation of dopamine may facilitate information within the ACC and participate in vocalization and speech as well as have emotional implications.

Furthermore, recent genetic study has indicated that the forkhead-domain gene (*FOXP2*) is mutated in the case of severe speech and language disorder (Lai et al. 2001). It is therefore interesting to determine how *FOXP2* is expressed in the ACC during the processes of speech and language. Because humans and chimpanzees are 98.7% identical in their genomic DNA sequences (Enard et al. 2002; The Chimpanzee Sequencing and Analysis Consortium 2005), a comparative study of the gene expression patterns of *FOXP2* in the ACC of great apes and humans will also be of importance.

4 Spindle Neurons in the ACC

4.1 Discovery of Spindle Neurons

The presence of spindle neurons in the human cingulate gyrus was first reported by Betz (1881). The neurons are spindle shaped and abundant in layer V. Rose (1927) observed these kinds of neurons in the cingulate cortex of the common chimpanzee. Recently, Nimchinsky et al. (1995) precisely analyzed these unique neurons in the human ACC using modern neuroanatomical methods. The neurons are demonstrated in layer Vb of the ACC (area 24), and the presence of a single apical and basal dendrite creates the shape of a spindle. The axon descends toward the white matter. The neurons do not contain any of the calcium-binding proteins parbalbumin, calbindin, and calretinin, which are neuronal markers for a subpopulation of GABAnergic neurons. These observations suggest that the spindle neurons are projection neurons and may be modified pyramidal neurons. Notably, the neurons are fairly vulnerable in the patients of Alzheimer's disease, with a loss of approximately 60%.

Nimchinsky et al. (1999) have reported that spindle neurons are present in the ACC of great apes such as bonobos (*Pan paniscus*), common chimpanzees (*Pan troglodytes*), gorillas (*Gorilla gorilla gorilla*), and orangutans (*Pongo pygmaeus*) but are absent in gibbons (*Hylobates lar*) and in other primate species and mammals. Moreover, the volume of spindle neurons varies with brain volume, and the neurons are most abundant in humans and decline in density as follows: bonobos > common chimpanzees > gorillas > orangutans. With regard to the function of spindle neurons, an interesting speculation regards their participation in vocalization, speech, and communication, which are highly developed in humans and great apes (Matsuzawa 2001). Further functional study of these unique neurons is necessary in future experiments.

4.2 Development of Spindle Neurons

After obtaining a postmortem brain from a chimpanzee fetus (stillbirth in July 1998, male, 1.5 kg; the father was Akira and the mother was Ai) at embryonic day 224 (gestation to approximately embryonic day 230) from our institute, we attempted to determine whether spindle neurons are present in the ACC during the embryonic stage (Hayashi et al. 2001a). The brain was fixed with 4% paraformaldehyde and 0.5% glutaraldehyde. The region of the ACC (Brodmann's area 24b) was sectioned frontally at 40 μm, and the sections were stained with 0.1% cresyl violet for Nissle staining.

Figure 2 shows a spindle neuron and a pyramidal neuron in layer Vb of the fetal ACC. The diameters of the spindle and pyramidal neurons were approximately 10 to 15 μm and 10 to 20 μm, respectively. The distributions of Nissle-stained cells and spindle neurons in the ACC are shown in Fig. 3. The spindle neurons constituted 5.3% ± 1.3% (±SD) of the total population of neurons in

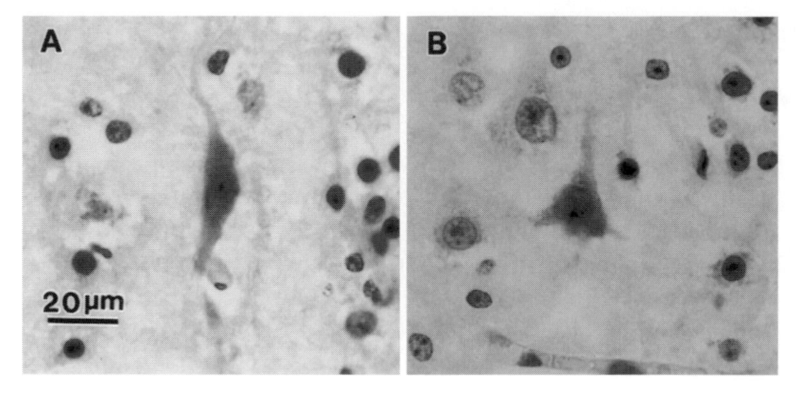

Fig. 2. A spindle neuron (**A**) and a pyramidal neuron (**B**) in the ACC of a chimpanzee fetus (embryonic day 224, stillbirth, male)

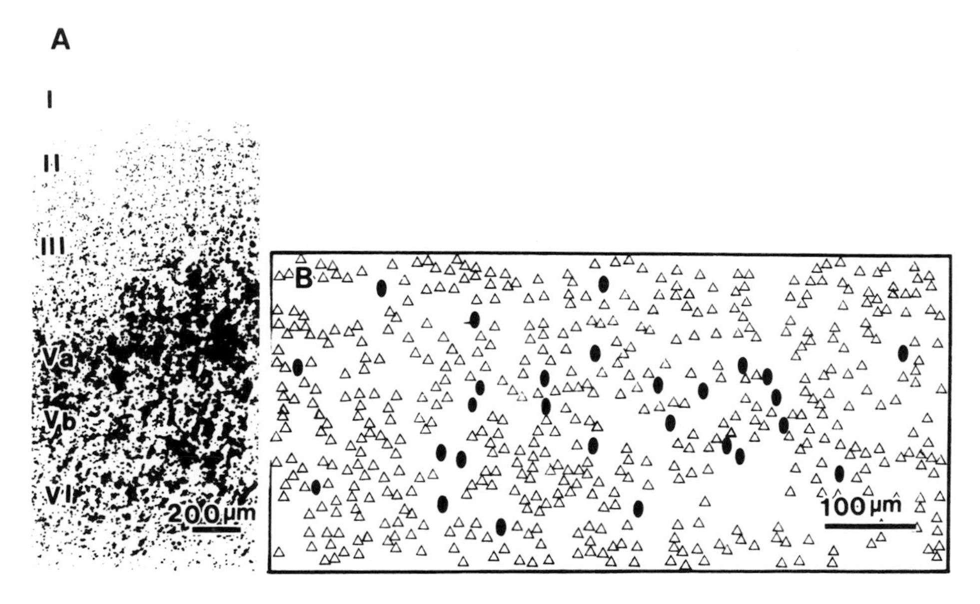

Fig. 3. **A** Distribution of Nissle-stained cells in the ACC of the chimpanzee fetus. **B** Location and distribution of the spindle neurons in layer Vb of the ACC. *Black ellipses* show the position of the spindle neurons; *triangles* indicate the position of Nissle-stained neurons. (From Hayashi et al. 2001a, with permission)

layer Vb. The spindle neurons occurred in clusters of two to three in layer Vb. Interestingly, in humans, spindle neurons have not been detected in late-term fetal brain or at birth, but first appear at 4 months after birth (Allman et al. 2001, 2002). The late appearance of the neurons in humans may result from the later development of the human brain compared to that of chimpanzees. We have recently observed spindle neurons in an adult female chimpanzee (Fig. 4). The neurons were present with 5.2% ± 0.6% (±SD) of the total population of neurons

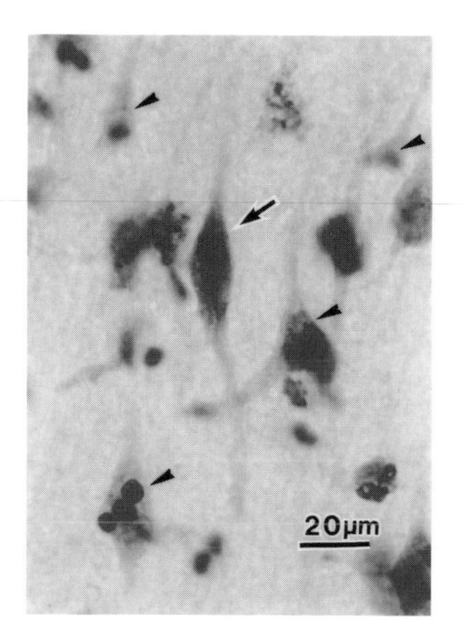

Fig. 4. A spindle neuron (*arrow*) and pyramidal neurons (*arrowheads*) in the ACC of adult chimpanzee (38 years old, female)

in layer Vb, which agrees with the value in the chimpanzee fetus in the present study.

4.3 Spindle Neurons and Brain-Derived Neurotrophic Factor

To date, we have been interested in the molecular mechanisms of the development and aging of the primate brain, particularly the cerebral cortex, which has expanded during evolution (Hayashi 1992, 1996, 2002). Within various signal molecules, we have focused on neurotrophic factors (neurotrophins), which are nerve growth factor family molecules such as BDNF, NT-3, and NT-4/5. Among them, brain-derived neurotrophic factor (BDNF) has been shown to influence various aspects of the development of the vertebrate nervous system, including neuronal survival and differentiation, the growth of axons and dendrites, and the formation and maintenance of synapses (Bibel and Barde 2000; McAllister 2001; Thoenen 2000).

Mori et al. (2004) have reported high levels of BDNF in the hippocampus and association cortices of adult macaque monkeys. In addition, the highest levels in the cerebral cortex occur at 2 months postnatal, which correlates with the time of synapse formation. In contrast, during the aging process, significant decreases in BDNF mRNA and protein in the monkey CNS have been reported (Hayashi et al. 1997, 2001b). These results suggest that BDNF controls the development, maintenance, and aging of the primate CNS. The recent most important finding is that BDNF is involved in hippocampal episodic memories in humans (Egan et al. 2003). We have, thus, examined whether the spindle neurons in the fetal chimpanzee are BDNF immunoreactive.

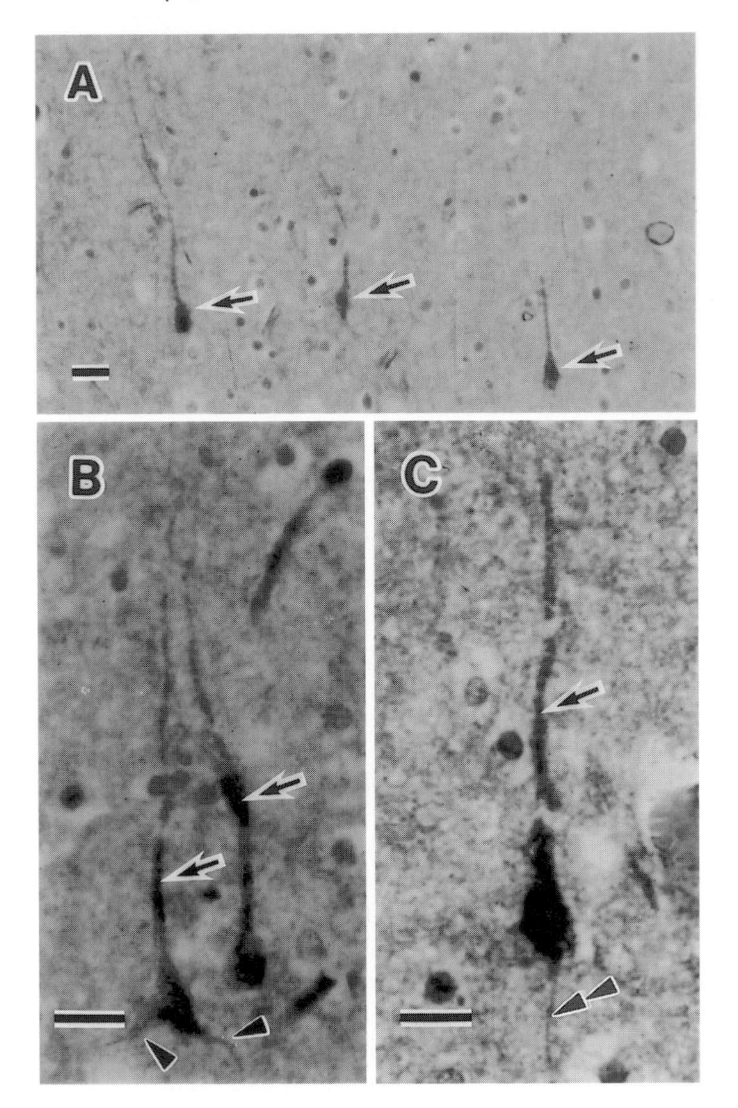

Fig. 5. **A** Brain-derived neurotrophic factor (BDNF)-immunoreactive pyramidal neurons (*arrows*) in layer Vb of the ACC of the chimpanzee fetus. *Arrows* in **B** and **C** indicate the apical dendrites; *arrowheads* in **B** point to the basal dendrites; *double arrow* in **C** indicates an axon. *Bars* 20 μm. (From Hayashi et al. 2001a, with permission)

The 40-μm sections from the ACC of the fetus were incubated for 48 h with rabbit polyclonal antibody (BDNF-N20, sc546; Santa Cruz Biotechnology, Santa Cruz, CA, USA). The immunoreactivity was visualized by the avidin-biotin-complex peroxidase method using an avidin-biotin-complex kit (Vectastain Elite ABC kit; Vector Laboratories, Burlingame, CA, USA).

As indicated in Fig. 5, BDNF-immunoreactive pyramidal neurons were found to exist in layer Vb of the ACC. Besides the cell body, the apical and basal dendrites and axons were BDNF immunopositive. These results correspond to the previous finding that BDNF is both anterogradely and retrogradely transported in the CNS (Mufson et al. 1999). Furthermore, BDNF was detected only in the pyramidal neurons, and there were no BDNF-immunoreactive spindle neurons. In the cerebral cortex of adult humans and macaque monkeys, BDNF has primarily been observed in pyramidal neurons (Kawamoto et al. 1999; Ferrer et al. 1999). In addition, the BDNF content is significantly reduced in the CNS of patients with Alzheimer's disease (Connor et al. 1997; Ferrer et al. 1999; Peng et al. 2005). Therefore, spindle neurons are possibly vulnerable in patients with Alzheimer's disease because they are BDNF immunonegative and also because of the decline of BDNF in the CNS in these patients.

Recently, Kawai et al. (2004) has reported that a chimpanzee fetus is capable of associative learning mediated by its auditory system. Furthermore, Takeshita et al. (see Chapter 2, this volume) have observed limb movements in a chimpanzee fetus using four-dimensional ultrasonography. These findings indicate the neuronal system has already been functioning during the embryonic stage of this animal. Furthermore, our present study (Hayashi et al. 2001a) has shown that the characteristics of the presence of both spindle neurons and BDNF containing pyramidal neurons in the ACC have already been determined intrinsically in the embryonic chimpanzee. Interestingly, gene expression of BDNF is increased by various stimuli such as learning (Kesslak et al. 1998; Tokuyama et al. 2000), environmental enrichment (Falkenberg et al. 1992), and maternal care (Liu et al. 2000). In contrast, immobilization stress has been shown to decrease BDNF expression in the rat hippocampus (Smith et al. 1995). These results strongly suggest that the intrauterine environment is of particular importance for the gene expression of BDNF, which may affect development of the fetal CNS.

5 Conclusion

In the present chapter, I have described the presence of spindle neurons in the ACC of the full-term chimpanzee fetus. The distributions of the neurons were found to be similar to those found in the adult stage, suggesting that the generation of neurons in the ACC may be genetically determined during the embryonic stage. The most important functions of this brain area are its involvement in attention, error recognition, and communication. A recent MRI study has also shown the possibility of ACC in personality (Pujol et al. 2002). Moreover, the area is postulated to be of importance in the maturation of self-regulatory behavior and emotional control from early childhood to adulthood (Posner and Rothbart 1998). These intelligent functions are highly developed in humans and great apes (Matsuzawa 2001). Therefore, in the future, it will be of particular interest to clarify how the spindle neurons participate in these higher brain functions and how these neurons grow and degenerate during ontogenetic development.

Acknowledgments

The author thanks Drs. M. Itoh, T. Mori, and K. Shimizu for their contributions and assistance. Drs. H. Endo and K. Yoshiwara generously provided the post-mortem brain of an adult chimpanzee used in this study. I sincerely thank Dr. T. Matsuzawa for his critical reading of the text. This work was supported by Grants-in-Aid for Scientific Research on Priority Areas-Advanced Brain Science (no 15016056), and for the Biodiversity Research of the 21st Century COE (A14) from the Ministry of Education, Culture, Sports, Science, and Technology of Japan, and the Core-to-Core Program HOPE by the Japan Society for the Promotion of Science (JSPS).

References

Allman JM, Hakeem A, Erwin JM, Nimchinsky E, Hof P (2001) The anterior cingulate cortex. The evolution of an interface between emotion and cognition. Ann N Y Acad Sci 935: 107–117

Allman J, Hakeem A, Watson K (2002) Two phylogenetic specializations in the human brain. Neuroscientist 8:335–346

Barrett J, Pike GB, Paus T (2004) The role of the anterior cingulate cortex in pitch variation during sad affect. Eur J Neurosci 19:458–464

Barris RW, Schuman HR (1953) Bilateral anterior cingulate gyrus lesions: syndrome of the anterior cingulate gyri. Neurology 3:44–52

Betz W (1881) Ueber die feinere Structur der Gehirnrinde des Menschen. Zentralbl Med Wiss 19:193–195, 209–213, 231–234

Bibel M, Barde YA (2000) Neurotrophins: key regulators of cell fate and cell shape in the vertebrate nervous system. Genes Dev 14:2919–2937

Brodmann K (1909) Vergleichende Lokalisationslehre der Groshirnrinde. Barth, Leipzig

Bush G, Luu P, Posner MI (2000) Cognitive and emotional influences in anterior cingulate cortex. Trends Cogn Sci 4:215–222

Connor B, Young D, Yan Q, Faull RL, Synek B, Dragunow M (1997) Brain-derived neurotrophic factor is reduced in Alzheimer's disease. Mol Brain Res 49:71–81

Devinsky O, MorrelL MJ, Vogt BA (1995) Contributions of anterior cingulate cortex to behaviour. Brain 118:279–306

Egan MF, Kojima M, Callicott JH, Goldberg TE, Kolachana BS, Bertolino A, Zaitsev E, Gold B, Goldman D, Dean M, Lu B, Weinberger DR (2003) The BDNF val66met polymorphism affects activity-dependent secretion of BDNF and human memory and hippocampal function. Cell 112:257–269

Enard W, Khaitovich P, Klose J, Zollner S, Hissig F, Giavalisco P, Nieselt-Struwe K, Muchmore E, Varki A, Ravid R, Doxiadis GM. Bontrop RE, Paabo S (2002) Intra- and interspecific variation in primate gene expression patterns. Science 296:340–343

Falkenberg T, Mohammed AK, Henriksson B, Persson H, Winblad B, Lidefors N (1992) Increased expression of brain-derived neurotrophic factor mRNA in rat hippocampus is associated with improved spatial memory and enriched environment. Neurosci Lett 138:153–156

Ferrer I, Marin C, Rey MJ, Ribalta T, Goutan E, Blanco R, Tolosa E, Marti E (1999) BDNF and full-length and truncated TrkB expression in Alzheimer disease. Implications in therapeutic strategies. J Neuropathol Exp Neurol 58:729–739

Gaspar P, Berger B, Febvret A, Vigny A, Henry JP (1989) Catecholamine innervation of the human cerebral cortex as revealed by comparative immunohistochemistry of tyrosine hydroxylase and dopamine-beta-hydroxylase. J Comp Neurol 279:249–271

George MS, Ketter TA, Parekh PI, Horwitz B, Herscovitch P, Post RM (1995) Brain activity during transient sadness and happiness in healthy women. Am J Psychiatry 152:341–351

Hayashi M (1992) Ontogeny of some neuropeptides in the primate brain. Prog Neurobiol 38:231–260
Hayashi M (1996) Neurotrophins and the primate central nervous system: a minireview. Neurochem Res 21:739–747
Hayashi M (2002) Development and aging of the primate brain—from the viewpoint of neuroactive molecules. Curr Topics Neurochem 3:155–165
Hayashi M, Yamashita A, Shimizu K (1997) Somatostatin and brain-derived neurotrophic factor mRNA expression in the primate brain: decreased levels of mRNAs during aging. Brain Res 749:283–289
Hayashi M, Itoh M, Shimizu K (2001a) The spindle neurons are present in the cingulate cortex of chimpanzee fetus. Neurosci Lett 309:97–100
Hayashi M, Mitsunaga F, Ohira K, Shimizu K (2001b) Changes in BDNF-immunoreactive structures in the hippocampal formation of the aged macaque monkey. Brain Res 918:191–196
Hutchison WD, Davis KD, Lozano AM, Tasker RR, Dostrovsky JO (1999) Pain-related neurons in the human cingulate cortex. Nature Neurosci 2:403–405
Jurgens U, Ploog D (1970) Cerebral representation of vocalization in the squirrel monkey. Exp Brain Res 10:532–554
Kawai N, Morokuma S, Tomonaga M, Horimoto N, Tanaka M (2004) Associative learning and memory in a chimpanzee fetus: learning and long-lasting memory before birth. Dev Psychobiol 44:116–122
Kawamoto Y, Nakamura S, Kawamata T, Akiguchi I, Kimura J (1999) Cellular localization of brain-derived neurotrophic factor-like immunoreactivity in adult monkey brain. Brain Res 821:341–349
Kesslak JP, So V, Choi J, Cotman CW, Gomez-Pinilla F (1998) Learning upregulates brain-derived neutrophic factor messenger ribonucleic acid: a mechanism to facilitate encoding and circuit maintenance? Behav Neurosci 112:1012–1019
Lai CSL, Fisher SE, Hurst JA, Vargha-Khadem F, Monaco AP (2001) A forkhead-domain gene is mutated in a severe speech and language disorder. Nature (Lond) 413:519–523
Liu D, Diorio J, Day JC, Francis DD, Meaney MJ (2000) Maternal care, hippocampal synaptogenesis and cognitive development in rats. Nat Neurosci 3:799–806
MacDonald AW III, Cohen JD, Stenger VA, Carter CS (2000) Dissociating the role of the dorsolateral prefrontal and anterior cingulate cortex in cognitive control. Science 288:1835–1838
Matsuzawa T (2001) Primate foundations of human intelligence: a view of tool use in nonhuman primates and fossil hominids. In: Matsuzawa T (ed) Primate origins of human cognition and behavior. Springer, Tokyo, pp 3–25
McAllister AK (2001) Neurotrophins and neuronal differentiation in the central nervous system. Cell Mol Life Sci 58:1054–1060
Mori T, Shimizu K, Hayashi M (2004) Differential expression patterns of TrkB ligands in the macaque monkey brain. NeuroReport 15:2507–2511
Mufson EJ, Kroin JS, Sendera TJ, Sobreviela T (1999) Distribution and retrograde transport of trophic factors in the central nervous system: functional implications for the treatment of neurodegenerative diseases. Prog Neurobiol 57:451–484
Nimchinsky EA, Vogt BA, Morrison JH, Hof PR (1995) Spindle neurons of the human anterior cingulate cortex. J Comp Neurol 355:27–37
Nimchinsky EA, Gilissen E, Allman JM, Perl DP, Erwin JM, Hof PR (1999) A neuronal morphologic type unique to humans and great apes. Proc Natl Acad Sci USA 96:5268–5273
Pardo JV, Pardo PJ, Janer KW, Raichle ME (1990) The anterior cingulate cortex mediates processing selection in the Stroop attentional conflict paradigm. Proc Natl Acad Sci USA 87:256–259
Paus T (2001) Primate anterior cingulate cortex: where motor control, drive, and cognition interface. Nat Rev Neurosci 2:417–424
Paus T, Koski L, Caramanos Z, Westbury C (1998) Regional differences in the effects of task difficulty and motor output on blood flow response in the human anterior cingulate cortex: a review of 107 PET activation studies. NeuroReport 9:R37–R47

Peng S, Wuu J, Mufson EJ, Fahnestock M (2005) Precursor form of brain-derived neurotrophic factor and mature brain-derived neurotrophic factor are decreased in the pre-clinical stages of Alzheimer's disease. J Neurochem 93:1412–1421

Perry DW, Zatorre RJ, Petrides M, Alivisatos B, Meyer E, Evans AC (1999) Localization of cerebral activity during simple singing. NeuroReport 10:3453–3458

Petersen SE, Fox PT, Posner MI, Mintun M, Raichle ME (1988) Positron emission tomographic studies of the cortical anatomy of single-word processing. Nature (Lond) 331:585–589

Pool KD, Devous MD, Freeman FJ, Watson BC, Finitzo T (1991) Regional cerebral blood flow in developmental stutterers. Arch Neurol 48:509–512

Posner MI, Rothbart MK (1998) Attention, self-regulation, and consciousness. Philos Trans R Soc Lond B 353:1915–1927

Pujol J, Lopez A, Deus J, Cardoner N, Vallejo J, Capdevila A, Paus T (2002) Anatomical variability of the anterior cingulate gyrus and basic dimensions of human personality. NeuroImage 15:847–855

Rainville P, Duncan GH, Price DD, Carrier B, Bushnell C (1997) Pain affect encoded in human anterior cingulate but not somatosensory cortex. Science 277:968–971

Rogers MA, Kasai K, Koji M, Fukuda R, Iwanami A, Nakagome K, Fukuda M, Kato N (2004) Executive and prefrontal dysfunction in unipolar depression: a review of neuropsychological and imaging evidence. Neurosci Res 50:1–11

Rose M (1927) Gyrus limbicus anterior und Regio retrosplenialis (Cortex holoprotoptychos quinquestratificatus): Vergleichende Architektonik bei Tier und Menschen. J Psychol Neurol 35:5–217

Ross ED, Stewart RM (1981) Akinetic mutism from hypothalamic damage: successful treatment with dopamine agonists. Neurology 31:1435–1439

Sanders GS, Gallup Jr GG, Heinsen H, Hof PR, Schmitz C (2002) Cognitive deficits, schizophrenia, and the anterior cingulate cortex. Trends Cogn Sci 6:190–192

Smith MA, Makino S, Kvetnansky R, Post RM (1995) Stress and glucocorticoids affect the expression of brain-derived neurotrophic factor and neurotrophin-3 mRNA in the hippocampus. J Neurosci 15:1768–1777

The Chimpanzee Sequencing and Analysis Consortium (2005) Initial sequence of the chimpanzee genome and comparison with the human genome. Nature (Lond) 437:69–87

Thoenen H (2000) Neurotrophins and activity-dependent plasticity. Prog Brain Res 128: 183–191

Tokuyama W, Okuno H, Hashimoto T, Xin Li Y, Miyashita Y (2000) BDNF upregulation during declarative memory formation in monkey inferior temporal cortex. Nat Neurosci 3: 1134–1142

Vogt BA, Pandya DN (1987) Cingulate cortex of the rhesus monkey: II. Cortical afferents. J Comp Neurol 262:271–289

Vogt BA, Pandya DN, Rosene DL (1987) Cingulate cortex of the rhesus monkey: I. Cytoarchitecture and thalamic afferents. J Comp Neurol 262:256–270

Vogt BA, Finch DM, Olson CR (1992) Functional heterogeneity in cingulate cortex: the anterior executive and posterior evaluative regions. Cerebral Cortex 2:435–443

5
Descent of the Larynx in Chimpanzees: Mosaic and Multiple-Step Evolution of the Foundations for Human Speech

Takeshi Nishimura

1 Introduction

Extant humans are unique in sharing language, which has doubtlessly contributed to the unfolding of humanity and civilizations, and which has a complex and multilayered configuration not found in any other animal. The evolutionary emergence of this human capacity has long been debated in many scientific fields, including linguistics, information science, and neuroscience (see Christiansen and Kirby 2003). These studies have evaluated how preexisting biological and cognitive foundations, for example, the capacity for imitation and learning, might have led to the emergence of language. The evolution of these foundations per se has been also examined in nonhuman animals based on biological evolutionary concepts, using methods in cognitive science, ethology, and neuroscience (see other chapters in this volume). Thus, the evolution of language remains one of the most enigmatic issues in studies on human evolution and is a challenge that attracts many scholars.

The evolution of human vocalizations, viz speech, has attracted much interest for understanding the evolution of language. No human groups lack verbal speech whereby concepts can be communicated, although some groups lack media such as writing. Human speech shares a distinct feature in that humans can regularly utter several phonemes—including vowels and consonants—sequentially and rapidly in a single exhalation. Humans are uniquely endowed with this faculty. In contrast, nonhuman mammals usually utter a single phoneme in a single exhalation, although there may be gradual changes in amplitude and pitch. Here, it must be noted that speech per se is not the same as language and does not necessarily reflect the high intelligence of humans. Although speech is just a kind of vocalization, this sophisticated feature of human speech allows us to turn much information that is encoded by language in the brain into sounds and to communicate it with others rapidly and efficiently. It is the best media for this efficient exchange of information, which is one of the functions of language and which has been essential for the

Laboratory of Physical Anthropology, Department of Zoology, Graduate School of Science, Kyoto University, Kitashirakawa Oiwake-cho, Sakyo-ku, Kyoto 606-8502, Japan

unfolding of human uniqueness. Even if language and speech arose independently in the human lineage, understanding the evolution of speech will thus shed light on the evolution of the language with which we are endowed today.

Language per se leaves no fossil or archaeological trace, which has prevented morphologists and paleoanthropologists from examining the issue of language evolution directly. However, these disciplines have made major contributions to the scientific understanding of when speech might have evolved (Lieberman and Crelin 1971; Lieberman et al. 1972; Laitman et al. 1979; Laitman and Heimbuch 1982; Arensburg et al. 1989; Kay et al. 1998; MacLarnon and Hewitt 1999). Speech can be explained in acoustic terms, and its unique acoustic feature is achieved by sophisticated manipulation of the physical foundations: the vocal apparatus. Human speech makes use of the same peripheral machinery as other mammalian vocalizations. However, several important and unique anatomical modifications enable humans to produce the sounds of speech (Lieberman et al. 1969; Lieberman 1984; Fitch and Hauser 1995; Fitch 2000). Such modifications have been evaluated in comparative studies of extant primates and fossil humans, and most of them are believed to have arisen primarily with advantages for speech (Lieberman 1984). It is generally believed that the anatomical foundations for speech arose during a single evolutionary shift. Many paleoanthropological and morphological studies have been based on this concept and have involved searches for "a" morphological basis for the faculty of speech. This is often the case for studies on the evolution of language, in that the biological or cognitive foundations of human language are believed to have arisen primarily with advantages to language, and not to any other faculties. The distinct unique feature of language could have led to this simplistic view. In this chapter, I review the developmental changes in the vocal apparatus of chimpanzees. I also challenge these traditional views and propose a new concept: a mosaic and multiple-step model of the morphological foundations for the evolution of speech and their secondary adaptations for speech in the human lineage.

2 Anatomical Foundations of Speech

Humans and nonhuman mammals principally make use of the same machinery for speech and vocalization: the lungs for generating sound power, the larynx and vocal tract for phonation and articulation, and the ears for perception. Together, speech physiology and acoustic theory reveal that humans share a unique anatomy of the vocal apparatus that underlies its sophisticated manipulation for speech production. The interested reader can consult Chiba and Kajiyama (1941), Fant (1960), Stevens (1998), or Titze (1994) for the details of the theory; see Lieberman and Blumstein (1988) or Fitch and Hauser (2003) for an intuitive description. Based on these works, the physiology and anatomical foundations for speech are as follows.

2.1 Physiology of Vocalization

Sounds uttered from the mouths of mammals, including humans, are classified into two main types: voiced and unvoiced sounds. The former are accompanied by vibrations of the vocal folds of the glottis, producing, for example, vowels or the pant-hoots of chimpanzees. The latter are not associated with vocal fold vibration and include, for example, consonants or smacking sounds. The voiced sounds form the platform for vocal communication. These sounds are essentially produced by a common set of physiological and anatomical mechanisms in most mammal vocalizations, including human speech (Fig. 1): exhalation from the lung, phonation in the glottis of the larynx, and articulation in the supralaryngeal vocal tract (hereafter, SVT).

The power for creating sounds is via the exhaled airflow produced by compression of the pulmonary volume (lungs). The airflow reaches the glottis of the larynx, which is composed of bilateral vocal folds. The airflow runs up through the narrow channels between the vocal folds, causing them to vibrate. The vibrations produce sequential air puffs, which comprise the sound sources (Fig. 1). These are conventionally called "laryngeal sounds" or "glottal sources," but they are not sounds heard by us. This physiological mechanism is termed "phonation," and it determines most of the tonal characters such as intensity, loudness, and fundamental frequency ("pitch").

The SVT, from the glottis to the lips, functions as the resonator for the laryngeal sounds to generate voiced sounds with some bands of the formant frequencies (see Fig. 1). This mechanism is called "articulation." The distribution pattern of the formants defines the phoneme of the voiced sounds heard by us, including, for example, the different kinds of vowels in human speech. These are

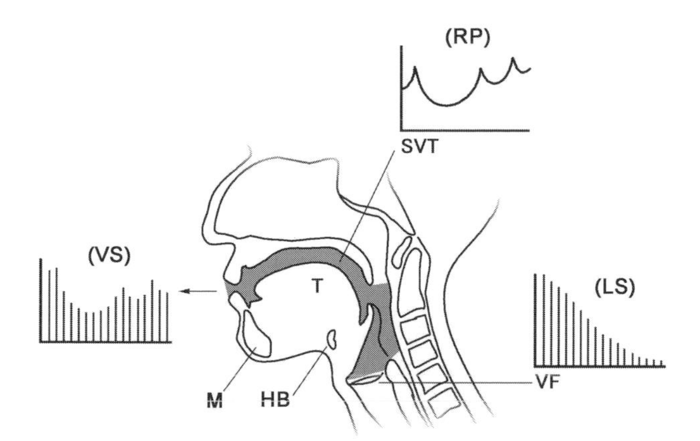

Fig. 1. Diagram for the acoustic theory in vocalizations. The sound sources, laryngeal sounds (*LS*), are produced by the vibrations of vocal folds (*VF*) in the larynx. The supralaryngeal vocal tract (*SVT*, colored in *gray*) functions as the resonator for the laryngeal sounds, to generate the voiced sounds (*VS*) uttered from the mouth. The resonant properties (*RP*) of the tract are a function of the sequential cross-sectional areas of the tract. The areas are modified by the movements of the tongue (*T*), hyoid bone (*HB*), and mandible (*M*)

determined by the resonant properties of the SVT, and these properties are a function of the sequential cross-sectional areas of the tract (Fig. 1). The areas are modified by the movements of the tongue, the hyoid, and the jaw. Nonhuman mammals have constraints to sequential and rapid modifications in the cross-sectional areas of the SVT (Lieberman et al. 1969; Lieberman 1984; Fitch 2000), although they can change the phonetic parameters—amplitude and pitch—in a single exhalation. In contrast, humans are endowed with the faculties for extensive manipulation to change the distribution of the formant frequencies sequentially even in a short single exhalation, forming complex phonemes.

2.2 Two-Tube System of the Vocal Tract

Sequential and rapid modifications of the shape of the SVT are indispensable to the production of human speech, but this is not the case in the vocalizations of nonhuman mammals. Although some neurological modifications underlie such sophisticated manipulations in humans, anatomical constraints on the SVT also prevent nonhuman mammals from such faculties (Lieberman et al. 1969; Lieberman 1984; Fitch 2000).

The SVT in most mammals, including humans, is principally composed of two cavities: the horizontal oral cavity extending from the lips to the velum and the vertical pharyngeal cavity from the velum to the glottis (see Fig. 2). In nonhuman mammals, the oral cavity is very long, but the pharyngeal cavity is much shorter or small, and the epiglottis, which is attached to the thyroid cartilage of the laryngeal skeleton, is locked to the velum to prevent the latter cavity from facing the movable tongue (Fig. 2a) (Negus 1949; Wind 1970; Lieberman 1984; Laitman and Reidenberg 1993; Dyce et al. 1996). The tongue is long in the horizontal direction. Although this anatomy allows the oral cavity to function as a single resonator, it prevents the pharyngeal cavity from doing much in that capacity (Lieberman 1984; Fitch 2000; Fitch and Hauser 2003). In addition, tongue anatomy restricts the extent to which nonhuman mammals can efficiently modify even the surface shape of the oral cavity (Takemoto and Ishida 1995). Thus, the single-tube system imposes physical constraints on the range of vocal behavior possible in nonhuman mammals.

Humans, in contrast, are endowed with a unique anatomical foundation: the "two-tube system" of the SVT. Humans form equally long oral and pharyngeal cavities in adults (Fig. 2c) (Lieberman 1984; Crelin 1987; Zemlin 1988). The human epiglottis is separated from the velum which secures the long oropharyngeal region facing the dorsal surface of the tongue (Lieberman 1984; Crelin 1987; Zemlin 1988). The tongue is round to fit this configuration, and its internal musculature makes the surface highly mobile (Takemoto 2001). In anatomical terms, these features allow the shapes of the oral and pharyngeal cavities to be modified sequentially, rapidly, and semiindependently of each other. It means that this system facilitates humans producing complex resonance property,

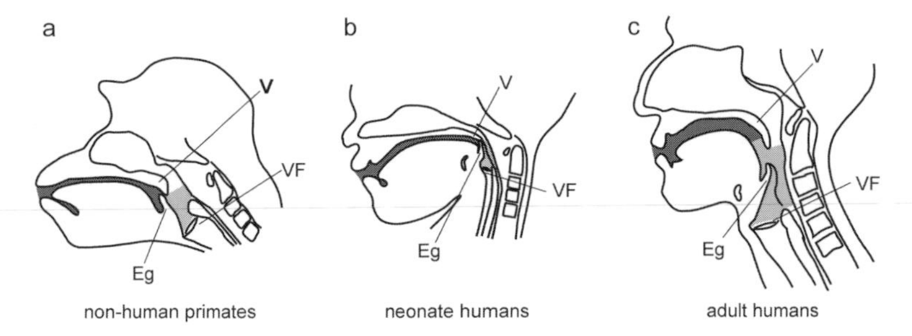

Fig. 2. Midsagittal diagrams of the head and neck in nonhuman primates, human neonates, and adults. **a** Nonhuman primates. The oral cavity (*dark gray*) is very long, but the pharyngeal cavity (*light gray*) is much shorter. Their epiglottis (*Eg*) is locked to the velum (*V*). **b** Neonate humans. The SVT configuration is similar to that in nonhuman primates. **c** Adult humans. The SVT forms the two-tube configuration with equally long oral and pharyngeal cavities. The epiglottis is separated from the velum which secures the long oropharyngeal region. *VF*, vocal fold

giving an acoustical advantage to produce a wide range of acoustically different phonemes even in a short single exhalation, sequentially and rapidly. Thus, the two-tube system contributes greatly to speech production in humans: it acts as the "linguistic hardware."

2.3 Descent of the Larynx in Humans

Some important anatomical constraints of nonhuman mammals preclude the two-tube system required for true speech. In humans, this system is believed to depend on the lower position of the larynx to the palate along the neck and on the flatter face relative to any other mammal (Lieberman 1987). These features form longer vertical pharyngeal and shorter horizontal oral cavities in humans, respectively. However, in human neonates the larynx is positioned close to the palate as with other mammals, even though the face is flat (Fig. 2b) (Wind 1970; Crelin 1973, 1987; Zemlin 1988). Although this high position of the larynx makes little vertical pharyngeal space possible at birth, the human larynx descends rapidly relative to the palate in the infant and early juvenile periods (Wind 1970; Crelin 1973, 1987; Zemlin 1988), to arrive at the adult level of the pharyngeal cavity at about 9 years of age (Lieberman DE et al. 2001). This descent also makes the epiglottis descend relative to the velum, thereby lengthening the oropharyngeal region (Crelin 1973, 1987; Sasaki et al. 1977). Thus, in human ontogeny, this descent of the larynx primarily contributes to development of the two-tube system with a great advantage to speech development.

The laryngeal skeleton, composed of the laryngeal cartilages, is suspended from the hyoid bone and the hyoid is in turn suspended from the mandible and cranial base, through various muscles and ligaments (Fig. 3) (Crelin 1987; Zemlin 1988; Williams 1995). They are not directly articulated with any skeletal

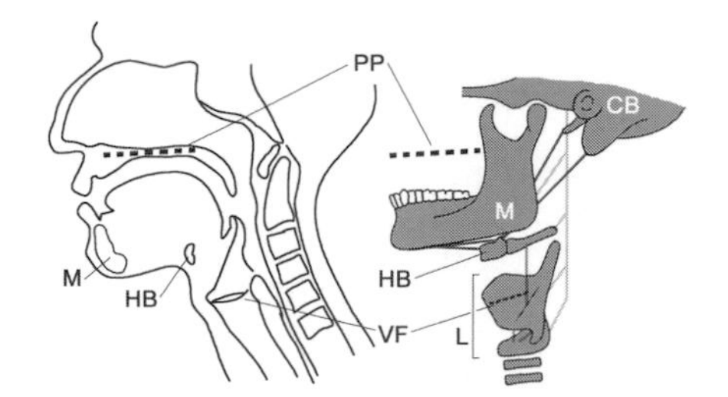

Fig. 3. Schema of the hyo–laryngeal complex and functional related structures. The laryngeal skeleton (*L*) is suspended from the hyoid (*HB*), and the hyoid is in turn suspended from the mandible (*M*) and cranial base (*CB*), through various muscles and ligaments (*gray lines*). *PP*, palatal plane; *VF*, vocal folds

structures, which means that anatomically the descent of the larynx is accomplished both through the descent of the laryngeal skeleton relative to the hyoid and through that of the hyoid relative to the palate (Fig. 3). This process has been evaluated using X-ray photographs and magnetic resonance imaging (MRI) in humans (Westhorpe 1987; Flügel and Rohen 1991; Fitch and Giedd 1999; Vorperian et al. 1999; Lieberman DE et al. 2001). According to these studies, in infancy there is a rapid double descent of the human laryngeal skeleton relative to the hyoid and of the hyoid relative to the palate. Both these processes then slow down to continue until 9 years of age. In early infancy (by 18 months of age), human infants develop to the point where they can utter several different phonemes, including vowels and consonants, sequentially in a single exhalation, although the full acoustical features of speech are achieved later (Oller 1980; Stark 1980). Thus, the great dual descent of the laryngeal skeleton and hyoid likely underlie speech development in early infancy and later.

3 Development of the Vocal Apparatus in Chimpanzees

Human speech is essentially attributable to the unique anatomy of the speech apparatus. While the face stays flat in growth, the laryngeal position descends rapidly in infancy, and this contributes greatly to the final placement of the human SVT in human "ontogeny." However, these facts do not support the "evolutionary" hypothesis that the descent arose to establish the two-tube system, leading to the origin of speech. This issue has been resolved by comparative studies on development in vocal apparatus anatomy between humans and their close phyletic relatives using new medical imaging methods.

3.1 Approaches to Unveil the Hidden Apparatus

There have been few comparative studies on developmental changes in the morphology and physiology of the vocal apparatus in humans and nonhuman mammals (Taylor et al. 1976; Flügel and Rohen 1991). As already described, the vocal apparatus is composed mainly of cartilaginous skeleton, muscles, and ligaments and is surrounded by the bony elements of the cranium and mandible.

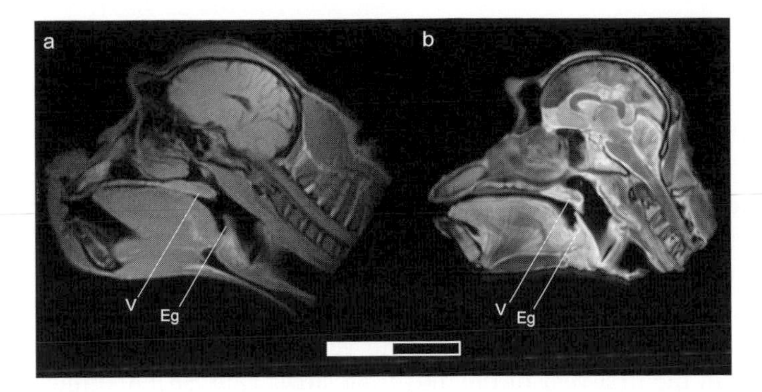

Fig. 4. Magnetic resonance images of a living subject (**a**) and embalmed cadaver (**b**) of adult male chimpanzees. The latter shows shrinkage and distortion in the soft tissues; for example, in the latter, the hyo-laryngeal complex is slightly pulled up relative to the palate, so the epiglottis (*Eg*) is touching the velum (*V*). The skin had been removed in this cadaver specimen. *Scale* 5 cm

Historically, this hampered anatomical and physiological examination of the intact vocal apparatus in living subjects. For example, radiography and cineradiography had been used conventionally for this kind of study, but these methods are not useful for imaging soft tissues or objects surrounded by bony structures. In addition, X-ray exposure restricts detailed and repetitive examinations on living subjects. Most studies approached this challenge by gross anatomical dissection and radiography of cadavers. However, these approaches are often associated with the potential risk of shrinkage and distortion in the tissues, which inevitably influence the results (Fig. 4). These technical limitations severely constrained comparative studies of the anatomical and physiological development of the vocal apparatus.

Newer medical imaging techniques such as computed tomography (CT) and MRI have now been introduced to the fields of physical anthropology and primatology and have allowed significant breakthroughs (Nishimura et al. 1999). These modalities provide tomograms for a region of interest in living subjects with no destruction of tissue and with minimum invasiveness. In particular, MRI has a great advantage for imaging the soft tissues with no risk of X-ray exposure to subjects, and it permits repetitive scans. Thus, these new imaging techniques allow us to evaluate the developmental changes in the vocal apparatus longitudinally using the same living subjects. They have contributed and will contribute greatly to our understanding of the processes in humans and nonhuman animals.

3.2 Descent of the Larynx in Living Chimpanzees

Although there are a few studies on the developmental changes of the vocal apparatus in some nonhuman mammals, it remains mostly unclear when, how, and why the unique features in humans evolved. Using MRI, we have succeeded in evaluating the developmental changes in the SVT shape in chimpanzees and in making a detailed comparison with established human data (Lieberman DE et al. 2001), which has largely solved this enigmatic issue (Nishimura et al. 2003; Nishimura 2005).

In one of the studies (Nishimura et al. 2003), three chimpanzee infants named Ayumu, Cleo, and Pal, at the Primate Research Institute of Kyoto University, Japan, were scanned in the first 2 years of life. This study showed that their larynx also descended relative to the palate to change the proportions of the SVT during infancy, as in human infants. In chimpanzees, the vertical pharyngeal cavity (SVT_V) grows rapidly in the first year of life and then slows (Fig. 5b). A similar developmental pattern is observed in human infants (Fig. 5b). Thus, the larynx rapidly descends relative to the palate during early infancy in both humans and chimpanzees. During that period, this descent of the chimpanzee larynx also makes the epiglottis descend relative to the velum, and it completely loses contact with the velum in early infancy, as in humans (Fig. 5c). Nevertheless, there are some differences between chimpanzees and humans in the changes in the spatial relations between the palate, hyoid, and laryngeal skeleton. In humans, descent of the larynx is achieved by the descent of the laryngeal skeleton relative to the hyoid bone (Fig. 5d) and by the descent of the hyoid relative to the palate, even in infancy (Fig. 5e). In contrast, in chimpanzee infants, it is caused primarily by the descent of the laryngeal skeleton relative to the hyoid bone (Fig. 5d), and it is not accomplished by a descent of the hyoid relative to the palate (Fig. 5e). Thus, although chimpanzees and humans show similar growth patterns in the SVT_V during infancy, they share the descent of the laryngeal skeleton relative to the hyoid, but not the descent of the hyoid in that period (Nishimura et al. 2003).

In the chimpanzees, the horizontal oral cavity (SVT_H) grows similarly to that in humans during infancy, although the length per se is longer in chimpanzees than in humans (see Fig. 5b). In both chimpanzees and humans, the growth of the SVT_H is slower than that of the SVT_V in that period (Fig. 5f). Thus, in chimpanzee infants, despite no descent of the hyoid, the descent of the laryngeal skeleton results in a proportional change to the SVT similar to that in human infants; the SVT develops toward a configuration where the SVT_V length is equal to that of the SVT_H (Fig. 5f; Nishimura et al. 2003).

The MRI study (Nishimura et al. 2003) did not show developmental changes of the SVT in juvenile chimpanzees. On the other hand, another MRI study (Nishimura 2005), using a cross-sectional ontogenetic series of embalmed specimens, showed that although in the juvenile period the SVT_V grows slightly, the SVT_H grows greatly in chimpanzees compared with humans (Fig. 6a); this clearly differentiates the proportion of the SVT in chimpanzees from that in humans (Fig. 6b). Thus, descent of the chimpanzee larynx may depend primarily on the descent of the laryngeal skeleton relative to the hyoid, and the lack of the descent of the hyoid in chimpanzees leads to the different architecture of the SVT in

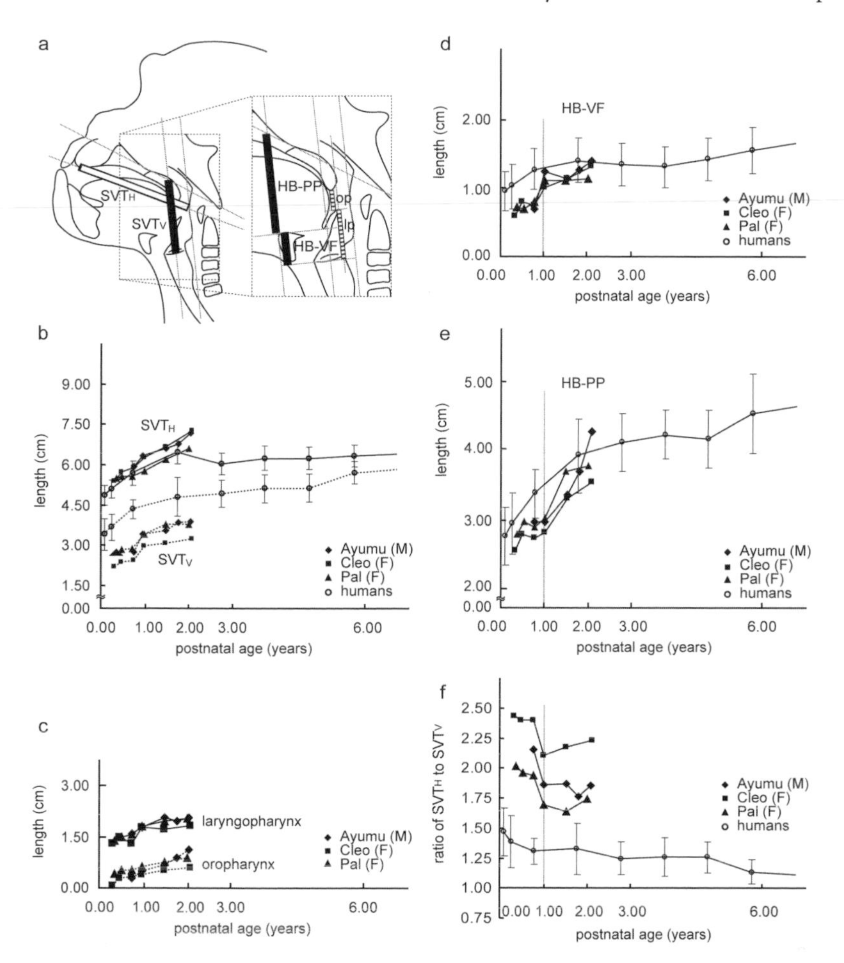

Fig. 5. Developmental changes in the shape of the SVT in living chimpanzees and humans. **a** Diagram of the dimensions of the SVT. **b** Growth of the SVT_H (*continuous line*) and the SVT_V (*dotted line*). **c** Growth of the laryngopharyngeal (*lp; continuous line*) and oropharyngeal (*op; dotted line*) parts of the vertical pharyngeal cavity. **d** Distance from the hyoid to the vocal folds (*HB–VF*). **e** Distance from the hyoid to the palate (*HB–PP*). **f** Age-related changes in the ratio of SVT_H to SVT_V lengths. See Lieberman DE et al. (2001) and Nishimura et al. (2003) for detailed explanations of methods and for definitions of the dimensions. [Measurements in chimpanzees and humans are from Nishimura et al. (2003) and Lieberman DE et al. (2001), respectively, with permission]

humans and chimpanzees (Nishimura et al. 2003; Nishimura 2005). Unfortunately, artifacts of embalming precluded the study by Nishimura (2005) from examining the developmental changes in the spatial relations between the palate, hyoid, and laryngeal skeleton in the juvenile period. Nevertheless, we plan to continue to use MRI to examine living subjects and to evaluate this issue, which should provide valuable information for discussing the proximate mechanisms leading to the unique conformation of the human SVT.

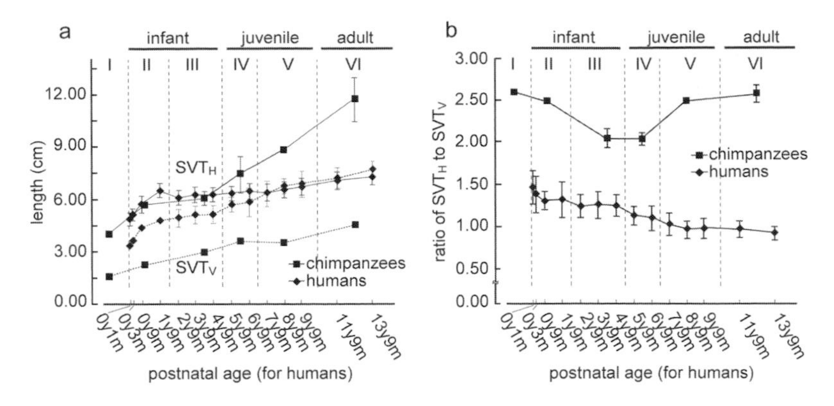

Fig. 6. Developmental changes in the shape of the SVT in a cross-sectional ontogenetic series of embalmed specimens of chimpanzees and living humans. **a** Growth of the SVT_H (*continuous line*) and the SVT_V (*dotted line*). **b** Age-related changes in the ratio of SVT_H to SVT_V lengths. *Roman numerals* represent the dental ages defined in Nishimura (2005). See Fig. 5 for a diagram of the dimensions of the SVT, and Lieberman DE et al. (2001) and Nishimura (2005) for detailed explanations of methods and for definitions of the dimensions. [Modified from Nishimura 2005. Measurements in chimpanzees and humans are from Nishimura (2005) and Lieberman DE et al. (2001), respectively, with permission]

3.3 Evolution of the Descent of the Larynx and the Two-Tube System

The new imaging methods succeeded in showing that chimpanzee infants share with human infants a rapid descent of the laryngeal skeleton relative to the hyoid, to change the proportions of the SVT. Thus, the latest this developmental descent must have arisen is in the last common ancestor of chimpanzees and humans. According to the anatomical examination of the hyo-laryngeal complex (Nishimura 2003a), all nonhuman hominoids have at least one feature in common with humans: the laryngeal skeleton is well separated from and assured of mobility independent of the hyoid (Fig. 7a,b). By contrast, both Old World and New World monkeys have anatomical features that contrast sharply with hominoids: the laryngeal skeleton is locked into and tied tightly with the hyoid so that the hyo-laryngeal complex acts as a functional unit (Fig. 7c). Anatomically, the hominoid type of the complex is likely to develop by the descent of the laryngeal skeleton relative to the hyoid, which strongly suggests that this descent probably arose in a common ancestor of all extant hominoids (Nishimura 2003a; Nishimura et al. 2003), although the developmental changes in the SVT are yet to be explored in other hominoids apart from chimpanzees and humans.

The rapid descent of the hyoid relative to the palate has not been identified in nonhuman primates to date, which may mean that this descent arose in the human lineage, in combination with modifications in the development of the mandible (Nishimura 2005). The hyoid is tightly anchored to the mandible by

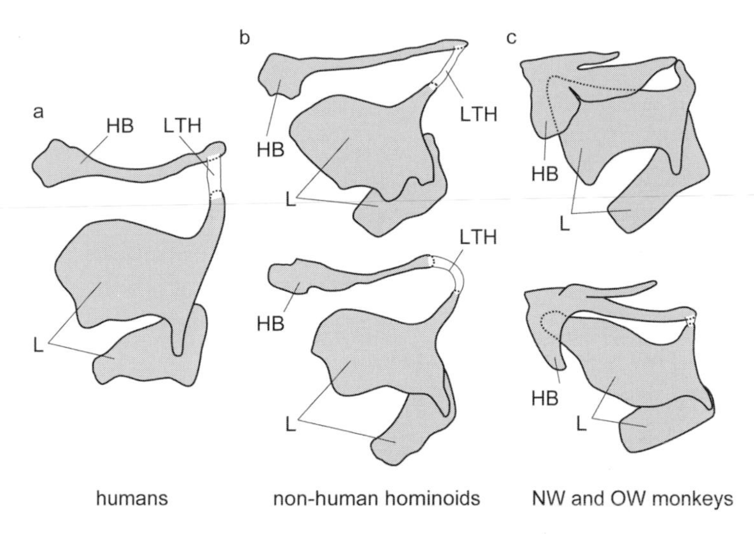

Fig. 7. Schema of the hyo-laryngeal complex from the left perspective. **a** Human. The laryngeal skeleton (*L*) is well separated from and connected through the long lateral thyrohyoid ligament (*LTH*) with the hyoid (*H*). There is a wide space between them, and the former is allowed to move independently of the latter. **b** White-handed gibbon, *Hylobates lar* (*top*), and chimpanzee, *Pan troglodytes* (*bottom*). All nonhuman hominoids have the feature in common with humans. **c** White-fronted capuchin, *Cebus albifrons* (*top*), and stump-tailed macaque, *Macaca arctoides* (*bottom*). The laryngeal skeleton is locked into and tied tightly with the hyoid. The hyo-laryngeal complex acts as a functional unit, to restrict the independence of their movements. *NW*, New World; *OW*, Old World. [Modified from Nishimura (2003), with permission]

muscles, ligaments, and membranes (see Fig. 3; Crelin 1987; Zemlin 1988; Williams 1995). This morphological restriction possibly maintains the spatial relationship between the two during growth in humans. In fact, in humans the hyoid descends relative to the palate, along with the superior–inferior growth of the mandibular ramus (Lieberman DE and McCarthy 1999; Lieberman DE et al. 2001). Taking into consideration similar anatomical characteristics in this region in chimpanzees (Swindler and Wood 1973), this process may also be the case for chimpanzees (Nishimura 2005). Mandibular ramus height is shorter relative to the horizontal dimension of the mandible body—including the ramus width—in adult chimpanzees than in adult humans (Swindler and Wood 1973; Johnson et al. 1976; Aiello and Dean 1998). This adult configuration is explained mostly by remodeling of bone on the surface of the mandible during growth (Enlow and Harris 1964; Atkinson and Woodhead 1968; Johnson et al. 1976; Bromage 1992; see also Enlow 1990). Thus, if the hyoid descent is not shared by chimpanzees, mandibular growth may underlie the evolution of the hyoid descent, resulting in the unique two-tube system of the SVT in the human lineage (Nishimura 2005).

In conclusion, the descent of the larynx arose at least in part before the divergence of human from chimpanzee lineages. The descent of the laryngeal skeleton relative to the hyoid, which may occur principally during infancy, arose at

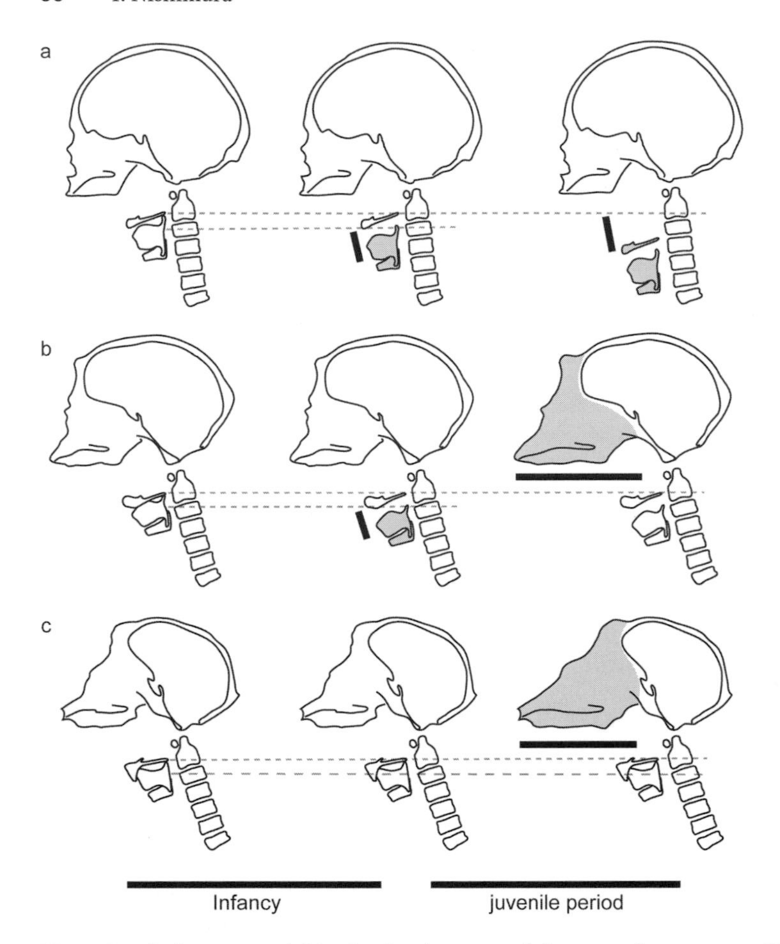

Fig. 8. Evolutionary model in the development of the two-tube system. **a** Humans. The laryngeal skeleton and hyoid descend relative to the hyoid and palate, respectively. The former and latter occur principally during the infant (*left*) and juvenile (*right*) periods, respectively. The face remains flat in growth. **b** Nonhuman hominoids. The laryngeal skeleton descends relative to the hyoid principally during infancy, but the hyoid position is kept at the newborn level. Nevertheless, in the juvenile period, the greater growth of the face develops the long horizontal oral cavity to form the single-tube system. **c** New World and Old World monkeys. The position of the laryngeal skeleton and hyoid bone are kept at the newborn level. In the juvenile period, the greater growth of the face develops the long horizontal oral cavity to form the single-tube system

latest in the last common ancestor of the extant hominoids (Fig. 8). On the other hand, the descent of the hyoid relative to the palate may have arisen in the human lineage. If it is true, the descent of the larynx must have evolved in at least two steps, one in the common ancestor of extant hominoids and one in the human lineage (Fig. 8; Nishimura 2003a,b, 2005; Nishimura et al. 2003), which means that the two-tube system of the human SVT did not evolve in a single step in human evolution.

4 Functional Adaptations for Speech Foundations

The vocal apparatus is associated with various functions, such as breathing, deglutition, and air-trapping in locomotion, besides phonation and articulation (Negus 1949; Lieberman 1984; Williams 1995; Hayama 1996; Schwenk 1999). The descent of the larynx at least partially arose before the divergence of human lineage, so it evolved with some advantages to functions other than speech. Thus, it seems likely that its evolutionary progression would have been affected inevitably by some of these associated activities. It is therefore improbable that any single selective advantage could have accounted for laryngeal descent. Unfortunately, there is insufficient information to elucidate these factors, for example, the comparative physiology of feeding, swallowing, or respiration (see Schwenk 1999). I now survey possible modifications in swallowing and acoustic physiologies that may have been caused by spatial rearrangements in the hyo-laryngeal apparatus and discuss functional adaptations underlying the descent of the larynx.

4.1 Modifications in the Swallowing Mechanism

The larynx comprises the orifice of the trachea, and it originally appeared during the evolution of vertebrate animals with pulmonary respiration, to shut off the approach into the trachea to prevent aspiration: accidental entrance of swallowed food or liquid boluses into the trachea (Negus 1949; Wind 1970; Harrison 1995). In mammals, the pharynx and larynx evolved to transfer the swallowed food or liquid bolus from the ventral oral cavity to the dorsal esophagus and the breathed air from the dorsal nasal cavity to the ventral trachea (Negus 1949; Wind 1970; Smith 1992; Harrison 1995). Thus, spatial rearrangements in the hyo-laryngeal complex may have affected or been affected by modifications in the physiological mechanism of swallowing.

In humans, the descent of the larynx makes the epiglottis lose contact with the velum in early infancy, between 4 and 6 months of age (Crelin 1973; Sasaki et al. 1979). This conformational change anatomically increases the risk of accidental aspiration during swallowing (Lieberman 1984; Laitman and Crelin 1980). Accompanying this conformational change, the adult mode of swallowing develops in human infants to decrease this risk (Sasaki et al. 1979). In this process, the hyoid ascends, the larynx approximates the hyoid, the epiglottis bends, and the laryngeal orifice closes (Ekberg 1982, 1986; Ekberg and Sigurjónsson 1982). In humans, when a food or liquid bolus enters through the pharynx into the esophagus, the laryngeal skeleton always moves anterosuperiorly toward the hyoid (Fig. 9); this applies stress to the connective tissue between the hyoid and epiglottis, which stress rotates the epiglottis back from an upright to a transverse position (Fink et al. 1979; Ekberg and Sigurjónsson 1982; Vandaele et al. 1995; Reidenbach 1997). This movement enables the epiglottis to close the laryngeal orifice (see Fig. 9). These mechanisms in humans ensure that

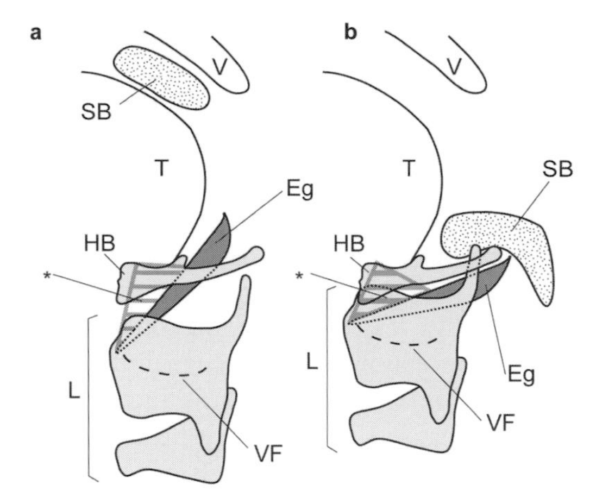

Fig. 9. Human adult mode of swallowing: **a** respiratory phase; **b** pharyngeal stage of deglutition. The laryngeal skeleton (*L*) moves anterosuperiorly towards the hyoid (*HB*), which applies stress to the connective tissue (*, *shaded region*) between the hyoid and epiglottis (*Eg*). This stress rotates the epiglottis back from an upright (**a**) to a transverse (**b**) position, to ensure that the swallowed bolus (*SB*) passes over the rotated epiglottis to the esophagus. *T*, tongue; *V*, velum; *VF*, vocal fold

the swallowed bolus of food or liquid passes over the rotated epiglottis to the esophagus without causing choking. This mode of swallowing requires an adequate space and independent mobility between the laryngeal skeleton and the hyoid. The laryngeal skeleton descends relative to the hyoid to secure the space and weaken the physical linkages between them. Thus, this descent possibly contributed greatly to the development of the human adult mode of swallowing (Nishimura 2003a).

In nonhominoid primates, the larynx does not descend relative to the palate, so the epiglottis always touches the velum and remains in the intranarial position even in the adult (Negus 1949; Geist 1965; Laitman et al. 1977; Crompton et al. 1997). Although the mechanism is controversial (Negus 1949; Laitman et al. 1977; Larson and Herring 1996; Crompton et al. 1997; Dyce et al. 1996), in these animals a masticated food or liquid bolus is believed to pass through deep lateral channels—the piriform recesses—on either side of the laryngeal orifice (Fig. 10). Even while swallowing, their epiglottis is kept erect to ensure that the masticated bolus flows from the oral cavity to the recesses: in some cases, it acts like as a snorkel to the nasal cavity for maintaining breathing (Fig. 10). It is possibly caused, in part, by morphological constraints on their hyo-laryngeal complex. The laryngeal skeleton is very close to and tied with the hyoid, and this prevents independent mobility relative to the hyoid (Nishimura 2003a). In fact, this mode of swallowing is observed in human infancy, but is modified to the adult mode along with the descent of the larynx (Sasaki et al. 1979). Thus, in nonhominoid anthropoids, lack of the descent of the laryngeal skeleton restricts any flexible mobility of the laryngeal skeleton and hyoid, as observed in the human adult mode of swallowing (Nishimura 2003a).

The descent of the laryngeal skeleton relative to the hyoid has been identified in chimpanzees and is probably shared with other hominoids (Nishimura et al. 2003). It results in a conformation of the hyo-laryngeal complex in the nonhuman hominoids that is similar to that in humans (Nishimura 2003a). Thus, even nonhuman hominoids share the physical basis for the human adult mode of

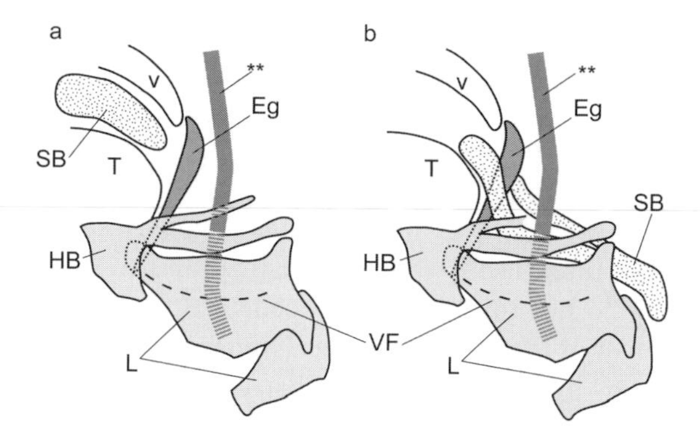

Fig. 10. Nonhuman mode of swallowing: **a** respiratory phase; **b** pharyngeal stage of deglutition. The laryngeal skeleton (*L*) is locked to the hyoid (*HB*), and the epiglottis (*Eg*) touches the velum (*V*) and remains in the intranarial position even while swallowing. The epiglottis ensures that a swallowed bolus (*SB*) passes from the oral cavity through deep piriform recesses on either side of the laryngeal orifice. It acts like a snorkel to maintain a breathing tract (**, *gray line*) from the nasal cavity to trachea. *T*, tongue; *VF*, vocal fold

swallowing and they possibly swallow in a similar manner. However, there is no information on the mode of deglutition in nonhuman hominoids that would allow us to evaluate the foregoing interpretation. There have been many studies on the movement of the epiglottis in nonhuman mammals (Laitman et al. 1977; Larson and Herring 1996; Crompton et al. 1997) and on the transfer of food through the oral cavity, using cineradiography or X-ray television (Franks et al. 1984; Thexton and McGarrick 1989; German et al. 1992; Hiiemae et al. 1995). However, to my knowledge there are no detailed examinations on the interaction of the hyoid and laryngeal skeleton while swallowing in nonhuman mammals. Thus, such information is expected to help us understand the functional background of variations in the hyo-laryngeal configuration and the evolution of the descent of the larynx.

4.2 Physical Basis for Phonation and Articulation

Observations on the descent of the larynx in chimpanzees also shed light on the evolution of the physical foundations of phonation and articulation (Nishimura 2003b; Nishimura et al. 2003). While speaking, humans can utter sequential phonemes planned in the brain, even in a single short exhalation. In humans, the laryngeal sounds depend only slightly on the resonance properties of the SVT (Chiba and Kajiyama 1942; Fant 1960; Fitch 2003). Moreover, phonation and articulation must be elaborately and flexibly modified for human speech (Lieberman 1984; Fitch 2000, 2003). Thus, humans can declaim a given vocal line at many different pitches. The larynx is the physical foundation for phonation, for example, modification of pitch, and the hyoid provides the basis for the tongue movements that participate in articulation (see the second section in this chapter). The descent of the laryngeal skeleton relative to the hyoid ensures that the physical linkages between them are weakened (Nishimura 2003a). Thus, this descent probably contributes to increased flexibility between the activities of phonation and articulation.

The independence of phonation and articulation applies to the vocalizations of nonhuman mammals, including primates, in terms of bioacoustics (Fitch 1997, 2003; Fitch and Hauser 2003). Intentional utterance also applies to them (Hihara et al. 2003). However, they do not imply that the acoustical features of the both are modified flexibly in most of their vocalizations (Hauser et al. 1993). This rigidity is possibly caused in part by morphological constraints on the hyo-laryngeal complex. The laryngeal skeleton and the hyoid are tightly connected to each other in the nonhominoid anthropoids, namely the Old World and New World monkeys, indicating that they function physically as a single unit (Nishimura 2003a). This connection means that the movements of and within the laryngeal skeleton are tightly linked to movements of the tongue. Thus, the lack of descent of the laryngeal skeleton relative to the hyoid anatomically restricts any flexible modifications to those laryngeal sounds and resonant properties that act independently of each other (Nishimura 2003b; Nishimura et al. 2003).

On the other hand, nonhuman hominoids share the physical basis for increased flexibility between the activities of phonation and articulation, to a degree similar to that seen in humans (Nishimura 2003b; Nishimura et al. 2003). Thus, it seems that their vocalizations reflect this property. However, there is little information on the bioacoustics of phonation and articulation in the vocalizations of nonhuman hominoids. Primate vocalizations have often been studied in terms of social context, contributing to studies of their communication (Cheney and Seyfarth 1990; Hauser 1996; Simmons et al. 2002). The independence or interactions of phonation and articulation in nonhuman primates remain a matter for empirical or preliminary study. In the future, artificial synthesis or mathematical simulation techniques that can be used to study their vocalized sounds bioacoustically (see Fitch 2003) will elucidate the physical constraints on their vocalizations and clarify the evolution of the physical bases of phonation and articulation.

4.3 Evolution of the Morphological Foundations for Speech

The descent of the larynx used to be an enigmatic developmental phenomenon. Insufficient comparative information led to the assumption that the descent evolved as an adaptation for speech in a single shift in the human lineage, in combination with flattening of the face and flexure of the cranial base. Communication by verbal language is a unique derived feature of humans and has been essential for the unfolding of many unique aspects of human behavior. This concept has tended to encourage a simplistic view, which has formed a basis for paleoanthropological studies on the origin and evolution of human speech. The evolution of the morphological basis for human speech has been regarded as synonymous with the evolution of the developmental descent of the larynx. In this, the "unique" morphological features marking the low position of the larynx have been examined through comparisons with extant primates. However, recent studies using imaging techniques have redefined this concept: the descent of the larynx did not arise in a single step in the human lineage. The descent of the laryngeal skeleton identified in humans and chimpanzees may confer an advantage for the development of the human adult mode of swallowing. If this suggestion is correct, then the first step of the laryngeal descent evolved through a developmental alternation of the swallowing mechanism at the latest in the last common ancestor of humans and chimpanzees, and possibly of all hominoids. It may have arisen as an adaptation to changes in hominoid diet and increases in body size, and secondarily facilitated the flexible controls of phonation and articulation (Nishimura 2003; Nishimura et al. 2003).

It remains unclear when the second phase, descent of the hyoid relative to the palate, arose. If it was in the human lineage, the emergence of the descent of the hyoid may have been related to modifications in mandibular development, such as decreases and increases in mandibular length and height, respectively (Nishimura 2005). These modifications may have accompanied modifications to facial architecture. The latter could underlie another contribution to the evolution of the two-tube system of SVT in humans with a shorter horizontal oral cavity. The evolutionary trend to flatten the face was an early feature of the human lineage (Klein 1989; Aiello and Dean 1998; Fleagle 1999). The australopithecines, however, retained an ape-like face with a robust masticatory apparatus (Rak 1983; Asfaw et al. 1999). During the divergence of genus *Homo* from contemporaneous Pliocene australopithecines, the face became peculiarly flattened with a gracilization of the masticatory system (Klein 1989; Wood 1992; Walker and Leakey 1993; Fleagle 1999; Stedman et al. 2004; Wood and Collard 2004). This rearrangement in the physiognomy of *Homo* possibly marked a distinct shift in human adaptive strategies involving innovations of dietary behavior (Wood 1992; Lieberman DE et al. 2004; Wood and Collard 2004). These strategies could have entailed changes toward a diet requiring less robust mastication, including meat eating, and improvements in food manipulation or in tool use for food preparation. Thus, such modifications associated with nutritional functions may have coevolved with the selective advantages of face flattening, with anatomical consequences for the evolution of hyoid descent and the shorter oral cavity of the SVT in the *Homo* lineage.

In summary, the morphological foundations underlying the separate mechanisms that facilitate speech in modern humans may have evolved under different selection pressures, possibly originally directed toward advantages unrelated to speech. Thus, they may have been secondarily advantageous for human speech. If this hypothesis is correct, the mosaic and multiple-step evolution of the SVT anatomy may have been a preadaptive set of functional modifications leading to the evolution of speech (Nishimura et al. 2003; Nishimura 2005).

5 Summary and Conclusions

I have surveyed the evolution of morphological foundations for speech to argue that morphological approaches to this issue are a necessary part of the study on the evolution of language, both linguistic and psychological. In the second section, I reviewed the physiology of speech and the anatomy of the human SVT. Humans have a two-tube system of the SVT, which underlies elaborate acoustical manipulations for speech production. The unique anatomy develops through the rapid descent of the larynx in the infant and early juvenile periods. In the third section, studies on the development in the shape of chimpanzee SVT were surveyed. New medical imaging techniques such as MRI showed that chimpanzees share the descent of the larynx with humans, at least partially. This finding strongly suggests the multiple—perhaps two-step—evolution of the two-tube system. In the fourth section, I proposed a new concept for the evolution of the morphological foundations for speech, based on separate and functionally significant modifications in spatial configuration between the laryngeal skeleton, hyoid, and palate. Selection pressures with advantages unrelated to speech have possibly played important roles in the evolutionary modifications of the laryngeal apparatus, and these may have been secondarily advantageous to speech in the human lineage. This concept may pave the way to better understanding of the evolution of speech. I argue that if we overvalue only one feature as a clue to the evolution of the speech, the evolutionary trail of speech will be misrepresented. A multidisciplinary approach to studies of the various aspects of speech should contribute greatly to our understanding of the evolution of the biological foundations of speech in the long mammalian and primate evolution. Speech evolved not only in its physical foundations, but also by the coevolution and separate evolution of the physiological and neurological foundations for the many systems that contribute to speech production. Many of these separate biological foundations of speech are considered to have evolved independently under different selection pressures unrelated to speech, and this may be also the case for the evolution of language.

This chapter may help to provide us with new concepts on the evolution of speech in terms of its morphological basis: the linguistic hardware. I hope it will also stimulate future studies on the evolution of other biological bases of speech and language.

Acknowledgments

This work was supported by a Research Fellowship of the Japan Society for the Promotion of Science (JSPS) for Young Scientist (no. 16000326, to T.N.), by Grants-in-Aid for the Biodiversity Research of the 21st Century Center of Excellence (no. A14, to Kyoto University), and for Specially Promoted Research (no. 12002009, to Professor Tetsuro Matsuzawa) from the Ministry of Education, Culture, Sports, Sciences, and Technology of Japan, and by JSPS core-to-core program HOPE (no. 15001, to Professor Tetsuro Matsuzawa). I gratefully thank for the comments of the editors on the earlier version of the manuscript.

References

Aiello L, Dean C (1998) An introduction to human evolutionary anatomy. Academic Press, New York

Arensburg B, Tiller AM, Vandermeersch B, Duday H, Schepartz LA, Rak Y (1989) A middle Palaeolithic human hyoid bone. Nature (Lond) 338:758–760

Asfaw B, White T, Lovejoy O, Latimer B, Simpson S, Suwa G (1999) *Australopithecus garhi*: a new species of early hominid from Ethiopia. Science 284:629–635

Atkinson PJ, Woodhead C (1968) Changes in human mandibular structure with age. Arch Oral Biol 13:1453–1463

Bromage TG (1992) The ontogeny of *Pan troglodytes* craniofacial architectural relationships and implications for early hominoids. J Hum Evol 23:235–251

Cheney DL, Seyfarth RM (1990) How monkeys see the world: inside the mind of another species. University of Chicago Press, Chicago

Chiba T, Kajiyama M (1942) The vowel: its nature and structure. Tokyo-Kaiseikan, Tokyo

Christiansen MH, Kirby S (eds) (2003) Language evolution. Oxford University Press, New York

Crelin ES (1973) Functional anatomy of the newborn. Yale University Press, New Haven

Crelin ES (1987) The human vocal tract. Vantage Press, New York

Crompton AW, German RZ, Thexton AJ (1997) Mechanisms of swallowing and airway protection in infant mammals (*Sus domesticus* and *Macaca fascicularis*). J Zool (Lond) 241: 89–102

Dyce KM, Sack WO, Wensing CJG (1996) Textbook of veterinary anatomy, 2nd edn. Saunders, Philadelphia

Ekberg O (1982) Closure of the laryngeal vestibule during deglutition. Acta Otolaryngol 93:123–129

Ekberg O (1986) The normal movements of the hyoid bone during swallow. Invest Radiol 21:408–410

Ekberg O, Sigurjónsson SV (1982) Movement of the epiglottis during deglutition. A cineradiographic study. Gastrointest Radiol 7:101–107

Enlow DH (1990) Facial growth, 3rd edn. Saunders, Philadelphia

Enlow DH, Harris DB (1964) A study of the postnatal growth of the human mandible. Am J Orthod 50:25–50

Fant G (1960) Acoustic theory of speech production. Mouton, The Hague

Fink BR, Martin RW, Rohrmann CA (1979) Biomechanics of the human epiglottis. Acta Otolaryngol 87:554–559

Fitch WT (1997) Vocal tract length and formant frequency dispersion correlate with body size in rhesus macaques. J Acoust Soc Am 102:1213–1222

Fitch WT (2000) The evolution of speech: a comparative review. Trends Cogn Sci 4:258–267

Fitch WT (2003) Primate vocal production and its implications for auditory research. In: Ghazanfar AA (ed) Primate audition: ethology and neurobiology. CRC Press, Boca Raton, pp 87–108

Fitch WT, Giedd J (1999) Morphology and development of the human vocal tract: A study using magnetic resonance imaging. J Acoust Soc Am 106:1511–1522

Fitch WT, Hauser M (1995) Vocal production in nonhuman primates: acoustics, phylogeny, and functional constraints on "Honest" advertisement. Am J Primatol 37:191–219

Fitch WT, Hauser M (2003) Unpacking "Honesty": vertebrate vocal production and the evolution of acoustic signals. In: Simmons AM, Popper AN, Fay RR (eds) Acoustic communication. Springer, New York, pp 65–137

Fleagle JG (1999) Primate adaptation and evolution, 2nd edn. Academic Press, San Diego

Flügel C, Rohen JW (1991) The craniofacial proportions and laryngeal position in monkeys and man of different ages (a morphometric study based on CT-scans and radiographs). Mech Aging Dev 61:65–83

Franks HA, Crompton AW, German RZ (1984) Mechanism of intraoral transport in macaques. Am J Phys Anthropol 65:275–282

Geist FD (1965) Nasal cavity, larynx, mouth, and pharynx. In: Hartman CG, Straus WL Jr (eds) The anatomy of the rhesus monkey. Hafner, New York, pp 189–209

German RZ, Crompton AW, Levitch LC, Thexton AJ (1992) The mechanism of suckling in two species in infant mammal: miniature pigs and long-tailed macaques. J Exp Zool 261:322–330
Harrison DFN (1995) The anatomy and physiology of the mammalian larynx. Cambridge University Press, Cambridge
Hauser MD (1996) The evolution of communication. MIT Press, Cambridge
Hauser MD, Evans CS, Marler P (1993) The role of articulation in the production of rhesus monkey (*Macaca mulatta*) vocalizations. Anim Behav 45:423–433
Hayama S (1996) Why does not the monkey fall from a tree? the functional origin of the human glottis (in Japanese with English summary). Primate Res 12:179–206
Hihara S, Yamada H, Iriki A, Okanoya K (2003) Spontaneous vocal differentiation of coo-calls for tools and food in Japanese monkeys. Neurosci Res 45:383–389
Hiiemae KM, Hayenge AM, Reese A (1995) Patterns of tongue and jaw movement in a cinefluorographic study of feeding in the macaque. Arch Oral Biol 40:229–246
Johnson PA, Atkinson PJ, Moore WJ (1976) The development and structure of the chimpanzee mandible. J Anat 122:467–477
Kay RF, Cartmill M, Michelle B (1998) The hypoglossal canal and the origin of human vocal behavior. Proc Natl Acad Sci U S A 95:5417–5419
Klein RG (1989) The human career: human biology and cultural origins. The University of Chicago Press, Chicago
Laitman JT, Crelin ES (1980) Developmental change in the upper respiratory system of human infants. Perinatol Neonatol 4:15–22
Laitman JT, Heimbuch RC (1982) The basicranium of Plio-Pleistocene hominids as an indicator of their upper respiratory systems. Am J Phys Anthropol 59:323–343
Laitman JT, Reidenberg JS (1993) Specialization of the human upper respiratory and upper digestive system as seen through comparative and developmental anatomy. Dysphasia 8:318–325
Laitman JT, Crelin ES, Conlogue GJ (1977) The function of the epiglottis in monkeys and man. Yale J Biol Med 50:43–48
Laitman JT, Heimbuch RC, Crelin ES (1979) The basicranium of fossil hominids as an indicator of their upper respiratory system. Am J Phys Anthropol 51:15–34
Larson JE, Herring SW (1996) Movement of the epiglottis in mammals. Am J Phys Anthropol 100:71–82
Lieberman DE, McCarthy RC (1999) The ontogeny of cranial base angulation in humans and chimpanzees and its implications for reconstructing pharyngeal dimensions. J Hum Evol 36:487–517
Lieberman DE, McCarthy RC, Hiiemae KM, Palmer JB (2001) Ontogeny of postnatal hyoid and larynx descent in humans. Arch Oral Biol 46:117–128
Lieberman DE, Krovitz GE, Yates FW, Devlin M, St Claire M (2004) Effects of food processing on masticatory strain and craniofacial growth in a retrognathic face. J Hum Evol 46:655–677
Lieberman P (1984) The biology and evolution of language. Harvard University Press, Cambridge
Lieberman P, Blumstein SE (1988) Speech physiology, speech perception, and acoustic phonetics. Harvard University Press, Cambridge
Lieberman P, Crelin ES (1971) On the speech of Neanderthal man. Ling Inq 2:203–222
Lieberman P, Crelin ES, Klatt DH (1972) Phonetic ability and related anatomy of the newborn and adult human, Neanderthal man, and the chimpanzee. Am Anthropol 74:287–307
Lieberman PH, Klatt DH, Wilson WH (1969) Vocal tract limitations on the vowel repertoires of rhesus monkey and other nonhuman primates. Science 164:1185–1187
MacLarnon AM, Hewitt GP (1999) The evolution of human speech: the role of enhanced breathing control. Am J Phys Anthropol 109:341–363
Negus VE (1949) The comparative anatomy and physiology of the larynx. Heinemann, London
Nishimura T (2003a) Comparative morphology of the hyo-laryngeal complex in anthropoids: two steps in the evolution of the descent of the larynx. Primates 44:41–49
Nishimura T (2003b) Studies on the ontogenetic changes in the shape of the vocal tract in chimpanzees. D.Sc. dissertation, Kyoto University, Kyoto

Nishimura T (2005) Developmental changes in the shape of the supralaryngeal vocal tract in chimpanzees. Am J Phys Anthropol 126:193–204
Nishimura T, Kikuchi Y, Shimizu D, Hamada Y (1999) New methods of morphological studies with minimum invasiveness (in Japanese with English summary). Primate Res 15:259–266
Nishimura T, Mikami A, Suzuki J, Matsuzawa T (2003) Descent of the larynx in chimpanzee infants. Proc Natl Acad Sci U S A 100:6930–6933
Oller DK (1980) The emergence of the sounds of speech in infancy. In: Yeni-Komshian GH, Kavanagh JF, Ferguson CA (eds) Child phonology, vol 1. Production. Academic Press, New York, pp 93–112
Rak Y (1983) The australopithecine face. Academic Press, New York
Reidenbach MM (1997) Anatomical considerations of closure of the laryngeal vestibule during swallowing. Eur Arch Otorhinolaryngol 254:410–412
Sasaki CT, Levine PA, Laitman JT, Crelin ES (1977) Postnatal descent of the epiglottis in man: a preliminary report. Arch Otolaryngol 103:169–171
Schwenk K (ed) (1999) Feeding. Academic Press, San Diego
Simmons AM, Popper AN, Fay RR (eds) (2002) Acoustic communication. Springer, New York
Smith KK (1992) The evolution of the mammalian pharynx. Zool J Linn Soc 104:313–349
Stark RE (1980) Stages of speech development in the first year of life. In: Yeni-Komshian GH, Kavanagh JF, Ferguson CA (eds) Child phonology, vol 1. Production. Academic Press, New York, pp 73–92
Stevens KN (1998) Acoustic phonetics. MIT Press, Cambridge
Stedman HH, Kozyak BW, Nelson A, Thesier DM, Su LT, Low DW, Bridges CR, Shrager JB, Minugh-Purvis N, Mitchell MA (2004) Myosin gene mutation correlates with anatomical changes in the human lineage. Nature (Lond) 428:415–418
Swindler DR, Wood CD (1973) An atlas of primate gross anatomy. University of Washington Press, Seattle
Takemoto H (2001) Morphological analyses of the human tongue musculature for three-dimensional modeling. J Speech Lang Hear Res 44:95–107
Takemoto H, Ishida H (1995) A functional and evolutional biology of human speech: part 2, the vocal tract and the pronunciation of vowels of humans and chimpanzees (in Japanese with English summary). In: Okada M (ed) Grant-in-aid report (B1) "Development of new methods for evaluating the air-trapping during activities," pp 157–165
Taylor EM, Sutton D, Lindeman RC (1976) Dimensions of the infant monkey upper airway. Growth 40:69–81
Thexton AJ, McGarrick JD (1989) Tongue movement in the cat during the intake of solid food. Arch Oral Biol 18:153–356
Titze IR (1994) Principles of voice production. Prentice Hall, Englewood Cliffs
Vandaele DJ, Perlman AL. Cassell MD (1995) Intrinsic fibre architecture and attachments of the human epiglottis and their contributions to the mechanism of deglutition. J Anat 186:1–15
Vorperian HK, Kent RD, Gentry LR, Yandell BS (1999) Magnetic resonance imaging procedures to study the concurrent anatomic development of vocal tract structures: preliminary results. Int J Pediatr Otorhinolaryngol 49:197–206
Walker A, Leakey RE (eds) (1993) The Nariokotome *Homo erectus* skeleton. Harvard University Press, Cambridge
Westhorpe RN (1987) The position of the larynx in children and its relationship to ease of intubation. Anaesth Intens Care 15:384–388
Williams PL (ed) (1995) Gray's anatomy, 38th edn. Churchill Livingstone, New York
Wind J (1970) On the phylogeny and the ontogeny of the human larynx. Wolters-Noordhoff, Groningen
Wood B (1992) Origin and evolution of the genus *Homo*. Nature (Lond) 355:783–790
Wood B, Collard M (2004) The changing face of genus *Homo*. Evol Anthropol 8:195–207
Zemlin WR (1988) Speech and hearing science: anatomy and physiology, 3rd edn. Prentice Hall, Englewood Cliffs

6
Understanding the Growth Pattern of Chimpanzees: Does It Conserve the Pattern of the Common Ancestor of Humans and Chimpanzees?

Yuzuru Hamada[1] and Toshifumi Udono[2]

1 Introduction

Chimpanzees are the species most closely related to humans. It is, therefore, considered a priori that the majority of differences including growth patterns between humans and chimpanzees were acquired in the lineage of humans (hominins) after their divergence. Fossil studies have also supported this idea that the growth pattern of living great apes is conservative, inherited from mid-Miocene ancestors. We examined the growth characteristics of chimpanzees based on growth pattern analyses on linear dimensions and fat deposits.

Although the adolescent growth spurt is not found in chimpanzees, the same growth mechanism functions both in chimpanzees and in humans. The slower growth in infancy and adolescence is one of the characteristics of chimpanzees. The long infantile period is the second growth characteristic of chimpanzees, but the longer growth period, the characteristic that great apes share, is the product of a scaling trend, that is, the consequence of their greater size. The age change pattern of fat deposit in chimpanzees is similar to that of macaques and much different from that of humans. Substantial fat begins to be deposited in chimpanzees, especially in females, starting from the adolescent stage, in preparation for reproduction. These growth characteristics of the chimpanzee are explained by their traditional way of rearing the immatures; that is, the mother almost exclusively rears and nutritionally supports the infant and juvenile. The growth pattern of humans is much attributable, on the other hand, owes much to their novel rearing system, that is, support from the father and other kin, and to their highly efficient food-acquiring ability.

[1]Primate Research Institute, Kyoto University, 41 Kanrin, Inuyama, Aichi 484-8506, Japan
[2]Kumamoto Primate Park, Sanwa Kagaku Kenkyusho, 990 Ohtao, Misumi-cho, Uki, Kumamoto 869-3201, Japan

2 The Phylogenetic Position and Growth Pattern of Chimpanzees

Chimpanzees (*Pan troglodytes* and *Pan paniscus*) are, among living animals, those species most closely related to humans, and it is estimated by genetic analyses that humans and chimpanzees diverged from their last common ancestor about 6 million years ago. However, there are great differences between them in various biological characters. Thus, chimpanzees have been classified as a member of the family Pongidae together with gorillas and orangutans, separate from the Hominidae (humans and their ancestors). Although a gradistic view (anagenesis) was once pervasive, whereby phylogeny was thought to have proceeded following successive stages of monkeys (cercopithecoids and ceboids), apes, and then humans, in the order of lower (primitive) to higher (advanced) taxa, this simplistic interpretation has received much criticism.

It is considered a priori that the differences between humans and chimpanzees—for example, bipedalism, encephalization, vocal communication (language), production and use of tools, and life history characteristics, including growth patterns—were acquired in the lineage of humans (hominins) after they diverged from their last common ancestor. Although authorities have different ideas on the evolutionary order of appearance of these characters, they do agree that this suite of characters is unique to humans. Included in growth characters are features such as the immature neonate (especially in motor performance, e.g., clinging to the mother), a long growth period, and the presence of adolescence and the adolescent growth spurt (Bogin 1999; Tanner et al. 1990). On the other hand, it is tacitly agreed that the greater portion of characters found in chimpanzees represent retention from the last common ancestor or even more remote ancestors, for example, the locomotor pattern of knuckle-walking (Richmond and Strait 2000). If this is true, then these characters are symplesiomorphic and only very few chimpanzee characters are autapomorphic, which may strengthen the idea of retention of primitive characters in the growth of chimpanzees. Although it is possible that only humans crossed the Rubicon to evolve their unique (and advanced) suite of characters, this is, of course, an open question, too open for the present discussion.

The precise aims of this discussion, then, are as follows:

What factors are responsible for the chimpanzee growth pattern: phylogenetic inertia, ecology, life history, or body size?

Do chimpanzees have a special way of rearing immature offspring, especially in regard to the nutritional supply to immatures?

3 Growth Studies on Chimpanzees

The fundamental physical growth patterns in chimpanzees, as described by Schultz (1969), Yerkes and his colleagues (Grether and Yerkes 1940; Nissen and Riesen 1964), Gavan (1971), Watts (1985; Watts and Gavan 1982), Leigh (1992,

1994, 1996; Leigh and Shea 1996), and Hamada and his colleagues (Hamada et al. 1996, 2002; Hamada et al. 2003a), encompass increases in body size, eruption of teeth, and skeletal and reproductive maturation. Comparisons between the growth patterns of macaques (Hamada et al. 1999), chimpanzees, and humans will help to shed light on the uniqueness of chimpanzee growth and aid in reconstructing the chimpanzees' evolutionary history.

3.1 Phylogenetic Consideration of the Growth Pattern of Chimpanzees

The growth patterns of fossil hominoids have been reconstructed by examination of the order of tooth eruption and enamel microstructure in such taxa as *Australopithecus africanus*, *Afropithecus*, and *Sivapithecus* (Kelley 2002; Anemone 2002). The hominoids had already acquired a life history as prolonged as those of extant great apes in the middle of the Miocene, and it is considered that the pattern would have evolved just after the divergence of hominoids from the primitive catarrhines (Kelley 2002). Slower growth may have been characteristic both of the great apes and even of the ancestors of the hominins for a rather long period. The much more prolonged life history of modern humans most probably evolved in the latest period of the Pliocene (in older *Homo erectus*, as old as about 1.5 million years ago; Kelley 2002).

These results from fossil studies present a rather gradistic view of evolution wherein only hominins would have advanced in their growth pattern rather recently. Although the great apes, including chimpanzees, are considered to have retained traditional growth patterns, that of chimpanzees should be reevaluated from various aspects.

3.2 Life History Analyses and the Growth Pattern

Life history, including growth, is the target of adaptation and is determined by ecology, population dynamics, body size, and other biological factors (Stearns 1992). Thus, the growth pattern is also considered to be shaped by these factors. Some researchers have pointed out that many life history parameters are strongly related to mortality in adulthood and that such parameters exhibited a trade-off relationship (Charnov 1993). In those species that have a lower mortality rate in adulthood, the growth period (represented by the age at which females first become pregnant) tends to be longer, that is, offspring mature later. The number of offspring produced over a lifetime also exhibits a similar relationship.

In life history studies, the trends, that is, the general relationships between parameters, are first sought for by statistical analyses using data taken from various species, and species specificity is recognized as a deviation from the trend. Thus, a given species is characterized by both trend and deviation. Scaling analysis, for example, is one of such analyses. I first characterize the growth pattern of chimpanzees in general.

3.3 Characteristics of Chimpanzee Growth

Growth patterns were compared by relative values (deviation from the scaling function, or trend), which were expressed by the percentage of real value against the estimated value from the function using the data published by Harvey et al. (1987). Table 1 lists the relative values in humans, chimpanzees, macaques (cercopithecoid), capuchins (ceboid), and lemurs (lemuroid). We did not use brain size for the calculation of the function because the inclusion of human data would have distorted the scaling relationship too much. However,

Table 1. Relative values of growth parameters in primates

	Humans	Chimpanzees	Macaques	Capuchins	Lemurs
Body mass, male/female (kg)	47.9/40.1	41.6/31.1	11.7/9.1	2.86/2.10	2.90/2.50
Intrauterine life (days)	+13 (267)	0 (228)	−10 (170)	+11 (160)	−10 (135)
Neonatal body mass (g)	+44 (3300)	−3 (1756)	+15 (503)	+70 (248)	−24 (88.2)
Age at weaning (days)	−30 (1095)[a]	+64 (1460)	−8 (182)	+22 (270)[b]	−22 (105)
Age at reproductive maturity (months)	+49 (198)	+1 (118)	−19 (60)	+31 (43.1)[c]	−28 (10.0)
Neonatal brain mass (g)	+80 (384)	−25 (128)	+13 (54.5)[d]	+46 (29.0)[e]	−23 (25.6)
Adult brain mass (g)	+205 (1,250)	+25 (410)	+24 (109)	+120 (71.0)	−14 (25.6)
Life span (in years)	+30 (60)	+4 (44.5)	−11 (25.0)[f]	+78 (40.0)	+6 (27.1)

Relative growth parameters are calculated by the division of original values by the estimated values using the scaling function [in percent, from the data of Harvey et al. (1987); the original data are shown in parentheses]

Representative species for macaques, capuchins, and lemurs are Japanese macaques, tufted capuchins, and ring-tailed lemurs, respectively; where data for these species were not obtained, those of closely related species were used, as indicated by footnotes b through f

[a]Age at weaning was taken from Bogin (1999)

[b]*Cebus albifrons*

[c]*Lemur fulvus*

[d]*Macaca mulatta*

[e]*Cebus capucinus*

[f]*Macaca nemestrina*

the relative value of human brain size was calculated using the scaling function.

In humans, other than the age at weaning (about 3 years), which was much less than the estimated age, all characters showed great and positive relative values, meaning a longer intrauterine life, greater neonatal body mass and brain size, later reproductive maturation, and an adult brain mass three times greater than that estimated from body mass. Life span is a little longer than the estimation. Capuchins showed similarity in many characters, that is, the relative values of all characters are positive, meaning that they grow slowly both in pre- and postnatal periods and they have a greater brain size at birth and in adulthood than the value estimated from the scaling function. Lemurs, on the other hand, with the exception of life span, showed the opposite, that is, rapid growth and a smaller brain. Cercopithecoids showed a distinctive combination of relative values. Their intrauterine period, age at weaning and at reproductive maturation, and life span were smaller than the respective estimates, but their brain mass is greater. In spite of a shorter pregnancy, they have a larger neonate that grows rapidly and matures earlier. Chimpanzees showed small relative values, that is, intrauterine life, neonatal body mass, age at reproductive maturation, and life span are all close to those estimated from the scaling function. It is only the age at weaning (about 4 years) that showed a greater deviation. Postnatal brain growth is rather large in chimpanzees, as shown by the relative body mass of –25 at birth and +25 in adulthood.

Apes and humans were then compared (Table 2). Gorillas were unique in that relative values are negative in all characters. Gibbons, except for neonatal body mass, showed the opposite, that is, their relative values are positive, especially the age at reproductive maturation. Thus, gibbons have similar life history parameters to capuchins. Orangutans appeared similar to chimpanzees, but they have smaller neonates and are reproductively mature earlier. The growth characteristic of smaller neonates is considered to be shared by all the great apes.

From these comparisons, each of the great apes showed specific growth patterns. The "prolonged life history" of hominoids is considered to be the consequence of larger body size, that is, the general trend in primates. The generic specificities are considered to be the product of their distinctive ecology and social system (Leigh 1994; Leigh and Shea 1996).

3.4 Growth Stages and the Growth Period

Several life stages have been noticed in humans. Schultz compared the duration of life stages between various primate taxa (Schultz 1969). The diagram of Schultz (1969, Fig. 57, p. 149) has been referred to as the standard of comparative primate growth studies (Fleagle 1999). Growth stages of humans were applied to nonhuman primates (for macaques, Hamada et al. 1999; for chimpanzees, Hamada et al. 1996); that is, infant, juvenile, adolescent, and adult. These stages were determined by such developmental phenomena as weaning,

Table 2. Relative values of growth parameters in hominoids

	Humans	Chimpanzees	Orangutans	Gorillas	Gibbons[a]
Body mass, male/female (kg)	47.9/40.1	41.6/31.1	69.0/37.0	160.0/93.0	5.70/5.30
Intrauterine life (days)	+13 (267)	0 (228)	+11 (260)	−3 (256)	+14 (205)
Neonatal body mass (g)	+44 (3300)	−3 (1756)	−19 (1728)	−58 (2110)	−4 (410.5)
Age at weaning (days)	−30 (1095)	+64 (1460)	+12 (1095)	−4 (1583)	+58 (730)
Age at reproductive maturity (months)	+49 (198)	+1 (118)	−34 (84)	−62 (78)	+91 (108)
Neonatal brain mass (g)	+80 (384)	−25 (128)	−14 (170)	−49 (227)	+33 (50.1)
Adult brain mass (g)	+205 (1250)	+25 (410)	+8 (413)	−40 (505.9)	+33 (107.7)
Life span (years)	+30 (60)	+4 (44.5)	+11 (50.0)	−33 (39.3)	+23 (31.5)

Relative values of growth parameters are calculated by the division of original values by the estimated values using the scaling function [in percent, from the data of Harvey et al. (1987)]
[a]Gibbons are represented by the white-handed gibbon (*Hylobates lar*)

Table 3. Duration of growth stages

	Infancy	Juvenile	Adolescence
Macaques			
F	0–0.5 (7)	0.5–3.5 (43)	3.5–7.0 (50)
M	0–0.5 (6)	0.5–4.5 (50)	4.5–8.0 (44)
Chimpanzees			
F	0–4.0 (27)	4.0–9.0 (33)	9.0–15.0 (40)
M	0–4.0 (29)	4.0–9.0 (36)	9.0–14.0 (36)
Humans			
F	0–3.0 (15)	3.0–12.0 (45)	12.0–20.0 (40)
M	0–3.0 (15)	3.0–14.0 (55)	14.0–20.0 (30)

Data from various sources, in years; numbers in parentheses indicate relative duration in percentage

reproductive maturation, and somatic maturation (the cessation of growth in body mass or linear dimensions). The duration of each stage was compared between macaques (Japanese macaque, *Macaca fuscata*), chimpanzees, and humans (Table 3). Macaque growth is characterized by a strikingly short infantile period and similar duration of juvenile and adolescent periods. In humans, the infantile and adolescent periods are characteristically short and the juvenile period is long. The three stages have almost the same duration in chimpanzees, and a definitely longer infancy should be emphasized. Therefore, although the hominoids appear, at first, to have inherited a prolonged life history from mid-Miocene ancestors (Kelley 2002), it is important to determine which stages have been prolonged between chimpanzees (infantile) and humans (juvenile). As some researchers have cautioned (Watts 1985), precocial and rapid macaque growth should not be regarded as the ancestral state for hominoids but rather as one of the diverged characters specific to cercopithecoids.

The details of growth patterns in linear dimensions, such as stature or trunk length, are considered to reflect growth stages (Bogin 1999); this means that age at changes of pace (velocity), accelerations and decelerations, come close to the boundary of the stages and that each of the growth stages has a specific growth velocity and velocity change. I examine the growth curve of chimpanzee next in relationship to growth stages.

3.5 Growth Pattern of Crown–Rump Length in Chimpanzees

The growth curve (distance) of crown–rump length (CRL) is shown in Fig. 1. The diagram shows plotting of cross-sectional CRL data and the smoothed

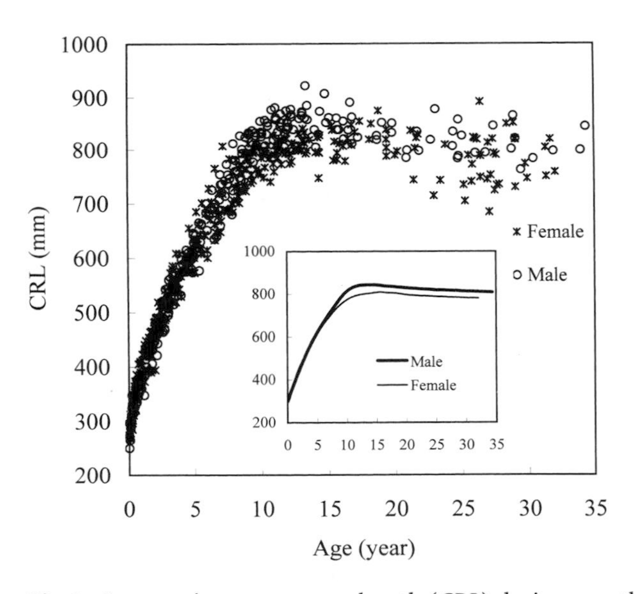

Fig. 1. Increase in crown–rump length (*CRL*) during growth in chimpanzees for 253 males (age range, 0.0–34.4 years) and 336 females (age range, 0.0–32.0 years). Average curves cross sectionally obtained by the Loess smooth algorithm (Math-Soft, 4.0) are superimposed. Note that the sex difference, which is slight even in adults, starts to be significant from about 7 years of age

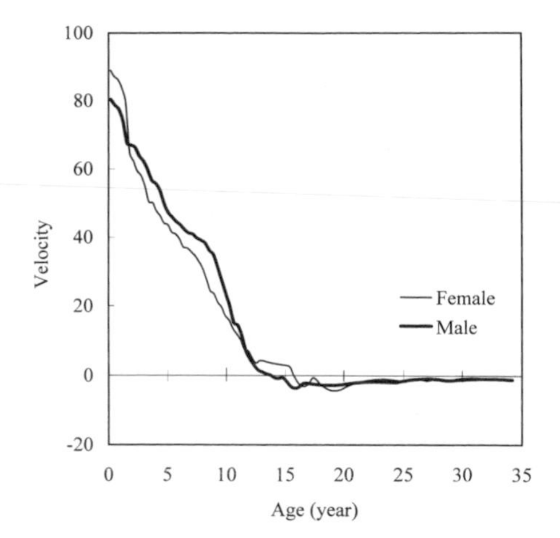

Fig. 2. Velocity (pseudo-velocity) curves of crown–rump length growth in chimpanzees. Velocity was calculated from the smoothed distance values

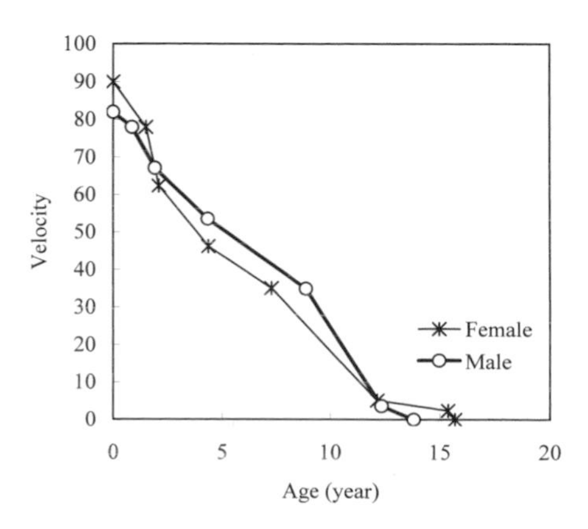

Fig. 3. Growth stages determined from growth velocity curves

average curve obtained from the Loess smooth algorithm (S-Plus 4.0; MathSoft Co. ltd.). Neonatal CRL, approximately 300 mm, does not show any sex difference, but the sex difference becomes greater from the age of 5.5 years. The maximum CRL, 845 mm in males and 810 mm in females, is attained at the age of 13.8 years in males and 15.7 years in females. The sex difference is not great even in adults, and females may stop growing a little later than males.

The distance growth curves appear simple in the two sexes, that is, it appears that there are not any changes of pace but, rather, a slow deceleration over the whole growth period. However, the velocity curve (pseudo-velocity; Fig. 2) revealed small but significant changes of pace, which suggests the possibility of staging. By the inclusion of longitudinal data taken from three infants born in 2000, pseudo-velocity curves were obtained for these subjects over the period from birth to the age of 5 years, when the velocity greatly changed. From these velocity curves, we could pinpoint where the pace of growth changed.

The growth of linear dimensions in chimpanzees was found to be composed of six phases as follows (Fig. 3):

1. A neonatal rapid growth phase where the velocity rapidly decreases
2. A midinfantile great deceleration phase
3. A moderate deceleration phase in the latter half of the infantile period
4. A slow deceleration phase in the juvenile period
5. An adolescent great deceleration phase
6. A slow deceleration, finally stopping the growing phase

Phases 1 to 3 correspond to the infantile period of 0 to 4 years. Phase 4 corresponds to the juvenile period. Phases 5 and 6 correspond to the adolescent phase, and at around the beginning of phase 5 reproductive maturation is attained, that is, females experience menarche and males show rapid development of the testes.

The growth curve of stature in humans (see Fig. 2.5, p. 69, in Bogin 1999) shows a remarkable change of pace, which facilitates the staging of the whole growth period. Infancy is the period of rapid growth and rapid deceleration, the juvenile period is that of slow and steady growth, and adolescence is that of abrupt acceleration and deceleration soon after the velocity curve attains its peak, which is followed by the termination of growth. The human growth curve, thus, looks unique among primates, and some authorities would ascribe a special evolution of growth and life history to the hominins (Bogin 1999). However, on the basis of the presence of change of pace at the end of phase 4, where the velocity greatly decreases and chimpanzees experience the beginning of reproductive maturation (puberty), Hamada et al. (1996) suggested that there are not definite differences in growth patterns between humans and chimpanzees.

There is a hypothesis that human growth is composed of two cycles (see Fig. 12 in Hamada and Udono 2002), and the velocity curve shown above is composed of two logistic function curves. The peak of the first cycle is located in the prenatal period and is higher than the second one in which the peak of velocity locates at around puberty. In postnatal life, the first cycle covers from birth to the midjuvenile period, and the second covers from the midjuvenile period to the cessation of growth. The appropriate choice of parameters, which determines the parabolic curves, can express the growth curve of humans and nonhuman primates, including chimpanzees. This two-cycle model appears applicable to the body mass growth of various anthropoid species in which pubertal accelerations are found (Leigh 1996). The differences in the growth patterns between humans and chimpanzees are thus considered to be caused by the parameters, that is, the difference is quantitative in nature. In humans, the curve of the second cycle is sharp with a high peak compared to that of chimpanzees. The question to be addressed is the background biology, which has induced the height of peaks and the sharpness (great acceleration and deceleration) of the growth cycles, the rapid growth both in infancy and adolescence in humans, and the longer infancy with a slower increase in chimpanzees (Goodall 1986).

4 The Answer May Be Found in the Nutrient Supply System (Parenting): The Economy of Supply and Consumption of Nutrients in Growing Individuals

Growing individuals need nutrients for basal metabolism, activity (physical and other), heat production at assimilation, and growth (Malina 1987). Growth shows plasticity in accordance with the quantity of nutrient supplied, that is, the individual may grow fast or slow according to the quantity of foods consumed. However, there is a limitation to the plasticity, and individuals should mature (stop growing) with a smaller body than the optimal size, which is genetically determined if the nutrient supply is sufficient (reaction norms; Stearns 1992). To increase adaptive success, the parent or parents should care sufficiently for offspring to survive long enough to reproduce themselves. It is beneficial in competition with conspecifics if offspring grow bigger. The mother supplies various resources including nutrients both prenatally and postnatally (lactation), safety, and support for juveniles in foraging. Adolescent and older offspring live independently from their mother and access food resources according to their social status, which they try to promote.

A surplus of supplied nutrients will be partly deposited in the form of fat. The mother, while she is taking care of her current offspring, tries to deposit fat in preparation for the next offspring, and thus tries to control the nutrient supply to the current dependent offspring. The current offspring, however, requests as much food as possible from its mother. Therefore, a compromise must be reached between the two demands, and the first priority of maternal nutrient supply is to minimize the surplus to the infant.

4.1 Age Change of Fat Deposit

Fat deposit is regarded as an indicator of the nutritional condition of an animal, lean versus fatty. Indirect measures of fat deposit are skinfold thickness or physique indices (e.g., body mass index, Rohler index). Direct measures are obtained by such methods as weighing dissected fat and dual-energy X-ray absorptiometry (DXA) (Roche et al. 1996). The results obtained from macaques (*Macaca fuscata*) using DXA (Hamada et al. 2003b) showed that macaques adopt the first option of nutrient supply mentioned earlier and that infant and juvenile macaques do not deposit any significant amount of fat, 4% or less of body mass. After reproductive maturation, adolescent macaques start to deposit a more substantial amount of fat. The immature human shows a strikingly distinctive pattern of age change in fat deposit (Kuzawa 1998). A baby is born with about 15% fat content, and the fat deposit is rapidly increased to 25% to 27% of body mass within half a year. From that peak, relative fat mass gradually decreases to about 15% at the age of 5 years. In the juvenile period, the relative fat mass is maintained but at reproductive maturation it is increased again to attain adult values of 15% to 20% in males and 20% to 25% in females. The age change pattern of fat deposit in humans suggests the second option of nutritional supply to offspring, that is, a higher relative fat deposit from birth and an especially higher deposit during early infancy. The question then is which option does the chimpanzee take?

4.2 Age Change of Fat Deposit in Chimpanzees

As far as we are aware, at present, there are not any reports on age changes in fat deposits in chimpanzees (Fig. 4), because access to immature chimpanzees is limited. Although limited in number, our colleagues and we have accumulated DXA measurements taken from immatures (5 years or younger) and adults (10 years or older). Figure 5 shows the preliminary results of age changes in fat deposits in chimpanzees.

Age changes in fat mass (relative to body mass, in percent) displayed a similar trajectory to that found in macaques. Chimpanzees of 5 years or younger do not

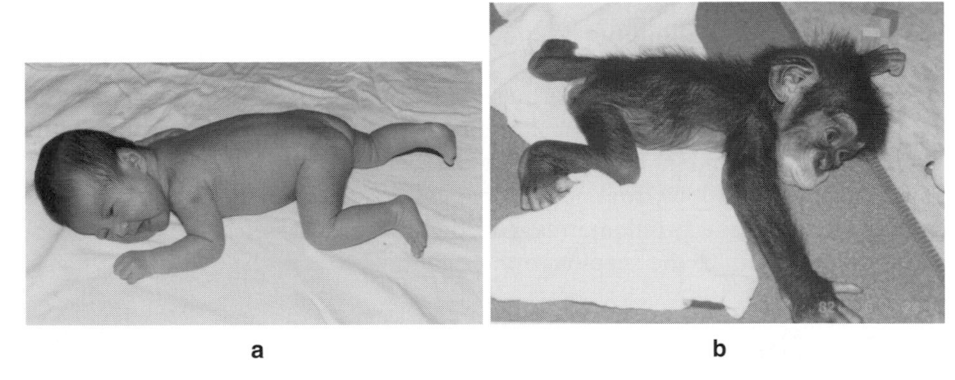

Fig. 4. **a** A human infant at the age of 30 days after the birth (Photograph taken by T. Matsuzawa). **b** A chimpanzee infant (Reo) at the age of 38 days after the birth (Photograph taken by T. Matsuzawa)

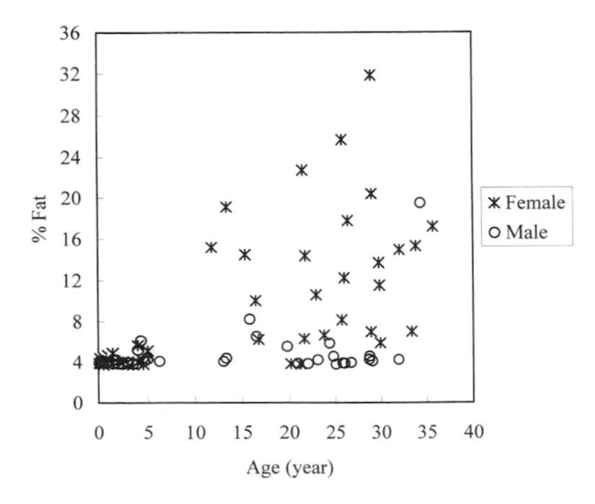

Fig. 5. Plotting of fat mass [in percent (%) relative to body mass] obtained using dual-energy X-ray absorptiometry (DXA) for 57 females (30 subjects < 10 years and 27 subjects > 10 years of age) and 38 males (18 subjects < 10 years and 20 subjects > 10 years of age)

show deposits of more than 4% fat mass, and slightly higher values of 5% or 6% were recorded only sporadically in both sexes. Thus, substantial fat mass was not deposited in chimpanzees of 5 years or younger, even though they were reared in a favorable environment.

Subadult and adult chimpanzees, of 10 years or older, showed substantial fat deposits, but a considerable number of chimpanzees showed the minimum deposit of fat mass (4% or less), especially males. Basic statistics for fat deposits in male chimpanzees of 12 years or older are 3.87%, 4.23%, and 7.73% for the 10th, 50th, and 90th percentiles, respectively. Thus, the majority of adults deposited a substantially smaller amount of fat, with the maximum being 19%. Adult females deposit more fat than males, as shown by their basic statistics (6.07%, 13.00%, and 21.57% for the 10th, 50th, and 90th percentiles, respectively, and the maximum value was 32%).

As the DXA data were limited, age changes in skinfold thickness, which has long been one of the indicators of fat deposit, were analyzed. Here we used the total thickness at the abdominal (at the level of navel), back (subscapular), and suprailiac regions, measured using a skinfold caliper (Eiken type). Figure 6 shows the plots for the data and the smoothed curves connecting the 10th, 50th,

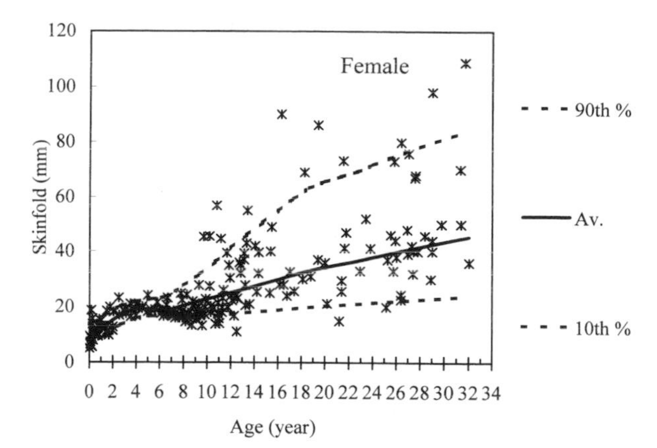

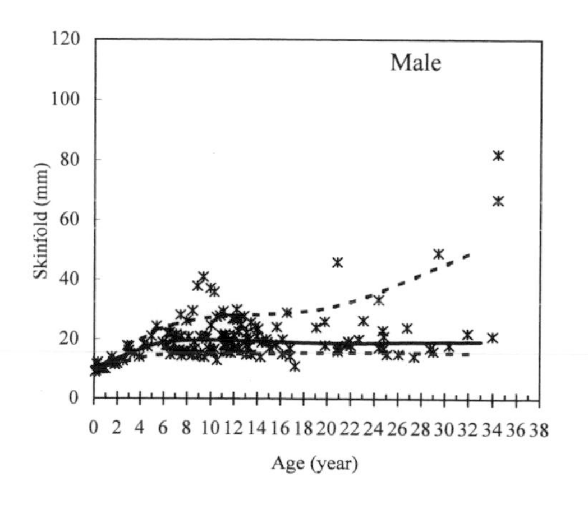

Fig. 6. Age change in the skinfold thickness for females (*top*) and males (*bottom*): 179 females (85 subjects < 10 years and 94 subjects > 10 years of age) and 135 males (53 subjects < 10 years and 82 subjects > 10 years of age). *Curves* connecting the 10th, 50th, and 90th percentiles of age classes are superimposed using the Loess smooth algorithm. From birth to about 5 years of age in females and to about 7 years of age in males, skinfold thickness gradually increases with age. These increases result not from the accumulation of subcutaneous fat but from the increase of skin thickness itself. After 8 years of age, skinfold steadily increases with age in females. However, that in adult males does not show significant increase, meaning that the majority of adult males do not show subcutaneous fat

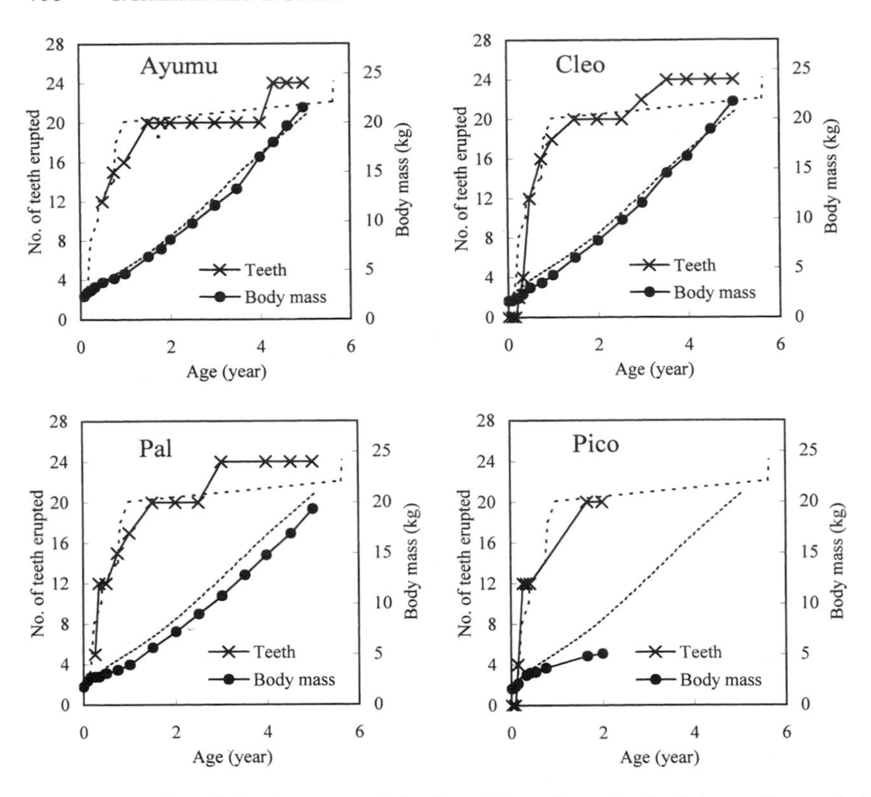

Fig. 7. Growth and development of the four infants born in the Primate Research Institute, Kyoto University, were compared with the average body mass growth curve and dental development (eruption of deciduous teeth and the first molar, *dotted line*). In each panel, *painted circle* and *X connected by line* express body mass and numbers of teeth erupted, respectively

and 90th percentile values of each age class. In females, the curve of the 50th percentile showed that skinfold rapidly increased to about 20 mm from birth to 4 years. The thickness of 20 mm was maintained until the age of 7 years and then gradually increased to finally become 40 mm. Curves of the 10th and 90th percentiles are parallel with that of the 50th percentile until the age of 7 years. From 7 years of age, the curve of the 90th percentile deviated more and more from that of the 50th percentile to finally attain 80 mm. The curve of the 10th percentile increased slowly to finally attain 25 mm. In males, curves of the 10th and 50th percentiles were maintained at almost 15 mm in the former and 20 mm in the latter, meaning that the subcutaneous fat deposit is nil or minimum in the majority of adult male chimpanzees. The curve of the 90th percentile slowly increased from 7 to 20 years of age to attain only 30 mm. Thus, sex difference is significant in subcutaneous fat deposits of adult chimpanzees.

The fat data for infantile period were mainly taken from four infants born in the Primate Research Institute, Kyoto University. Their growth and development are shown in Figure 7.

The results obtained from macaques showed that skinfold thickness correlated significantly well with the percent fat mass obtained using DXA, meaning that skinfold is a good indicator of fat deposit. It is probably also true in chimpanzees, and although the data are limited, the skinfold and percent fat mass correlated well in adults ($R^2 = 0.59$ in 25 females, $R^2 = 0.94$ in 27 males). Based on these analyses, it appears that chimpanzees also adopt the first option of maternal nutrition supply. The longer infantile period in chimpanzees during which the mother supplies nutrient to infants by lactation is a characteristic unique to chimpanzees.

The age change pattern of fat deposit appears to be species specific and may have implications for the life history of a given species. The level of parental care (rearing) is a part of reproductive adaptation. There should be an age-specific nutritional supply to immature offspring, which may be reflected in the age-specific fat deposit. We found there are two modes of age-specific nutrition supply, the first being that adopted by macaques and chimpanzees and second being unique to humans. Infants and juveniles of the former have the minimum fat deposit, meaning that the mother does not supply a surplus, only that amount necessary to support the maintenance and growth of the offspring. After reproductive maturation, although the subadult is not fully mature in body, they try to better their social status, which determines the quality and quantity of food acquired. Reproductively active females try to obtain nutrients efficiently and accumulate fat for reproduction. As males do not invest much energy in offspring, they do not have to accumulate as much fat as females. Female chimpanzees devote as long as 4 years to rearing their infants, and their interbirth interval is, therefore, great (Goodall 1986). The rearing system of chimpanzees, in that only the mother takes care of and supplies nutrient to the immature over such an extended period, may have caused this unique life history in chimpanzees.

The second option in the pattern of age change in fat deposits, which humans have adopted, appears unique, not only in the greater accumulation of fat at any growth stage but also by a much higher deposit of fat in infancy (especially in early infancy) and adolescence. Taking the growth pattern of CRL into consideration, that is, the very rapid growth in infancy, and that the amount of nutrient supplied is much more than that required to sustain rapid growth, it seems that the greater fat deposit is the byproduct of a guarantee to rapid growth. The greater amount of nutrient cannot be supplied by the mother alone, and the interbirth interval in humans is much shorter than that of chimpanzees. Therefore, humans must have a different way of rearing, which depends on immatures acquiring food with much greater efficiency than chimpanzees or macaques.

Individuals that help the mother in the rearing of immatures have been documented in some anthropoids, the marmosets being the popular example, in which the father and elder siblings help the mother to care for babies. Colobines show allomothering behavior, in which neonates have a different pelage color (Kohda 1985). In these primates, however, food is supplied to neither the infants nor the mothers. In modern humans, however, helping is much developed, and even juveniles are supplied with nutrients. Fathers share foods with family members (Kaplan et al. 2000), and the contributions of grandparents (intergenerational resource transfer; Lee 2003) have also been suggested ("grandmother hypothesis"; Hawkes et al. 2003). Bogin (1999) also advocated the "baby-sitter hypothesis" in which elder siblings give substantial aid to their mother in rearing infants. Such systematic helping would be a prerequisite to the second option of nutrition supply, which is reflected in the greater amount of fat deposit and the shorter interbirth interval.

The characteristically rapid growth of humans is found not only in the body but also in the brain (Rice 2002; Leigh 2004; Hamada et al. 2004), which requires a continuous and greater supply of nutrient per unit of mass. The greater fat deposit found in infants is considered to guarantee this supply. Brain size shows a higher correlation with many life history parameters, such as age of dental eruption (Smith et al. 1994). It is not always easy to show the direct cause–effect relationship between brain size (cognitive ability) and growth patterns. The duration of the growth period is explained by such hypotheses as "needing to learn," "brain growth constraint," "juvenile risk," and the "brain constraint," which has proved to be the most feasible (Ross and Jones 1999). This hypothesis is based on the fact that the developing brain requires a higher energy supply than other organs, which coincides well with the discussion already described. The role of the brain (cognitive ability) in the rearing system (nutrient supply) is a topic for future consideration.

5 Conclusions

We have tried to address the questions of which characters are unique to chimpanzees as found in their growth patterns and what are the factors contributing to the production of these characters. Based on the fact that living great apes have diverse growth characteristics, the growth pattern of living hominoids, which is sometimes expressed simply as "prolonged life history," may not be a retention from the mid-Miocene ancestors. The prolonged life history derives from their greater size as a consequence of scaling trends. It is the relatively immature neonate, the character shared by great apes, that needs to be explained. The unique growth characteristic of chimpanzees is their long infancy, which may have been ultimately produced by their traditional system of infant rearing, that is, in which only the mother takes care of and supplies nutrients to the infant without any assistance.

Based on the two-cycle model of postnatal growth, the difference in growth pattern between chimpanzees and humans is explained by differences in the parameters describing the two cycles. The growth pattern difference is shaped by the species-specific system of rearing immatures, that is, the manner of nutrient supply to immatures, which is reflected in age changes in fat deposit. The chimpanzee or macaque mother does not supply the amount of nutrient required by their infant or juvenile to accumulate much fat. Immature human individuals receive a greater amount of nutrients and they accumulate fat deposits, which may guarantee their very fast body and brain growth in infancy. This development is supported by the unique social system in humans whereby the father and other kin provide help to the mother in ensuring the infant's nutrition supply. Brain and cognitive ability appear to contribute much in shaping the rearing system, which is a topic for future investigation. The growth pattern of chimpanzees could well be understood in terms of its rearing system, in which only the mother contributes substantially in rearing her immature offspring.

Acknowledgments

The authors thank Dr. Matsuzawa, who kindly invited us to write a review. We also thank all staff of the Primate Park of Sanwa Kagaku-Kenkyuso and the Primate Research Institute, Kyoto University, especially those taking care of chimpanzees. This study was supported by the Japan Society for Promotion of Science (no. 14204083).

References

Anemone RL (2002) Dental development and life history in hominid evolution. In: Minugh-Purvis N, McNamara KJ (eds) Human evolution through developmental change. Johns Hopkins Univ. Press, Baltimore, pp 249–280

Bogin B (1999) Patterns of human growth, 2nd edn. Cambridge University Press, Cambridge

Charnov EL (1993) Life history invariants. Oxford University Press, Oxford

Fleagle JG (1999) Primate adaptation and evolution, 2nd edn. Academic Press, San Diego

Gavan JA (1971) Longitudinal, postnatal growth in chimpanzee. Chimpanzees 4:46–102

Goodall J (1986) The chimpanzees of Gombe. Belknap Press/Harvard University Press, Cambridge

Grether WR, Yerkes RM (1940) Weight norms and relations for chimpanzee. Am J Phys Anthropol 27:181–197

Hamada Y, Udono T (2002) Longitudinal analysis of length growth in the chimpanzee (*Pan troglodytes*). Am J Phys Anthropol 118:268–284

Hamada Y, Udono T, Teramoto M, Sugawara T (1996) The growth pattern of chimpanzee: somatic growth and reproductive maturation in *Pan troglodytes*. Primates 37:277–293

Hamada Y, Hayakawa S, Suzuki J, Ohkura S (1999) Adolescent growth and development in Japanese macaques (*Macaca fuscata*): punctuated adolescent growth spurt by season. Primates 40:439–452

Hamada Y, Chatani K, Udono T, Kikuchi Y, Gunji H (2003a) A Longitudinal study on hand and wrist skeletal maturation in chimpanzees (*Pan troglodytes*), with emphasis on growth in linear dimensions. Primates 44:259–271

Hamada Y, Hayakawa S, Suzuki J, Watanabe K, Ohkura S (2003b) Seasonal variation in the body fat of Japanese macaques *Macaca fuscata*. Mamm Stud 28:79–88

Hamada Y, Udono T, Teramoto M, Hayasaka I (2004) Body, head, and facial growth: comparison between macaques (*Macaca fuscata*) and chimpanzee (*Pan troglodytes*) based on somatometry. Ann Anat 186:451–461

Harvey PH, Martin RD, Clutton-Brock TH (1987) Life histories in comparative perspective. In: Smuts BB, Cheney, DL, Seyfarth RM, Wrangham RW, Struhsaker TT (eds) Primate societies. University of Chicago Press, Chicago, pp 181–196

Hawkes K, O'Connell JF, Blurton Jones NG (2003) Human life histories: primate trade-offs, grandmothering, socioecology, and the fossil record. In: Kappeler PM, Pereira ME

(eds) Primate life histories and socioecology. University of Chicago Press, Chicago, pp 204–231
Kaplan H, Hill K, Lancaster J, Hurtado M (2000) A theory of human life history evolution: diet, intelligence, and longevity. Evol Anthropol 9:156–185
Kelly J (2002) Life-history evolution in Miocene and extant apes. In: Minugh-Purvis N, McNamara KJ (eds) Human evolution through developmental change. Johns Hopkins Univ Press, Baltimore, pp 223–248
Kohda M (1985) Allomothering behaviour of New and Old World monkeys. Primates 26:28–44
Kuzawa CW (1998) Adipose tissue in human infancy and childhood: an evolutionary perspective. Yearb Phys Anthropol 41:177–209
Lee RD (2003) Rethinking the evolutionary theory of aging: transfers, not births, shape senescence in social species. Proc Natl Acad Sci U S A 100:9637–9642
Leigh SR (1992) Patterns of variation in the ontogeny of primate body size dimorphism. J Hum Evol 23:27–50
Leigh SR (1994) Ontogenetic correlates of diet in anthropoid primates. Am J Phys Anthropol 94:499–522
Leigh SR (1996) Evolution of human growth spurts. Am J Phys Anthropol 101:455–474
Leigh SR (2004) Brain growth, life history, and cognition in primate and human evolution. Am J Primatol 62:139–164
Leigh SR, Shea BT (1996) Ontogeny of body size variation in African apes. Am J Phys Anthropol 99:43–65
Malina RM (1987) Nutrition and growth. In: Nutritional anthropology. Liss, New York, pp 173–196
Nissen HW, Riesen AH (1964) The eruption of the permanent dentition of chimpanzee. Am J Phys Anthropol 22:285–294
Rice SH (2002) The role of heterochrony in primate brain evolution. In: Minugh-Purvis N, McNamara KJ (eds) Human evolution through developmental change. Johns Hopkins University Press, Baltimore, pp 154–170
Richmond BG, Strait DS (2000) Evidence that humans evolved from a knuckle-walking ancestor. Nature (Lond) 404:382–385
Roche AF, Heymsfield SB, Lohman TG (eds) (1996) Human body composition. Human Kinetics, Champaign
Ross C, Jones K (1999) Socioecology and the evolution of primate reproductive rates. In: Lee PC (ed) Comparative primate ecology. Cambridge University Press, Cambridge, pp 73–110
Schultz AH (1969) The life of primates. Weidenfeld and Nicolson, Hampshire
Smith BH, Crummett TL, Brandt KL (1994) Ages of eruption of primate teeth: a compendium for aging individuals and comparing life histories. Yearb Phys Anthropol 37:177–231
Stearns SC (1992) The evolution of life histories. Oxford Univ Press, Oxford
Tanner JM, Wilson ME, Rudman CG (1990) Pubertal growth spurt in the female rhesus monkey: relation to menarche and skeletal maturation. Am J Hum Biol 2:101–106
Watts ES (1985) Adolescent growth and development of monkeys, apes and humans. In: Watts ES (ed) Nonhuman primate models for human growth and development. Liss, New York, pp 41–65
Watts ES, Gavan JA (1982) Postnatal growth of nonhuman primates: the problem of the adolescent spurt. Hum Biol 54:53–70

7 The Application of a Human Personality Test to Chimpanzees and Survey of Polymorphism in Genes Relating to Neurotransmitters and Hormones

Miho Inoue-Murayama[1], Emi Hibino[1], Tetsuro Matsuzawa[2], Satoshi Hirata[3], Osamu Takenaka[2], Ikuo Hayasaka[4], Shin'ichi Ito[1], and Yuichi Murayama[5]

1 Introduction

1.1 Evaluation of Chimpanzees' Characters and Survey of Their Genetic Background

Among nonhuman primates, the chimpanzee (together with the bonobo) is the species closest to humans, and its high intelligence is reflected in social behavior, tool use, and language learning (Whiten et al. 1999; Matsuzawa 2003). Because the individual characters of chimpanzees are full of variety, an analytical method that is able to objectively evaluate their behavioral traits may be useful for understanding their social behavior and interactions. Relationships have been reported between polymorphism of the genes regulating neurotransmitters or hormones and human personality. The study of the association of these genes with the personality of chimpanzees and comparison of the results with that of humans may be considered to be useful for elucidating human evolution. In this study, for the first step to conduct such a comparison, we evaluated personality and genetic polymorphism in chimpanzees.

We adopted the human personality questionnaire Yatabe-Guillford (YG) Personality Inventory for self-scoring and evaluated 11 adult chimpanzees at the Primate Research Institute, Kyoto University, based on the answers of three evaluators to 120 questions for each chimpanzee. To evaluate reliability of the test, 11 humans were also scored by the same methods, and the scores were compared with those by self. The average percentage of coincidence was 76.5%. The chim-

[1]Faculty of Applied Biological Sciences, Gifu University, 1-1 Yanagito, Gifu 501-1193, Japan
[2]Primate Research Institute, Kyoto University, 41 Kanrin, Inuyama, Aichi 484-8506, Japan
[3]Great Ape Research Institute, Hayashibara Biochemical Laboratories, 952-2 Nu, Tamano, Okayama 706-0316, Japan
[4]Kumamoto Primate Park, Sanwa Kagaku Kenkyusho, 990 Ohtao, Misumi-cho, Uki, Kumamoto 869-3201, Japan
[5]National Institute of Animal Health, Tsukuba, Ibaraki 305-0856, Japan

panzees showed significantly higher scores on the scales of "selfish," "impulsive," "nervous," and "sensible" than those of the human subjects. The individual difference observed in the chimpanzees was most remarkable in "leadership qualities." We then analyzed polymorphism of neurotransmitter- and hormone-related genes in chimpanzees. In 4 loci (dopamine receptor D4, androgen receptor, serotonin transporter, and estrogen receptor-beta), 3, 4, 7, and 6 alleles were observed, respectively, in 11 chimpanzees. Based on our preliminary results, a wide-range assessment of more loci with more individuals will provide a better characterization of genes affecting chimpanzee personality.

1.2 Understanding the Personality of Chimpanzees

Individual difference in the personality of chimpanzees is also remarkable, and a few studies of variation in personality within this species have been published (Lilienfeld et al. 1999; Weiss et al. 2000). Various studies have shown associations between differences in human personality traits and variations of neurotransmitter- and hormone-related genes. For example, the long allele of the variable number of tandem repeat (VNTR) in the exon 3 region of the dopamine receptor D4 gene is associated with a behavioral trait (novelty seeking) in humans (Benjamin et al. 1996). In humans, the roles of androgen have been identified in the regulation of sexuality, aggression, cognition, emotion, and personality (Rubinow and Schmidt 1996). The study of the association of these genes with the personality of chimpanzees and comparison of the results with those of humans may be considered to be useful for elucidating human evolution. To conduct such a comparison, evaluation of personality and genetic polymorphism in chimpanzees is essential. However, in the previous studies on chimpanzees, methods for personality testing were different from those applied to humans, and the relationships with genotypes were not investigated in chimpanzees. In primates, relationships with genotypes were previously studied in the rhesus monkey (Champoux et al. 2002; Miller et al. 2001). However, in these reports, methods for personality testing were quite different from those applied to humans.

In this study, as a first step to conduct the comparison of the genetic basis of personality between chimpanzees and humans, we attempted to evaluate personality and genetic polymorphism of chimpanzees. We applied a human personality questionnaire to chimpanzees because this test is very precise to express the personality of each individual, and the result of the relationship between personality and genotype is comparable with the previous reports of humans. Therefore, we adopted the human personality questionnaire Yatabe-Guillford (YG) Personality Inventory for self-scoring and evaluated 11 adult chimpanzees at the Primate Research Institute (PRI), Kyoto University, based on the answers of three evaluators to 120 questions for each chimpanzee. Because this test was originally based on self-scoring, to evaluate the reliability of scaling chimpanzees, 11 humans were also scored by the same methods and the scores were compared with those by self.

In chimpanzees, reports of polymorphic loci to survey the association with personalities as in humans are not sufficient. The afore-mentioned dopamine receptor D4 was monomorphic in chimpanzees (Inoue-Murayama et al. 1998). Also, in the promoter region of the serotonin transporter gene (Lesch et al. 1996), no variation was detected in chimpanzees (Inoue-Murayama et al. 2000, 2001). Polymorphism within chimpanzees was reported in the intron 2 region of the dopamine receptor D4 gene (hereafter referred to as *DRD4in2*) (Shimada et al. 2004) and the first exon of the androgen receptor gene (*AR*) (Choong et al. 1998). In this study, to increase the information of polymorphic loci, we newly surveyed the intron 2 region of the serotonin transporter gene (*STin2*) and the intron 6 region of the estrogen receptor gene (*ERβ*) in chimpanzees. We genotyped 4 loci (*DRD4in2, AR, STin2, ERβ*) in a total of 54 unrelated individuals kept in PRI and in the Kumamoto Primates Park (KPP), Sanwa Kagaku Kenkyusho. Genotypes of the 11 personality-scaled chimpanzees at PRI were also surveyed.

2 Methods

2.1 Subjects

Eleven chimpanzees (3 males and 8 females; average age, 27.4 ± 6.6 years) kept at PRI were scored for their personality (Table 1). They were all *Pan troglodytes verus*, except for Pendesa, who was a hybrid of *P. t. verus* and *P. t. schweinfurthii*. Gon was the father of Reo, Popo, and Pan; Puchi was the mother of Popo and Pan; and Reiko was the mother of Reo. Popo and Pan were human reared. For the purpose of comparison, 11 humans (6 men and 5 women; average age, 30.7 ± 8.5 years) were scored on the personality scales and included in the analysis (see Table 1). To survey allele distribution at each locus in chimpanzees, we further genotyped 47 unrelated chimpanzees at KPP (*P. t. verus*: 21 males and 26 females).

2.2 Scoring Personality Traits

For each chimpanzee, three evaluators were randomly chosen from 11 researchers who knew the personality of all chimpanzees well. The evaluators answered "yes" or "no" to 120 questions of the YG Personality Inventory (Japan Institute for Psychological Testing, http://www.sinri.co.jp/index.html). This test, which has been designed for self-scoring, consists of 10 questions for each of 12 different personality scales (see Table 1). The answer "?" is used only for questions that are difficult to decide. The final answer for each question was decided by the majority of answers by three evaluators. For example, when two evaluators answered "yes" and one answered "no," the final decision was "yes." When the answers of three evaluators were "yes," "no," and "?," the final decision was "?." In the questionnaires, the word "person" or "people" was replaced by "chim-

Table 1. Scores of YG Personality Inventory decided by majority of answers by three evaluators

			Chimpanzees										
Personality scale	P[a]	Name Sex (age[b])	Gon M (37[c])	Puchi F (37[c])	Reiko F (36[c])	Akira M (27[c])	Mari F (27[c])	Ai F (26[c])	Pendesa F (26[c])	Chloe F (22)	Reo M (21)	Popo F (21)	Pan F (19)
Depressed	0.74183		5	0	3	0	3	2	0	2	2	8	2
Impulsive	0.00008***		2	12	12	12	9	9	4	12	12	13	8
Feeling inferior	0.75533		12	0	2	4	4	0	0	2	10	15	0
Nervous	0.01146*		8	6	5	8	8	8	8	16	14	7	8
Subjective	0.37682		4	4	4	6	5	4	2	12	4	8	6
Selfish	0.00001***		6	11	10	4	11	11	2	16	10	14	5
Aggressive	0.21472		0	8	12	8	8	16	14	16	8	4	11
Active	0.01801*		4	0	2	4	2	10	11	12	2	2	8
Optimistic	0.38963		4	10	8	14	3	8	14	14	3	4	8
Sensible	0.04186*		20	16	16	18	15	8	18	10	16	16	18
Leadership qualities	0.97605		2	14	18	16	8	16	20	10	10	0	4
Sociable	0.06675		6	6	20	16	5	8	18	6	6	2	2

Each of the 12 different personality scales consists of 10 questions; the respective scores for answers "Yes", "?", and "No" were "2", "1", and "0", or "0", "1", and "2", depending on each questionnaire

[a]Scores of chimpanzees and humans were compared by one-way ANOVA, *: $P < 0.05$, ***: $P < 0.001$

[b]Ages at the time of personality test in September 2003

[c]Ages were estimated

panzee (s)." For example, "I like to be a friend of many people" was replaced to "I like to be a friend of many chimpanzees."

To confirm the reliability of evaluation of the chimpanzees, 11 humans (evaluators of the above study) were, as a control, also scored by the same methods, and scores by themselves and those by three randomly chosen other evaluators were compared. Based on the answers, scores for 12 personality scales were analyzed. A dendrogram of the subject 22 individuals was drawn using a principal component analysis (PCA) based on the scores of the 12 personality scales.

2.3 Analysis of Genetic Polymorphism

DNA was extracted from peripheral blood using phenol-chloroform. Four primer sets were used: 5′-GCCATCAGCGTGGACAGGT-3′ and 5′-CGTCGTTGAGGCCGCACAGCAC-3′ for *DRD4in2*; 5′-TCTGGCGCTTCCCCTACATAT-3′ and 5′-TGTTCCTAGTCTTACGCCAGTG-3′ for *STin2*; 5′-TCCAGAATCTGTTCCAGAGCGTGC-3′ and 5′-GCTGTGAGGGTTGCTGTTCCTCAT-3′ for *AR;* and 5′-GGTAAACCATGGTCTGTACC-3′ and 5′-AACAAAATGTTGAATGAGTGGG-3′ for *ERβ*. For amplification of *DRD4in2* and *ERβ*, we used 50 ng DNA in 10 μl reaction mixture containing 0.5 μM of each primer, 0.5U of *LA Taq* polymerase, GC buffer I (TaKaRa, Shiga, Japan), and 400 μM of each dNTP. After an initial incubation at 95°C for 2 min, polymerase chain reaction (PCR) amplification was performed for 35 cycles consisting of 95°C for 30 s, 65°C for 1 min, and 74°C for 2 min, followed by a final extension at 74°C for 10 min. For

Table 1. *Continued*

			Humans														
Range	Average	SD	A	B	C	D	E	F	G	H	I	J	K	Range	Average	SD
0–8	2.5	2.4	0	4	2	0	4	0	0	6	4	3	8	0–8	2.8	2.7
2–13	9.5	3.6	2	2	5	0	2	1	4	8	0	5	4	0–8	3.0	2.5
0–15	4.5	5.4	0	0	2	2	6	3	0	14	0	0	14	0–14	3.7	5.4
5–16	8.7	3.3	4	0	4	2	6	0	0	15	2	4	8	0–15	4.1	4.4
2–12	5.4	2.7	0	4	5	4	4	4	4	8	6	6	4	0–8	4.5	2.0
2–16	9.1	4.3	0	2	2	2	2	0	2	4	0	0	0	0–4	1.3	1.4
0–16	9.5	4.9	14	2	8	6	12	3	6	6	8	11	2	2–14	7.1	4.0
0–12	5.2	4.3	20	10	16	2	10	13	16	2	18	11	4	2–20	11.1	6.3
3–14	8.2	4.4	8	4	16	0	10	8	8	0	4	10	3	0–16	6.5	4.8
8–20	15.5	3.6	4	12	15	8	12	20	16	0	6	16	12	0–20	11.0	6.0
0–20	10.7	6.7	20	12	9	12	0	18	20	2	8	16	2	0–20	10.8	7.3
2–20	8.6	6.3	18	16	13	8	12	18	20	5	8	20	10	5–20	13.5	5.3

amplification of *AR* and *STin2*, the annealing temperature was decreased to 60°C and 55°C, respectively. The PCR products were separated by electrophoresis on 1.5% agarose gel and were then extracted from the gel and directly sequenced using the dye termination method and a ABI 3100 DNA Sequencer (Perkin-Elmer, Applied Biosystems Division, Foster City, CA, USA). Allele frequency distributions of four genes were surveyed among the total of 54 unrelated chimpanzees (*P. t. verus*, 23 males and 31 females) at PRI and KPP.

3 Results and Discussion

3.1 Personality Test

Table 1 shows the scores on 12 personality scales of the YG Personality Inventory for the 11 chimpanzees and 11 humans at PRI. Humans are indicated by the letters A to K. In humans, the average percentage (and range) of coincidence between evaluation by self and by non-self (the final answer decided by majority of answers by three evaluators) was 76.5 (71.8–83.6), indicating that non-self evaluation is more than 70% reliable for all scales (Table 2). On the other hand, the percentage of perfectly coincident answers among the three evaluators was not very high in either humans (average, 42.9%) or chimpanzees (average, 48.0%). In chimpanzees, a relatively low percentage of coincidence was observed in the scales "selfish" (31.8%) and "impulsive" (36.4%). These scales include ques-

Table 2. Percentage of coincident answers

Personality scale	Human: final and self[a]	Human: three evaluators[b]	Chimpanzee: three evaluators[b]
Depressed	75.5	45.5	52.7
Impulsive	71.8	40.9	36.4
Feeling inferior	77.3	50.9	54.5
Nervous	77.3	40.0	42.7
Subjective	77.3	42.7	54.5
Selfish	83.6	57.3	31.8
Aggressive	76.4	40.0	50.0
Active	76.4	32.7	49.1
Optimistic	81.8	37.3	48.2
Sensible	74.5	33.6	59.1
Leadership qualities	73.6	46.4	53.6
Sociable	72.7	47.3	42.7
Average	76.5	42.9	48.0

[a]Percentage of coincident answers between the final answer decided by the majority of answers by three evaluators and the answer by self
[b]Percentage of perfectly coincident answers among three non-self evaluators

tions that seemed to be difficult to evaluate appropriately for non-self evaluators, such as, "I feel most chimpanzees are lazy without someone watching them" and "I cannot summarize my thinking." On the other hand, the average ratio of perfect coincidence of three evaluators was relatively high for the "sensible" (59.1%), "feeling inferior" (54.5%), and "subjective" (54.5%) scales. However, the tendency was not the same in humans.

In a dendrogram based on PCA of the scores of the 12 personality scales, 11 chimpanzees and 11 humans were categorized into three large clusters (Fig. 1). The three clusters were characterized as high scores in "depressed" and "feeling inferior" for the first cluster, "implusive" and "selfish" for the second cluster, and "active" and "sociable" for the third cluster. In the first, second, and third clusters, 2, 6, and 3 chimpanzees were included, indicating the diverse personalities within chimpanzees.

A different tendency was observed between chimpanzees and humans; 8 of 11 chimpanzees belonged to the first and the second clusters, whereas 9 of 11 humans belonged to the third cluster. In the comparison of scores between chimpanzees and humans by a one-way analysis of variance (ANOVA), chimpanzees showed significantly higher scores on the scales of "selfish," "impulsive" ($P < 0.001$), "nervous," and "sensible" ($P < 0.05$), and lower on the "active" scale ($P < 0.05$). These results may indicate the species difference in temperament between chimpanzee and humans. Or, this difference might be caused by the passive condition under which the chimpanzees were kept in their enclosure. Comparison of the result in this study with that of chimpanzees kept in the other institutes

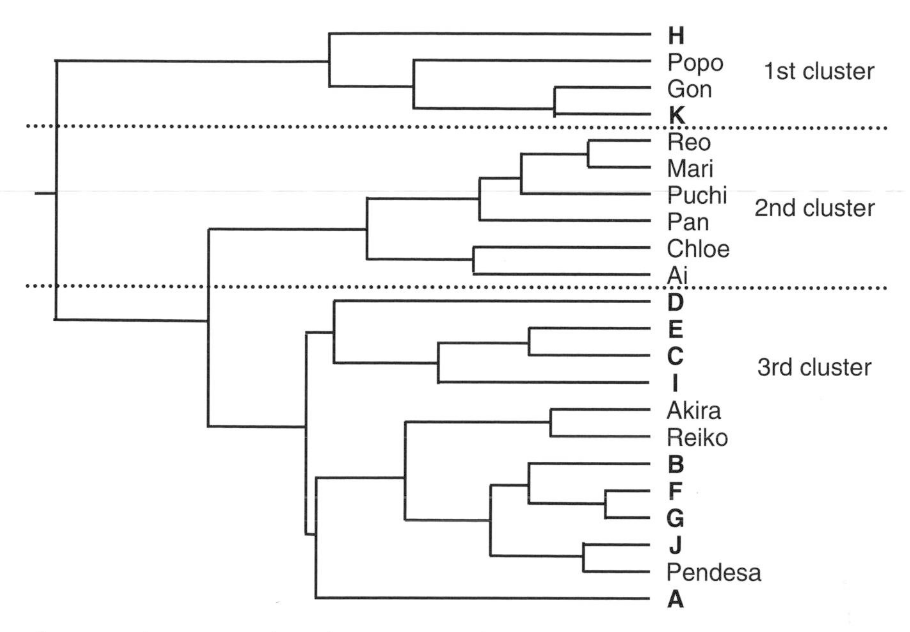

Fig. 1. Dendrogram resulting from a principal component analysis (PCA) of the scores of the 12 personality scales of the YG Personality Inventory for 11 chimpanzees and 11 humans: 22 individuals were divided into three large clusters

and with those living in the wild would be necessary to fully describe the temperament of chimpanzees. A common tendency across chimpanzees and humans was also found. The largest individual differences were observed in "leadership qualities" (0–20) in both chimpanzees and humans. Leadership qualities may be a good indicator for individual differences of animals living in groups with a complex social structure with hierarchies.

In this study, by applying the method for human personality scaling to chimpanzees, the result enabled us to describe precise individual difference within chimpanzees and to compare the scores between chimpanzees and humans. However, some questions were difficult to answer for non-self evaluators. Modifying questionnaires to more concrete ones may raise the percentage of coincidence among evaluators. Also, three evaluators for each chimpanzee were chosen randomly from 11 researchers without considering their intimacy with each chimpanzee. Considering the experience of each evaluator with each chimpanzee may lead to a more precise description of the personality. To assess the power of this method, other traits reflecting personality should also be scored and the result should be compared with the present results. For instance, the frequency of grooming in PRI chimpanzees was related to a high score for leadership qualities ($P < 0.05$; data not shown). Scoring some traits such as variable levels of hyperactivity of rhesus monkeys (Miller et al. 2001) would also be a possibility.

3.2 Analysis of Genetic Polymorphism

In chimpanzees, the polymorphism of *STin2* and *ERβ* was reported for the first time in this study. Polymorphism had already been reported for *DRD4in2* (Shimada et al. 2004) and *AR* (Choong et al. 1998). However, 7 alleles with 15, 16, 19, and 24–27 repeats of glutamine in *AR* were newly found in this study. Allele frequency distributions are shown in Table 3. In the polymorphic regions of *DRD4in2, STin2, AR*, and *ERβ*, 3, 3, 12, and 7 alleles, respectively, were observed in the 54 unrelated individuals. The expected heterozygosity (Nei and Roychoudhury 1974) was the highest in *AR* (0.828) and the lowest in *STin2* (0.230). The sequences of the alleles can be obtained from the DDBJ/EMBL/GenBank nucleotide sequence database with the accession numbers AB194967, AB194968, and AB194971 for alleles *18, 19*, and *25* of *STin2* and AB194972–AB194975 for alleles *194, 198, 202*, and *206* of *ERβ*.

In the 11 chimpanzees at PRI, 3, 4, 7, and 6 alleles were observed for *DRD4in2, STin2, AR*, and *ERβ* (Table 4). In *STin2*, the allele *23* (accession number: AB194969) was observed only in one individual. Polymorphic loci found in this study will become good markers for elucidating their effect on the personality of chimpanzees.

In humans, the dopamine receptor D4 gene includes many variable sites, and the function of *DRD4in2* is unknown (Shimada et al. 2004). Another locus, *STin2*, includes a VNTR of 15- to 17-bp units, and 3 alleles with 9, 10, and 12 repeats have been reported in humans (Ogilvie et al. 1998). Repeat number affects reporter gene expression in transgenic mice (MacKenzie and Quinn 1999), and the 12-repeat allele is related with bipolar affective disorder (Collier et al. 1996). In the present study, repeat numbers of chimpanzees were 18–25 and were greatly different from those in humans. The efficiency of signal transduction should also be surveyed in chimpanzees for comparison with humans.

In hormonal receptors, relationships between repeat number and sexual function have been reported in humans (Chamberlain et al. 1994; Westberg et al. 2001). In the *AR*, increased repeat number of a polyglutamine tract in the protein has been associated with decreased transactivation activity in vitro (Chamberlain et al. 1994). Reports indicate that repeat number may influence serum testosterone level (Westberg et al. 2001), aggressiveness (Jönsson et al. 2001), and cognition problems (Yaffe et al. 2003). Furthermore, the repeat number of a (CA)n microsatellite in human *ERβ* (Tsukamoto et al. 1998) affects serum androgen level (Westberg et al. 2001). In this study, we found that the *AR* and *ERβ* loci were highly polymorphic in chimpanzees, showing expected heterozygosities of 0.828 and 0.780, respectively. It would also be very interesting to investigate the effect of the genotypes of these loci on reproductive functions in chimpanzees.

In humans, polymorphisms of many other genes have been reported in relationship to personality. Recently, the chimpanzee genome sequence has been reported (The Chimpanzee Sequencing and Analysis Consortium 2005), and information about orthologous genes in chimpanzees is increasing. It will be

Table 3. Allele frequencies at four loci in 54 chimpanzees (*Pan troglodytes verus*)

	Dopamine receptor D4		Serotonin transporter		Androgen receptor		Estrogen receptor β	
Gene	*DRD4in2*		*STin2*		*AR*		*ERβ*	
Abbreviation of loci	Allele	Frequency	Allele	Frequency	Allele	Frequency	Allele	Frequency
	104	0.231	*18*	0.120	*15*	0.048	*180*	0.074
	111	0.722	*19*	0.870	*16*	0.012	*194*	0.278
	114	0.046	*25*	0.009	*18*	0.048	*198*	0.213
					19	0.349	*200*	0.009
					20	0.060	*202*	0.296
					21	0.012	*204*	0.028
					22	0.145	*206*	0.102
					23	0.120		
					24	0.108		
					25	0.048		
					26	0.036		
					27	0.012		
He[a]	0.426		0.230		0.828		0.780	

Allele names indicate sizes of polymerase chain reaction (PCR) products for an insertion/deletion polymorphism (*DRD4in2* and *ERβ*) or repeat number for tandem repeats (*STin2* and *AR*); alleles found in 11 individuals at the Primate Research Institute (PRI) are underlined

[a]Expected heterozygosity: $He = n(1 - \Sigma q_i^2)/(n - 1)$

Source: Nei and Roychoudhury (1974)

Table 4. Genotypes of chimpanzees in PRI

	DRD4in2		*STin2*		*AR*		*ERβ*	
Gon	*111*	*111*	*18*	*19*	*18*	—	*194*	*198*
Puchi	*104*	*104*	*19*	*19*	*20*	*24*	*198*	*202*
Reiko	*104*	*111*	*19*	*19*	*19*	*22*	*180*	*202*
Akira	*111*	*111*	*18*	*19*	*22*	—	*198*	*202*
Mari	*111*	*111*	*19*	*19*	*20*	*24*	*198*	*198*
Ai	*104*	*111*	*19*	*19*	*15*	*25*	*194*	*202*
Pendesa	*104*	*111*	*19*	*23*	*18*	*24*	*180*	*206*
Chloe	*111*	*114*	*19*	*25*	*19*	*23*	*198*	*200*
Reo	*111*	*111*	*19*	*19*	*22*	—	*180*	*198*
Popo	*104*	*111*	*18*	*19*	*18*	*20*	*198*	*202*
Pan	*104*	*111*	*19*	*19*	*18*	*24*	*194*	*198*

necessary to analyze these genes in chimpanzees and investigate a possible relationship with personality so that a comparison across the two species can be made.

In Japan, a total of 349 chimpanzees are maintained in 57 institutes and zoos (Tama Zoological Park 2005). Based on our preliminary results, a wide-range assessment of chimpanzees using a modified questionnaire will provide a better characterization of their personality and allow us to investigate the effect of genes with higher statistical power. Such results may be used in captive breeding by typing personalities and applying this information, for instance, when new groups need to be formed. A better understanding of temperamental differences between chimpanzees and humans is also likely to help when designing appropriate conservation measures for chimpanzees in the wild.

Acknowledgments

We thank Ms. Y. Ueda, Gifu University, for her technical assistance. We are indebted to the staff of Primate Research Institute, Kyoto University, and Great Ape Research Institute, Hayashibara Biochemical Laboratories, for their help in elucidating the personalities of their chimpanzees. We are also grateful to Dr. J.R. de Ruiter, Durham University, for valuable comments on the manuscript. This study was funded by a grant (2001–2003) from the National Agriculture and Bio-oriented Research Organization (NARO) of Japan, the Cooperation Research Program of the Primate Research Institute, Kyoto University, MEXT (#12002009, #16002001), and JSPS-HOPE. For the blood sampling of chimpanzees, we have complied with the ethical standards in the treatment of animals with guidelines laid down by the Primate Society of Japan.

References

Benjamin J, Li L, Patterson C, Greenberg BD, Murphy DL, Hamer DH (1996) Population and familial association between the D4 dopamine receptor gene and measures of Novelty Seeking. Nat Genet 12:81–84

Chamberlain NL, Driver ED, Miesfeld RL (1994) The length and location of CAG trinucleotide repeats in the androgen receptor N-terminal domain affect transactivation function. Nucleic Acids Res 22:3181–3186

Champoux M, Bennett A, Shannon C, Higley JD, Lesch KP, Suomi SJ (2002) Serotonin transporter gene polymorphism, differential early rearing, and behavior in rhesus monkey neonates. Mol Psychiatry 7:1058–1063

Choong CS, Kemppainen JA, Wilson EM (1998) Evolution of the primate androgen receptor: a structural basis for disease. J Mol Evol 47:334–342

Collier DA, Arranz MJ, Sham P, Battersby S, Vallada H, Gill P, Aitchison KJ, Sodhi M, Li T, Roberts GW, Smith B, Morton J, Murray RM, Smith D, Kirov G (1996) The serotonin transporter is a potential susceptibility factor for bipolar affective disorder. Neuroreport 7:1675–1679

Inoue-Murayama M, Takenaka O, Murayama Y (1998) Origin and divergence of tandem repeats of primate D4 dopamine receptor genes. Primates 39:217–224

Inoue-Murayama M, Niimi Y, Takenaka O, Okada K, Matsuzaki I, Ito S, Murayama Y (2000) Allelic variation of the serotonin transporter gene polymorphic region in apes. Primates 41:267–273

Inoue-Murayama M, Niimi Y, Takenaka O, Murayama Y (2001) Evolution of personality-related genes in primates. In: Miyoshi K, Shapiro CM, Gaviria M, Morita Y (eds) Contemporary neuropsychiatry. Springer, Tokyo, pp 425–428

Jönsson EG, von Gertten C, Gustavsson JP, Yuan QP, Lindblad-Toh K, Forslund K, Rylander G, Mattila-Evenden M, Asberg M, Schalling M (2001) Androgen receptor trinucleotide repeat polymorphism and personality traits. Psychiatr Genet 11:19–23

Lesch KP, Bengel D, Heils A, Sabol SZ, Greenberg BD, Petri S, Benjamin J, Muller CR, Hamer DH, Murphy DL (1996) Association of anxiety-related traits with a polymorphism in the serotonin transporter gene regulatory region. Science 274:1527–1531

Lilienfeld SO, Gershon J, Duke M, Marino L, de Waal FB (1999) A preliminary investigation of the construct of psychopathic personality (psychopathy) in chimpanzees (*Pan troglodytes*). J Comp Psychol 113:365–375

MacKenzie A, Quinn J (1999) A serotonin transporter gene intron 2 polymorphic region, correlated with affective disorders, has allele-dependent differential enhancer-like properties in the mouse embryo. Proc Natl Acad Sci U S A 96:15251–15255

Matsuzawa T (2003) The Ai project: historical and ecological contexts. Anim Cogn 6:199–211

Miller GM, De La Garza RD II, Novak MA, Madras BK (2001) Single nucleotide polymorphisms distinguish multiple dopamine transporter alleles in primates: implications for association with attention deficit hyperactivity disorder and other neuropsychiatric disorders. Mol Psychiatry 6:50–58

Nei M, Roychoudhury AK (1974) Sampling variances of heterozygosity and genetic distance. Genetics 76:379–390

Ogilvie AD, Russell MB, Dhall P, Battersby S, Ulrich V, Smith CA, Goodwin GM, Harmar AJ, Olesen J (1998) Altered allelic distributions of the serotonin transporter gene in migraine without aura and migraine with aura. Cephalalgia 18:23–26

Rubinow DR, Schmidt PJ (1996) Androgens, brain, behavior. Am J Psychiatry 153:974–984

Shimada MK, Inoue-Murayama M, Ueda Y, Maejima M, Murayama Y, Takenaka O, Hayasaka I, Ito S (2004) Polymorphism in the second intron of dopamine receptor D4 gene in humans and apes. Biochem Biophys Res Commun 316:1186–1190

Tama Zoological Park (2005) Internal studbook of common chimpanzee, *Pan troglodytes*

The Chimpanzee Sequencing and Analysis Consortium (2005) Initial sequence of the chimpanzee genome and comparison with human genome. Nature 437:69–87

Tsukamoto K, Inoue S, Hosoi T, Orimo H, Emi M (1998) Isolation and radiation hybrid mapping of dinucleotide repeat polymorphism at the human estrogen receptor beta locus. J Hum Genet 43:73–74

Weiss A, King JE, Figueredo AJ (2000) The heritability of personality factors in chimpanzees (*Pan troglodytes*). Behav Genet 30:213–221

Westberg L, Baghaei F, Rosmond R, Hellstrand M, Landen M, Jansson M, Holm G, Björntorp P, Eriksson E (2001) Polymorphisms of the androgen receptor gene and the estrogen receptor β gene are associated with androgen levels in women. J Clin Endocrinol Metab 86:2562–2568

Whiten A, Goodall J, McGrew WC, Nishida T, Reynolds V, Sugiyama Y, Tutin CE, Wrangham RW, Boesch C (1999) Cultures in chimpanzees. Nature (Lond) 399:682–685

Yaffe K, Edwards ER, Lui LY, Zmuda JM, Ferrell RE, Cauley JA (2003) Androgen receptor CAG repeat polymorphism is associated with cognitive function in older men. Biol Psychiatry 54:943–946

Part 3
Communication and Mother–Infant Relationship

8 Evolutionary Origins of the Human Mother–Infant Relationship

Tetsuro Matsuzawa

1 Introduction

This chapter aims to speculate on the evolutionary origins of the human mother–infant relationship by providing evidence from fieldwork as well as laboratory work with nonhuman primates, particularly chimpanzees. In short, a form of mother–infant relationship characterized by clinging and embracing is common to all primate, especially simian, species with grasping hands and feet. Such ventro-ventral contact provides the basis for further extensions of the relationship, such as various forms of face-to-face communication shared by humans and chimpanzees, including mutual gaze and smiling. Uniquely in humans, mother and infant often rely on vocal exchanges as a result of increased physical separation. The stable supine posture of human infants facilitates the free movement of limbs and fingers and manual gestures. As a result, humans have developed a complex system of communication using multimodal signals. The following section outlines a possible evolutionary scenario for the human mother–infant relationship.

2 Evolutionary Stages of Mother–Infant Relationships

Picture in your mind the image of a human mother embracing her infant. She is looking into the eyes of her baby, smiling, and talking softly. Such a scene represents a common everyday encounter. However, it is also a particularly enlightening example, highlighting some unique features of the human mother–infant relationship.

The human mind is a product of evolution, much like the human body and society. Characteristics of the ways in which human mothers and their offspring interact can be no exception. This chapter proposes a model consisting of successive evolutionary stages for the emergence of the human mother–infant relationship (Table 1). I outline five distinctive stages, each with a different point of evolutionary origin: mammalian, primate, simian, hominoid, and *Homo*.

Primate Research Institute, Kyoto University, 41 Kanrin, Inuyama, Aichi 484-8506, Japan

Table 1. Evolutionary stage model of mother–infant relationship

Stage	Year[a]	Unique behavior	Number of species
Mammalian	65	Provision of milk	4,500
Primate	50	Clinging by the infant	200
Simian	40	Embracing by the mother	80
Hominoid	5	Mutual gaze and smiling	2
Homo	2	Vocal exchange and manual gestures	1

[a]Year represents the onset of each stage in millions of years before the present

2.1 Mammalian Origin

The first stage of the human mother–infant relationship has its origins in the period of mammalian adaptation and divergence. Parental care directed toward offspring is a common practice among birds and mammals. Parental care itself may have emerged in the common ancestor of mammals, birds, and some dinosaurs around 200 million years ago.

There are presently about 4,500 extant species of mammals. They are characterized by the nursing of offspring through the provision of milk—a feature that sets them apart from birds, reptiles, amphibians, and other animals. Researchers envisage the common ancestor of living mammals as a small ground-dwelling animal active during the night at a time in Earth's history when dinosaurs dominated the land. The mammalian mother–infant relationship may have been established around 65 million years ago, after the extinction of the dinosaurs, and during the subsequent diversification of the mammalian lineage.

2.2 Primate Origin

The second stage of the human mother–infant relationship derives from primate origins. Primates differentiated from the common mammalian ancestor, shifting from nocturnal and terrestrial habits to a diurnal and arboreal lifestyle. As part of their ecological adaptation, the tips of the four limbs were transformed to allow the grasping of objects. Primates were once known as "*Quadrumana*"—creatures with four hands for manipulation.

Primates' vision also improved over the course of evolution. It is now characterized by a clear tendency for good three-dimensional depth perception as well as color vision. Let us picture a primate moving through trees in the daytime: the hands grasp the branches firmly and the eyes discriminate objects within the environment with depth and color. These characteristics contribute enormously to the individual's survival.

Thanks to the shape of the hands and as a result of the necessity for transportation, primate infants began to cling to their mothers by grasping her body hair. Later, the mothers came to embrace their infants. Note that infant cats, dogs,

Fig. 1. An infant ring-tail lemur clings to its mother. (Photo by Tetsuro Matsuzawa)

horses, or cows cannot cling to their mothers, who in turn cannot embrace their offspring, simply because they have no grasping hands.

The evolutionary model needs to discriminate the stage of clinging from the stage of both clinging and embracing. Prosimians are thought to closely resemble living primates' common ancestor. Most prosimians are arboreal, but they are not diurnal: active during the night, they feed on insects. Some prosimians (e.g., the aye-aye, *Daubentonia madagascariensis*, and the ruffed lemur, *Varecia variegata*) often leave their infants in nests, while others (such as the greater galago, *Otolemur crassicaudatus*, and the western tarsier, *Tarsius bancanus*) leave them clinging to branches. These species transport their infants by mouth (oral carrying), much as do cats, dogs, and many other mammals (Ross 2002).

Clinging is likely to have emerged within primate evolution from the mammalian ancestor. Primate infants began to cling to their mothers as a result of the appearance of grasping hands. In this early stage, primate mothers would not generally have embraced their infants: it was the infants themselves who were responsible for clinging during transports by the mother. Some prosimians, such as the ring-tail lemur (*Lemur catta*), as well as some New World monkeys, such as the cotton-top tamarin (*Saguinus oedipus*), are characterized by this form of "clinging without embracing" mother–infant relationship (Fig. 1).

2.3 Simian Origin

In the third stage, mothers began supporting or embracing the infants already clinging to their fur (Fig. 2). Such embrace is common practice among humans, apes, Old World monkeys, and some New World monkeys (the "simians," in

Fig. 2. A Japanese monkey mother embraces her infant as the infant clings to her. (Photo by Michio Nagamine)

contrast with prosimians). This third stage would thus have emerged with the appearance of the simian lineage from the primate ancestor.

The simians have a variety of common characteristics. Although prosimians, as well as many mammals, possess only very limited or rudimentary color vision, simians are color sensitive (De Valois and Jacobs 1968). Humans, apes, and Old World monkeys (catarrhines) are all trichromats. It has been suggested that New World monkeys (platyrrhines) show color vision polymorphisms (Jacobs and Rowe 2004), that is, individuals are either trichromats or dichromats. One important difference between trichromats and dichromats is the fact that dichromats are unable to discriminate monochromatic from white light over a small portion of the spectrum, the neutral point. A recent study has shown that dichromats may have an advantage in shape recognition in dim light (Saito et al. 2005). Nevertheless, simians inhabit the same colorful world that we humans perceive.

In comparison with prosimians, simians are also larger in body size. Heavy infants require additional support; this may be one of the reasons why mothers began to put their hands on the back of their clinging infants, thereby forming the basis of the embrace.

The important point to note, therefore, is that clinging comes first and embracing second. This model is supported by observations of events after chimpanzee births. In the case of captive chimpanzee mothers, almost half fail to raise their own infants (see Chapter 1, this volume). The major difficulty is with the initial establishment of the clinging–embracing bond. Although the mother's embrace inevitably follows once the infant first begins to cling, a commonly seen problem among captive-born mothers is that following delivery they refuse to approach their infants and to allow them to cling in the first place.

2.4 Hominoid Origin

The fourth stage of human mother–infant relationships may derive from the period of hominoid adaptation, particularly the common ancestor of humans and chimpanzees. Among humans and chimpanzees, a new level of mother–infant relationship has evolved in addition to the clinging–embracing characteristics common to most primates; these are mutual gaze and smiling.

Japanese monkey mothers (*Macaca fuscata*) seldom look into the eyes of their infants. Direct gaze in general has negative connotations among, for example, Japanese monkeys; in human terms, it is comparable to a hostile stare and carries the meaning of a mild threat. If you continue to look directly into the eyes of a monkey, they will eventually open their mouth to threaten you back, or grimace, show their teeth in a submissive manner, even begin to scream. Other Old World monkeys also exhibit hostile staring. Baboons (*Papio* spp.) repeatedly blink their eyes to display the white portions of the eye lids: this sparkling effect communicates agonistic intention.

However, in the case of humans and chimpanzees, direct gaze can have two different meanings: hostility and affection. In the context of the latter, it is often referred to as eye-to-eye contact or mutual gaze (Fig. 3), and is frequently accom-

Fig. 3. Mutual gaze in a chimpanzee mother and her infant. The mother Ai is lifting up Ayumu at the age of 51 days. (Photo by Tomomi Ochiai)

panied by smiling in both humans and chimpanzees. Smiling modifies the meaning of the direct gaze and conveys information about the actor's affectionate state of mind. This kind of gaze is so far known to occur only in hominoids, particularly in humans and chimpanzees.

2.5 *Homo* Origin

What is uniquely human in the context of mother–infant relationships? To provide an answer to this question, we need to bear in mind two important prerequisites. First, we need to explore details of chimpanzee mother–infant relationships because chimpanzees are our closest living relatives. Second, we need to speculate on evolutionary trends within fossil hominids—on the changes that occurred during the transition from the apelike Australopithecines to the *Homo* species. The first point is more fully described in the next section; the present section focuses on the second point.

Australopithecines arose from the common ancestor of humans and chimpanzees 5 million years ago. *Australopithecus afarensis* had roughly the brain volume of a chimpanzee. So, why is it classified as a hominid? The main reasons are upright posture and bipedal locomotion, features that set *A. afarensis* apart from its precursors. Australopithecines are thought to have used various kinds of tools, including stone tools that resemble those of living chimpanzees. Chimpanzees modify grass stems, branches, and other perishable material and use stones to crack nuts, but they do not intentionally modify the stones. Presumably australopithecines did at least as well as chimpanzees, but not until *Homo* are there signs that stone was deliberately modified to form tools. Lithic technology is characterized by the manufacture of stone tools through the use of other stone tools.

Homo spp. appeared about 2 million years ago, as did lithic technology (Asfaw et al. 1999). Thus, this was the time when hominids took up tool-making and evolved from the small-brained, apelike australopithecines into the first members of the *Homo* genus. Well-preserved specimens of *Homo* appear at around 2 million years ago in East Africa, mainly at Olduvai Gorge (Tanzania), where the remains of *Homo habilis* have been unearthed, and Koobi Fora (Kenya), where two species are present, a *habilis*-like species and the larger *Homo rudolfensis*. A couple of hundred thousand years after these two early *Homo* species arrived on the scene, the first more modern looking species, *Homo ergaster*, with long legs, shortened forearms, short face, and cranial capacity greater than 800 cm^3, appears in the record (Chapter 1, Fig. 1).

The emergence of *Homo* was truly a turning point in the evolution of humans. The increase in brain volume was accompanied by the appearance of tool manufacture and complex communication involving facial-gestural signals and vocalizations. These changes likely went hand in hand with changes in society: division of labor for hunting and gathering, cooperative hunting, meat-sharing, sexual roles, and helpers looking after close kin. In parallel to shifts in the nature of socio-material life, the mother–infant relationship also underwent a transformation. Most importantly, there was an increase in the occurrence of physical separation between mothers and their infants.

Homo mothers did not hold their infants constantly: the clinging–embracing relationship common to primates had gone. Mothers laid their infants down; when the infants needed help, they cried. Mothers began talking to their infants in face-to-face situations. One of the unique features of human infants is that they can remain stable in the supine posture. Ape infants (I have so far observed this in chimpanzees and orangutans) are unstable in the supine posture: when placed on their back, they move their contralateral limbs simultaneously (left arm and right leg, or right arm and left leg), and start to whimper (Fig. 4). They cannot roll over until the age of 2 months. Monkey infants (I have observed six species of macaque infants, i.e., *Macaca fuscata, Macaca fascicularis, Macaca mulatta, Macaca nemestrina, Macaca arctoides,* and *Macaca radiata*) are also unstable in the supine posture but can roll over to the prone posture by themselves from just after birth. Chimpanzee and orangutan infants are able to roll over at the age of about 2 months.

Being stable in the supine posture freed human infants' hands from grasping and from supporting the body. Hands could instead be used for various other actions such as extension toward objects, holding, pinching, pointing, and touching. Thus, the mother–infant relationship in humans can be characterized by mutual-gaze, vocal exchange, and manual-gestural signs in the context of face-to-face communication. The fifth and final stage of human mother–infant

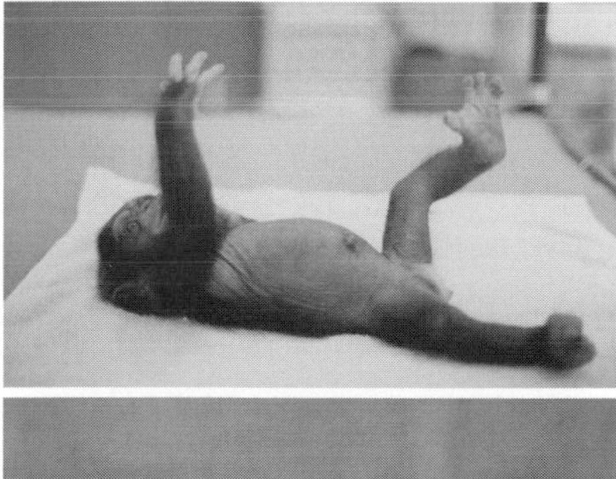

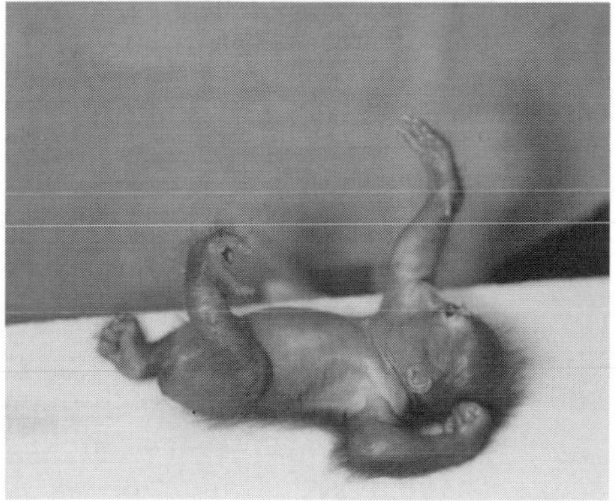

Fig. 4. Ape infants (*top:* chimpanzee; *bottom:* orangutan) are unstable when they are laid down in the supine posture. They start moving the contralateral limbs simultaneously (left arm and right leg, or right arm and left leg) and whimper. (Photo provided by H. Takeshita and T. Matsuzawa)

relationships has its origin in the emergence of *Homo* species. I return to these unique characteristics in the final section.

3 Early Development of Communication in Chimpanzees

My colleagues and I have been working with three pairs of mother–infant chimpanzees since 2000. The project has focused on sociocognitive development in chimpanzees from just after birth (see Matsuzawa, Chapter 1, this volume; Matsuzawa and Nakamura, 2004; Tomonaga et al. 2004; see Tomonaga, Chapter 12, this volume) and has utilized a novel and unique research method we have called participation observation. Unlike in previous studies, the chimpanzee infants are not reared by human surrogates; they are being reared by their biological mothers. All three mother–infant pairs live in a community of captive chimpanzees of three generations, within an enriched environment. This setting has provided the infants with a socioecological environment that resembles that found in the wild in many important ways. Human investigators, who have established long-term relationships with the chimpanzee mothers, participate in the everyday lives of the subjects.

This section focuses on a series of studies that have illuminated similarities and differences between mother–infant relationships in humans and chimpanzees. The two species share many common characteristics of face-to-face communication from the moment of birth. Through a review of our recent studies of chimpanzees using participation observation, I address topics such as neonatal smiling, neonatal imitation, neonatal face recognition, and neonatal vocalization in response to sounds. I use the word "neonate" to refer to infants from birth up to the age of 2 months.

3.1 Visual-Facial Communication

Chimpanzee infants as well as human infants exhibit neonatal smiling (Mizuno et al. 2006). They spontaneously smile while they sleep (Figs. 5, 6). The eyes remain closed, implying that the smile is not directed toward any specific individual in the infant's surroundings. The tendency in infants to smile seems to be innate: without any explicit stimuli, they spontaneously perform the behavior. Neonatal smiling disappears around 2 months of age, to be later replaced by social smiling. In contrast with neonatal smiling, during social smiling the eyes remain open, and the gesture is directed toward the individual in front of the infant. As the behavior first appears around 3 months of age, it is also known as "3-months-old smiling" (Fig. 7). Both humans and chimpanzees go through similar developmental changes in the occurrence of neonatal and social smiling (Fig. 8, Mizuno et al. 2006; see also Tomonaga, Chapter 12, this volume).

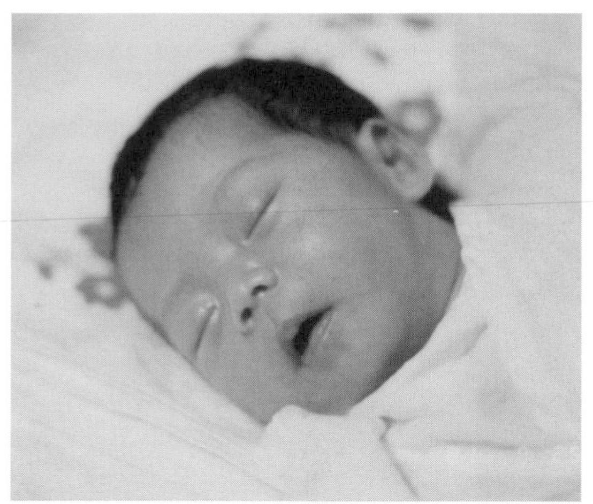

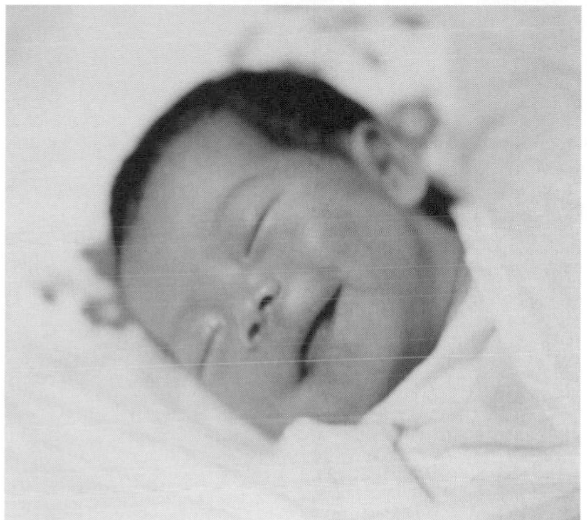

Fig. 5. Neonatal smiling in a human infant YM, 11 days after birth. (Photo by Matsuzawa)

Chimpanzees also exhibit neonatal imitation comparable to that shown by humans (Myowa-Yamakoshi et al. 2004; see also Myowa-Yamakoshi, Chapters 9 and 14, this volume). In our tests, a human experimenter performed one of three facial gestures in front of infant chimpanzee subjects—tongue protrusion, mouth opening, and lip protrusion—while we video-recorded the infants' facial expressions. The procedure was exactly the same as that used in Meltzoff and Moore's (1997) original study with human infants. The results showed that chimpanzee infants during their first 2 months of life (1–8 weeks) imitated two of the three facial expressions: tongue protrusion and lip protrusion. However, facial imitation disappeared over the following months (9–16 weeks). These findings reveal essential similarities between human and chimpanzee infant development.

Human infants are also known to recognize their mothers' face from a very early age (Johnson and Morton 1991). We investigated face recognition by

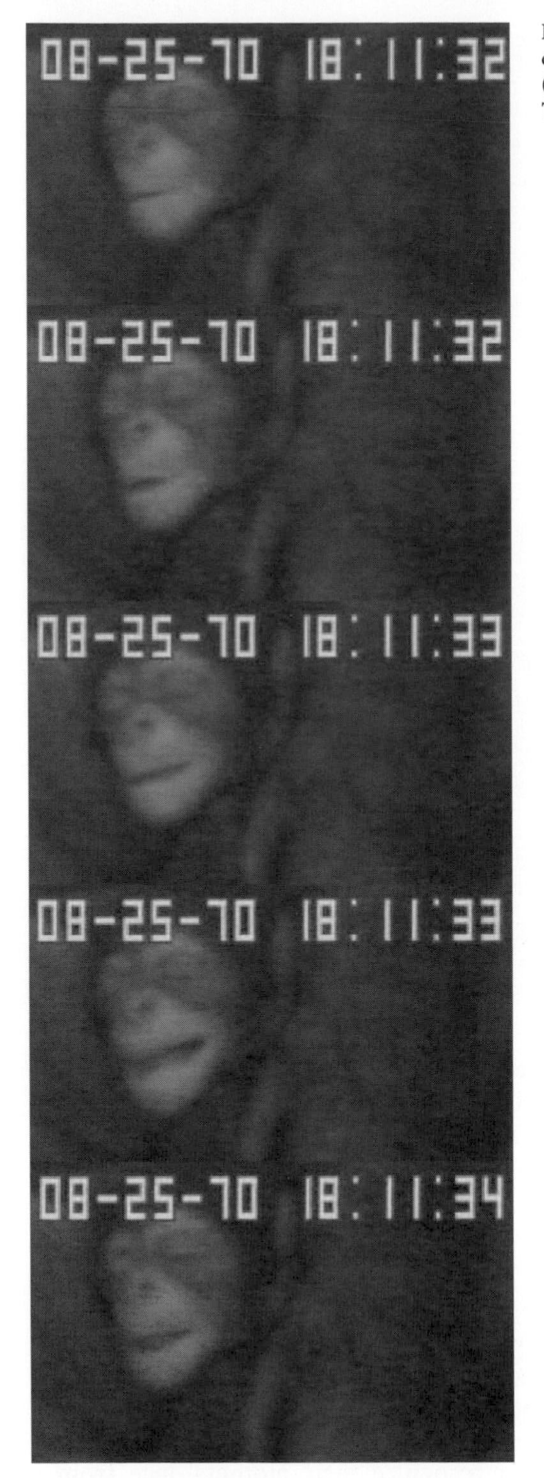

Fig. 6. Neonatal smiling in an infant chimpanzee Pal, 16 days after birth. (Photo provided by Y. Mizuno, H. Takeshita, and T. Matsuzawa)

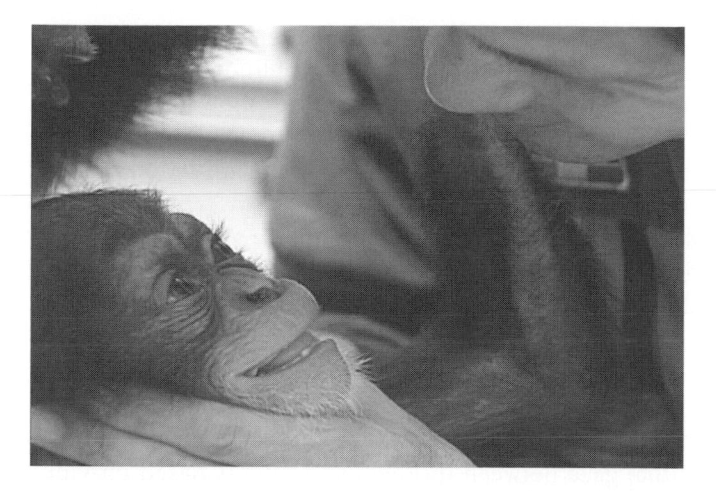

Fig. 7. Social smiling toward a human tester. Ayumu, at 3 months of age. (Photo provided by Chukyo-TV)

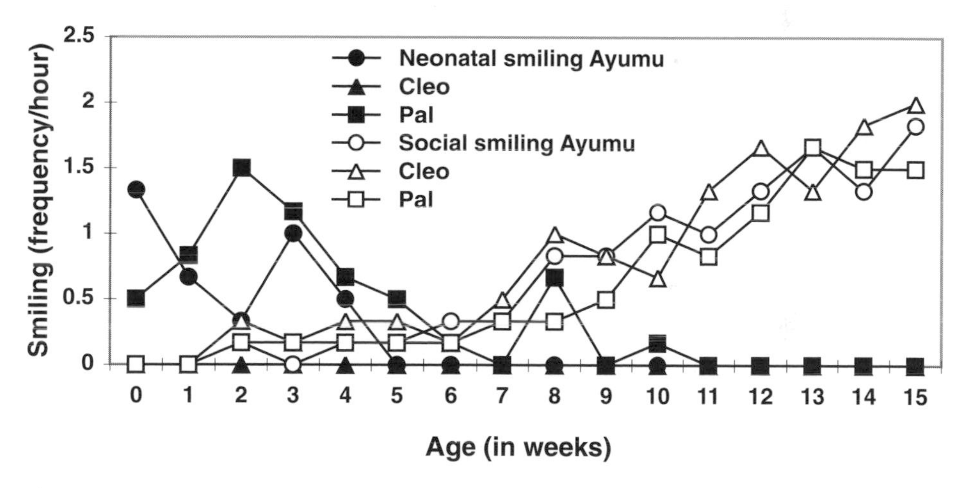

Fig. 8. Neonatal smiling and social smiling measure in the first 4 months of life in three infant chimpanzees reared by the mothers. (From Mizuno et al. 2006)

infant chimpanzees, using pictures of the three mother chimpanzees (Myowa-Yamakoshi et al. 2005; see Tomonaga, Chapter 12, this volume). As a control, we used a composite picture consisting of a computer-generated "average" face based on the three mothers' photographs. Each photo was cropped around the contours of the individual's face and mounted onto a small video camera pointed toward the infants' face, recording their gaze. The face stimulus was moved around slowly in front of the subjects, and we measured how long the infants' gaze remained fixed on the images. We found no clear differences in infants' gaze toward the mother's versus the control face up to 1 month of age. Then, between 1 and 2 months, a preference emerged and infants began to look longer at photos of their mother's face. After the age of 2 months, the tendency disappeared. We concluded that the infants may have developed the ability to discriminate their mothers when they were around 1 month old and thereafter showed particularly affectionate responses to these faces.

The eyes are thought to play an important role in the discrimination of faces. We analyzed gaze recognition in infant chimpanzees (Myowa-Yamakoshi et al. 2003) through the so-called preferential looking method. We showed the infants two pictures of faces side by side, in which only the state of the eyes was varied. For example, we pitted a face with open eyes against one with the eyes closed, or one with direct gaze against another with averted gaze. The results showed that chimpanzees preferred to look at faces in which the eyes were open, as well as those with direct gaze, from at least 2 months of age. This finding strongly implies that chimpanzee infants are sensitive to gaze and gaze direction.

We also analyzed mutual gaze between mothers and infants based on video recordings of daily interactions (Bard et al. 2005). The occurrence of mutual gaze began to increase at around 2 months of age. Thus, many developmental changes co-occur at the age of 2 months, coinciding with the disappearance of neonatal smiling and its subsequent replacement by social smiling.

Taken together, many parallels can be drawn between human and chimpanzee infant development. Neonates of both species exhibit smiling beginning right after birth. Both species are sensitive to the mother's face and possess an innate tendency to imitate facial expressions. They are also sensitive to the direction of gaze and show mutual gaze with smiling. Communication in face-to-face settings thus seems to be a characteristic shared by humans and chimpanzees. Furthermore, developmental changes progress along very similar paths between the two species.

3.2 Auditory-Vocal Communication

Through our participation observation method, we explored various features of chimpanzee infant vocalization and auditory perception, uncovering some contrasts with humans. Chimpanzee infants emit a vocalization called the "staccato" (a kind of "ho-ho-ho-ho" sound) in calm, relaxed situations. In a pilot study (Matsuzawa and Nakashima, unpublished data), we examined our infant subjects' responses to others' vocalizations. The principal subject was an infant male called Ayumu. A human tester (the present author) produced in front of the chimpanzee infant one of seven distinct vocalizations: pant-hoot, pant-grunt, food-grunt, laughter, staccato, whimper, or calling the infant's name. Data combined for the first 5 months of Ayumu's life showed that when the tester emitted a pant-hoot, the infant replied with his staccato 86.9% of the time. The pant-hoot is a long-distance communication call in chimpanzees. Ayumu's behavior parallels findings from the wild, where infant chimpanzees also emit staccato in response to pant-hoots by their mothers as well as other members of their community.

When the tester emitted one of the other five vocalizations taken from the chimpanzee vocal repertoire, the infant replied by staccato 5% to 34% of the time (5.3% for whimper, 8.9% for food-grunt, 11.1% for laughter, 20.5 % for pant-grunt, and 33.7% for staccato). In other words, other chimpanzee vocalizations are not as effective as the pant-hoot, but can nevertheless stimulate the infant's response. However, Ayumu never responded to the sound of his name "Ayumu" being called. He did not emit the staccato, nor did his facial expression change in response to his name. Interestingly, there was a facial expression elicited by the chimpanzee voice. The voiced "Laughter" resulted in the play face, that is, a relaxed open-mouth face, in Ayumu in 40.7% of the trials, while the play face was not elicited by the other voices, in 0% to 7.2% of trials. It must be noted that the laughter inevitably accompanied the play face so that the reply of play face is a form of exchange of social smiling, as already described.

Bear in mind at this point that the same person produced each of the vocalizations, succeeding in eliciting the staccato by mimicking the chimpanzee vocal repertoire but failing to do so using human speech. This result strongly suggests that chimpanzee infants are primed to respond to chimpanzee vocalizations but not to sounds typically made by humans. Imagine the reverse experiment, using human infants; clearly, they would show the opposite tendency!

4 What Is Uniquely Human in Mother–Infant Relationships?

This chapter has discussed human mother–infant relationships with special reference to its evolutionary origins. The study of chimpanzees has revealed that the two species share many common characteristics, which include face-to-face communication beginning immediately after birth. So, what is uniquely human? What are the major differences between the two species?

Let us look more closely at the mother–infant relationship in chimpanzees. In clear contrast to humans, chimpanzees show continuous physical contact. In most cases, for the first 3 months of the infant's life, mother and offspring are in physical contact 24 hours a day: the infant clings to the mother, and the mother in turn embraces the infant. However, in humans, mother and infant are separated from each other for the largest part of the day. Human infants cry to attract the mother's attention, and mothers often reply vocally instead of actually embracing them.

Although some mother–infant interactions such as mutual gazing and smiling have their root in humans' and chimpanzees' common ancestor, hominization (the evolution of humans) was accompanied by an increasing tendency for physical separation between mother and infant. In exchange, communication through facial, gestural, and vocal signals has proliferated. The onset of such new types of interactions may have occurred in parallel with other changes such as the division of roles within society, complex tool manufacture, and subtle multimodal communication. The interplay among these factors may have fueled some form of autocatalytic process, ultimately resulting in the emergence of *Homo* about 2 million years ago. With face-to-face communication serving as the foundation, infants may have begun to manipulate objects in relation to their mothers and acquired the ability to imitate, all in the context of the triadic relationship of self–others–object that characterizes human intelligence.

Mutual gaze and smiling can be regarded as meta-communication phenomena in that they enable the modification of the meaning of a signal by the addition of a second signal. If accompanied by smiling, direct gaze can convey an affectionate message. Facial information can modify the recognition of auditory stimuli in what is known as McGurk's effect (McGurk and MacDonald 1976). Even if exactly the same vocal sound is presented, it can be recognized by the listener as either "ba" or "da" depending on visual cues provided by lip movement. Refined visual perception can thus affect the recognition of vocal sounds. Such multimodal communication can be identified not only in signal perception but also in signal production. Human infants have a strong tendency to accompany their own vocalizations by the synchronous movement of limbs and fingers. Facial gestures, vocalizations, and manual movement emerge together in human infants. The three different media of communication complement each other to modulate the original meaning of the signals, thus producing the complexity seen in human face-to-face communication.

This chapter has postulated a possible evolutionary scenario for the emergence of the human mother–infant relationship. Similar to all other mammals, human mothers suckle their infants. The primate ancestor, diurnal and arboreal, developed grasping hands. This preadaptation gave rise to the uniquely primate relationship of clinging, then embracing. Based on continuous ventro-ventral physical contact, mutual gazing and smiling developed in the common ancestors of humans and chimpanzees. Then, the emergence of *Homo* brought with it an interruption of continuous physical contact between mother and infant, a physical separation that was likely related to changes in social and material intelligence. Mutual gaze and smiling are also preadaptations. In addition, human mother–infant physical separation resulted in the facilitation of vocal exchange, and the stable supine posture of the infants provided freedom of limb movement. Through these changes, humans developed a unique way of communication, incorporating multiple sources of signals such as facial expressions, manual gestures, and vocalizations.

Acknowledgments

The present study was supported by grants from the Ministry of Education, Science, and Culture in Japan (#12002009 and #16002001), from the biodiversity research of the 21COE (A14), and from the JSPS core-to-core program, HOPE. I thank the staff and students at the Section of Language and Intelligence, Primate Research Institute, Kyoto University, and the keepers and veterinarians who took care of the chimpanzees at the institute. Thanks are also offered to Hideko Takeshita, Yuu Mizuno, Noe Nakashima, Masako Myowa-Yamakoshi, Masaki

Tomonaga, and Masayuki Tanaka for allowing me to use the materials of the collaborative research. I also thank Dr. Dora Biro for her editing of the English text.

References

Asfaw B, White T, Lovejoy O, Latimer B, Simpson S, Suwa G (1999) *Australopithecus garhi*: a new species of early hominid from Ethiopia. Science 284:629–635

Bard K, Myowa-Yamakoshi M, Tomonaga M, Tanaka M, Costall A, Matsuzawa T (2005) Group differences in the mutual gaze of chimpanzees (*Pan troglodytes*). Dev Psychol 41:616–624

DeValois R, Jacobs G (1968) Primate color vision. Science 162:533–540

Jacobs G, Rowe M (2004) Evolution of vertebrate color vision. Clin Exp Optom 87(4–5):206–216

Johnson MH, Morton J (1991) Biology and cognitive development: the case of face recognition. Blackwell, Oxford

Matsuzawa T, Nakamura M (2004) Caregiving: mother infant relations in chimpanzees. In: Bekoff M (ed) Encyclopedia of animal behavior. Greenwood Press, Westport, CT, pp 196–203

McGurk H, MacDonald J (1976) Hearing lips and seeing voices. Nature (Lond) 264:746–748

Meltzoff AN, Moore MK (1977) Imitation of facial and manual gestures by human neonates. Science 198:75–78

Mizuno Y, Takeshita H, Matsuzawa T (2006) Behavior of infant chimpanzees during the night in the first four months of life: smiling and suckling in relation to arousal levels. Infancy 9:215–234

Myowa-Yamakoshi M, Tomonaga M, Tanaka M, Matsuzawa T (2003) Preference for human direct gaze in infant chimpanzees (*Pan troglodytes*). Cognition 89:B53–B64

Myowa-Yamakoshi M, Tomonaga M, Tanaka M, Matsuzawa T (2004) Imitation in neonatal chimpanzees (*Pan troglodytes*). Dev Sci 7:437–442

Myowa-Yamakoshi M, Yamaguchi M, Tomonag, M, Tanaka M, Matsuzawa T (2005) Development of face recognition in infant chimpanzees (*Pan troglodytes*). Cogn Dev 20:49–63

Ross C (2002) Park or ride?: Evolution of infant carrying in primates. Int J Primatol 22:749–771

Saito A, Mikami A, Kawamura S, Ueno Y, Hiramatsu C, Widayati KA, Suryoboroto B, Teramoto M, Mori Y, Nagano K, Fujita K, Kuroshima H, Hasegawa T (2005) Advantage of dichromats over trichromats in discrimination of color-camouflaged stimuli in nonhuman primates. Am J Primatol 67:1–12

Tomonaga M, Tanaka M, Matsuzawa T, Myowa-Yamakoshi M, Kosugi D, Mizuno Y, Okamoto S, Yamaguchi M, Bard K (2004) Development of social cognition in infant chimpanzees (Pan troglodytes): face recognition, smiling, gaze, and the lack of triadic interactions. Jpn Psycol Re 39:253–265

9
Development of Facial Information Processing in Nonhuman Primates

Masako Myowa-Yamakoshi

1 Introduction

The face provides significant information about the social lives of human and nonhuman primates. For example, following the gaze of others can help individuals perceive the location of important components of the environment, such as food and predators, and can facilitate certain kinds of social interaction among group mates (Langton et al. 2000; Tomasello et al. 1998). Several researchers have suggested that, from an evolutionary perspective, primates may have a specialized neural system within the brain that is devoted to gaze processing (Baron-Cohen 1994, 1995; Langton et al. 2000; Perret and Emery 1994; Perret et al. 1992).

Developmental evidence that supports this claim exists. Developmental psychologists have discovered that human infants preferentially look at human faces. Empirical studies on the development of face/nonface discrimination in human infants using visual preference techniques have demonstrated that even newborns preferentially track facelike patterns over nonface patterns (Goren et al. 1975; Johnson et al. 1991; Macchi Cassia et al. 2001).

Are human beings unique among primates in their ability to process facial information from just after birth? In this chapter, I present our recent findings on the early development of facial information processing in nonhuman primates. We observed two infant species—a lesser ape, the gibbon (*Hylobates agilis*), and our closest evolutionary relative, the chimpanzee (*Pan troglodytes*). We first focused on the ability of gibbon and chimpanzee infants to recognize individual faces. Specifically, we investigated the time around which they might be able to recognize individual faces, especially most familiar (caregivers') faces. Further, we examined the gaze sensitivity of infant gibbon and chimpanzees to explore how and when they would perceive gaze direction. Finally, we explored the early cognitive mechanism underlying facial information processing and its adaptive significance from both phylogenetic and ontogenetic perspectives.

School of Human Cultures, The University of Shiga Prefecture, 2500 Hassaka-cho, Hikone, Shiga 522-8533, Japan

2 Development of Facial Recognition in Human Infants

Attempting to explain the phenomenon that human neonates prefer looking at facelike stimuli rather than nonfacelike stimuli, Johnson (1990) suggested that a subcortical visual pathway involving the superior colliculus controls the tracking of moving facelike stimuli during the first month. Johnson and Morton (1991) have named this primary mechanism CONSPEC. CONSPEC operates from birth, and its functioning rapidly declines within the first month. A second mechanism, which they named CONLERN, is believed to be acquired at around 6 to 8 weeks. Johnson and Morton (1991) proposed that there is a developmental shift in processing from the subcortical visual pathway to the second mechanism that appears in plastic cortical visual pathways. This second mechanism is believed to enable the recognition of individual faces.

However, several experimental studies have produced evidence that is not consistent with Johnson and Morton's (1991) two-process theory. For example, Pascalis et al. (1995) demonstrated that 4-day-old neonates look longer at their mothers' faces than at a stranger's face (Field et al. 1984; Bushnell et al. 1989).

Explaining this disagreement, Johnson and de Haan (2001) suggested that, during the first few weeks after birth, face recognition is mediated by an early hippocampus-based preexplicit memory (Nelson 1995). The hippocampal system is believed to form an accurate representation of the memory of the visual stimuli, independent of whether they are facelike or nonfacelike. According to this hypothesis, newborns might be able to discriminate between individual faces by memorizing the shape of their individual features (Simion et al. 2002).

On the other hand, this type of face processing during the neonatal period is believed to differ from that during adulthood. de Haan et al. (2001) suggested that once higher cortical areas begin to mediate, they would relate one memorized face to another. Through prolonged experience with faces, this cortical area would enable infants to mentally form average prototypic representations of faces after 6 to 8 weeks. These representations guide infants to encode new faces in terms of the way in which they deviate from the prototype. This process allows them to discriminate between individual faces (Valentine 1991).

The ability to form a prototypic representation of the face seems to develop only after 6 to 8 weeks, with the emergence of a functional cortical system for face processing. This view is supported by de Haan et al. (2001). They familiarized 1- and 3-month-old infants to four individual faces and then tested whether they recognized a computer-generated image of a face composed of the average of the four faces and one of the exact individual faces. They found that both 1- and 3-month-old infants were able to recognize familiar individual faces. However, only the 3-month-old infants looked longer at the familiar face than the average face.

3 When Do Nonhuman Primates Begin to Recognize "Individual" Faces?

Several theoretical models and the empirical evidence supporting them have significantly facilitated our understanding of this domain. However, when and how human infants begin to discriminate individual faces still seems unclear. Comparative studies to determine the phylogenic origin of human face recognition may help reveal answers to these questions. We assessed the developmental changes in face recognition by gibbon and chimpanzee infants using the "preferential-looking" paradigm that measured the infants' eye and head tracking of moving stimuli. The infants were shown several different photographs of faces, including that of the mother (caregiver) of each infant. We speculated that if they were able to discriminate the most familiar faces, they would look preferentially at one photograph rather than another.

3.1 Facial Recognition in Infant Gibbon

Myowa-Yamakoshi and Tomonaga (2001a) investigated the face recognition ability of an infant male gibbon aged 4 to 5 weeks. He had been reared by human caregivers since he was less than 2 weeks old because his biological mother had provided inadequate maternal care. The stimuli consisted of three gray-scale photographs: a familiar human's face (caregiver), an unfamiliar human's face (stranger), and an unfamiliar conspecific's face (gibbon). One session was conducted for each pair of faces: (a) caregiver versus stranger, (b) caregiver versus gibbon, and (c) stranger versus gibbon.

As soon as the gibbon fixated on the stimuli presented in front of his face, one was moved slowly to his left and the other to his right, both at a rate of approximately 9°/s. This procedure was repeated five times per session and performed on 4 days of each week. For each stimulus, the gibbon's preference was scored according to whether he fixated on the stimulus with an eye or with a head turn of more than approximately 60°.

We found that by 4 weeks of age the gibbon was able to discriminate between a familiar (human caregiver) face and unfamiliar (human stranger and gibbon) faces. Moreover, the gibbon discriminated between different unfamiliar individuals' faces (human stranger versus gibbon). On the whole, the human faces elicited more attention than the gibbon face (Fig. 1). No significant preferential differences were observed between the three sets of stimuli during the experimental period. The fact that the gibbon was able to recognize human faces within only 1 week of being reared by humans was noteworthy. The gibbon might have already developed the ability to discriminate between different individual faces by at least 4 weeks of age.

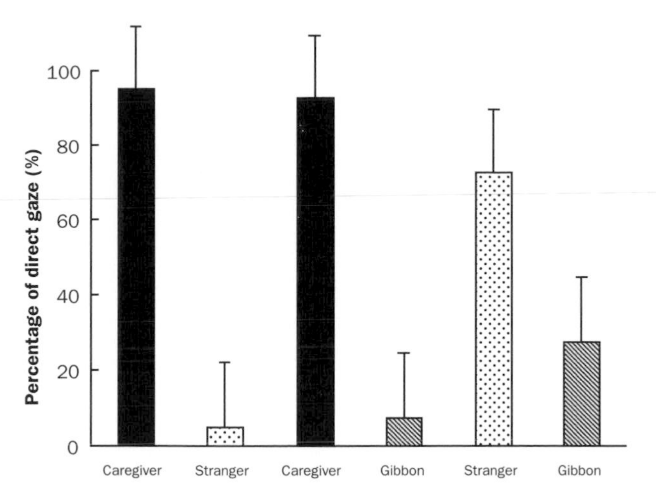

Fig. 1. The mean percentage of the preference score (plus standard error) for the gibbon's gazing at each of the three faces. [From Myowa-Yamakoshi and Tomonaga (2001a)]

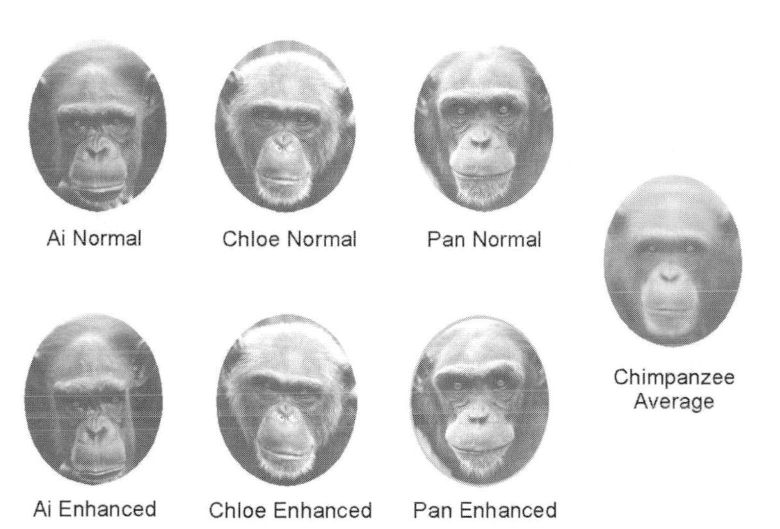

Fig. 2. The seven face photographs used in Myowa-Yamakoshi et al. (2005). Ai, Chloe, and Pan are the mothers of the three infant chimpanzees. [From Myowa-Yamakoshi et al. (2005)]

3.2 Facial Recognition in Infant Chimpanzees

We also investigated the ability of three infant chimpanzees, aged 1 to 18 weeks, to recognize others' faces (Myowa-Yamakoshi et al. 2005). They were reared by their biological mothers, who had participated in several cognitive experiments in the Primate Research Institute, Kyoto University. They had also participated in a variety of tests related to the development of cognitive abilities (Matsuzawa 2003; Tomonaga et al. 2004).

We prepared the following three photographs: (a) a normal face of the chimpanzee's mother, (b) an enhanced face of the chimpanzee's mother, and (c) an average face that was generated from that of 11 chimpanzees (Fig. 2). The tester

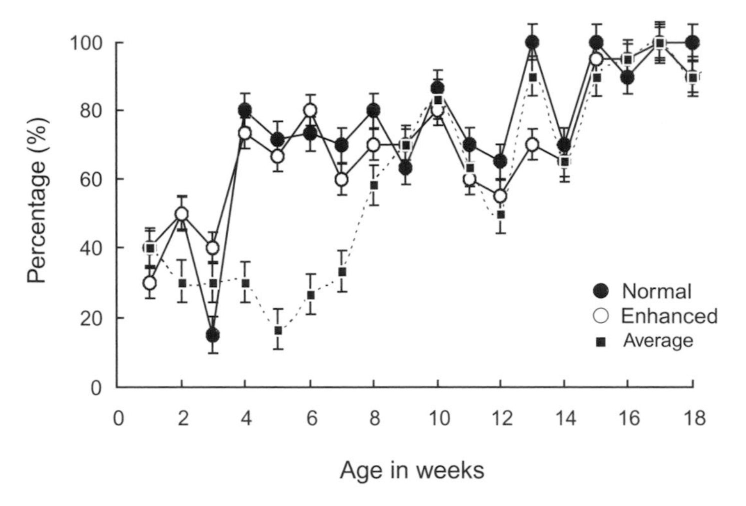

Fig. 3. Developmental change as a mean percentage of the tracking score for each of the three conditions of the mother's face for each week of age (plus standard error). [From Myowa-Yamakoshi et al. (2005)]

positioned the stimulus directly in front of the chimpanzee's face at a distance of approximately 30 cm. As soon as the chimpanzee fixated on the stimulus, it was slowly moved to one side (right or left) at a rate of approximately 18°/s. This procedure was defined as 1 trial and was repeated five times for each side. Each session consisted of 10 trials (5 trials × 2 sides) per stimulus.

Results showed that, before 4 weeks of age, the chimpanzees showed few tracking responses and no differential responses among the three photos. Between 4 and 8 weeks of age, they paid a greater amount of attention to their mother's faces, both the normal and enhanced, in comparison to the average face. On the other hand, from 8 weeks onward, they again showed no differences, but exhibited frequent tracking responses (Fig. 3).

Moreover, we investigated the same chimpanzees' ability to discriminate among individuals of another species, humans, using the same procedure. The stimuli were (a) a normal face of the most familiar human, (b) an enhanced face of the most familiar human, and (c) an average face created from pictures of Japanese humans (half were males and half were females). However, in contrast to the results observed for chimpanzees' faces, we did not note any clear evidence indicating the ability to discriminate between human faces during the testing period.

4 Mechanism Underlying the Development of Facial Recognition in Nonhuman Primates

In conclusion, human infants are not unique among primates in their ability to pay attention to faces after birth. Both gibbon and chimpanzee infants were able to distinguish each of their caregivers' faces from other faces from very shortly after birth.

We may consider these findings regarding nonhuman primates according to the theoretical models that cognitive developmental research has offered. In the infant gibbon, we did not find any developmental changes in the preference for the presented faces during the experimental period. When we began the experiment, the gibbon was 4 weeks old and already showed a strong preference for the caregiver's face over those of others. As of now, we are unable to conclude when and how the gibbon might have developed the ability to recognize individual faces. To verify the developmental models of facial recognition, we need to investigate gibbons' capacity for face recognition beginning much sooner after birth (in the "true" neonatal period).

However, our results may still have implications for the model. Although the gibbon had witnessed human faces only up to 2 weeks, he showed a preference for whole human-type faces rather than the gibbon face. There is a possibility that he might have already formed a prototype of faces based on human facial features before 4 weeks of age.

On the other hand, the chimpanzees began to show a preference for their mothers' faces at around 4 weeks of age. According to Johnson and Morton's (1991) CONSPEC/CONLERN theory, the findings are open to two interpretations: (1) it was not until 4 weeks of age that the chimpanzees might have accurately memorized their mother's faces based on the shape of individual facial features, and (2) at 4 weeks, the chimpanzees might have already formed a prototype of faces and discriminated the mother's faces based on this prototype.

When debating these interpretations, it is important to consider that the chimpanzees' face-to-face interaction with mothers and infants was much less than that of humans (Bard et al. 2005). It is reasonable to state that the restriction of visual experience in chimpanzees would cause a lag in memorizing the mother's face. We assume that, at around 4 weeks of age, the chimpanzees began to recognize their mothers' faces by memorizing local information such as the shape of the facial elements.

Furthermore, the same may be said about the process by which a prototypic representation of faces is formed in chimpanzees. The chimpanzees had fewer opportunities to look at other chimpanzees' faces during the first few weeks of life. From 0 to 3 weeks of age, we separated each mother–infant pair from other chimpanzees because the mothers seemed very cautious about other chimpanzees who might peek at or touch their infants. As a result, the emergence of the formation of a prototype of faces would have also been later than that in humans. We assume that, at around 8 weeks, when the average face also became attractive to the infants, they might begin to form an average prototypic representation through extensive visual experience with other chimpanzees' faces.

Our findings have another important implication. The early ability to recognize faces may develop flexibly, depending upon the surrounding faces to which the infant was exposed since birth. Nelson (1995) proposed that the ability to recognize faces developmentally "narrows" with extensive experience in processing the most frequently observed faces. As a result, human faces can be gradually represented as a human face-specific prototype. This hypothesis is supported by Pascalis et al. (2002), who demonstrated that 6-month-old infants, but not adults and 9-month-old infants, could discriminate between faces of another species (monkey). In contrast, adults are worse at recognizing unfamiliar individuals of monkeys, compared to their own species (Goldstein and Chance 1980).

From birth, human infants engage in face-to-face interactions with their caregivers. Through such social interaction, the nursery-reared gibbon should have had a considerable number of visual experiences with the faces of human caregivers. Interestingly, the gibbon preferred human faces over a conspecific's face, irrespective of whether the human faces were familiar. It is possible that nonhuman primates reared in a human environment might form a more human face-specific prototype than their own species face prototype through their daily social interactions with humans.

5 Development of Gaze Perception in Human Infants

Of the different facial elements, the eyes seem to be the most important for animals from the perspective of survival. Several studies have revealed that human infants are extremely sensitive in their perception of eyes and eyelike stimuli. By the time infants are 4 months old, they are able to discriminate between direct and averted gazes (Vecera and Johnson 1995; Farroni et al. 2000).

Two hypotheses have been proposed regarding the onset of this ability. The first is that eye-direction processing is the product of an "encapsulated" innate module (Baron-Cohen 1994, 1995). Baron-Cohen (1994) has named this neural module devoted to processing gaze information the "eye direction detector" (EDD). According to this view, human infants can automatically detect others' gaze from just after birth. As evidence, Batki et al. (2000) demonstrated that neonates even younger than 2 days old looked longer at a photograph of a face in which the eyes were open than they did at a photograph of the same face with the eyes shut.

The second hypothesis is that this ability gradually emerges over the first few months of life (Vecera and Johnson 1995; Farroni et al. 2000). Vecera and Johnson (1995) revealed that infants did not discriminate gaze any more efficiently when intact eyes were presented in the context of a scrambled face than they did when the eyes were presented in the context of an intact face. In contrast with the modular hypothesis, they insisted that gaze information is not processed completely isolated from facial information.

The second view that there is a developmental change in processing gaze direction is based on the structural hypothesis proposed by Johnson and Morton (CONSPEC/CONLERN theory; Johnson and Morton 1991). According to the two-process theory, the emergence of CONLERN around 2 months of age is believed to enable the processing of information associated with whole facial features, including facial expressions, identities, and gaze direction, through each individual's experience of faces. They suggested that such cortical maturation would cause a developmental change in the processing of gaze direction.

6 Development of Gaze Perception in Nonhuman Primates

To verify this theoretical controversy, it seems important to explore whether gaze sensitivity is developmentally influenced by the surrounding facial context. We investigated the gaze sensitivity of gibbon and chimpanzee infants from the following two aspects: (1) whether they could discriminate between direct and averted gaze, and (2) whether gaze direction is influenced by the surrounding facial context.

6.1 Perceiving Eye Gaze in Infant Gibbon

Myowa-Yamakoshi and Tomonaga (2001b) investigated the ability of processing eye gaze in the same gibbon observed in Myowa-Yamakoshi and Tomonaga (2001a). The experiment was conducted 2 days per week during 2 to 6 weeks of age, using a two-choice preferential-looking paradigm. The stimuli consisted of three facial types (upright, inverted, and scrambled) of black-and-white line drawings. Each face type had eyes with direct and averted gazes. One session was conducted for each combination of two faces. Each session consisted of one of the three types of conditions: (a) upright face with directed gaze versus upright face with averted gaze, (b) inverted face with directed gaze versus inverted face with averted gaze, and (c) scrambled face with directed gaze versus scrambled face with averted gaze. The procedure followed was identical to that used in Myowa-Yamakoshi and Tomonaga (2001a).

Results revealed that, by 2 weeks of age, the gibbon preferentially looked at the face with a directed gaze rather than that with an averted gaze, irrespective of the different face contexts (Fig. 4). The gibbon's sensitivity to detect eye gaze was not influenced by the context of the faces unlike that of human infants.

6.2 Perceiving Eye Gaze in Infant Chimpanzees

We also observed the same chimpanzees studied by Myowa-Yamakoshi et al. (2005) (Myowa-Yamakoshi et al. 2003). The chimpanzees were tested approximately 1 day per week between 10 and 32 weeks of age, using a two-choice preferential-looking paradigm. We created six conditions using the color photographs of a human female (Fig. 5):

Condition 1: Open (frontal view) versus closed (frontal view)
Condition 2: Direct (frontal view) versus averted (frontal view)
Condition 3: Direct (frontal view) versus averted (three-fourths view)

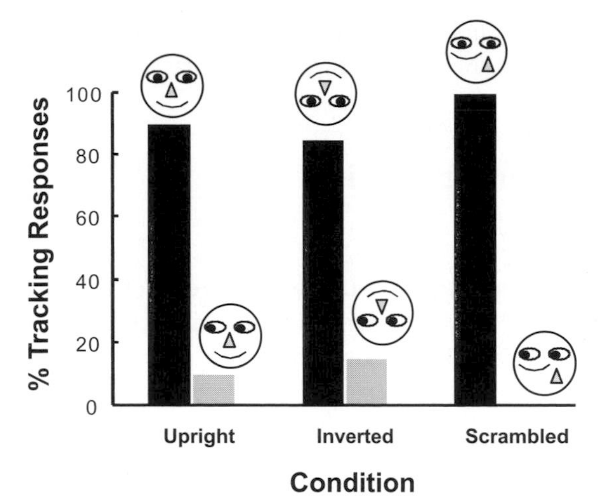

Fig. 4. The mean percentage of the preference score for the gibbon's gazing at each of the six stimuli. [From Myowa-Yamakoshi and Tomonaga (2001b)]

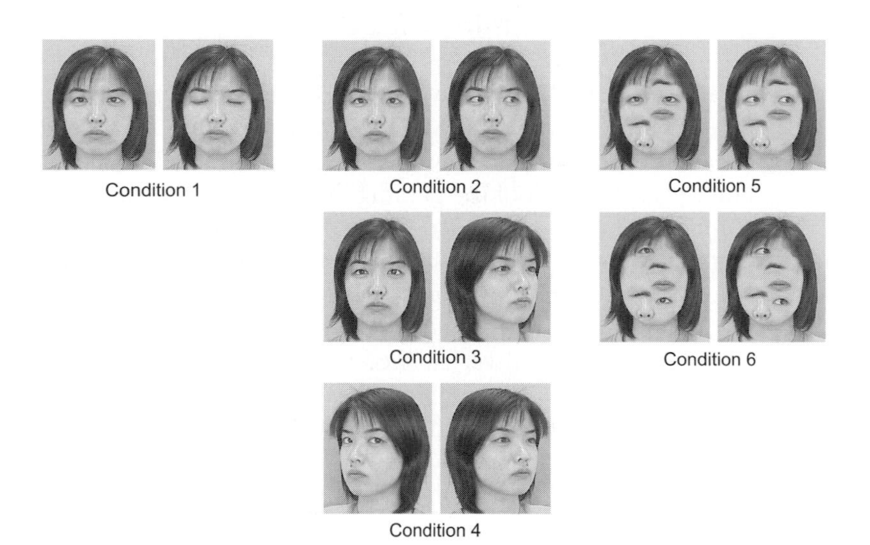

Fig. 5. The ten photographs used in the study by Myowa-Yamakoshi et al. (2003). [From Myowa-Yamakoshi et al. (2003)]

Condition 4: Direct (three-fourths view) versus averted (three-fourths view)
Condition 5: Direct (frontal view, scrambled with intact eyes) versus averted (frontal view, scrambled with intact eyes)
Condition 6: Direct (frontal view, all features scrambled) versus averted (frontal view, all features scrambled)

The chimpanzee was shown each pair consisting of the two faces. Each session consisted of 12 trials (6 conditions × 2 sides), and each chimpanzee took part in one session per day. A trial started when the chimpanzee spontaneously looked at the stimulus and lasted for 15 s after the stimulus had been presented (Fig. 6).

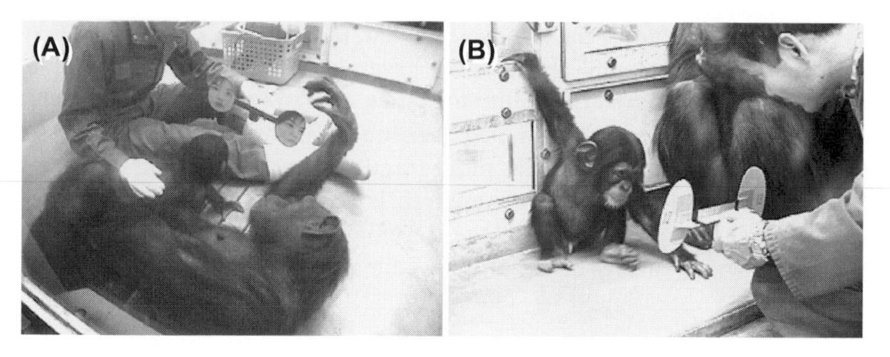

Fig. 6. Experimental situation for two of the chimpanzees, Pal (**A**) and Ayumu (**B**). The chimpanzees, facing the human tester, were shown two face photographs. [From Myowa-Yamakoshi et al. (2003)]

We found that chimpanzees preferred looking at faces with the eyes open or directed gaze than at faces with the eyes closed or averted gaze (conditions 1 through 4) throughout the experimental period. However, in the context of scrambled faces, the chimpanzees did not look differently at the faces with direct and averted gazes (conditions 5 and 6). These findings suggest that chimpanzees' gaze perception may be influenced by normal facial configurations.

7 How Do Nonhuman Primates Develop the Ability to Detect Gaze?

Both gibbons and chimpanzees paid considerable attention to the eyes and gaze direction beginning shortly after birth. However, whether they possess the ability to perceive gaze at birth is still unknown. We found no consistent developmental changes in their gaze perception during the experimental periods. Considering the differences in the developmental time course in primates, our gibbons and chimpanzees seemed to have already passed the neonatal stage when we started our experiments. There is room for argument regarding the existence of an innate gaze module in neonates.

However, our results have implications for the controversy regarding the development of gaze perception in humans. In the context of scrambled faces, the chimpanzees did not look differently at the faces with direct and averted gazes. This finding is consistent with the hypothesis proposed by Vecera and Johnson (1995) that the development of gaze perception may depend upon prolonged exposure to faces during the first few months of life.

On the other hand, gaze perception in the gibbon was not influenced by the surrounding facial context, in contrast to human and chimpanzee infants. It is possible that there may be species differences among nonhuman primates. However, we cannot deny the possibility that the gibbon showed gaze processing that was dependent on facial context after the experimental period ended at

7 weeks of age. Some studies have suggested that, in primates, this ability may show a gradual developmental change over several years of life. For example, Ferrari et al. (2000) examined the ability to follow gaze in juvenile (2 to 6 years) and adult (more than 6 years) pig-tailed macaques (*Macaca nemestrina*). The results showed that gaze following was more frequent in adults than in juveniles.

Further longitudinal observational research is required to confirm whether there exist interspecies differences in the developmental process of gaze processing across a wider variety of nonhuman primates.

8 Why Is Facial Expression So Important for Great Apes?

Finally, I would like to emphasize that, at least in great apes, whole facial configuration has a strong effect on gaze processing. I assume that facial expression as well as gazing must play a significant role in mother–infant communication in great apes. Facial expression is rather important for survival during infancy, considering their neonatal immaturity compared with other primates. During the course of communication, human caregivers expose their infants to several facial expressions in an "exaggerated" mode, such as raising their eyebrows and opening their mouths wider, to attract the infants' attention. Similarly, infants are attracted to their caregivers' changeable and attractive faces and react to them. The early ability to orient to faces is believed to enable infants to form such face-to-face interactions as soon as possible after birth. Through face-to-face interaction, infants may attract the attention of caregivers for as long as possible. As a result, their opportunities for receiving care would increase much more, as compared with the opportunities they would gain by simply crying (Myowa-Yamakoshi et al. 2005).

Chimpanzees, both in the wild and in captivity, also engage in face-to-face interactions with mutual gazing during their first 3 months of life (Bard et al. 2005; Plooij 1984). Evidence for this can be seen in the phenomenon of facial imitation in chimpanzee neonates (Bard and Russell 1999; Myowa 1996; Myowa-Yamakoshi et al. 2004). Moreover, chimpanzees seem to discriminate their mothers' faces in face-to-face interaction by the time they are 1 month old.

9 Conclusion

Taken together, the data show that humans as well as gibbons and chimpanzees have the ability to process facial information from shortly after birth. Although the eyes should have been the most significant facial element in evolution, gaze perception in great apes may not be simply a product of an innate module that is automatically processed. Rather, the ability may depend upon prolonged exposure to faces and may develop through face-to-face social interaction during the first few months of life. Few experimental studies have investigated early cognitive development in nonhuman primates. Further developmental and comparative studies will help reveal the relationship between species-specific biological foundations and the effect of postnatal experience in the development of face processing.

Acknowledgments

The author thanks T. Matsuzawa, M. Tomonaga, M. Tanaka, M.K. Yamaguchi, H. Takeshita, T. Sato, M. Inaba, Y. Kuniyoshi, S. Hirata, N. Nakashima, A. Ueno, Y. Mizuno, S. Okamoto, M. Uchikoshi, T. Imura, M. Hayashi, K. Kumazaki, N. Maeda, A. Kato, K. Matsubayashi, G. Hatano, O. Takenaka, K.A. Bard, J.R. Anderson, K. Yuri, and M. Sakai for their help throughout the projects. The research reported here and preparation of the manuscript were financially supported by Grants-in-Aid for Scientific Research from the Japan Society for the Promotion of Science (JSPS), and the Ministry of Education, Culture, Sports, Science and Technology (MEXT) (nos. 12002009 and 16002001 to T. Matsuzawa, 13610086 and 16300084 to M. Tomonaga, 09207105 to G. Hatano, 10CE2005 to O. Takenaka, 16203034 to H. Takeshita, 16683003 to M. Myowa-Yamakoshi), the MEXT Grant-in-Aid for the 21st Century COE Programs (A2 and D2 to Kyoto University), the research fellowship to M. Myowa-Yamakoshi from JSPS for Young Scientists (no. 3642), the JSPS core-to-core program HOPE, the JSPS Grant-in-Aid for Creative Scientific Research, "Synthetic Study of Imitation in Humans and Robots" to T. Sato, and the Cooperative Research Program of the Primate Research Institute, Kyoto University.

References

Bard KA, Russell CL (1999) Evolutionary foundations of imitation: Social cognitive and developmental aspects of imitative processes in non-human primates. In: Nadel J, Butterworth G (eds) Imitation in infancy. Cambridge University Press, Cambridge, pp 89–123

Bard KA, Myowa-Yamakoshi M, Tomonaga M, Tanaka M, Quinn J, Costall A, Matsuzawa T (2005) Group differences in the mutual gaze of chimpanzees (*Pan troglodytes*). Dev Psychol 41:616–624

Baron-Cohen S (1994) How to build a baby that can read minds: cognitive mechanisms in mindreading. Curr Psychol Cogn 13:513–552

Baron-Cohen S (1995) Mindblindness: an essay on autism and theory of mind. MIT Press, Cambridge

Batki A, Baron-Cohen S, Wheelwright S, Connellan J, Ahluwalia J (2000) Is there an innate module? Evidence from human neonates. Infant Behav Dev 23:223–229

Bushnell IWR, Sai F, Mullin JT (1989) Neonatal recognition of the mother's face. Br J Dev Psychol 35:294–328

de Haan M, Johnson MH, Maurer D, Perrett DI (2001) Recognition of individual faces and average face prototypes by 1- and 3-month-old infants. Cogn Dev 16:659–678

Farroni T, Johnson MH, Brockbank M, Simon F (2000) Infants' use of gaze direction to cue attention: the importance of perceived motion. Visual Cogn 7:705–718

Ferrari PF, Kohler E, Fogassi L, Gallese V (2000) The ability to follow eye gaze and its emergence during development in macaque monkeys. Proc Natl Acad Sci U S A 97:13997–14002

Field TM, Cohen D, Garcia R, Greenberg R (1984) Mother-stranger face discrimination by the newborn. Infant Behav Dev 7:19–25

Goldstein AG, Chance JE (1980) Memory for faces and schema theory. J Psychol 105:47–59
Goren C, Sarty M, Wu P (1975) Visual following and pattern discrimination of face-like stimuli by newborn infants. Pediatrics 56:544–549
Johnson MH (1990) Cortical maturation and the development of visual attention in early infancy. J Cogn Neurosci 2:81–95
Johnson MH, de Haan M (2001) Developing cortical specialization for visual-cognitive function: the case of face recognition. In: McClelland JL, Seigler RS (eds) Mechanisms of cognitive development: behavioral and neural perspectives. Erlbaum, Mahwah, NJ, pp 253–270
Johnson MH, Morton J (1991) Biology and cognitive development: the case of face recognition. Blackwell, Oxford
Johnson MH, Dziurawiec S, Ellis H, Morton J (1991) Newborns' preferential tracking of face-like stimuli and its subsequent decline. Cognition 40:1–19
Langton SRH, Watt RJ, Bruce V (2000) Do the eyes have it? Cues to the direction of social attention. Trends Cogn Sci 4:50–59
Macchi Cassia V, Simion F, Umiltà C (2001) Face preference at birth: the role of an orienting mechanism. Dev Sci 22:892–903
Matsuzawa T (2003) The Ai project: historical and ecological contexts. Anim Cogn 6:199–211
Myowa M (1996) Imitation of facial gestures by an infant chimpanzee. Primates 37:207–213
Myowa-Yamakoshi M, Tomonaga M (2001a) Development of face recognition in an infant gibbon *(Hylobates agilis)*. Infant Behav Dev 24:215–227
Myowa-Yamakoshi M, Tomonaga M (2001b) Perceiving eye gaze in an infant gibbon (*Hylobates agilis*). Psychologia 44:24–30
Myowa-Yamakoshi M, Tomonaga M, Tanaka M, Matsuzawa T (2003) Preference for human direct gaze in infant chimpanzees (*Pan troglodytes*). Cognition 89:B53–B64
Myowa-Yamakoshi M, Tomonaga M, Tanaka M, Matsuzawa T (2004) Imitation in neonatal chimpanzees (*Pan troglodytes*). Dev Sci 7:437–442
Myowa-Yamakoshi M, Yamaguchi MK, Tomonaga M, Tanaka M, Matsuzawa T (2005). Development of face recognition in infant chimpanzees (*Pan troglodytes*). Cogn Dev 20:49–63
Nelson CA (1995) The ontogeny of human memory: a cognitive neuroscience perspective. Dev Psychol 31:723–738
Pascalis O, de Schnen S, Morton J, Deruelle C, Fabre-Grent M (1995) Mother's face recognition by neonates: a replication and an extension. Infant Behav Dev 18:79–95
Pascalis O, de Haan M, Nelson C (2002) Is face processing species-specific during the first year of life? Science 296:1321–1323
Perret DI, Emery NJ (1994) Understanding the intentions of others from visual signals: neurophysiological evidence. Cahiers Psychol Cogn 13:683–694
Perret DI, Hietanen JK, Oram MW, Benson PJ (1992) Organization and functions of cells responsive to faces in the temporal cortex. Philos Trans R Soc Lond B335:23–30
Plooij FX (1984) The behavioral development of free-living chimpanzee babies and infants. Monographs on infancy, vol 3. Ablex, Norwood
Simion F, Farroni T, Macchi Cassia V, Turati C, Barba BD (2002) Newborns' local processing in schematic facelike configurations. Br J Dev Psychol 50:465–478
Tomasello M, Call J, Hare B (1998) Five primate species follow the visual gaze of conspecifics. Anim Cogn 55:1063–1069
Tomonaga M, Tanaka M, Matsuzawa T, Myowa-Yamakoshi M, Kosugi D, Mizuno Y, Okamoto S, Yamaguchi MK, Bard KA (2004) Development of social cognition in chimpanzees (*Pan troglodytes*): face recognition, smiling, mutual gaze, gaze following and the lack of triadic interactions. Jpn Psychol Res 46:227–235
Valentine T (1991) A unified account of the effects distinctiveness, inversion, and race in face recognition. Q J Exp Psychol A 43:161–204
Vecera SP, Johnson MH (1995) Gaze detection and the cortical processing of faces: Evidence from infants and adults. Visual Cogn 2:59–87

10 Development of Joint Attention in Infant Chimpanzees

Sanae Okamoto-Barth[1] and Masaki Tomonaga[2]

1 General Introduction

1.1 Gaze Direction, Gaze Following, and Joint Attention

Humans are highly sensitive to the direction of gaze. Determining the precise direction of another's attention is an important ability. Gaze shifts provide salient information about the location of objects but may also function in complex forms of social cognition (Whiten 1997).

In our daily lives, a great deal of information is communicated by means of following another individual's gaze to specific objects and events. This behavioral sequence is called *gaze following* or *joint attention*. Gaze following/joint attention is characterized by one individual (X) following the direction of the attention of another individual (Y) attention to an object (Z) (an object of joint focus; Emery 2000). These terms are used interchangeably by researchers. Emery et al. (1997) suggested that gaze following and joint attention are different yet intimately related abilities (probably with different developmental and phylogenetic time courses). They defined gaze following as the ability of X to follow the direction of the gaze of Y to a position in space (not an object). Joint attention has the additional requirement that X follows the direction of Y's gaze to object Z that is the focus of Y's attention. Joint attention thus requires extra computation to process the object of attention not just the direction of gaze.

Emery (2000) similarly argued that there are subtle differences between joint attention and *shared attention* (see also Perrett and Emery 1994). Nevertheless, these two terms are also used interchangeably in the literature. Emery (2000) defined shared attention as a more-complex form of communication than joint attention. Shared attention requires that individuals X and Y each have knowledge of the directions of another individual's attention (or a method for checking that what the another individual is looking at is the same as what they are looking at).

[1]Department of Cognitive Neuroscience, Faculty of Psychology, Maastricht University, P.O. Box 616, 6200 MD Maastricht, The Netherlands
[2]Primate Research Institute, Kyoto University, 41 Kanrin, Inuyama, Aichi 484-8506, Japan

Determining the direction of another individual's attention is easier to establish from more salient visual cues, such as head or body orientation. Thus, attentional cues, or social gaze, are not only provided by the eyes. Perrett and his colleagues have suggested that the direction of the head and the orientation of the body may also provide important indicators of attention when the eyes are obscured or not clear or when the eyes are used for other purposes (Perrett et al. 1992; Perrett and Emery 1994). When all cues are available for processing, a hierarchy of importance exists whereby the eyes provide more important cues than the head and the head is a more important cue than the body.

Research with human infants suggests that sensitivity to gaze shifts occurs very early in infancy. From 3 months of age, infants are able to discriminate changes in an adult's eye direction (Hains and Muir 1996). The ability to follow gaze has been demonstrated most successfully in studies with human infants. In most of those experiments, the general procedure involves the experimenter (or the mother of the infant) sitting face-to-face with the infant. After making eye contact, the experimenter shifts her gaze to a particular location or object. Infant's responses in this task have a specific developmental trajectory. However, the age at which an infant first follows another's gaze is controversial, ranging from 3 months to 18 months (Scaife and Bruner 1975; Butterworth and Cochran 1980; Butterwort and Jarrett 1991; Corkum and Moore 1995; D'Entremont et al. 1997). These conflicting results may be due to methodological, conceptual, or definitional differences. Nevertheless, before 12 months of age, human infants follow their mother's gaze but do not direct their attention to the object of her attention. At around 12 months, infants begin to follow their mother's gaze toward particular objects in their visual field, and at around 18 months they can direct their attention to objects outside of their visual field.

Joint attention is considered an early social cognitive ability leading to later developments associated with mental state attribution (e.g., theory of mind, deception, perspective taking; cf. Baron-Cohen 1995; Tomasello 1995, 1999). However, there are some accounts that do not need to attribute understanding other's mental states which apply to young infants. Young infants are primed from an early age to look in the direction that others are looking (cf. D'Entremont et al. 1997). When they do so, they often see interesting (and rewarding) objects and events. Hence, infants may learn to use gaze direction as a cue to where such rewarding events are located (Corkum and Moore 1995, 1998; Moore 1999). In this view, gaze is merely a discriminative stimulus for the general direction in which an attractive event might be encountered, and, once encountered, search should presumably cease. Butterworth's account of gaze following stressed the innate properties of this behavior. He suggests that young infants are hardwired to follow the gaze direction of others. They are held to terminate search at the first salient object in their scan path (Butterworth and Cochran 1980; Butterworth and Jarrett 1991). In contrast to Moore's conception (1999), there is no learning on the infant's part and no expectation of finding an event.

Although previous studies have suggested that infants are innately sensitive to eye gaze and gaze direction, it is still an open question whether gaze following (or joint attention) is, in fact, an innate ability or a learning effect from daily experience.

However, gaze is not limited to information from the eyes as simple stimuli. Gaze is one cue that is often used to determine the focus of another individual's direction of attention. Of course, eye gaze is not the only attentional cue. The orientation of the whole head, body, and hand (pointing) are similarly good indicators of attention and interest and are used in our daily interactions with others. Especially, pointing is considered as an important component of joint attention as an indicator of particular objects, location, or events. The age of 15 months is the breaking point for the comprehension of pointing (Desrochers et al. 1995).

1.2 Gaze Following and Joint Attention in Nonhuman Primates

Gaze following is also found to occur in a number of nonhuman primates. The use of gaze shifts as social cues has various evolutionary advantages. For instance, gaze shifts may index the location of predators, potential mates, or food sources. Several field studies suggest that primates can follow the gaze of conspecifics (Chance 1967; Menzel and Halperin 1975; Whiten and Byrne 1988). However, in field studies, it is difficult to identify which object or event is looked at two individuals by means of gaze following. For instance, individuals may come to fixate on the same object because the object is inherently interesting even if they do not follow gaze. Such interpretational confounds can be effectively excluded in laboratory studies. In fact, various studies have demonstrated that many primate species follow the gaze direction of conspecifics to objects (chimpanzees, mangabeys, macaques; Tomasello et al. 1998; Emery et al. 1997). The general procedure in primate studies is the same as in studies with humans (see earlier). Furthermore, primates (especially apes) follow the gaze of nonconspecific individuals (e.g., a human experimenter). They do this even when the target is located above and/or behind them (Itakura 1996; Povinelli and Eddy 1996, 1997). Itakura (1996) studied the ability of various species of prosimians, monkeys, and apes to follow a human experimenter's gaze. Only the ape (orangutan and chimpanzee) responded at above-chance levels. Neither Old nor New World Monkeys (i.e., brown lemur, black lemur, squirrel monkey, brown capuchin, whiteface capuchin, stump-tailed macaque, rhesus macaque, pigtailed macaque, and tonkean macaque) responded at above-chance levels.

The clearest evidence for the ability to follow gaze in nonhuman primates comes from laboratory work on great apes, in particular, studies with chimpanzees. Although these studies have shed some light on the topic, they have left many questions unanswered. For instance, *how* do chimpanzees follow the other's gaze? *Which* cues are important for gaze following? *Why* do chimpanzees follow other's gaze? And *when* do chimpanzees start to follow other's gaze?

Previous studies have investigated *how* chimpanzees follow another's gaze. Povinelli and Eddy (1996), for example, installed an opaque barrier in a testing room to obstruct the chimpanzee subjects' line of sight. In cases where the experimenter looked to an object next to the barrier (outside the immediate line of sight of the subject), chimpanzees followed the experimenter's line of sight around the barrier to the unseen object. This ability may be important when trying to extrapolate information from another's attention, specifically, when the focus of attention is out of sight. In another paradigm using a distracter in the subjects' visual field, Emery et al. (1997) reported that monkeys bypassed looking at the first interesting object in their line of sight and followed the demonstrator's gaze to the target object; following gaze geometrically. Tomasello et al. (1999) have reported similar results in chimpanzees. These results suggest that subjects do not reflexively follow gaze to the first available object within their view but actively track the gaze of others geometrically to localize objects or locations others are attending to.

A number of other studies have demonstrated that primates use a variety of cues to track the focus of another's attention (e.g., pointing, head orienting, gazing without head orientation). For example, in one study, Perrett and his colleagues investigated *which* cue(s) primates use to direct their own attention by measuring eye movements of monkeys during presentation of head and body cues, and head only and head with eyes cues (Lorincz et al. 1999). They found that the subjects used the information from the head more readily than the body. They also appeared to follow gaze cues when the demonstrator's head was oriented toward the subjects.

In a series of experiments, Povinelli and his colleagues (1996, 1997) have investigated *why* chimpanzees follow the experimenter's gaze. The main question is whether they attribute mental states to other individuals when they follow their gaze direction (not automatically). They suggested that chimpanzees can follow an experimenter's gaze but do not use that information to learn about objects in the world or about the "mental state" of the individual providing the gaze cues. As with human infants, it is still an open question whether chimpanzees use gaze as indicators of mental states (as mentioned earlier, however, the question for human infants is at what age do they start to attribute mental states to other individuals' gaze).

Previous studies with chimpanzees, however, have tested adult (or juvenile) subjects and described the ability to follow gaze. A longitudinal study of infant chimpanzees that measures the frequency of gaze following in the course of development may yield important clues as to the ontogeny and evolution of this behavior. Such a study would, in principle, address the question of *when* (e.g., a developmental time course) infant chimpanzees start to follow another's gaze. A series of longitudinal experiments evaluated the development of joint attention in an infant chimpanzee to address this important question in both human and chimpanzee development.

The present longitudinal series of studies were conducted to clarify the emergence (study 1) and the development (study 2) of the ability in chimpanzees to follow experimenter-given cues, such as gazing and pointing.

2 The Emergence of Joint Attention

2.1 Introduction

By the end of their first year, human infants are sensitive to information specifying where others are looking. Butterworth and his colleagues (Butterworth and Cochran 1980; Butterworth and Jarrett 1991; Butterworth 1991) propose a naturalistic approach of joint attention with three successive mechanisms that develop in human infants between the ages of 6 and 18 months. At 6 months, infants progress gradually from responding to the head movements of others to orienting in the same general direction within their visual field (*ecological mechanism*). At this age, however, infants terminate their search at the first salient object in their scan path. By 12 months, infants are able to localize the particular object at which the other is looking (*geometric mechanism*). They also found that infants establish joint attention to objects within their visual field before they do so for objects outside their visual field. That is, infants younger than 18 months cannot yet represent their whole environment, some region of which might be visible to another person. By 18 months, infants can follow the other person's gaze into space that is outside their own initial visual field (*representational mechanism*). On the other hand, Corkum and Moore, who advocate an empirical and parsimonious approach, assumed that when infants follow an adult's gaze, they often see interesting objects and events and hence learn to use the gaze direction of others as a cue to where such events might be located (Corkum and Moore 1995, 1998; Moore 1999). That is, social learning drives joint attention, although learning is constrained by certain causal and social sensitivities.

Human infants progressively develop the ability for gaze following between 6 and 18 months of age. In contrast to the huge amount of human infant literature, there are very few studies on joint attention/gaze following in young nonhuman primates. Myowa-Yamakoshi and Tomonaga (2001) reported that between 1 and 6 weeks of age an infant gibbon (*Hylobates agilis*) preferred to look at a schematic face with direct gaze rather than averted gaze. Ferrari et al. (2000) assessed in juvenile and adult pig-tailed macaques (*Macaca nemestrina*) the ability to follow the eye gaze of an experimenter. The juvenile monkeys were not able to orient their attention on the basis of eye cues alone. Chimpanzee infants are also sensitive to faces that are looking at them (Myowa-Yamakoshi et al. 2003). Tomasello et al. (2001), however, reported that chimpanzees less than 3 to 4 years old do not look outside their own visual field when using the experimenter's head-turn cue.

As a first step of the present longitudinal study, we conducted the traditional gaze-following paradigm to clarify the emergence of the ability in chimpanzees to follow experimenter-given cues such as gazing and pointing.

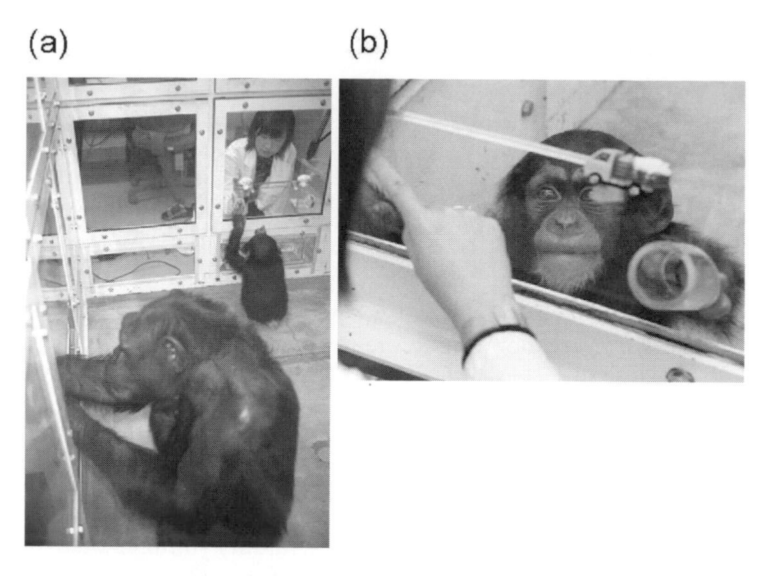

Fig. 1. **a** Ayumu is performing the experimental task. His mother worked at her own tasks in the same experimental booth. **b** Ayumu is looking at the target object, at which the experimenter is pointing. (From the *Mainichi* Newspaper, Japan, with permission)

2.2 Methods

We tested a male chimpanzee infant, Ayumu, from 6 months to 13 months of age. He had participated in various kinds of cognitive tasks in a face-to-face situation after birth (Matsuzawa 2003; Myowa-Yamakoshi et al. 2003, 2004, 2005). During the experiment, an experimenter faced the infant (Fig. 1) and gave four types of social cues [tap, point, head turn, glance (without head turn)] to one of two objects that were placed in front of the infant (Fig. 2). The subject was given food rewards independently of his responses in the first three conditions, so that his responses to the objects were not influenced by the rewards. In the glance condition, irrespective of the subject's response no food reward was presented. This condition was considered as the test for following the eye-gaze cues. We measured the infant's following responses to the target object from video recordings. The subject also received three kinds of control (non-cued) trials, corresponding to each of the four types of social cues described earlier.

2.3 Results

Figure 3 shows that the infant started to follow the experimenter-given cues (tap, point, and head turn) to the target object in front of him from around 9 months old. By the age of 13 months, the subject showed reliable following responses to the object that was indicated by the glance cue. Furthermore, additional tests clearly showed that the subject's performance was controlled by the "social"

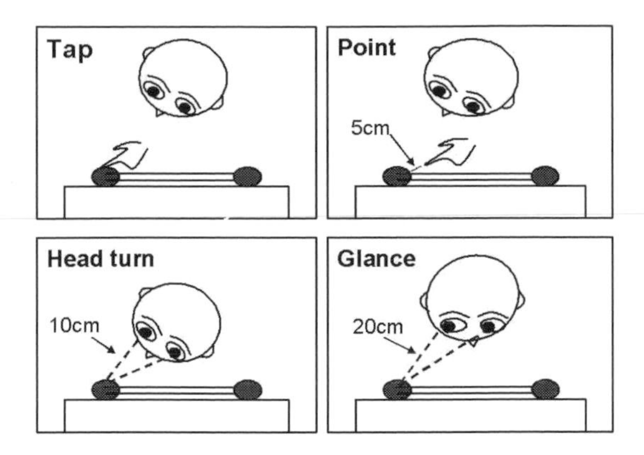

Fig. 2. The four types of cue conditions. (Okamoto et al. 2002a)

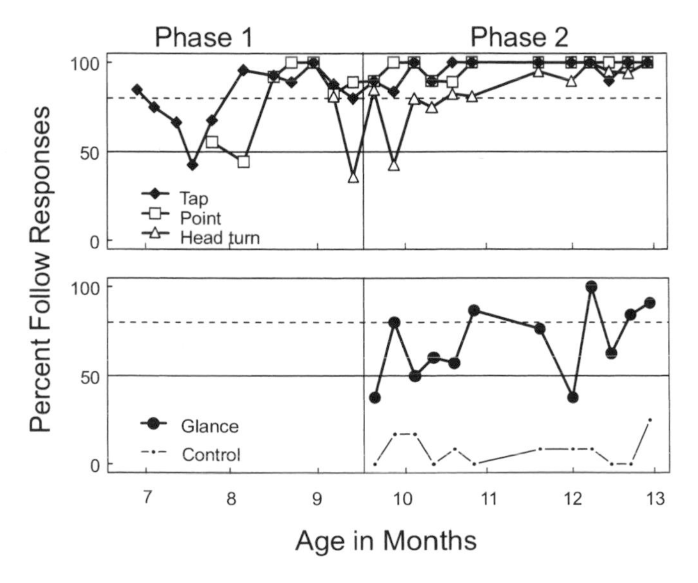

Fig. 3. The percentage of gaze following as a function of the subject's age in months. (Okamoto et al. 2002a)

properties of the experimenter-given cues but not by the nonsocial, local-enhancing peripheral properties.

2.4 Discussion

The results of study 1 clearly demonstrate that a chimpanzee infant less than 1 year old can reliably follow the social cues by a human experimenter to shift his attention to a target object. By the age of 13 months, the subject showed reliable following responses to the object that was indicated by the glance cue. This behavior was not controlled by the nonsocial peripheral property (or local enhancement) of the experimenter-given cues.

It has previously been reported that infant chimpanzees less than 3 to 4 years old do not use head-turn cues (Tomasello et al. 2001). Contrary to this result, our subject reliably used "head turn" as a cue at 11 months and "glance" at 13 months of age. One possible reason for this inconsistency might be differences in the experimental settings. In the study of Tomasello et al., the experimenter looked up to the sky or the ceiling, whereas in the present study, the experimenter looked at the toy as a target object in the subject's visual field. A more important point is that these procedures focus on different aspects of joint attention. As mentioned in the introduction, Butterworth and Jarrett (1991) reported three successively emerging mechanisms of joint attention in human infants from 6 to 18 months. In the present study, we used a specific object within the subject's visual field as a target. This procedure may investigate the "ecological" or "geometric mechanism" of joint attention in their terminology. The Tomasello et al. study, on the other hand, used no specific objects, but subjects were required to move their heads or bodies. This procedure may require the "representational mechanism." Tomasello et al. (1999) reported that chimpanzees follow the gaze direction of other individuals to specific locations geometrically, in much the same way as human infants do. Our experiment used only the two specific locations in the subject's visual field. This procedure might be insufficient for distinguishing between the ecological and geometric mechanisms (see Butterworth and Jarrett 1991). To address problems concerning the underlying mechanisms of gaze following in chimpanzee infants, we need further experimental manipulations, for example, a greater number of specific locations for the target.

In the present experiment, the infant followed the glance cues without explicit differential reinforcement training by 13 months of age. Povinelli and Eddy (1996) reported that 5- to 6-year-old chimpanzees responded appropriately both to head turns and eye movement alone. Itakura and Tanaka (1998) also reported that adult chimpanzees can use eye movement as a cue in object-choice tasks. However, with the exception of the present study, there has been no evidence that infant chimpanzees can use eye movements alone as a cue. Human infants at 9 months of age are unable to shift attention by glance cues without head turning. At 12 and 14 months of age, half the subjects can shift their attention using glance cues (Butterworth and Jarret 1991; Lempers 1979; but see also Corkum and Moore 1995). These chronological ages apparently correspond to the chimpanzee infant in the present study. However, it is well known that the speed of body growth and perceptual development in chimpanzees is faster than that of humans. We need further studies both in humans and chimpanzees to draw clear conclusions concerning the onset of gaze-following abilities.

As a next step of the investigation of this ability, we might have to focus on the developmental aspect, which includes the "representational mechanism" period of joint attention. In humans, 12-month-olds do not follow gaze to objects behind them but 18-month-olds do (Butterworth and Jarrett 1991). From which age on do infant chimpanzees look back to the target? To clarify this question, we conducted a further study focusing on the spatial representation in joint attention. Moreover, we assessed whether other factors such as the characteristic of target objects affect the infant's following responses.

3 The Development of Joint Attention

3.1 Introduction

As joint attention has an important role for development in social animals, it is also important to understand how the actions of others elicit infants' joint attention. Along the lines of Corkum and Moore's empirical and parsimonious account in another study with human subjects, pointing cues elicited more episodes of joint attention than looking alone, and distinctive and complex targets elicited more episodes of joint attention than identical targets (Deák et al. 2000). The authors also found that infants looked more at front than at back targets, but there was also an effect of magnitude of head turn. They also suggested that human infants' joint attention to targets behind them is affected by the distinctiveness and complexity (i.e., interesting) of the targets. Thus, environmental factors also affect the infant's joint attention.

Study 2 was conducted to clarify the ability of the infant chimpanzee (Ayumu) to follow experimenter-given cues to targets outside his visual field, the "representational mechanism" in Butterworth's terminology. Moreover, we manipulated two factors to investigate what affects the chimpanzee's joint attention to objects outside his visual field: incentive and subject's memory of targets.

3.2 Methods

In the present study, from 13 months old, Ayumu was tested to look at one of two identical object pairs, which an experimenter indicated by pointing and head turning (Fig. 4). The object pairs were set in front of or behind the subject

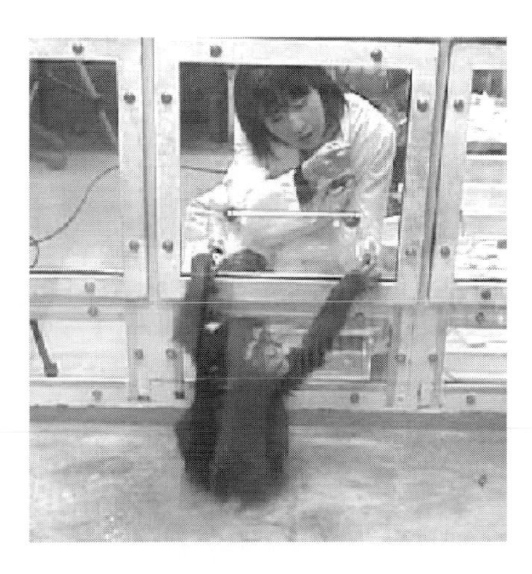

Fig. 4. Ayumu is looking at the target object behind him, which the experimenter is pointing at (from the view of camera 2). (Okamoto et al. 2004)

(Fig. 5). We administered a preliminary test phase and a main test phase. In the preliminary test phase, each session was composed of baseline trials with four cues from study 1 to the close object (tap, point, head turn, and glance). We could use these cues as baseline trials for letting the infant stay in front of the experimenter because the following responses to these cues have been fully established in study 1. To assess the emergence of the following response to the distant target, "looking back," we introduced a "distant-pointing" cue to a distant target in some of the trials. In the main test phase, we used moving or stationary objects as the distant targets. Moreover, the experimenter manipulated a computer at the onset of each block of trials. In the main test, two factors were manipulated: incentive and subject's memory of targets.

The main test consisted of four phases. *Phase 1*: two stationary and identical toys were presented as the distant objects. *Phase 2a*: two identical computer screens showing moving screen savers served as the distant objects. After activating the screen savers, the experimenter indicated one of the distant objects by giving one of the experimental cues. *Phase 2b*: the computers with stationary screen savers served as targets. The computers were not activated. The experimenter indicated one of the distant objects by distant-pointing cues according to the predetermined time setting as in phase 2a. Sessions of phases 2a and 2b were alternated. *Phase 3*: instead of the experimental cues, control cues (see those in study 1) were used with a moving screen saver. *Phase 4*: This phase is the same as phase 2a. It was considered as the recovery condition for investigating the effect of the cues. In the latter three phases, the experimenter manipulated the computer in front of the subject at the onset of each block of trials.

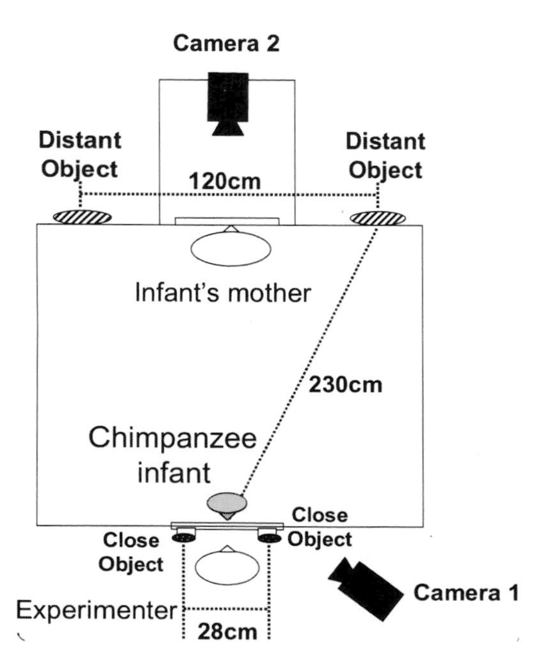

Fig. 5. Experimental setting (aerial view). (Okamoto et al. 2004)

3.3 Results

Figure 6 shows that the percentages of "looking-back" responses to the distant object by following the distant-pointing cue. The results show that by the age of 20 months the infant reliably began to follow the experimenter's distant-pointing cues and looked back to the target behind him. Moving targets elicited more responses than stationary targets, and the subject showed more following responses after having seen the experimenter manipulating the computer.

Although Ayumu often turned his head or body to the left side (71.7% of total looking-back responses), the side he turned his head to and the side the experimenter pointed to matched in 80.4% of total looking-back responses. One more important result to be noted is, however, that the subject did not look at the experimenter again once he looked back behind himself.

3.4 Discussion

In the present study, by the age of 21 months, the infant chimpanzee reliably followed the experimenter's cues and looked back to the target behind him.

Our previous study 1 (Okamoto et al. 2002a) and the present study clearly indicate that mechanisms of joint attention also emerge successively in an infant chimpanzee as in human infants (cf. Butterworth and Cochran 1980; Butterworth and Jarrett 1991). Moreover, factors such as the distinctiveness of targets also influenced the chimpanzee's joint attention, as in human infants (Deák et al. 2000). The comparison of results between phase 1 and phase 2 suggest that the subject's looking-back behavior was facilitated by seeing the experimenter manipulating the computer. Some episodic memory of targets being manipulated may influence joint attention in the sense of increased expectancy of a subsequent event. Furthermore, the attractiveness of subsequent events also affected the subject's response (moving versus stationary targets, both of which were manipulated by the experimenter; phases 2a versus 2b). In phase 2, we introduced 2a (moving target) first, and then the 2b (stationary target) conditions alternately. It is possible that the mean responses of 2a are higher than that of 2b because the more interesting condition was presented first and the conditions were not presented in a random order. However, as seen in Fig. 6, the main tendency that the looking-back response is higher in condition 2a than in the adjacent condition 2b is preserved. This tendency indicates that distinctive (e.g., attractive, interesting) targets elicited more looking-back responses than identical ones. Moreover, we stress that the looking-back behavior only occurred when the experimenter indicated the targets, even though the response rate decreased gradually. In other words, only the experimenter's gesture was used as a trigger to look back to the target behind the chimpanzee infant. Additionally, because the side that the subject looked back to often matched the side the experimenter pointed to, we can suggest that the subject's responses might represent a "representational mechanism."

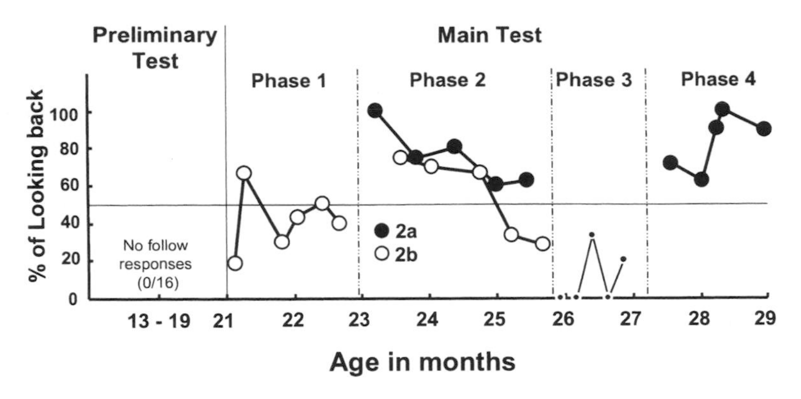

Fig. 6. The percentage of "looking back" responses in the distant-point condition as a function of session and the subject's age. (Okamoto et al. 2004)

Looking-back responses according to the experimenter's gaze cues are apparently evidence for a "representational mechanism" in chimpanzee infants. As noted in the Results section, however, the subject showed a lack of subsequent behavioral sequences after looking back, which are observed commonly in human infants. That is, the subject did not look at the experimenter again after looking back at the target. This result implies that there must be some substantial difference in "joint attention" between these two species. Thus, the question arises: are the present results really evidence for "joint attention" in chimpanzee infants? Our chimpanzee infant followed the experimenter's gaze but did not interact with her. Carpenter et al. (1998) and Tomasello (1999) described the three main types of joint attentional interaction as check attention (9–12 months), follow attention (11–14 months), and direct attention (13–15 months). According to their distinction, our chimpanzee infant followed the experimenter's attention but did not direct her attention to the external object. Similarly, Emery (2000) noted that joint attention is different from shared attention. "Joint attention" is the same as "gaze following" except that there is a focus of attention (such as an object). "Shared attention," on the other hand, is a combination of mutual attention and joint attention, where the focus of attention of individuals A and B is on the object of joint focus and each other. Thus, "triad relationship-based" joint attention is an important component of social cognitive skills in human infants older than 1 year. On the basis of Emery's definition, the present results suggest that the chimpanzee infant and human experimenter jointly attended to the object behind the infant but did not share their attention with each other. More strictly speaking, the experimenter's behavior was merely a trigger for the subject to initiate appropriate reactions such as searching behind him. As in previous studies, it remains unresolved whether chimpanzees attribute referential intent and visual experience to the signaler or merely follow gaze direction geometrically to specific locations (e.g., Tomasello et al. 1999).

4 General Discussion

4.1 Similarities and Differences of Joint Attention in Chimpanzees and Humans

The present studies found similarities in certain levels of joint attention between human and chimpanzee infants. Two consecutive studies clearly showed that an infant chimpanzee did follow social cues (e.g., tapping, pointing, and head turning) from around 9 months of age. The infant began to reliably follow eye gaze at 13 months. Although, for the gaze cues, a nondifferential reinforcement design was introduced to avoid any type of shaping, it is still possible that the infant may have learned to follow human gaze by "generalizing" from the other cues. Starting at 21 months of age, the infant looked back to targets located behind him, even when there was a distracter in front of him. In this study, infant chimpanzees exhibited the looking-back response, which Butterworth and Jarrett (1991) have interpreted as evidence of a "representational mechanism" in 18-month-old human infants. Although there are some developmental differences about the onset of each "level" of joint attention, on the surface, the development of joint attention in human and chimpanzee infants appears to be highly similar.

However, unlike human infants, the chimpanzee infant in our study failed to look at the experimenter after following her gaze to an object located behind him. This triadic interaction between mother, child, and object of interest has been widely reported in the human developmental literature. For instance, human infants look at the mother and the object alternately. Especially, the situation of study 2 in which the target object was manipulated (movement) might trigger this interaction in triadic relationships between mother and human infant. The absence of this interaction in the infant chimpanzee we observed suggests a fundamental difference in joint attention of humans and chimpanzees.

There might be a potential explanation of the difference of joint attention between the two species. In humans, a number of qualitative changes in social communication occur at around 9 months of age (Carpenter et al. 1998). Human infants, at 6 months, interact "dyadically" with objects or with a person in a turn-taking (or reciprocally exchanging) sequence. However, they do not interact with the person who is manipulating objects (Tomasello 1999). From 9 months on, they start to engage in "triadic" exchanges with others. Their interactions involve both objects and persons, resulting in the formation of a referential triangle of infant, adult, and object to which they share attention (Rochat 2001; Tomasello 1999). In contrast, Tomonaga et al. (in press) reported that chimpanzee infants never showed an object or gave an object to a caregiver. Such actions are taken to be indicative of referential communication in triadic relationships in humans. One exceptional case of these actions was reported for an 18-month-old nursery-raised chimpanzee by Russell et al. (1997). Yet, unlike human infants who eventually develop an ability to establish this referential communication in triadic relationships, it might be possible that chimpanzees never reach this cognitive milestone.

4.2 Prerequisites for Joint Attention

The first appearance of triadic relationship has been treated as an important qualitative change in the development of human infants. The following components are necessary to establish a triadic relationship: infant, adult, and a particular object or event. Previous studies have investigated gaze-following abilities in human infants and nonhuman primates and have used several types of particular objects as targets. Some characteristic of target objects such as distinctiveness or attractiveness elicited the infant's gaze following behavior in both humans and chimpanzees [Deák et al. 2000; present study 2 (Okamoto et al. 2004)]. These results might be supported by studies that demonstrated that "expectancies" about the appearance of a target elicited more gaze-following responses.

Okamoto-Barth and Kawai (in press) investigated how the expectancy of a target that appeared inconsistently affects attention to gaze-direction cues in human adults. The results showed clear differences between the performances of two groups tested in trials with a consistently and inconsistently appearing target. In trials where the target appeared inconsistently, the subjects disengaged attention when no target appeared because they expected the trial to be a catch trial (no target).

Apes also performed differently in trials where targets appeared consistently or inconsistently. Okamoto-Barth et al. (unpublished data) manipulated the presence of a target in a gaze-following task with all four great ape species (chimpanzees, bonobos, orangutans, and gorillas). When there was no target, the apes showed fewer following responses.

For humans' daily joint attention episodes, it is possible that a receiver of social attention cues has the expectancy and prediction that a sender may/must be looking at a particular object. Such expectancies and predictions facilitate social communication. However, when there is no particular object or event, the receiver would search for it or wonder what the sender was looking at. In this case, receivers might look back at the sender to gather more contextual information.

A behavior referred to as "checking back" typically accompanies gaze following. This "checking-back" behavior has been observed in adult chimpanzees in instances of apparent uncertainty (Call et al. 1998). However, as mentioned earlier, the chimpanzee infant in the present study did not show this behavior. We interpreted the absence of "checking back" in the infant chimpanzee we studied as evidence against "triad relationship-based" joint attention (or "shared attention"). However, there may be an essential difference between "checking back" and "shared attention." Checking back could be completed without an understanding of mental states. However, checking back cannot be accomplished without the ability to track gaze geometrically. A number of studies have demonstrated that chimpanzees can effectively track an individual's gaze geometrically, even when there is a distracter between the subject and the target (e.g., Tomasello et al. 1999). Yet it does not follow from these results that subjects that track gaze are interpreting them mentalistically. After following the sender's gaze, does the receiver need a terminal point (target object or event)? Although this is still an open question, it is clear that the presence or expectancy of a target affects the receiver's attention to a potential target location. For example, consider the preparation for new environmental information that was observed in human adults presented with a schematic face whose eyes were directed to a given location (Okamoto-Barth and Kawai, in press). However, the use of this cue does not imply mental-state attribution in the triadic relationships, but it is more dependent on reflexive orienting.

Is it impossible for the chimpanzee to engage in real joint (shared) attention, that is, "triad relationship-based joint attention"? In a dyad interaction, infant chimpanzees do show "mutual gaze" when interacting with their mother from early infancy (cf. Bard et. al. 2005; Okamoto et al. 2002b). Furthermore, similar to a report on human infants (Deák et al. 2000), the present studies suggested that some environmental factors influence joint attention to objects outside the visual field of chimpanzee infants. In the future, we should conduct more detailed comparative examinations concerning the developmental changes from dyad- to triad-based interactions involving eye gaze and factors affecting these interactions. Such studies will provide a clearer idea of visual communication including joint attention and the understanding of social-cognitive abilities in nonhuman primates.

Acknowledgments

The present study was supported by grants from the Japan Society of Promotion of Science (JSPS) and the Ministry of Education, Culture, Sports, Science, and Technology (MEXT), Japan (#12002009, #13610086, #16002001), MEXT 21st Century COE Program (A14), and the cooperative research program of the Primate Research Institute (PRI), Kyoto University. Preparation of the manuscript was supported by the Faculty of Psychology, Department of Cognitive Neuroscience, Maastricht University, The Netherlands. Parts of this manuscript were submitted to Maastricht University by Sanae Okamoto-Barth in partial fulfillment of the requirements for a doctorate degree in psychology. We wish to express thanks to M. Tanaka, N. Kawai, K. Ishii, S. Itakura, K.A. Bard, J.R. Anderson, and T. Matsuzawa for their support and advice given in the course of the study. Thanks are also due to J. Barth and F. Subiaul for their suggestions on early versions of this manuscript. We are grateful to the staff of the Center for Human Evolution Modeling Research, PRI, for their daily care of the chimpanzees.

References

Baron-Cohen S (1995) Mindblindness: an essay on autism and theory of mind. MIT Press, Cambridge
Bard KA, Myowa-Yamakoshi M, Tomonaga M, Tanaka M, Quinn J, Costall A, Matsuzawa T (2005) Group differences in the mutual gaze of chimpanzees (*Pan troglodytes*). Dev Psychol 41:616–624
Butterworth G (1991) The ontogeny and phylogeny of joint visual attention. In: Whiten A (ed) Natural theories of mind. Blackwell, Oxford, pp 223–232
Butterworth G, Cochran E (1980) Towards a mechanism of joint visual attention in human infancy. Inter J Behav Dev 3:253–272
Butterworth G, Jarrett N (1991) What minds have in common is space: spatial mechanisms serving joint visual attention in infancy. Br J Dev Psychol 9:55–72
Call J, Hare BA, Tomasello M (1998) Chimpanzee gaze following in an object-choice task. Anim Cogn 1:89–99
Carpenter M, Nagell K, Tomasello M (1998) Social cognition, joint attention, and communicative competence from 9 to 15 months of age. Monogr Soc Res Child Dev 63 (4, Serial No. 255)
Chance MRA (1967) Attention structure as a basis of primate rank orders. Man 2:503–518
Corkum V, Moore C (1995) The development of joint attention in infants. In: Moore C, Dunham PJ (eds) Joint attention: its origins and role in development. Erlbaum, Hillsdale, NJ, pp 61–85
Corkum V, Moore C (1998) The origins of joint visual attention in infants. Dev Psychol 34:28–38
D'Entremont B, Hains SMJ, Muir DW (1997) A demonstration of gaze following in 3- to 6-month-olds. Infant Behav Dev 20:569–572
Deák GO, Flom RA, Pick AD (2000) Effects of gesture and target on 12- and 18-month-olds' joint visual attention to objects in front of or behind them. Dev Psychol 36:511–523.
Desrochers S, Morissette P, Ricard M (1995) Two perspective on pointing in infancy. In Moore C, Dunham PJ (eds) Joint attention: its origins and role in development. Erlbaum, Hillsdale, NJ, pp 85–102
Emery NJ (2000) The eyes have it: the neuroethology, function and evolution of social gaze. Neurosci Biobehav Rev 24:581–604
Emery NJ, Lorincz EN, Perrett DI, Oram MW, Baker CI (1997) Gaze following and joint attention in rhesus monkeys (*Macaca mulatta*). J Comp Psychol 111:286–293
Ferrari PF, Kohler E, Gallese V (2000) The ability to follow eye gaze and its emergence during development in macaque monkeys. Proc Natl Acad Sci U S A 97:13997–14002
Hains SMJ, Muir DW (1996) Infant sensitivity to adult eye direction. Child Dev 67:1940–1951
Itakura S (1996) An exploratory study of gaze-monitoring in non-human primates. Jpn Psychol Res 38:174–180
Itakura S, Tanaka M (1998) Use of experimenter-given cues during object-choice tasks by chimpanzees (*Pan troglodytes*), an orangutan (*Pongo pygmaeus*) and human infants (*Homo sapiens*). J Comp Psychol 112:119–126
Lempers JD (1979) Young children's production and comprehension of nonverbal deictic behaviors. J Genet Psychol 135:93–102
Lorincz EN, Baker CI, Perrett DI (1999) Visual cues for attention following in rhesus monkeys. Cahiers Psychol Cogn 18:973–1003
Matsuzawa T (2003) The Ai project: historical and ecological contexts. Anim Cogn 6:199–211
Menzel EW, Halperin S (1975) Purposive behavior as a basis for objective communication between chimpanzees. Science 189:652–654
Moore C (1999) Gaze following and the control of attention. In Rochat P (ed) Early social cognition. Erlbaum, Mahwah, NJ, pp 241–256
Myowa-Yamakoshi M, Tomonaga M (2001) Perceiving eye gaze in an infant gibbon (*Hylobates agilis*). Psychologia 44:24–30
Myowa-Yamakoshi M, Tomonaga M, Tanaka M, Matsuzawa T (2003) Preference for human direct gaze in infant chimpanzees (*Pan troglodytes*). Cognition 89:B53–B64

Myowa-Yamakoshi M, Tomonaga M, Tanaka M, Matsuzawa T (2004) Imitation in neonatal chimpanzees (*Pan troglodytes*). Dev Sci 7(4):437–442

Myowa-Yamakoshi M, Yamaguchi MK, Tomonaga M, Tanaka M, Matsuzawa T (2005) Development of face recognition in infant chimpanzees (*Pan troglodytes*). Cogn Dev 20:49–63

Okamoto S, Tomonaga M, Ishii K, Kawai N, Tanaka M, Matsuzawa T (2002a) An infant chimpanzee (*Pan troglodytes*) follows human gaze. Anim Cogn 5:107–114

Okamoto S, Kawai N, Sousa C, Tanaka M, Tomonaga M, Ishii K, Matsuzawa T (2002b) Visual interaction between mother and infant chimpanzee. 18th Annual conference of Primate Society of Japan, Tokyo, Japan. Prim Res 18:394

Okamoto-Barth S, Kawai N (in press) The role of attention in the facilitation effect and another "inhibition of return". Cognition

Okamoto S, Tanaka M, Tomonaga M (2004) Looking back: the "representational mechanism" of joint attention in an infant chimpanzee (*Pan troglodytes*). Jpn Psychol Res 46:236–245

Perrett DI, Emery NJ (1994) Understanding the intentions of others from visual signals: neurophysiological evidence. Cahiers Psychol Cogn 13:683–694

Perrett DI, Hietanen JK, Oram MW, Benson PJ (1992) Organization and functions of cells responsive to faces in the temporal cortex. Philos Trans Biol Sci 335:23–30

Povinelli DJ, Eddy TJ (1996) Chimpanzees: joint visual attention. Psychol Sci 7:129–135

Povinelli DJ, Eddy TJ (1997) Specificity of gaze-following in young chimpanzees. Br J Dev Psychol 15:213–222

Rochat P (2001) The infant's world. Harvard University Press, Cambridge, MA

Russell CL, Bard KA, Adamson LB (1997) Social referencing by young chimpanzees (*Pan troglodytes*). J Comp Psychol 111:185–193

Scaife M, Bruner JS (1975) The capacity for joint visual attention in the infant. Nature (Lond) 253:265–266

Tomasello M (1995) Joint attention as social cognition. In: Moore C, Dunham PJ (eds) Joint attention: its origins and role in development. Erlbaum, Hillsdale, NJ, pp 61–85

Tomasello M (1999) The cultural origins of human cognition. Harvard University Press, London

Tomasello M, Call J, Hare B (1998) Five primate species follow the visual gaze of conspecifics. Anim Behav 55:1063–1069

Tomasello M, Hare B, Agnetta B (1999) Chimpanzees, *Pan troglodytes*, follow gaze direction geometrically. Anim Behav 58:769–777

Tomasello M, Hare B, Fogleman T (2001) The ontogeny of gaze following in chimpanzees, *Pan troglodytes*, and rhesus macaques, *Macaca mulatta*. Anim Behav 61:335–343

Tomonaga M, Myowa-Yamakoshi M, Mizuno Y, Okamoto S, Yamaguchi MK, Kosugi D, Bard KA, Tanaka M, Matsuzawa T (in press) Chimpanzee social cognition in early life: comparative-developmental perspective. In: Wasserman EA, Zentall TR (eds) Comparative cognition: experimental explorations of animal intelligence. Oxford University Press, New York

Whiten A (1997) Evolutionary and developmental origins of the mindreading system. In Langer J, Killen M (eds) Piaget: evolution and development. Erlbaum, London

Whiten A, Byrne RW (1988) The manipulation of attention in primate tactile deception. In: Byrne RW Whiten A (eds) Machiavellian intelligence: social expertise and the evolution of intellect in monkeys, apes and humans. Oxford University Press, Oxford, pp 211–223

11
Food Sharing and Referencing Behavior in Chimpanzee Mother and Infant

Ari Ueno

1 Introduction

Youngsters of omnivorous mammals have to learn about food in the time course of their development. In their developmental course, they learn which objects are edible and adequate for their food among many potential food resources in their surrounding environment. Learning about food is an essential task for their survival.

Humans, one of the omnivorous mammals, make use of a variety of potential food resources and live in a wide range of environments. Such an ability to make use of various potential food resources is one of the hallmarks of the human species (Ungar and Teaford 2002). As one of the aspects presumably contributing to such an ability, we need to consider how people acquire a wide range of their food selection habits in their developmental course.

In humans, the acquisition process of food selection habits is affected by self-evaluation based on physiological consequences of ingestion and sensory-affective factors such as tastes (Mennella and Beauchamp 1997; Rozin 1976), but it is also much influenced by social factors (Birch 1987; Rozin 1988, 1996). When encountering a novel food, youngsters accept it more readily if they see other individuals ingest it than when this is not the case (Harper and Sanders 1975). Seeing another individual's choice among several kinds of food, youngsters even shift their own choice from the original one to converge with the other's choice (Birch 1980; Dunker 1938). Moreover, human caregivers regulate youngsters' experiences of foods by preparing their daily meals and actively teach them what and how to eat. In humans, several kinds of food-related interactions, including both passive and active interferences, seem to strongly influence the youngsters' food learning and the acquisition process of their food selection habits.

To understand human characteristics in the acquisition processes of food selection habits, comparative studies of nonhuman primates are very helpful. However, few studies have had comparative perspectives over species on this

Center for Evolutionary and Cognitive Sciences, Graduate School of Arts and Sciences, The University of Tokyo, 3-8-1 Komaba, Meguro-ku, Tokyo 153-8902, Japan

issue (see Chapter 16 of this volume). As well as in humans, physiological consequences of ingestion and sensory-affective factors such as tastes seem to be fundamental elements contributing to the acquisition of food selection habits in nonhuman primates. Indeed, nonhuman primates exhibit presumably affective responses to sweet and bitter tastes differently (Steiner et al. 2001), even from an early stage of life (Ueno et al. 2004). How about social influences? Do some kinds of food-related interactions, including passive and active interferences, as well as in those in humans, contribute to the youngsters learning about food and the acquisition process of their food selection habits?

In this chapter, I briefly introduce some reports that elucidate that food-related interactions provide some information about food and affect food selection in nonhuman primates. I then describe food-related interactions between a chimpanzee mother and infant (*Pan troglodytes*), phylogenetically the closest neighbor of humans. Comparing food-related interactions with those seen in other situations, I also discuss general patterns of mother–infant interactions in the chimpanzee.

2 Social Influences on the Acquisition Process of Food Selection Habits in Nonhuman Primates

In nonhuman primates, the degree of importance of social influence has been presumed to differ among species, depending upon their social structure and food habits (Cambefort 1981; Milton 1993). Although social influence on the acquisition process of food selection habits has long been emphasized in several species, it has scarcely been investigated systematically (Fragaszy and Visalberghi 1996; Visalberghi 1994; Visalberghi and Addessi 2003; see also Chapter 16 of this volume).

Under experimental settings, animals are reported to show interest selectively toward objects that were seemingly ingested by other human individuals (in rhesus macaque: Santos et al. 2001) and feed more when other individuals also feed than when alone or when other individuals do not feed (in capuchin monkey: Addessi and Visalberghi 2001; Fragaszy and Visalberghi 1996; Visalberghi and Addessi 2000, 2001, 2003). In the Tonkean macaque (Drapier et al. 2002), it has been reported that animals could obtain information about food ingested by other individuals via olfactory cues through muzzling behavior (sniffing of other individuals' mouth/nose closely). Moreover, in a study that was the only case in which clear evidence for social transmission of food selection has been proposed in nonhuman primates (mother–offspring pairs of Japanese macaque: Hikami et al. 1990), food-related interactions, such as looking at other feeding individuals, were pointed out to be important for the obtained results. All these reports evoke the need for focusing on food-related interactions to know more about the acquisition process of food selection habits in nonhuman primates.

3 Food-Related Interactions in the Chimpanzee

In the chimpanzee, social influence on the acquisition process of food selection habits has long been claimed (Nishida et al. 1983; Goodall 1986). Chimpanzees are known to be omnivorous and inhabit environments that vary greatly across season and habitat (Nishida et al. 1983). In particular, under such conditions in which the environment varies considerably across seasons and habitats, animals are assumed to benefit from the social transmission of food selection habits. Chimpanzee youngsters are often observed to exhibit food-related interactions with other feeding individuals, such as inspecting others who are feeding from a very close distance and trying to obtain the other's food. Such food-related interactions have been assumed to play some roles for youngsters' food learning (Goodall 1986).

In the following sections, I describe the details of food-related interactions exhibited by captive infant chimpanzees in two different experimental settings. First, the details of interactions that lead to food sharing between mother and infant (less than 2 years old) are described. Second, I report infant (at around 3 years old) response toward novel food and visual attention to the mother, presumable referencing behavior, under the condition in which an infant can freely explore food items together with the mother. The research was conducted in a series of longitudinal studies on the behavioral and cognitive development of chimpanzees (Matsuzawa 2002, 2003). Three infants, Ayumu (male), Cleo (female), and Pal (female), born in 2000 at the Primate Research Institute, Kyoto University, and their mothers, Ai, Chloe, and Pan, respectively, participated in the research.

3.1 Food Sharing Between Mother and Infant Chimpanzee

Among various kinds of food-related interactions, direct food transfer between individuals, so-called food sharing, has been reported to occur frequently between mother and offspring in particular (Goodall 1986; Hiraiwa-Hasegawa 1990a,b; McGrew 1975; Nishida and Turner 1996; Silk 1978). Incidents of food sharing have been reported in several observation fields. It has been suggested that food sharing facilitates an infant's learning of the diversity of adult foods available in his/her environment. Even though the function of food sharing has long been discussed, details of interactions that lead to food sharing and shared food parts have not been investigated systematically. In the wild, observational conditions might restrict the opportunity to see the whole event closely. Such an opportunity was available under conditions of "participant observation" (Matsuzawa 2002, 2003) in the Primate Research Institute, Kyoto University. To investigate the details of food-sharing behavior and the characteristic of shared food parts, Ueno and Matsuzawa (2004) passed one kind of food items to a chimpanzee mother and observed all interactions between the mother and her infant. In their study, various kinds of food items, including both novel and familiar

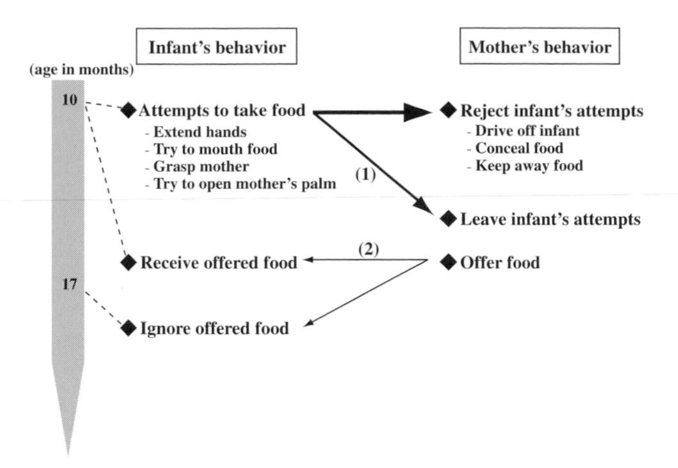

Fig. 1. Outline of infant and mother behaviors that affect food sharing immediately. The *arrows* and their size represent the direction of behavior and the approximate relative frequency of occurrence of each interaction, respectively. According to the preceding interactions, two patterns of food sharing were defined: infant-initiated sharing (*1*) and mother-initiated sharing (*2*). The *shaded vertical arrow* (*left side*) represents the developmental course of the infant's behavior. An infant's ignoring of offered food was first observed at the age of 17 months

foods for infants, were used. Three pairs of mother and infant were observed longitudinally from the age of 8 or 10 months up to 23 months.

Figure 1 represents the outline of mother–infant interactions observed in three pairs of mother and infant chimpanzees. Infants looked at their mothers from a very short distance and often attempted to take food from the mothers. In a majority of such cases, the mothers rejected the infants' attempts to reach to the food. In some cases, however, infants succeeded in obtaining part of the food from their mothers.

On the other hand, mothers sometimes offered a part of her food spontaneously without precedent infants' attempt to obtain food. Until the age of 16 months, infants received offered food parts in all such cases. However, at the age of 17 months, an infant first ignored offered pieces of food.

According to the precedent interactions, two patterns of food sharing were classified. A pattern of food sharing in which the infant attempted to obtain food from its mother and succeeded was labeled infant-initiated sharing. Another pattern of food sharing, in which mothers spontaneously offered a part of their food and infants received it, was labeled mother-initiated sharing. In total, infants obtained a part of food from their mothers through infant-initiated sharing in 34 cases and through mother-initiated sharing in 15 cases.

When we consider the palatability of shared food pieces, there were clear differences between that in infant-initiated and mother-initiated sharing (Table 1). In all three pairs, infants tended to obtain the same piece food as mothers were eating through infant-initiated sharing. In contrast, they could obtain only

Table 1. The number of incidents in which infants obtained either palatable or unpalatable food parts through infant- or mother-initiated sharing in the three mother–infant pairs

	Palatability	
Pattern of food sharing	Palatable	Unpalatable
Infant-initiated sharing	32	2
Mother-initiated sharing	0	15

unpalatable parts of foods, such as seed and calyx, through mother-initiated sharing. In other words, mothers spontaneously offered their infants only what they did not eat themselves. In all cases in which infants exhibited ignoring behavior, pieces of food offered to infants were unpalatable parts of familiar foods.

Overall, chimpanzee mothers seem to be reluctant to give their food to infants. On the other hand, infants express interest in food eaten by mothers. In the chimpanzee, infants, rather than mothers, appeared to be responsible for experiencing a variety of foods in the mothers' diet.

3.2 Referencing Behavior of Chimpanzee Infants in a Feeding Context

As already described, in the experimental setting for observing food sharing (Ueno and Matsuzawa 2004), infants exhibited interest in the mothers' food. In that situation, however, only a mother could have food at first, thus an infant could not have an opportunity to access the food freely and to explore it on his/her own. With improvement in locomotor ability, infants begin to move away from their mothers and are able to explore food resources in the surrounding environment on their own. Even in a situation in which infants can explore the food freely on their own, do they exhibit interest in their mothers and their mothers' food?

Ueno and Matsuzawa (2005) investigated this issue in two pairs of chimpanzee mother and infant (Ai–Ayumu and Pan–Pal) at around the age of 3 years. In an experimental setting in which mother and infant could explore food pieces freely together, their interactions and responses toward food were observed (Fig. 2). Both familiar and novel foods for infants were presented, and their responses were compared between familiar and novel foods.

In the results, infants' responses differed much between familiar and novel foods. As to novel foods, infants did not ingest them immediately but exhibited exploratory behavior, such as sniff-licking, first (Table 2, Fig. 3a). When encountering novel food, infants also tended to exhibit visual attention toward their mothers before ingesting/mouthing it (Table 3, Fig. 3b). With a familiar food,

Fig. 2. Mother and infant freely interact and explore food pieces allocated on the floor of an experimental booth. *Left* individual is a mother (Ai) and *right* one is an infant (Ayumu)

Table 2. The number of sessions in which the infants sniff-licked the food items first or immediately ingested them according to whether the food was familiar or novel for the infants

Response to food	Food items	
	Familiar	Novel
Sniff-lick first	1	11
Ingest immediately	34	0

Table 3. The number of sessions in which the infants paid attention to the mother or they did not according to whether they encountered familiar or novel food

Visual attention toward the mother	Food items	
	Familiar	Novel
Attention	0	7
No attention	36	4

however, infants immediately ingested it without any preceding exploratory behavior or visual attention toward their mothers.

From this research, it was elucidated that infant chimpanzees exhibited interest in their mothers/mothers' food even in a situation in which they could explore the food freely on their own. Incidents of infants' visual attention toward their mothers, which were observed only when they encountered novel food, remind us of referencing behavior that has been reported in other contexts in chimpanzee youngsters (Russel et al. 1997; Itakura 1995). The obtained results

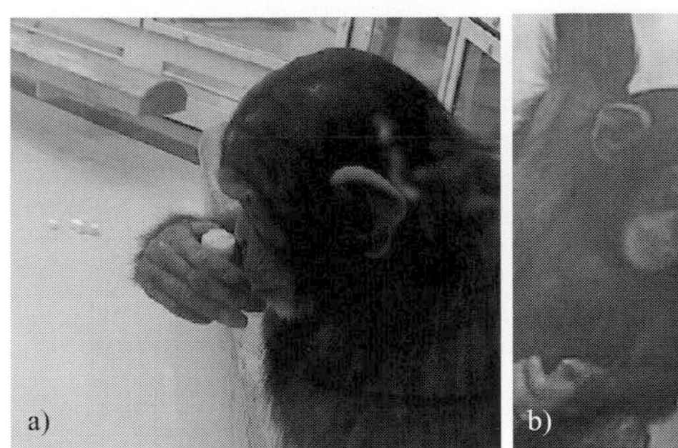

Fig. 3. **a** An infant did not ingest novel food immediately, but sniffed it first. **b** Before ingesting novel food on her own, an infant paid attention to her mother, who was ingesting it

imply that infant chimpanzees tend to hesitate to eat novel foods at first and refer to their mothers for some kind of cue. In contrast to the infants' active approach toward feeding mothers, mothers never interfered with infants' activity directly.

4 Insight from Food-Related Interactions Between Chimpanzee Mother and Infant

Through the observation of food-related interactions in two kinds of experimental settings, the following three points are clarified as characteristic of mother–infant interactions in chimpanzees. (1) Chimpanzee mothers do not actively interfere with infants' feeding activity. (2) In contrast, infants attend to feeding mothers/food eaten by mothers and even exhibit apparently referencing behavior toward mothers when encountering novel foods. (3) Infants, rather than mothers, seem to be responsible for experiencing and learning the diversity of adult foods in the chimpanzee.

An infant chimpanzee, the same individual as one of the subjects in Ueno and Matsuzawa's research (2004, 2005), was reported to start attending to an object that was indicated by various social cues, such as tapping, pointing, and glancing, by the age of 13 months (Okamoto et al. 2002; also see Chapter 10 of this volume). As to tapping and pointing cues, the infant began to attend to them successfully from 8 to 9 months of age. This finding might mean that subject infants of Ueno and Matsuzawa's research (2004, 2005) had already developed the cognitive ability to follow the direction of another individual's action by the time the research was conducted. Based on such development in social cognition, infants probably exhibited various kinds of food-related interactions with their mothers.

Throughout two kinds of observation on feeding situations, the characteristics of interaction between chimpanzee mother and infant seemed to be consistent: mothers do not actively interfere with infants' activity, and infants seem to be responsible for experiencing and learning the diversity of adult foods. Are such characteristics of the mother–infant interaction context specific? Hirata and Celli (2003) examined the role of mothers in infants' acquisition of tool use behavior in the laboratory. In this behavior acquisition process, infants frequently observed skillful mothers and finally began to use a tool the same as the mothers did. Moreover, infants selected the same kind of objects for tools as their mothers did. Explaining the obtained results, Hirata and Celli emphasized the importance of an infant's observation on mother's tool use from a very close distance. As in the research by Ueno and Matsuzawa (2004, 2005), active teaching by the mother was never observed. The same characteristic of mother–infant interactions has been also reported for tool use in the wild (Matsuzawa et al. 2001).

In the situations of both food learning and tool use (Ueno and Matsuzawa 2004, 2005; Hirata and Celli 2003), a direct goal of infants' activity is clearly attractive in itself for the infants, that is, food. When a direct goal of infants' activity is not food, do infants attend to their mothers and learn from them? Hayashi and Matsuzawa (2003) investigated development of object manipulation by chimpanzee infants. In their study, an infant's behavior was not reinforced by food rewards and infants' activity, manipulation of objects, did not link directly to obtaining food. They reported that their subjects began to exhibit a complex type of object manipulation, object–object combination, earlier than the other precedent studies. They elucidated that the early emergence of such a complex type of manipulation was due to the opportunity to observe the skillful mothers. Differing from the other studies, the infants were free to observe very closely that mothers displayed object–object combinations and free to access the objects manipulated by mothers in their study. A skillful mother never teached her infant, but infants were motivated to observe their mothers' behavior. Not merely motivated by food itself but based on an affectionate bond with their mother, chimpanzee infants might attend to their mothers and learn from them.

Learning about food is one of the essential tasks for an infant's survival, and thus is expected to be an important matter also for a mother. Even in such a fundamental situation, chimpanzee mothers do not interfere with infants' activity actively, but rather infants seemed to be responsible for solving their task. Such a characteristic of mother–infant interaction, which differs much from that in humans, might be very general in the chimpanzee.

Acknowledgments

This study was funded by grants from the Ministry of Education, Sciences, and Culture in Japan (#12002009 and #16002001), from the biodiversity research of the 21COE (A14), and from the JSPS Core-to-Core Program, HOPE. Preparation of the manuscript was financially supported by the Center for Evolutionary Cognitive Sciences at the University of Tokyo. I wish to thank Tetsuro Matsuzawa for his valuable comments on a draft of this article and generous guidance throughout this study. Thanks are also due to Masaki Tomonaga, Masayuki Tanaka and all the colleagues who participated in the research project on behavioral and cognitive development of infant chimpanzees for their support and advice. I also extend my thanks to the staff at the Section of Language and Intelligence, Social Behavior, and Ecology, Primate Research Institute, Kyoto University, and to the keepers and veterinarians who take daily care of the chimpanzees at the institute.

References

Addessi E, Visalberghi E (2001) Social facilitation of eating novel food in tufted capuchin monkeys (*Cebus apella*): input provided, responses affected, and cognitive implications. Anim Cogn 4:297–303

Birch LL (1980) Effects of peer models' food choices and eating behaviors on preschooler's food preferences. Child Dev 51:489–496

Birchi LL (1987) The acquisition of food acceptance patterns in children. In: Boakes R, Popplewell D, Burton M (eds) Eating habits. Wiley, New York, pp 107–130

Cambefort JP (1981) A comparative study of culturally transmitted patterns of feeding habits in the chacma baboon *Paio ursinus* and the vervet monkey *Cercopithecus aethiops*. Folia Primatol 36:243–263

Dunker K (1938) Experimental modification of children's food preferences through social suggestion. J Abnorm Soc Psychol 33:489–507

Drapier M, Chauvin C, Thierry B (2002) Tonkean macaques (*Macaca tonkeana*) find food sources from cues conveyed by group-mates. Anim Cogn 5:159–165

Fragaszy DM, Visalberghi E (1996) Social learning in monkeys: primate "primacy" reconsidered. In: Heyes CM, Galef BG (eds) Social learning in animals: the roots of culture. Academic Press, San Diego, pp 65–84

Goodall J (1986) The chimpanzees of Gombe: patterns of behavior. Harvard University Press, Cambridge

Harper LV, Sanders KM (1975) The effect of adults' eating on young children's acceptance of unfamiliar foods. J Exp Child Psychol 20:206–214

Hayashi M, Matsuzawa T (2003) Cognitive development in object manipulation by infant chimpanzees. Anim Cogn 6:225–233

Hikami K, Hasegawa Y, Matsuzawa T (1990) Social transmission of food preferences in Japanese monkeys (*Macaca fuscata*) after mere exposure or aversion training. J Comp Psychol 104:233–237

Hiraiwa-Hasegawa M (1990a) Role of food sharing between mother and infant in the ontogeny of feeding behavior. In: Nishida T (ed) The chimpanzees of Mahale Mountains. Tokyo University Press, Tokyo, pp 267–275

Hiraiwa-Hasegawa M (1990b) A note on the ontogeny of feeding. In: Nishida T (ed) The chimpanzees of Mahale Mountains. Tokyo University Press, Tokyo, pp 279–283

Hirata S, Celli M (2003) Role of mothers in the acquisition of tool use behaviours by captive infant chimpanzees. Anim Cogn 6:235–244

Itakura S (1995) An exploratory study of social referencing in chimpanzees. Folia Primatol 64:44–48

Matsuzawa T (2002) Chimpanzee Ai and her son Ayumu: an episode of education by master-apprenticeship. In: Bekoff M, Allen C, Burghart G (eds) The cognitive animal. MIT Press, Cambridge, pp 189–195

Matsuzawa T (2003) The Ai project: historical and ecological contexts. Anim Cogn 6:199–211

Matsuzawa T, Biro D, Humle T, Inoue-Nakamura N, Tonooka R, Yamakoshi G (2001) Emergence of culture in wild chimpanzees: education by master-apprenticeship. In: Matsuzawa T (ed) Primate origins of human cognition and behavior. Springer, Tokyo, pp 557–574

McGrew WC (1975) Patterns of plant food sharing by wild chimpanzees. In: Kondo S, Kawai M, Ehara A (eds) Contemporary primatology. Karger, New York, pp 304–309

Mennella JA, Beauchamp GK (1997) The ontogeny of human flavor perception. In: Beauchamp GK, Bartoshuk L (eds) Tasting and smelling. Academic Press, San Diego, pp 199–221
Milton K (1993) Diet and social organization of a free-ranging spider monkey population: the development of species-typical behavior in the absence of adults. In: Pereira ME, Fairbanks LA (eds) Juvenile primates. Oxford University Press, New York, pp 173–181
Nishida T, Turner L (1996) Food transfer between mother and infant chimpanzees of the Mahale Mountains National Park, Tanzania. Int J Primatol 17:947–968
Nishida T, Wrangham RW, Goodall J, Uehara S (1983) Local differences in plant-feeding habits of chimpanzees between the Mahale Mountains and Gombe National Park, Tanzania. J Hum Evol 12:467–480
Okamoto S, Tomonaga M, Ishii K, Kawai N, Tanaka M, Matsuzawa T (2002) An infant chimpanzee (*Pan troglodytes*) follows human gaze. Anim Cogn 5:107–114
Rozin P (1976) The selection of foods by rats, humans, and other animals. In: Lehrman D, Hinde RA, Shaw E (eds) Advances in the study of behavior, vol 6. Academic Press, New York, pp 21–76
Rozin P (1988) Social learning about food by humans. In: Zentall TR, Galef BG (eds) Social learning: psychological and biological perspectives. Erlbaum, Mahwah, NJ, pp 165–187
Rozin P (1996) The socio-cultural context of eating and food choice. In: Meiselman HL, MacFie HJH (eds) Food choice, acceptance and consumption. Blackie, London, pp 83–104
Russel CL, Adamson LB, Bard KA (1997) Social referencing by young chimpanzees (*Pan troglodytes*). J Comp Psychol 111:185–193
Santos LR, Hauser MD, Spelke ES (2001) Recognition and categorization of biologically significant objects by rhesus monkeys (*Macaca mulatta*): the domain of food. Cognition 82:127–155
Silk JB (1978) Patterns of food sharing among mother and infant chimpanzee at Gombe National Park, Tanzania. Folia Primatol 29:129–141
Steiner JE, Glaser D, Hawilo ME, Berridge KC (2001) Comparative expression of hedonic impact: affective reactions to taste by human infants and other primates. Neurosci Biobehav Rev 25:53–74
Ueno A, Matsuzawa T (2004) Food transfer between chimpanzee mothers and their infants. Primates 45:231–239
Ueno A, Matsuzawa T (2005) Response to novel food in infant chimpanzees: do infants refer to mothers before ingesting food on their own? Behav Process 68:85–90
Ueno A, Ueno Y, Tomonaga M (2004) Facial responses to four basic tastes in newborn rhesus macaques (*Macaca mulatta*) and chimpanzees (*Pan troglodytes*). Behav Brain Res 154:261–271
Ungar PS, Teaford MF (2002) Perspectives on the evolution of human diet. In: Ungar PS, Teaford MF (eds) Human diet: its origin and evolution. Bergin & Garvey, Westport, CT, pp 1–6
Visalberghi E (1994) Learning processes and feeding behavior in monkeys. In: Galef BG, Mainardi M, Valsecchi P (eds) Behavioral aspects of feeding: basic and applied research on mammals. Harwood, Chur, Switzerland, pp 257–270
Visalberghi E, Addessi E (2000) Seeing group members eating a familiar food enhances the acceptance of novel foods in capuchin monkeys, *Cebus apella*. Anim Behav 60:69–76
Visalberghi E, Addessi E (2001) Acceptance of novel foods in capuchin monkeys: do specific facilitation and visual stimulus enhancement play a role? Anim Behav 62:567–576
Visalberghi E, Addessi E (2003) Food for thought: social learning about food in feeding capuchin monkeys. In: Fragaszy DM, Perry S (eds) The biology of traditions: models and evidence. Cambridge University Press, Cambridge, pp 187–212

12
Development of Chimpanzee Social Cognition in the First 2 Years of Life

Masaki Tomonaga

1 Introduction

Many researchers have become interested in the development of social cognition in nonhuman primates since Premack and Woodruff (1978) proposed the concept of theory of mind. Theory of mind, the ability to infer another conspecific's mental state, was then elaborated on by developmental psychologists, and many experimental studies were conducted with human children using "false belief" tasks (Wimmer and Perner 1983). These studies clarified that theory of mind is emergent only after 4 or 5 years of age and that 3-year-old children do not show any clear evidence for the understanding of the other's false belief (Mitchell 1997). On the other hand, many researchers began trying to find the ontogenetic prerequisites for the 5-year-old's theory of mind in much younger children (Wellman 1992). At the same time, a group of primatologists in the mid-1980s proposed the hypothesis that human intelligence evolved to deal with the complexities of social living (Byrne and Whiten 1988; Whiten and Byrne 1997). This hypothesis, called the social intelligence hypothesis or Machiavellian intelligence hypothesis, was linked with progress in human developmental psychology. Since then, comparative developmental approaches to social cognition have been recognized as being important to the evolutionary understanding of human social cognition. Throughout the 1990s, findings on various aspects of social cognition in nonhuman primates (especially the great apes) accumulated, such as tactical deception, imitation, observational learning in cultural behavior including tool use, gaze following, perspective taking, empathy, social referencing, and false belief (Tomasello and Call 1997; Whiten and Byrne 1997).

Many of these studies, however, tested only juvenile or adult subjects. Although they revealed the great ape's abilities in social cognition, the developmental course of these abilities is still not well understood. Developmental studies of captive chimpanzees had been conducted since the 1930s, but most used human-raised (i.e., enculturated) chimpanzee infants (Kellogg and Kellogg 1933; Hayes 1951; Gardner and Gardner 1969). It is quite plausible that interactions between the human caregiver and the infant would modify the emergence of abilities in social cognition (cf. Russell et al. 1997). To truly understand the comparative development of social cognition in great apes, we need to investi-

Primate Research Institute, Kyoto University, 41 Kanrin, Inuyama, Aichi 484-8506, Japan

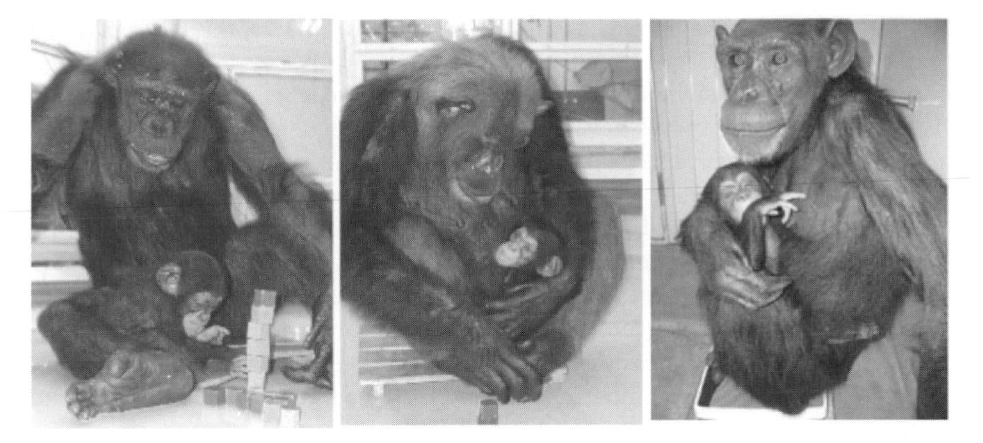

Fig. 1. Three mother–infant pairs of chimpanzees in the Primate Research Institute, Kyoto University. *Left*: Ai (mother) and Ayumu (male, born April 24, 2000); *center*: Chloe (mother) and Cleo (female, born June 19, 2000); *right*: Pan (mother) and Pal (female, born August 9, 2000). (All photographs by Primate Research Institute, Kyoto University)

gate the "natural" emergence of these abilities during the course of development. It was primarily for this purpose that the Primate Research Institute of Kyoto University (PRI) started a project of longitudinal study on chimpanzee development in 2000 (Matsuzawa 2002, 2003; Tanaka et al. 2002; Tomonaga et al. 2003, 2004). That year, three infants were born to chimpanzees at the PRI (Fig. 1), and each mother successfully held her baby, demonstrating good maternal competence (Bard 2002). Given the limitations imposed by captivity, we arranged as best we could the necessary conditions to facilitate the natural development of chimpanzees in regard to community and mother–infant bonds. Our research project ranges over various domains from physiological to cognitive aspects. In this chapter, we focus on cognitive development in the social domain on the basis of mother–infant bonds: recognition of the mother's face, mutual gaze, gaze following, and triadic interactions. These topics have recently been extensively discussed and are at the center of controversies concerning the evolutionary origin of primate cognition (Tomasello and Call 1997; Tomasello 1999). In the last part, we also discuss the relationship between early social cognition in the chimpanzees and their way of social transmission of cultural behaviors in the wild.

2 Emergence of Social Smiling: Changes at Around 2 Months of Age in Face Recognition and Neonatal Imitation

Chimpanzee infants before 2 months of age display capabilities quite similar to those of human infants. They recognize their mother's face and show clear matching facial responses to the human model of the facial expressions at around 1 month old.

For the chimpanzee infants as well as for human infants, the mother is the most familiar individual. Previous studies repeatedly revealed that human infants recognize their mother's face at around 1 month of age (Bushnell et al. 1989). We investigated the developmental changes of the recognition of their mother's face in chimpanzee infants from the first week of their life (see Chapter 9 by Myowa-Yamakoshi et al., this volume). We prepared two types of stimulus sets, the mother's face and a prototypic averaged chimpanzee face prepared by computer software based on the mother's face and those of the other members in the community of PRI. We set a photograph on the CCD camera, presented it in front of the infant's face, and moved the photograph slowly left or right repeatedly five times (the preferential tracking procedure: Bard et al. 1992; Johnson and Morton 1991). The number of tracking responses (eye movements or head turning) to each of the photographs was compared.

The results are summarized in Fig. 2A. We found three phases in the development of mother's face recognition in chimpanzees. At less than 1 month old, they showed very few tracking responses and no difference among the type of

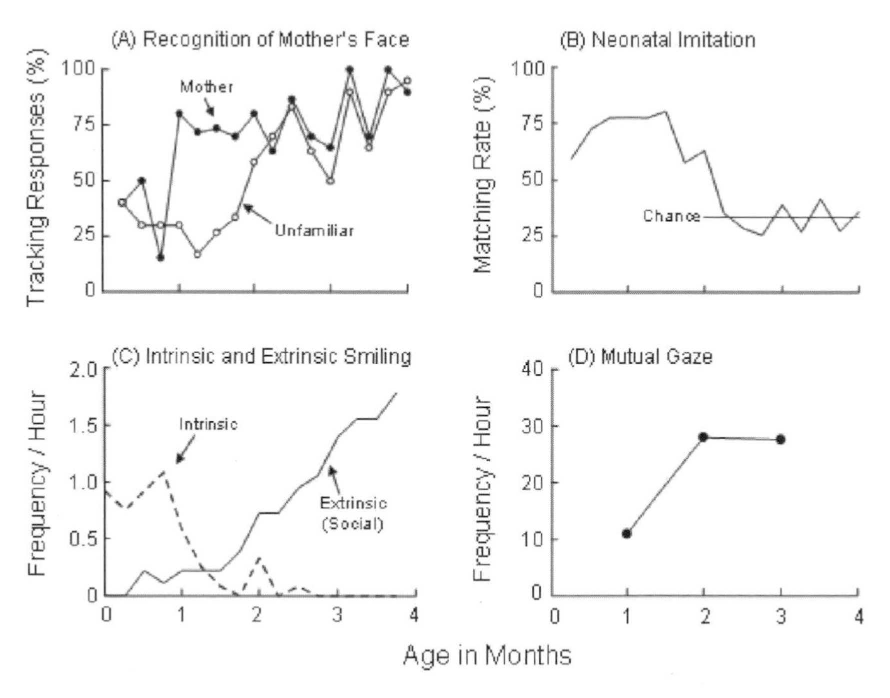

Fig. 2. Developmental changes in recognition of the mother's face and their relationship to other developmental changes. **A** Recognition of the mother's face. The data are averaged across the three infants. (Adapted from Myowa-Yamakoshi et al. 2005) **B** Neonatal imitation. The data are averaged across two infants and across three types of model gestures. *Horizontal line* at 33.3% shows chance performance. (Adapted from Myowa-Yamakoshi et al. 2004) **C** Developmental changes in smiling responses. *Vertical axis* indicates the mean number of occurrences per hour in which the facial responses were visible in the video recordings. (Adapted from Mizuno 2004) **D** Frequency of mutual gaze between mother and infants. (Adapted from Bard et al. 2005)

face. At 1 month of age, however, all the infants exhibited more responses to their mother's face than the averaged face. These results suggest that the infant chimpanzees recognized their mother's face at least at the age of 1 month and generally correspond to the results for human infants (Bushnell et al. 1989). In the later period at 2 months old, however, the infants increasingly preferred all types of faces. These preferences to all types of faces were also the case in the experiment with human faces and schematic faces (Kuwahata et al. 2003).

Myowa-Yamakoshi and her colleagues (2004; see also Chapter 14, this volume) tested the neonatal imitation of facial expressions in chimpanzees (Meltzoff and Moore 1977; Myowa 1996; Bard et al. 1992) and found that the matched facial-expression responses to the model facial expressions made by the human experimenter decreased to the chance level when the infants became 2 months old (Fig. 2B).

These developmental changes corresponded to the transition of two different types of smiling responses during this period. Mizuno et al. (in preparation; Mizuno 2004) found that the chimpanzee neonates clearly displayed spontaneous smiling responses, which occurred without external stimuli during rapid eye movement sleep, at least when the neonates were 0 day old. Furthermore, these spontaneous smiling responses decreased at 1 month of age, whereas extrinsic or social smiling responses caused by explicit stimulation of the infant, such as presenting the objects and face-to-face interactions, increased from 1 to 2 months of age (Fig. 2C).

Interestingly, chimpanzee infants older than 2 months began to display these extrinsic smiling responses to the faces during the face-recognition experiments (Fig. 3) and neonatal limitation experiments. All these results suggest that developmental changes occurred in chimpanzees from reflex-like (nonsocial) responses to social responses during the age of 1 to 2 months. The initial abilities of social cognition in chimpanzees emerge during this age period.

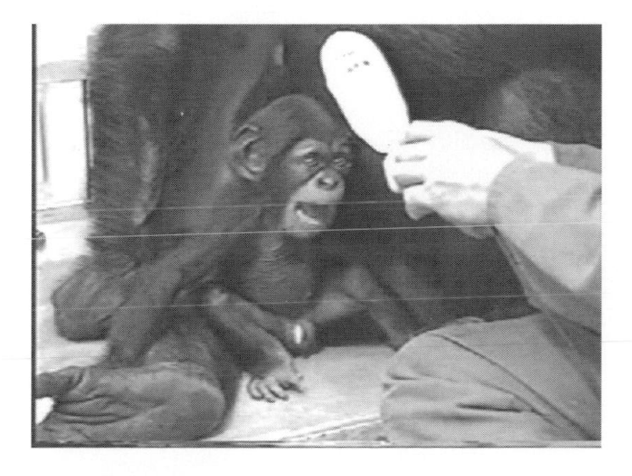

Fig. 3. Social smiling in response to a photograph of a face during the face-recognition experiment. (Photo courtesy of ANC Production)

3 Mutual Gaze: Other Signs of Developmental Changes at 2 Months

In parallel with the emergence of social reactions such as indifferent preferences to faces and social smiling in response to external stimuli, there are also developmental changes in the mutual gaze between the mother and infant (Fig. 4). Mutual gaze is defined as when both the mother and infant look at each other's face (Emery 2000). Okamoto-Barth (2005) reported that a mother chimpanzee looked at her infant's face during 6% of experimental time when the infant was in the first month of life and 24% when the infant was 4 months old, while the mother was working on an experimental computer-controlled task. The infant also increased the time spent looking at the mother's face in this situation from 3% to 12% of the experimental time. Mizuno (2004) measured the duration of mutual gaze in which both the mother and infant looked at each other's face at the same time, and found that it increased from 0 to 50 s/h from age 1 week to age 12 weeks.

We also conducted more detailed observations of development of mutual gaze in the natural setting (Bard et al. 2005). We video-recorded the behavior of mother and infant in the indoor living area from 2 to 12 weeks of infant age. The three mother–infant pairs increased the occurrences of mutual gaze from 11 times per hour when infants were 2 to 4 weeks old to 28 times per hour when infants were 10 to 12 weeks of age (Fig. 2D). This increase in mutual gaze corresponded to a decrease in cradling behavior by the mother: the frequency of mutual gaze is negatively correlated with that of physical contact between

Fig. 4. Mutual gaze between Ai (mother) and Ayumu (infant, at 1 month old). (Photo by Nancy Enslin)

mother and infant. The same tendency is also reported in human mother–infant pairs (LaVelli and Fogel 2002).

Presumably, the mother's responses were reinforced by the infant's behavioral change that occurred during 1 to 2 months of age and vice versa. In addition to these changes, it is quite plausible that the sensitivity to the other's gaze also changes during this period. To investigate this possibility, we tested the ability to discriminate gaze direction of the other's face in infant chimpanzees (Myowa-Yamakoshi et al. 2003; Fig. 5). In humans, neonates younger than 2 days old looked longer at a photograph of a face with the eyes open than the same face with the eyes closed (Batki et al. 2000). However, the majority of studies revealed that human infants discriminate eye-gaze direction when they are 3 to 4 months old (Farroni et al. 2000; Samuels 1985; Vecera and Johnson 1995). In nonhuman primates, there are very few reports on the development of discrimination of eye gaze direction. Myowa-Yamakoshi and Tomonaga (2001) reported that a nursery-raised agile gibbon infant showed preference to schematic directed-gaze face over to averted-gaze face when he was younger than 1 month old.

We tested the infants from 10 to 32 weeks of age using a forced-choice preferential looking procedure. We prepared various sets of photographs of human faces with directed and averted eye gaze and presented these directed- and averted-gaze faces to the infants for 15 s, then measured looking time for each photograph. All three infants looked longer at the directed-gaze faces than at the averted-gaze faces. These results indicate that the chimpanzee infants, at least around 2 months of age, clearly discriminate eye-gaze directions and, furthermore, prefer directed-gaze faces to averted-gaze faces. Mutual gaze in the mother–infant chimpanzees is established on the basis of the infant's preference for the directed-gaze face and is maintained by the mother's reaction toward the infant.

Bidirectional mother–infant interactions on the basis of mutual gaze may facilitate "primary intersubjectivity" (Trevarthen and Aitken 2001), defined as a dyadic social relationship maintained by mutual gaze between mother and infant chimpanzees. Emergence of social smiling and mutual gaze is the sign of

Fig. 5. Setting for the gaze-recognition experiment. The human experimenter presented a pair of photographs of direct- and averted-gaze faces to the infant. Looking behavior of the infant was recorded by the small CCD camera mounted at the center of the photographs. (Photo by Primate Research Institute, Kyoto University)

remarkable changes in early social-cognitive development also in humans; this change is frequently referred to as the "2-month revolution" (Rochat 2001). Our findings with chimpanzee infants clearly indicate that this 2-month revolution in social-cognitive development is shared by both humans and chimpanzees.

4 Formation of Attachment to the Mother

Chimpanzee infants as young as 1 month old initially discriminate the mother's face from others, and their social-cognitive abilities emerge from 1 to 2 months of age, as was evident in a decrease of reflex-like responsiveness. Based on these changes, they recognize another's eye-gaze direction, pay attention to directed-gaze faces, and engage in dyadic social interactions with the mother via mutual gaze. At 3 to 5 months of age, chimpanzees begin to interact with objects in a very simple manner, and show more complex, combinatorial manipulations at 8 to 9 months (Hayashi and Matsuzawa 2003). Infants also began to walk around the mother to explore the environment surrounding them; they walked away from their mother and then came back to the mother (Fig. 6A). Okamoto-Barth (2005) and Nakashima (2003) clearly showed that the rate of physical contact between chimpanzee mothers and infants decreased drastically at around 4

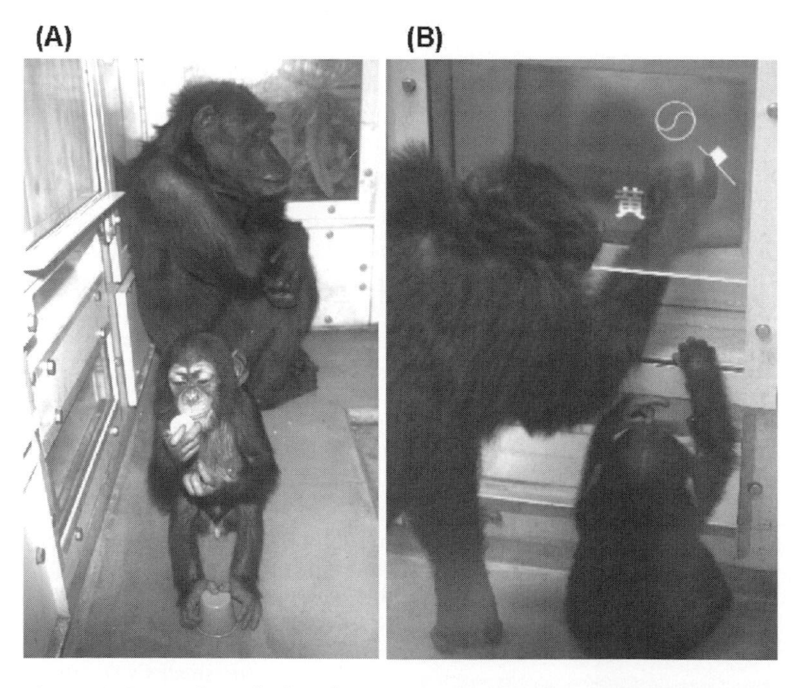

Fig. 6. **A** Ayumu is exploring the experimental booth away from his mother, Ai. **B** Ayumu is looking at his mother performing a computer-controlled cognitive task. (Photo courtesy of the *Mainichi* Newspaper, Japan)

months of age. The infants began to pay attention to physical objects around them and tried to manipulate them. Furthermore, they also paid close attention to the mother's behavior. This type of behavior characterized by the repeated explorations of the environment indicates that the mother becomes a secure base for the infants, and implies that the mother–infant attachment was established at around 4 to 6 months of age in chimpanzees (Ainsworth et al. 1978; Doran 1992). Sousa et al. (2003) reported that one infant chimpanzee, Ayumu, began to try to perform the computer-controlled task setup for his mother when he was 9 months old (Fig. 6B). This finding also suggests that the chimpanzee infants became more interested in the environment and the mother's behavior.

Human infants at 6 months also interact dyadically with objects or with a person in a turn-taking (or reciprocally exchanging) sequence. However, they do not interact with the person who is manipulating the objects (Tomasello 1999). At this age, the dyadic format of social interaction is prototypical. However, both human and chimpanzee infants then confronted the two mutually exclusive demands: exploration of the objects in the environment and physical contact with their mothers. Human infants come to solve this problem in a very sophisticated way; that is, they involve their mothers in a context of object manipulations, so-called triadic interactions. During triadic interactions in humans, gaze plays an important role again, as in the 2-month revolution, but in a different form: gaze following.

5 Gaze Following: Basis for Shared Attention?

The step toward the emergence of triadic interactions for the infants is to follow the gaze of others to direct their own attention, that is, gaze following. Here, gaze following refers to when an individual detects that another's gaze is not directed toward him and follows the line of sight of that other individual to a point or an object in space (cf. Emery 2000). Human infants at around 6 months old begin to follow the gaze direction of others, and this ability becomes more sophisticated during the course of development (Butterworth and Jarrett 1991; Moore and Dunham 1995). The ability to follow another's gaze has been intensively examined in various nonhuman primates from prosimians to great apes (see Emery 2000 for review), but there are few studies on gaze following from the comparative-developmental perspective (Ferrari et al. 2000; Tomasello et al. 2001). Okamoto et al. (2002; see also Chapter 10, this volume) tested the ability of an infant chimpanzee to follow a human experimenter's social cues, including gaze, longitudinally from 7 to 13 months of age.

In their experiment, the human experimenter positioned outside the experimental booth gave various types of cues to the infant, who was in the booth (Fig. 7A). The cues were directed to one of two identical objects, and consisted of tapping, pointing to it, head turning toward it, and only eyes directed to it. Three seconds after the presentation of the social cue, the experimenter delivered a

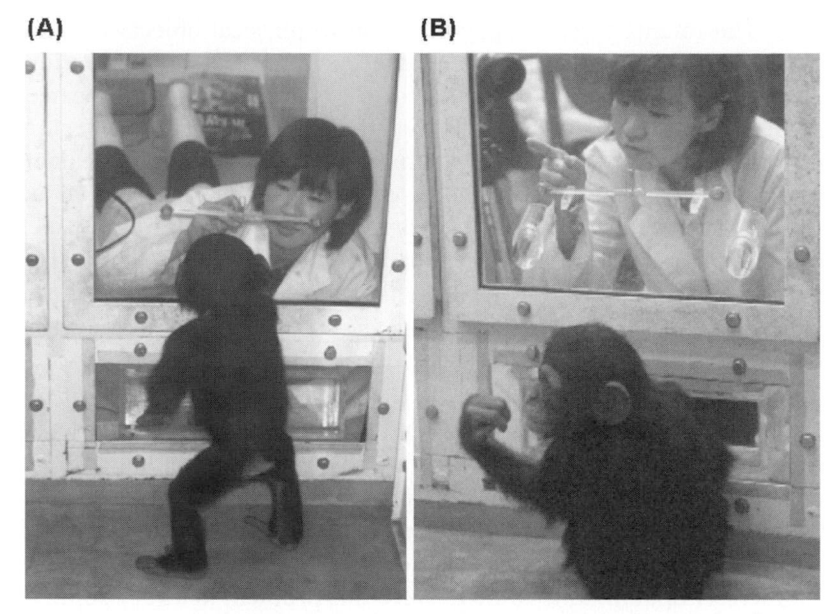

Fig. 7. **A** Ayumu at 1 year old is following the experimenter's social cue. **B** Ayumu at 2 years old is looking back following the experimenter's pointing. (Photo courtesy of the *Mainichi* Newspaper, Japan)

food reward to that side, irrespective of the infant's responses. We defined a following response as the subject's looking or approaching the side to which the experimenter attended before the delivery of food. The infant reliably followed the pointing cue before the age of 9 months, the head-turn cue by the age of 10 months, and the eye-gaze cue (without head movements) by 13 months.

These experiments clearly showed that an infant chimpanzee did follow social cues, including gaze at around 9 months of age. Although our experimental design was nondifferential reinforcement testing to avoid learning by differential reinforcement, the infant in fact might have "learned" to follow human gaze through the outcomes given by the experimenter. Nevertheless, his performance was constrained by the type of social cues, especially in the latter phase of the experiment: pointing was easiest and eye gaze was the most difficult. This constraint may result from the nature of social cues such as saliency to some degree, but we cannot rule out the possibility of developmental constraints. Okamoto et al. (2004; see also Chapter 10, this volume) report on the later changes in the same subject's gaze-following ability. The subject at 13 to 18 months old could not look back by following the human pointing to an object behind the subject, but did successfully when he was older than 20 months (Fig. 7B). Taken together with these results, gaze-following ability in chimpanzee infants seems to develop gradually and in a step-by-step manner, but in some aspects is delayed in comparison with humans.

6 Shared Attention and Triadic Interactions: Origin of Different Developmental Paths Between Humans and Chimpanzees

Humans exhibit a substantial change concerning social communications at around 9 months old (Carpenter et al. 1998; Ohgami 2002). As mentioned earlier, human infants at 6 months interact dyadically with objects or with a person. These dyadic interactions become triadic from around 9 months on (probably up to 12 months). Their interactions involve both objects and the person, resulting in the formation of a referential triangle of infant, adult, and the object upon which they share attention (Tomasello 1999). The propensity of the infant to look up toward the adult and then back to the object demonstrates that the infant is checking the joint visual attention of the other person (Rochat 2001). This behavior is called shared attention (Emery 2000). Shared attention is different from gaze following and emphasizes the role of communicative interactions via gaze (cf. Emery 2000). This is a decisive, critical development occurring at around 9 months of age. Some researchers refer to this change as "the 9-month revolution" (Tomasello 1999). The 9-month revolution appears on the basis of the primitive but necessary ability of gaze following and understanding the intention or goal-directedness of others and then becomes the basis for understanding the other's mental state (Baron-Cohen 1995; Tomasello 1999).

The 9-month revolution, however, does not seem to occur in chimpanzees, although the conclusion is still not decisive (Tomasello and Call 1997). Our studies provide, at present, no affirmative results. In an opportunistic observation, we tried to engage in triadic exchange with the chimpanzee infants using various kinds of objects, but they did not interact with humans in a reciprocating manner. When the human experimenter played with the infant chimpanzee at the age of 1 to 2 years using a towel, the infant displayed both social and solitary play with it, but did not engage in reciprocal exchange with us. In another case, we tried to reciprocate with the infant using a ball, but she "stole" the ball and started solitary play with it. She did give it back to the experimenter but only when the ball was exchanged for food (Tomonaga and Hayashi 2003). The chimpanzee infants never displayed "object showing" or "object giving," indicative of referential communication in a triadic relationship in human infants, as was found in an 18-month-old nursery-raised chimpanzee by Russell et al. (1997). Okamoto et al. (2004) also report that the infant did not look at the experimenter's face again having followed the human's pointing and looked back, which is one of the common behaviors of shared attention in human infants (Carpenter et al. 1998).

In addition to these naturalistic observations, we also conducted more controlled observations (Kosugi et al. 2003). We presented a novel animate-like object (a remote-controlled toy) to the mother–infant pairs when the infants were 1 and 2 years old and observed the mother–infant interactions. Initially,

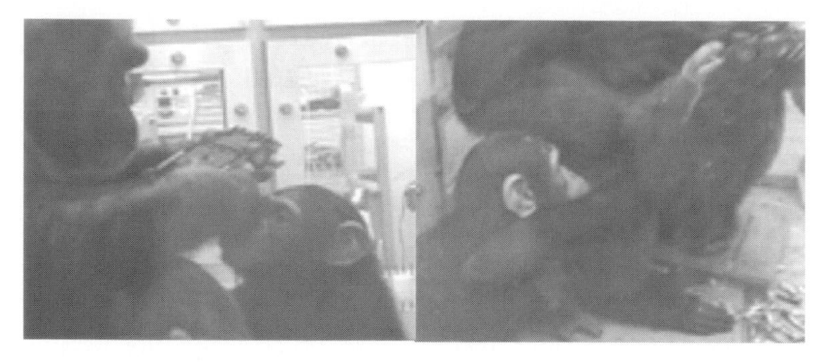

Fig. 8. Joint engagement in a triadic situation. Cleo (infant, 1 year old) manipulated the novel object (model car) held and manipulated by the mother. (Photo courtesy of Daisuke Kosugi)

the infants showed some fearful responses toward this novel object, such as withdrawing from the object and hiding themselves behind their mothers. When the infants manipulated the object, they always kept their unoccupied hand on the mother's body. After watching the mother manipulate the object, they often tried to touch it, and the mother never refused this kind of approach for searching. Such triadic interaction, or shared (joint) engagement, was frequently observed both when the infants were 1 and 2 years old (Fig. 8). However, when the infant manipulated the object, she seldom looked back to her mother, showed the object to her, or gave it to her. Similarly, the mothers did not display such showing or giving behaviors. These results suggest that the mother–infant interactions with an object were not based on shared attention, which may imply that the chimpanzee mother–infant pairs interacted without referential triadic relationships. However, there might be precursors for triadic interactions in chimpanzees. As described earlier, the infant chimpanzees showed fearful responses toward novel animate-like objects at first, and they did not manipulate them by themselves. Only after they had seen the mother manipulate the object or had participated in shared engagement did they actively try to manipulate it by themselves. This behavior can be interpreted as one type of "social referencing" (Feinman 1982; Sorce et al. 1985); that is, the infant obtained some information concerning the ambiguous object through watching and joining in on the manipulation of it by the mother as may be the case in the acquisition of novel food repertoire in chimpanzee infants. Ueno and Matsuzawa (2005) investigated the reactions to novel food items in a controlled setting. When the mother and infants (at around 3 years old) came into the experimental room where the novel food items were scattered in patches, the infants initially paid attention to the mothers' behavior (70% of episodes). Such kinds of referencing behaviors were never observed when the foods were familiar. It is still controversial whether the infant's reactions were homologous to the social referencing in humans, but it may be plausible that there are some ecological constraints on the emergence of triadic-like, shared attention based interactions in chimpanzees.

7 Lack of Triadic Interactions and "Education by Master-Apprenticeship"

Chimpanzees appear to lack complex triadic interactions, which are commonly observed in humans. This species-specific property of social interactions presumably affects the processes of acquisition of cultural or community-specific repertoire of behaviors in wild chimpanzees. In communities of wild chimpanzees, it is well known that they use tools and that there exist "cultural differences" in repertoires of tool-using behaviors among communities (Whiten et al. 1999). Many researchers agree that the chimpanzee learns tool-using behavior through the observation of the other's behavior, although the exact processes are still controversial. Matsuzawa et al. (2001) summarized the social learning process of chimpanzees using the term education by master-apprenticeship (cf. de Waal 2001). For a long period, the infants observe the adult's tool-using behaviors (especially those of their mother) closely and intensively and try those behaviors by themselves (cf. Hirata and Morimura 1999; Tonooka et al. 1997). Adults are relatively tolerant to being observed or to interrupted by the infants but do not actively teach them. Apparently, it is very rare that the mother and infant interact triadically in the tool use context.

This situation was also the case in our research project. To simulate tool-using behaviors in the wild, longitudinal studies were conducted on the acquisition of various cognitive skills in infant chimpanzees, such as tool-using behavior (Hirata and Celli 2003; see also Chapter 13, this volume) and computer-controlled tasks (Sousa et al. 2003). Both experiments reported that the infants watched the adults' behaviors very intensively and tried these target behaviors by themselves (Fig. 9). These responses by infants were in some part based on local/stimulus enhancement processes (cf. Inoue-Nakamura et al. 2003). However, the most important point is that the chimpanzee mother did not actively interact with the infant in these settings. That is, there is a lack of active teaching. There were very rare (or no) triadic interactions between the mother, infant, and tools. The mothers do not show the model, or guide, mold, punish, and praise the infant's behavior. The infants pay attention to the mother's behavior but it is seemingly not on the basis of shared attention. The ability of social (observational) learning in chimpanzees has been discussed on the basis of their ability to imitate the other's actions. However, the differences in social learning processes between humans and chimpanzees are not simply a result of these imitative abilities but of their social-communicative abilities.

8 Conclusion

In this chapter, we summarized part of an ongoing research project of cognitive development in infant chimpanzees with special reference to their social-cognitive abilities. Mother-raised chimpanzee infants seemingly lack the

Fig. 9. Social transmission of tool-using skills from the mother to the infant at 1.5 years of age. (Photo courtesy of Satoshi Hirata)

human-like ability for triadic social interactions. Early social cognitive development, such as face recognition and gaze recognition, however, is similar to that of humans. These abilities are the basis for dyadic social interactions. In the latter part of development in infancy, mother-raised chimpanzees diverge from the path taken by Western humans. As some researchers noted, triadic interactions may be required for more advanced cognitive abilities, such as for self-recognition and understanding of the other's mental state (Tomasello 1999; Tomasello and Call 1997). Based on Baron-Cohen's (1995) "mindreading system" model, detection of intentionality and eye direction is a prerequisite for shared attention, which is the base for theory of mind. The apparent inability of chimpanzees to understand the other's mental state (Call and Tomasello 1999; cf. Hirata and Matsuzawa 2001) therefore may be caused by the lack of shared attention.

Acknowledgments

This chapter is based on papers by the author (Tomonaga et al. 2004, in press a, b), with partial fulfillments. The research reported here and the preparation of the manuscript were financially supported by Grants-in-Aid for Scientific Research from the Japan Society for the Promotion of Science (JSPS), and the Ministry of Education, Culture, Sports, Science and Technology (MEXT) (nos. 07102010, 09207105, 10CE2005, 11710035, 12002009, 13610086, 14000773, 16002001, and 16300084), the MEXT Grant-in-Aid for the 21st Century COE Programs (A2 and D2 to Kyoto University), and the Cooperative Research Program of the Primate Research Institute, Kyoto University. I thank coauthors of the studies summarized in this chapter—M. Myowa-Yamakoshi, K.A. Bard, S. Okamoto-Barth, M. Tanaka, M.K. Yamaguchi, Y. Mizuno, and T. Matsuzawa—for their enthusiastic collaboration. Thanks are also due to O. Takenaka, G. Hatano, K. Fujita, S. Itakura, N. Kawai, C. Douke, M. Hayashi, T. Imura, C. Murai, N. Nakashima, T. Ochiai, R. Oeda, A. Ueno, M. Uozumi, T. Takashima, S. Hori, E. Nogami, Y. Fukiura, K. Kumazaki, N. Maeda, A. Kato, J. Suzuki, S. Goto, S. Watanabe, T. Kageyama, and K. Matsubayashi for their help throughout this research project.

References

Ainsworth M, Blehar M, Waters I, Wall S (1978) Patterns of attachment: a psychological study of the strange situation. Erlbaum, Hillsdale, NJ

Bard KA (2002) Primate parenting. In: Bornstein M (ed) Handbook of parenting, 2nd edn, vol 2. Biology and ecology of parenting. Erlbaum, Mahwah, NJ, pp 99–140

Bard KA, Myowa-Yamokoshi M, Tomonaga M, Tanaka M, Quinn J, Costal A, Matsuzawa T (2005) Cultural variation in the mutual gaze of chimpanzees (*Pan troglodytes*). Dev Psych 41:616–624

Bard KA, Platzman KA, Lester BM, Suomi SJ (1992) Orientation to social and nonsocial stimuli in neonatal chimpanzees and humans. Infant Behav Dev 15:43–56

Baron-Cohen S (1995) Mindblindness. MIT Press, Cambridge

Batki A, Baron-Cohen S, Wheelwright S, Connellan J, Ahluwalia J (2000) Is there an innate module? Evidence from human neonates. Infant Behav Dev 23:223–229

Bushnell IW, Sai F, Mullin JT (1989) Neonatal recognition of the mother's face. Br J Dev Psychol 7:3–15

Butterworth GE, Jarrett NLM (1991) What minds have in common is space: spatial mechanism serving joint visual attention in infancy. Br J Dev Psychol 9:55–72

Byrne RW, Whiten A (eds) (1988) Machiavellian intelligence: Social expertise and the evolution of intellect in monkeys, apes, and humans. Oxford University Press, NY

Call J, Tomasello M (1999) A nonverbal false belief task: the performance of chimpanzees and human children. Child Dev 70:381–395

Carpenter M, Nagell K, Tomasello M (1998) Social cognition, joint attention, and communicative competence from 9 to 15 months of age. Monogr Soc Res Child Dev 63(4):1–143

de Waal F (2001) The ape and the sushi master: cultural reflections of a primatologist. Basic Books, New York

Doran DM (1992) The ontogeny of chimpanzee and pygmy chimpanzee locomotor behavior: a case study of paedomorphism and its behavioral correlates. J Hum Evol 23:139–157

Emery NJ (2000) The eyes have it: the neuroethology, function and evolution of social gaze. Neurosci Biobehav Rev 24:581–604

Farroni T, Johnson MH, Brockbank M, Simon F (2000) Infants' use of gaze direction to cue attention: the importance of perceived motion. Vis Cogn 7:705–718

Feinman S (1982) Social referencing in infancy. Merrill Palmer Q 28:445–470

Ferrari PF, Kohler E, Fogassi L, Gallese V (2000) The ability to follow eye gaze and its emergence during development in macaque monkeys. Proc Natl Acad Sci U S A 97:13997–14002

Gardner RA, Gardner BT (1969) Teaching sign language to a chimpanzee. Science 165:664–672

Hayashi M, Matsuzawa T (2003) Cognitive development in object manipulation by infant chimpanzees. Anim Cogn 6:225–233

Hayes C (1951) The ape in our house. Harper, NY

Hirata S, Celli ML (2003) Role of mothers in the acquisition of tool use behaviour by captive infant chimpanzees. Anim Cogn 6:235–244

Hirata S, Matsuzawa T (2001) Tactics to obtain a hidden food item in chimpanzee pairs (*Pan troglodytes*). Anim Cogn 4:285–295

Hirata S, Morimura N (2000) Naive chimpanzees' (*Pan troglodytes*) observation of experienced conspecifics in a tool-using task. J Comp Psych 114:291–296

Inoue-Nakamura N, Myowa-Yamakoshi M, Hayashi M, Matsuzawa T (2003) Effect of stimulus enhancement on preference to objects in mother and infant chimpanzees. In: Tomonaga M, Tanaka M, Matsuzawa T (eds) Cognitive and behavioral development in chimpanzees: a comparative approach (in Japanese). Kyoto University Press, Kyoto, pp 254–257

Johnson MH, Morton J (1991) Biology and cognitive development: the case of face recognition. Blackwell, Oxford

Kellogg WN, Kellogg LA (1933) The ape and the child. McGraw-Hill, NY

Kosugi D, Murai C, Tomonaga M, Tanaka M, Ishida H, Itakura S (2003) Relationship between the understanding of causality in object motion and social referencing in chimpanzee mother-infant pairs: comparisons with humans. In: Tomonaga M, Tanaka M, Matsuzawa T (eds) Cognitive and behavioral development in chimpanzees: a comparative approach (in Japanese). Kyoto University Press, Kyoto, pp 232–242

Kuwahata H, Fujita K, Ishikawa S, Myowa-Yamakoshi M, Tomonaga M, Tanaka M, Matsuzawa T (2004) Development of schematic face preference in chimpanzee infants. In: Tomonaga M, Tanaka M, Matsuzawa T (eds) Cognitive and behavioral development in chimpanzees: a comparative approach (in Japanese). Kyoto University Press, Kyoto, pp 89–93

LaVelli M, Fogel A (2002) Developmental changes in mother-infant face-to-face communication. Dev Psychol 38:288–305

Matsuzawa T (2002) Chimpanzee Ai and her son Ayumu: An episode of education by master-apprenticeship. In: Bekoff M, Allen C, Gordon GM (eds) The cognitive animal. MIT Press, Cambridge, MA, pp 190–195

Matsuzawa T (2003) The Ai project: historical and ecological contexts. Anim Cogn 6:199–211

Matsuzawa T, Biro D, Humle T, Inoue-Nakamura N, Tonooka R, Yamakoshi G (2001) Emergence of culture in wild chimpanzees: education by master-apprenticeship. In: Matsuzawa T (ed) Primate origins of human cognition and behavior. Springer, Tokyo, pp 557–574

Meltzoff AN, Moore MK (1977) Imitation of facial and manual gestures by human neonates. Science 198:75–78

Mitchell P (1997) Introduction to theory of mind: Children, autism and apes. Edward Arnold Publishers, London, England

Mizuno Y (2004) Mother-infant communications in chimpanzee in the first four months (in Japanese). Doctoral thesis, University of Shiga Prefecture, Hikone, Japan

Moore C, Dunham PJ (eds) (1995) Joint attention: its origins and role in development. Erlbaum, Hillsdale, NJ

Myowa M (1996) Imitation of facial gestures by an infant chimpanzee. Primates 37:207–213

Myowa-Yamakoshi M, Tomonaga M (2001) Perceiving eye gaze in an infant gibbon (*Hylobates agilis*). Psychologia 44:24–30

Myowa-Yamakoshi M, Tomonaga M, Tanaka M, Matsuzawa T (2003) Preference for human direct gaze in infant chimpanzees (*Pan troglodytes*). Cognition 89:B53–B64

Myowa-Yamakoshi M, Tomonaga M, Tanaka M, Matsuzawa T (2004) Imitation in neonatal chimpanzees (*Pan troglodytes*). Dev Sci 7:437–442

Myowa-Yamakoshi M, Yamaguchi M, Tomonaga M, Tanaka M, Matsuzawa T (2005) Development of face recognition in infant chimpanzees (*Pan troglodytes*). Cogn Dev 20:49–63

Nakashima N (2003) Developmental changes of the responses to the vocalizations in infant chimpanzees (in Japanese). Master's thesis, Primate Research Institute, Kyoto University, Inuyama, Japan

Ohgami H (2002) The developmental origins of early joint attention behaviors (in Japanese). Kyushu Univ Psychol Res 3:29–39

Okamoto S, Tomonaga M, Ishii K, Kawai N, Tanaka M, Matsuzawa T (2002) An infant chimpanzee (*Pan troglodytes*) follows human gaze. Anim Cogn 5:107–114

Okamoto S, Tanaka M, Tomonaga M (2004) Looking back: the "representational mechanism" of joint attention in an infant chimpanzee (*Pan troglodytes*). Jpn Psychol Res 46:236–245

Okamoto-Barth S (2005) Gaze processing in chimpanzees and humans. Doctoral thesis, University of Maastricht, Maastricht, Netherlands

Premack D, Woodruff G (1978) Does the chimpanzee have a theory of mind? Behav Brain Sci 1:515–526

Rochat P (2001) The infant's world. Harvard University Press, Cambridge

Russell CL, Bard KA, Adamson LB (1997) Social referencing by young chimpanzees (*Pan troglodytes*). J Comp Psychol 111:185–193

Samuels CA (1985) Attention to eye contact opportunity and facial motion by three-month-old infants. J Exp Child Psychol 40:105–114

Sorce JF, Emde RN, Campos JJ, Klinnert MD (1985) Maternal emotional signaling: Its effect on the visual cliff behavior of 1-year-olds. Dev Psych 21:195–200

Sousa C, Okamoto S, Matsuzawa T (2003). Behavioral development in a matching-to-sample task and token use by an infant chimpanzee reared by his mother. Anim Cogn 6:259–267

Tanaka M, Tomonaga M, Matsuzawa T (2002) A developmental research project with three mother-infant chimpanzee pairs: A new approach to comparative developmental science (in Japanese). Jpn Psych Rev 45:296–308
Tomasello M (1999) The cultural origins of human cognition. Harvard University Press, London
Tomasello M, Call J (1997) Primate cognition. Oxford University Press, New York
Tomasello M, Hare B, Fogleman T (2001) The ontogeny of gaze following in chimpanzees, *Pan troglodytes*, and rhesus macaques, *Macaca mulatta*. Anim Behav 61:335–343
Tomonaga M, Hayashi M (2003) Object exchange between infant chimpanzees and humans. In: Tomonaga M, Tanaka M, Matsuzawa T (eds) Cognitive and behavioral development in chimpanzees: a comparative approach (in Japanese). Kyoto University Press, Kyoto, pp 153–157
Tomonaga M, Tanaka M, Matsuzawa T (eds) (2003) Cognitive and behavioral development in chimpanzees: a comparative approach (in Japanese). Kyoto University Press, Kyoto
Tomonaga M, Myowa-Yamakoshi M, Mizuno Y, Yamaguchi MK, Kosugi D, Bard KA, Tanaka M, Matsuzawa T (2004) Development of social cognition in infant chimpanzees (*Pan troglodytes*): face recognition, smiling, gaze and the lack of triadic interactions. Jpn Psychol Res 46:227–235
Tomonaga M, Myowa-Yamakoshi M, Mizuno Y, Okamoto S, Yamaguchi MK, Kosugi D, Bard KA, Tanaka M, Matsuzawa T (in press a) Chimpanzee social cognition in early life: comparative–developmental perspective. In: Wasserman EA, Zentall TR (eds) Comparative cognition: experimental explorations of animal intelligence. Oxford University Press, New York
Tomonaga M, Myowa-Yamakoshi M, Okamoto S, Bard KA (in press b) Development of gaze recognition in chimpanzees (*Pan troglodytes*). In: Fujita K, Itakura S (eds) Diversity of cognition: evolution, development, domestication, and pathology. Kyoto University Press, Kyoto
Tonooka R, Tomonaga M, Matsuzawa T (1997) Acquisition and transmission of tool use and making for drinking juice in a group of captive chimpanzees (*Pan troglodytes*). Jpn Psych Res 39:253–265
Trevarthen C, Aitken KJ (2001) Infant intersubjectivity: research, theory, and clinical applications. J Child Psychol Psychiatry 42:3–48
Ueno A, Matsuzawa T (2005) Response to novel food in infant chimpanzees: do infants refer to mothers before ingesting food on their own? Behav Process 68:85–90
Vecera SP, Johnson MH (1995) Gaze detection and the cortical processing of faces: evidence from infants and adults. Vis Cogn 2:59–87
Wellman HM (1992) The child's theory of mind. MIT Press, Cambridge, MA
Whiten A, Byrne RW (eds) (1997) Machiavellian intelligence II: Extensions and evaluations. Cambridge University Press, NY
Whiten A, Goodall J, McGrew WC, Nishida T, Reynolds V, Sugiyama Y, Tutin CEG, Wrangham RW, Boesch C (1999) Cultures in chimpanzees. Nature 399:682–685
Wimmer H, Perner, J (1983) Beliefs about beliefs: Representation and constraining function of wrong beliefs in young children's understanding of deception. Cognition 13:103–128

Part 4
Social Cognition: Imitation and Understanding Others

13
Chimpanzee Learning and Transmission of Tool Use to Fish for Honey

Satoshi Hirata

1 Culture in Nonhuman Primates

Since the early 1950s, Japanese primatologists have conducted research to examine culture in nonhuman animals (Hirata et al. 2001). Imanishi (1952) proposed that an important aspect of culture is learning from group members. In other words, if a species forms a group and this species has the ability to learn something from other group members, then this species may have created a culture. Another important aspect of culture is the transmission of behavior from one generation to the next, implying that a species with culture must live in a perpetual group. Some insects live in groups, but each group disappears after a certain period, meaning that behaviors cannot be transmitted to the next generation in any media other than genes. The same is true for any random members of the same organism that form a group for a random period. An intensive study begun in 1948 by Japanese primatologists revealed that Japanese monkeys live in perpetual groups. The scientists adopted a method unique at the time, which named individual monkeys and revealed, for example, dominance relationships and social structure. In Japanese monkey society, females remain in a natal group and males leave to join other groups; that is, each group is maintained by maternal lineage. After examining the results of this early study, Imanishi (1952) suggested that Japanese monkeys may have their own culture. Imanishi's suggestion spurred research conducted by colleagues and students, who studied cultural phenomena in nonhuman primates, as exemplified by the study of monkeys living in Koshima Islet that wash sweet potatoes (Kawamura 1954; Kawai 1965; Hirata et al. 2001).

Half a century has passed since then, and researchers have accumulated knowledge about the behaviors of several primate species. They agree that among all the primate species, chimpanzees have an immense behavioral repertoire and that each community of chimpanzees has a different behavioral repertoire which cannot be explained by ecological differences; that is, chimpanzees have their own cultures (McGrew 1992; Whiten et al. 1999; Matsuzawa 2003).

Great Ape Research Institute, Hayashibara Biochemical Laboratories, 952-2 Nu, Tamano, Okayama 706-0316, Japan

Long-term studies of nut cracking revealed a form of tool use among wild chimpanzees in parts of West Africa (Sugiyama and Koman 1979; Boesch et al. 1994); Matsuzawa et al. (2001) applied the term "master-apprenticeship" to characterize the chimpanzee cultural process. According to this description, a chimpanzee "master" skilled in a certain type of tool use does not actively teach the chimpanzee "apprentice," which is naive in the use of this tool, but through long-term repetitive observation of the master (which is supported by high levels of tolerance on the master's part, including allowing access to tools and edible materials obtained by tool use), acquires the skill.

To investigate in more detail how chimpanzees learn culture, Hirata and colleagues conducted two related studies on how a captive group of chimpanzees learned tool use (Hirata and Morimura 2000; Hirata and Celli 2003; see also Celli et al. 2004). They introduced the task of honey fishing, which simulated ant/termite fishing found in the wild (Goodall 1968; Nishida 1973; Paquette 1982). The first study involved adult chimpanzees, and the second involved mother–infant pairs. This chapter focuses on both these studies.

2 Timing of Observation and Use of Leftover Tools

To clarify cognitive mechanisms such as imitation, emulation, stimulus enhancement, and social facilitation (all of which underlie social transmission of behaviors), many researchers have studied animal behavior under experimental conditions in which the animals observed a model (Nagell et al. 1993; Whiten et al. 1996; Myowa-Yamakoshi and Matsuzawa 1999). In such experimental situations, human experimenters obliged an animal subject to observe actions performed by a model; in a natural wild setting, however, animals are never obliged to observe a model. If animals were permitted to behave freely, would they really observe a model, and when would they be motivated to observe a model? Hirata and Morimura (2000) performed an experiment designed to explore these questions.

As mentioned previously, the task was honey fishing. Researchers made a 5-mm-diameter hole in the transparent wall of an experimental booth and attached a bottle containing honey to the wall from the outside. Chimpanzees remained inside and had the opportunity to insert a slender and flexible object into the hole to dip honey. Twenty objects were scattered on the floor to make them available to the chimpanzees; these included a rubber tube, knobbly plastic string, wire, cotton string, chain, stick, spoon, and brush. Not all of these items could be used as a honey-fishing tool; 8 were larger than the 5-mm honey hole. There were 12 usable tools that could be inserted into the hole.

The study applied two conditions: the first was single-subject condition, in which three chimpanzees were tested individually; the second was a pair condition in which a chimpanzee naive to this task was paired with an experienced chimpanzee already having the skill (Fig. 1). Six naive–experienced chimpanzee pairs were tested using two honey-fishing sites.

Fig. 1. A naive chimpanzee observing an experienced partner

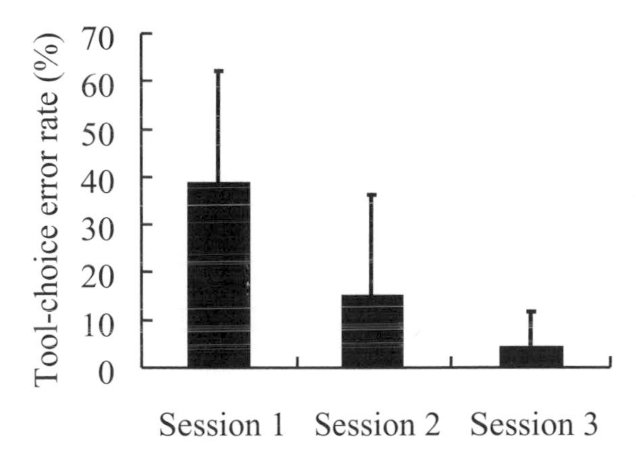

Fig. 2. Mean tool-choice error rate (+SD) of the six successful chimpanzees

Overall, adult chimpanzees learned by trial and error. Two of three chimpanzees tested during the single-subject condition mastered honey fishing and four of six chimpanzees tested along with experienced partners mastered the task. In their early attempts, they unsuccessfully used inappropriate objects such as a spoon or brush; Figure 2 shows the tool choice error rate, which represents the ratio of failed attempts with unusable tools. The six successful chimpanzees exhibited a similar gradual reduction in the rates of attempts with unusable tools. Thus, repeated experience taught them which objects were appropriate and efficient. Figure 3 shows the number of tool types used in a session; there was also a decrease in the number of types of tools used. Among the 20 kinds of objects, 2 were the most efficient: the rubber tube and knobbly plastic string. By the end of testing, the adult chimpanzees used the rubber tube and/or plastic string during most attempts.

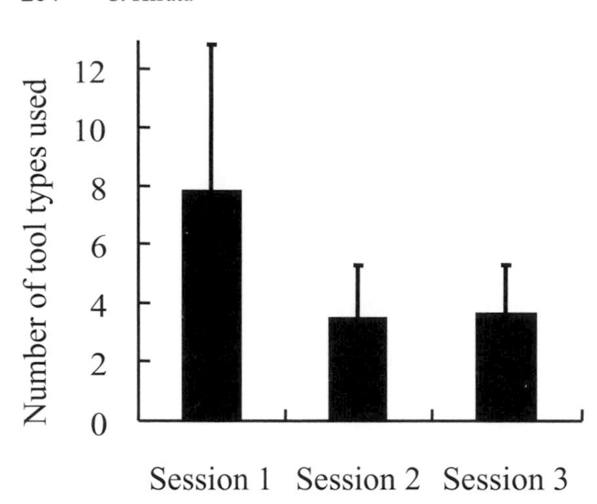

Fig. 3. Number of tool types used (+SD) over three sessions

When comparing tool use acquisition, there were no clear differences between chimpanzees that were tested alone or with experienced partners in terms of latency to first success or success rates over sessions. Given that an individual in the single-subject condition mastered this task quite quickly (within 10 min), it may be that adult chimpanzees found the task too simple for social learning to be a measurable advantage. For unsuccessful individuals, hand dexterity when using tools seemed to be more difficult than comprehending the task.

The main focus of the study involved the pair-condition learning process, in which subjects observed their partners. Most observation took place during the first session; for each pair, there tended to be a greater number of partner observations during the first session than in the second and third sessions. Of a total of 40 observation episodes, 34 involved a naive chimpanzee observing an experienced partner; 23 of these cases occurred during the first session. There were 7 cases during the second session involving two subjects; during the third session, there were 4 cases, all involving one subject.

To determine the motivation for chimpanzees to observe their partners, Hirata and Morimura (2000) examined events that preceded a chimpanzee's partner observation (Fig. 4). These preceding events were categorized according three patterns: immediate, failure, and success. Immediate refers to a naive chimpanzee that observed a partner before any attempt at tool use, failure refers to a chimpanzee that observed a partner after a personal failure, and success refers to a chimpanzee that observed a partner after a personal success.

There were 2 cases of naive chimpanzees observing experienced partners before a first attempt, 32 cases after a failure, and no cases after a success. There were 6 cases of experienced chimpanzees observing naive partners, all of which occurred after their own success; it is worth noting that there was no statistically significant difference between the number of observations after failure or success.

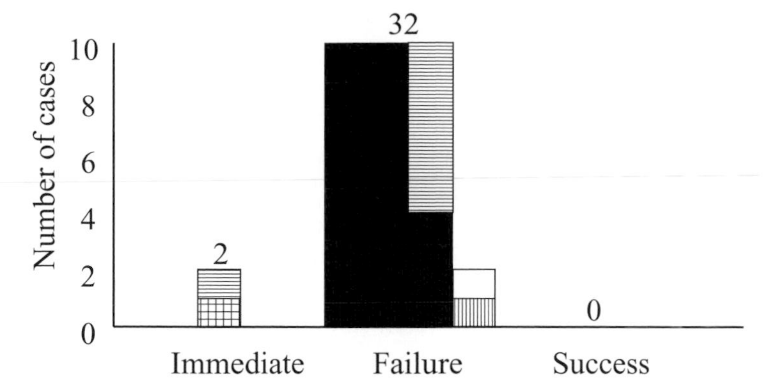

Fig. 4. Events before the naive chimpanzees' observations of the experienced partners. Different patterns indicate different individuals

Hirata and Morimura (2000) were able to determine another possible factor in the transmission of tool use: use of a partner's leftover tools. Four naive chimpanzees used their partners' leftover tools on ten occasions (two to four times per individual). Seven of the ten cases occurred during the first session; one of these was a case of "robbing," the active taking of a partner's tool. Nine cases occurred after an experienced individual exited the honey-fishing site after leaving a tool inserted in the hole, after which a naive chimpanzee took it. Naive chimpanzees had a 66% success rate when using a partner's leftover tool, whereas they only managed a 36% success rate during the same session when they used their own tool selection(s).

3 From Mother to Offspring

Ai, Chloe, and Pan were among the adult chimpanzees involved in the above study; all of them subsequently gave birth in 2000. Hirata and Celli (2003) took this opportunity to investigate how their babies, Ayumu, Cleo, and Pal, learned tool use from their mothers.

The researchers placed two pairs of mothers and infants together, and observed the combinations of Ai, Ayumu, Chloe, and Cleo; Ai, Ayumu, Pan, and Pal; and then Chloe, Cleo, Pan, and Pal twice a month. Each session lasted 40 min and involved four honey-fishing sites: two upper holes and two lower ones. Each pair was given 8 sets of the 20 kinds of tools scattered on the floor, which were identical to the materials used by Hirata and Morimura (2000).

In this situation, each infant was able to observe its own mother as well as another infant and its mother. This research allowed three modes of information transmission: mother–infant transmission, or "vertical transmission"; infant–infant transmission, or "horizontal transmission"; and infant–nonmother adult, or "diagonal transmission."

From a young age, infants observed their mothers and the other adult (Fig. 5); during observation, their faces almost touched. Ayumu observed his mother Ai, as well as Chloe; Cleo observed her mother Chloe, as well as Ai. Pan and Pal had a slightly different situation: the mother (Pan) was fairly skilled at honey fishing but was less motivated and quit fishing for honey after a few sessions. During almost all sessions, she remained in a room connected to the experimental area, about 5 m from the honey-fishing sites; this limited the opportunity for her daughter Pal to observe other individuals fishing for honey, because Pal tended to stay near her mother, away from the honey-fishing sites. However, as she grew, Pal gradually began to move farther away from her mother and was able to observe closely the honey fishing done by other individuals.

When the infants were 1 to 1.5 years old, researchers observed that their development allowed combinatory manipulations (Hayashi and Matsuzawa 2003). When Ayumu was just 1 year old, he extended a rubber tube toward a hole; at the time he was unable to insert it into the hole, but this was his first attempt. Cleo was 1.5 years old and Pal was 1.2 years old when they began to extend objects toward holes.

After infants reached 1.5 years, they began to extend objects toward holes with increasing frequency. They appeared to be trying to insert objects into holes, but their hands were still clumsy and they were unable to complete the task. Although they experienced continual failure, none gave up and they patiently kept trying. Figure 6 shows the cumulative number of honey-fishing attempts as a function of their age, before the infants' first success. The first attempts of all three infants occurred several months before their first success, and the number of attempts increased greatly in the month preceding the first success.

Fig. 5. An infant (Ayumu) observing his mother (Ai)

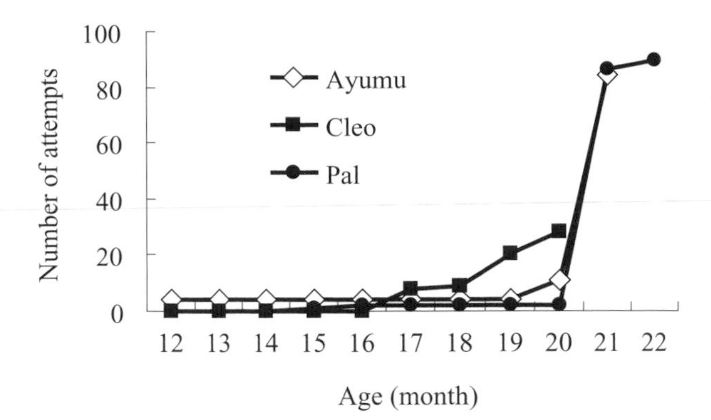

Fig. 6. Cumulative number of attempts before the infants' first success

Fig. 7. An infant (Ayumu) inserting a tool into a hole to fish for honey

On February 14, 2002, Ayumu first succeeded in dipping a tool (a plastic string) into honey; Cleo first succeeded using a plastic string on March 2, 2002; and Pal used a rubber tube in her first success on June 12, 2002. At the time of success, Ayumu was 1 year and 9 months old, Cleo was 1 year and 8 months old, and Pal was 1 year and 10 months old. These results show clearly that chimpanzee infants are able to use tools when they are just under 2 years old (Fig. 7).

Videotape analysis provided data on the length of time infants spent observing adult models, measured from the onset of the study until their first successes. Ayumu observed his mother fishing for honey 484 times for a total of 8,245 s; he also observed Chloe 146 times for a total of 3,143 s. Cleo observed her mother fishing for honey 353 times for a total of 5,619 s; she also observed Ai 16 times for a total of 491 s. Pal observed her mother 5 times for a total of 110 s; she also observed Chloe 30 times for a total of 516 s, and Ai 7 times for 65 s. Furthermore,

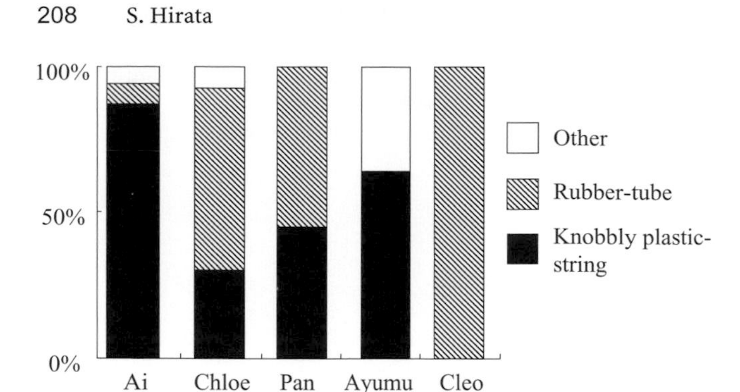

Fig. 8. Tools used by each individual when observed by infants (proportion of duration)

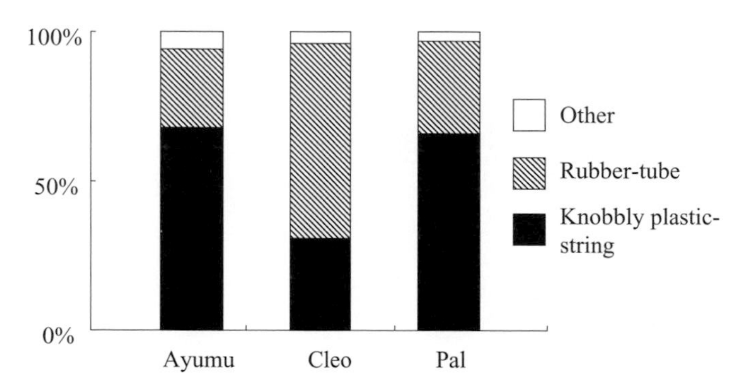

Fig. 9. Infants' observation of tools being used (proportion of duration)

she observed her peers Ayumu and Cleo, who succeeded earlier than she did, 7 times for a total of 64 s and 1 time for 27 s, respectively.

Figure 8 shows the tools each individual used when observed by infants. Two mothers, Ai and Chloe, were the main targets of observation by the three infants. Both selectively used 2 of the 20 tools: the rubber tube and knobbly plastic string. Ai's favorite tool was the plastic string and her second favorite was the rubber tube; Chloe's favorite was the rubber tube and her second favorite was the plastic string. Therefore, all three infants observed these two tools being used exclusively (Fig. 9). Subsequently, infants attempted to fish for honey by using these two kinds of tools, and their use of the adults' favorite tools led to their first successes.

Figure 10 shows the tools that each infant selected in attempts before the first success. Ayumu's first success was his 85th attempt, that is, he tried to insert objects into holes 84 times in vain. In all these attempts, Ayumu used only two kinds of objects: the plastic string and the rubber tube, never once selecting 1 of the 18 remaining objects. Pal exhibited similar behavior; 89% of the 89

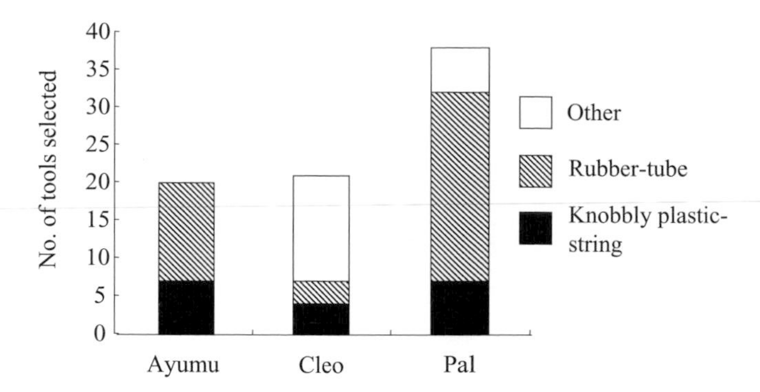

Fig. 10. Tools selected in attempts before the first success

attempts before her first success were done with the plastic string or rubber tube. Cleo exhibited slightly different behavior, making 28 attempts with eight different kinds of objects. She may have had an independent character, but the percentage of attempts she made using the plastic string or rubber tube was higher than random use.

Infants did not receive any reward during attempts before their first success; by definition, failure resulted in no honey. All three infants, especially Ayumu and Pal, continued to use the two objects, which were the same tools the adults selectively used to fish for honey. It is likely that the infants learned socially from their mothers and from other models what kinds of tools should be used.

In this sort of situation, human mothers would assist their infants by guiding their hands or verbally encouraging them; they might also perform the task for their babies, providing them with honey. Chimpanzee mothers, however, do not actively teach their offspring in these ways. This does not mean that chimpanzee mothers do nothing for their infants; infants often intervene in their mothers' activities by reaching for or trying to steal tools.

In these situations, adult response was twofold: the first was rejection, meaning that adults pushed away infants' hands to prevent them from reaching for tools; the second was allowing, meaning that adults allowed infants to steal tools and sometimes even stopped moving when they reached for tools (Fig. 11). Ai and Chloe both exhibited these two types of responses; sometimes they rejected and at other times they allowed. There were no clear differences in the reactions of adults to their own or to another's infant; they exhibited a similar degree of tolerance for each baby.

In a few cases, adults offered tools to infants (their own offspring or unrelated infants) after infants approached tools using their hands or mouths (Fig. 12). Interestingly, the tools offered by adults to infants almost never had honey on them; mothers only offered a tool to infants after they had already licked the honey from it, or before they had used the tool. These results do not indicate that mothers actively offered honey.

Fig. 11. A mother (Chloe) allowing her infant (Cleo) to lick the tool

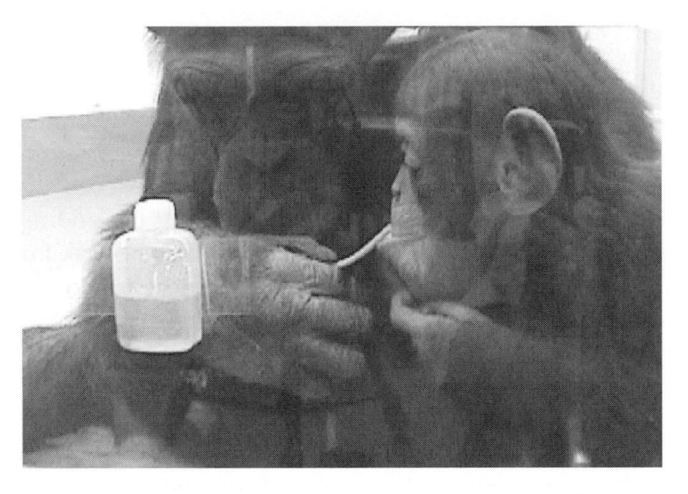

Fig. 12. A mother (Ai) giving the tip of a tool to her infant (Ayumu)

4 Chimpanzee Education and Learning by Observing

In both studies described previously in this chapter, chimpanzees learned the use of tools for honey fishing in social situations. In the first study, naive chimpanzees never observed experienced partners after a personal success; they only observed partners after a failure or before a first attempt. Therefore, partner observation was used efficiently to improve upon attempts to use a tool. Unskilled naive chimpanzees used tools left by experienced partners; this was another way that naive animals benefited from the activities of skilled partners.

In the wild, a similar scenario may occur; a skilled mother chimpanzee might abandon her tool, after which her infant might pick it up and try to use it. In this way, infants learn to use the same tool and behave in the same way as their mothers; when chimpanzees in the wild learn from group members, they not only observe models, but also use tools left by group members.

In the second study, all three infants first succeeded in fishing for honey using a tool when they were 20 to 22 months old. Nishida and Hiraiwa (1982) did not observe any wild chimpanzees at Mahale that were under 2 years of age ant fishing; the youngest individual they observed fishing for ants was 32 months old. There are several possible reasons for this age difference: our task may have been easier; the prolonged exposure to skilled individuals in our study may have had a facilitating effect; or the captive environment, in which there were many more opportunities to manipulate a variety of objects, may have accelerated manipulatory development. Detailed observations on the development of tool use in the wild would be necessary to evaluate these possibilities. Putting these issues aside, our study clearly showed that chimpanzee infants developed a tool-using skill, which entailed inserting an object into a hole to obtain an otherwise inaccessible food item, just before they reached 2 years of age.

Two infants selectively used 2 of the 20 kinds of available objects as tools in their attempts. These were the two tools predominantly used by adults, and the infants frequently observed them using these tools. Hirata and Celli (2003) simply recorded infants' attempts before their first success and never rewarded their selections, but the two infants consistently selected these two tools. It is likely that tool selectivity is transmitted from adults to infants; Tonooka (2001) observed both adult and infant chimpanzees at Bossou, Guinea, selecting a particular kind of leaf to use for drinking water. Infants may acquire an initial bias for a particular material through longitudinal observation of adults, leading to a local tradition in selecting certain tool materials.

Chimpanzee tool use is transmitted from mothers to offspring from one generation to the next. Infants have a strong motivation to learn, repeatedly observing skilled adult group members, and sometimes stealing tools used by adults. Their motivation to engage in the same activity as adults is a basis for their acquisition of similar tool use behavior. Adult chimpanzees do not actively assist infants by guiding their hands; even when they see infants failing at a task such as inserting a tool into a honey hole, they will not do anything in particular to help, or even encourage, infants. These initial findings based on captive chimpanzees have been supported by results from a recent study of wild chimpanzees' termite-fishing behavior at Gombe, Tanzania (Lonsdorf, 2006).

As infants learn by working near adults, adults often respond with tolerance; if infants reach for tools being used by adults, the adults may stop moving and allow infants to touch the tool, or even lick or steal the tool. In a few cases, adults may offer their tools to infants, so infants have opportunities to touch tools used by adults. Adult tolerance seems to support infants' motivation (Premack and Premack 2003). Because infants are not able to handle tools efficiently for a certain period after their birth, they are not rewarded during this period, but they still repeatedly observe adults and touch and mouth the tools adults use. Without such interaction between adults and infants, infants might lose their motivation and give up easily. The mother–infant interactions observed in this study, including the few cases of mothers providing tools without honey to infants, may represent steps in the evolution of teaching and may constitute an important step in the development of human culture.

Acknowledgments

The studies were financially supported by the Japan Society for the Promotion of Science (JSPS) for Young Scientists (No. 2926) to S. Hirata, the Core-to-Core Program HOPE by JSPS, and Grants-in-Aid for Special Promotion Research (#12002009, #16002001) of the Ministry of Education, Culture, Sports, Science and Technology to T. Matsuzawa. I greatly acknowledge T. Matsuzawa for his generous guidance throughout the studies, M. Tomonaga and M. Tanaka for their support and suggestions, and all the collaborators for their assistance in data collection and analysis.

References

Boesch C, Marchesi P, Marchesi N, Fruth B, Joulian F (1994) Is nut-cracking in wild chimpanzees a cultural behavior? J Hum Evol 26:325–338

Celli ML, Hirata S, Tomonaga M (2004) Socioecological influences on tool use behaviour of captive chimpanzees. Int J Primatol 25:1267–1281

Goodall JVL (1968) Behavior of free-living chimpanzees of the Gombe Stream area. Anim Behav Monogr 1:163–211

Hayashi M, Matsuzawa T (2003) Cognitive development in object manipulation by infant chimpanzees. Anim Cogn 6:225–234

Hirata S, Celli ML (2003) Role of mothers in the acquisition of tool-use behaviours by captive infant chimpanzees. Anim Cogn 6:235–244

Hirata S, Morimura N (2000) Naive chimpanzees' (*Pan troglodytes*) observation of experienced conspecifics in a tool-using task. J Comp Psychol 114:291–296

Hirata S, Watanabe K, Kawai M (2001) Sweet-potato washing revisited. In: Matsuzawa T (ed) Primate origins of human cognition and behavior. Springer, Tokyo, pp 487–508

Imanishi K (1952) The evolution of human nature (in Japanese). In: Imanishi K (ed) Ningen. Mainichi-shinbunsha, Tokyo, pp 36–94

Kawai M (1965) Newly acquired pre-cultural behavior of the natural troop of Japanese monkeys on Koshima Islet. Primates 6:1–30

Kawamura S (1954) On a new type of feeding habit that developed in a group of wild Japanese monkeys (in Japanese). Seibutsu-shinka 2:11–13

Lonsdorf EV (2006) What is the role of mothers in the acquisition of termite fishing behaviors in wild chimpanzees (*Pan troglodytes shweinfurthii*)? Anim Cogn 9:36–46

Matsuzawa T (2003) Koshima monkeys and Bossou chimpanzees: culture in nonhuman primates based on long-term researches. In: de Waal FBM, Tylack P (eds) Animal social complexity. Harvard University Press, Cambridge, pp 374–387

Matsuzawa T, Biro D, Humle T, Inoue-Nakamura N, Tonooka R, Yamakoshi G (2001) Emergence of culture in wild chimpanzees: education by master-apprenticeship. In: Matsuzawa T (ed) Primate origins of human recognition and behavior. Springer–Verlag, Tokyo, pp 557–574

McGrew WC (1992) Chimpanzee material culture. Cambridge University Press, Cambridge

Myowa-Yamakoshi M, Matsuzawa T (1999) Factors influencing imitation of manipulatory actions in chimpanzees (*Pan troglodytes*). J Comp Psychol 113:128–136

Nagell K, Olguin RS, Tomasello M (1993) Process of social learning in the tool use of chimpanzees (*Pan troglodytes*) and human children (*Homo sapiens*). J Comp Psychol 107: 174–186

Nishida T (1973) The ant-gathering behavior by the use of tools among chimpanzees of Mahale Mountains. J Hum Evol 2:357–370

Nishida T, Hiraiwa M (1982) Natural history of a tool-using behavior by wild chimpanzees in feeding upon wood-boring ants. J Hum Evol 11:73–99

Paquette D (1982) Discovering and learning tool-use for fishing honey by captive chimpanzees. J Hum Evol 7:17–30

Premack D, Premack A (2003) Original intelligence: unlocking the mystery of who we are. McGraw-Hill, New York

Sugiyama Y, Koman J (1979) Tool-using and making behavior in wild chimpanzees at Bossou, Guinea. Primates 20:513–524

Tonooka R (2001) Leaf-folding behavior for drinking water by wild chimpanzees (*Pan troglodytes verus*) at Bossou, Guinea. Anim Cogn 4:325–334

Whiten A, Custance D, Gomez JC, Teixidor P, Bard KA (1996) Imitative learning of artificial fruit processing in children (*Homo sapiens*) and chimpanzees (*Pan troglodytes*). J Comp Psychol 110:3–14

Whiten A, Goodall J, McGrew W, Nishida T, Reynolds V, Sugiayama Y, Tutin C, Wrangham R, Boesch C (1999) Culture in chimpanzees. Nature 399:682–685

14
How and When Do Chimpanzees Acquire the Ability to Imitate?

Masako Myowa-Yamakoshi

1 Introduction

Imitation is an important activity in humans, since a large amount of communicative and adaptive learning is based on reproducing others' skills. Numerous psychologists have mainly emphasized two aspects of the function of bodily imitation. The first is social learning, which contributes to adaptive skills in the human environment. Within the second year of life, human infants acquire the ability to perform a wide variety of novel actions (e.g., tool-using behaviors, symbolic gestures) by imitation (Abravanel and Gingold 1985; Meltzoff 1988). Moreover, imitation is considered to play a key role in supporting human cultural traditions by assisting in the transmission of knowledge and skills from one generation to the next (Matsuzawa et al. 2001; Tomasello et al. 1993a).

The second aspect is communication. Several researches on human infants have suggested that imitation plays an important role in developing social cognitive abilities. For example, the ability to imitate others is considered to be fundamental to the development of the normal theory of mind proposed by Premack and Woodruff (1978; Barresi and Moore 1996; Meltzoff and Gopnik 1993; Rogers and Pennington 1991) as well as self-awareness and the awareness of others (Meltzoff 1990). Further, imitation is considered to be a precursor to the capacity to represent symbols such as language (Piaget 1962; Werner and Kaplan 1963).

The foregoing two aspects appear to suggest the significant evolutionary advantage of this critical ability. The question regarding the extent to which nonhuman species are capable of imitation has important implications for the biological and phylogenetic foundations of human cognitive complexity (Visalberghi and Fragaszy 1990; Tomasello et al. 1993b; Whiten and Custance 1996). In this chapter, I compare the imitative abilities of humans (*Homo sapiens*) and our closest evolutionary relatives—chimpanzees (*Pan troglodytes*)—to explore the unique aspect of human cognition from an evolutionary perspective. Moreover, I also discuss the imitative ability from a developmental per-

School of Human Cultures, The University of Shiga Prefecture, 2500 Hassaka-cho, Hikone, Shiga 522-8533, Japan

spective. In other words, I have focused on the development of imitation in chimpanzees, comparing it with the development of the same ability in humans.

2 Factors Influencing Imitation in Adult Chimpanzees

Thus far, most experimental studies have shown that chimpanzees do not imitate a broad range of actions to the same degree as humans (Custance et al. 1995; Hayes and Hayes 1952; Nagell et al. 1993; Tomasello et al. 1987; Whiten et al. 1996). If this is the case, what kinds of actions are difficult/easy for chimpanzees to imitate? Myowa-Yamakoshi and Matsuzawa (1999) systematically investigated the factors that determine the degree of difficulty faced by chimpanzees in imitating human actions in a face-to-face situation.

The subjects of the study were five female nursery-reared chimpanzees, from 12 to 19 years of age, belonging to the Kyoto University Primate Research Institute (PRI). They had participated in several cognitive experiments (Matsuzawa 2003). Four pairs of objects were used as test stimuli. Each pair consisted of two objects that differed from each other and had no explicit relationship. Each session consisted of three conditions: (a) one object (O), (b) one object to self (O to S), and (c) one object to another (O to O). In the O condition, the chimpanzees watched the demonstrator manipulate one object (e.g., hitting the bottom of a bowl). In the O to S condition, the demonstrator manipulated one object at certain positions on his body (e.g., placing a bowl on his head). In the O to O condition, the demonstrator manipulated one object with respect to another (e.g., hitting a ball with a bowl). These three conditions involved many different motor patterns (e.g., hitting, pulling, pushing).

The human demonstrator and the chimpanzee sat face to face during the sessions. Before the start of the test, a pair of objects was presented to the chimpanzee for approximately 3 min of free play. During this time, the chimpanzee interacted with each of the objects in some way. The demonstrator then retrieved the objects and began to demonstrate an action. Each action was demonstrated two or three times to ensure that the chimpanzee paid attention to the action. After an action was demonstrated, the chimpanzee was handed the objects and was told "Do this!" (Fig. 1).

We conducted the test in two phases, depending on the chimpanzee's responses. In the first phase, we observed the chimpanzee's responses during the first attempt to determine whether she could reproduce the demonstrated action (imitation phase). If the experimenters judged that the chimpanzee was able to perform the action, the next action was demonstrated. If the chimpanzee was unable to perform the demonstrated action, we proceeded to the teaching phase, in which the demonstrator trained the chimpanzee to perform the action using verbal guidance, gestures, molding, and shaping by rewarding the chimpanzee with verbal praise and food reinforcements. The demonstrator then repeated the model trial to show the action again and handed the objects to the chimpanzee. When the experimenters judged that the chimpanzee could

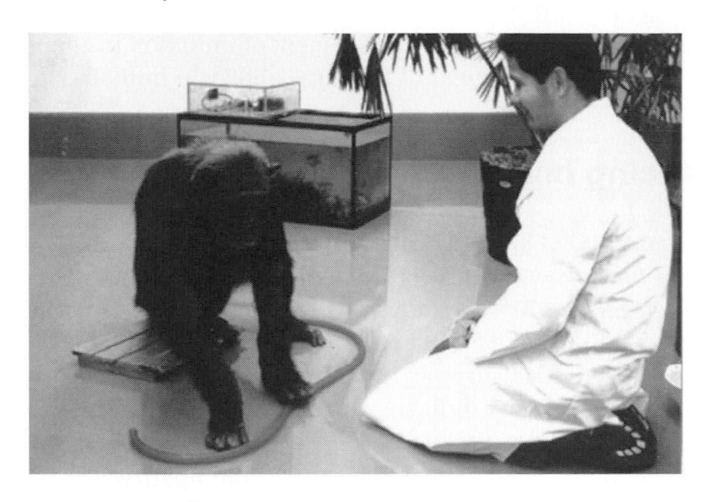

Fig. 1. A chimpanzee (Chloe) performing the demonstrated actions in the one-object condition (rolling the hose)

perform the action three times in succession, we proceeded to the next action. The trial began with the initial response of the chimpanzee and ended either when the chimpanzee successfully performed the demonstrated action or after the demonstrator had taught the action. Any one action was repeated a maximum of 20 times.

The chimpanzee's responses were videotaped, and the motor patterns involved in the chimpanzee's responses in each trial were identified as 1 of 23 mutually exclusive types. These motor patterns were classified into two main categories: (a) general motor patterns that had been observed in the free play involving manipulation and (b) nongeneral motor patterns that were not observed in the free play. To assess the level of difficulty to reproduce an action, we counted the total number of trials that were required for each chimpanzee to successfully perform each demonstrated action and compared the mean number of trials across the three conditions and two categories of motor patterns.

As a result, we arrived at three important findings. First, the chimpanzees found it more difficult to perform actions involving novel motor patterns as compared to performing actions involving familiar motor patterns (Fig. 2). It was noteworthy that the chimpanzees seldom reproduced demonstrated actions in the first attempt, even when these actions involved motor patterns that they had already acquired. Second, the chimpanzees found it easiest to perform actions in the O to O condition. On the other hand, single-object manipulations were the most difficult to reproduce (Fig. 3). It seems likely that the chimpanzees focused on the direction in which objects were manipulated to acquire visual cues for reproducing the demonstrator's actions. Third, we found some very specific types of errors in the imitative task. The chimpanzees persistently repeated actions that they were taught in a previous session and also continued

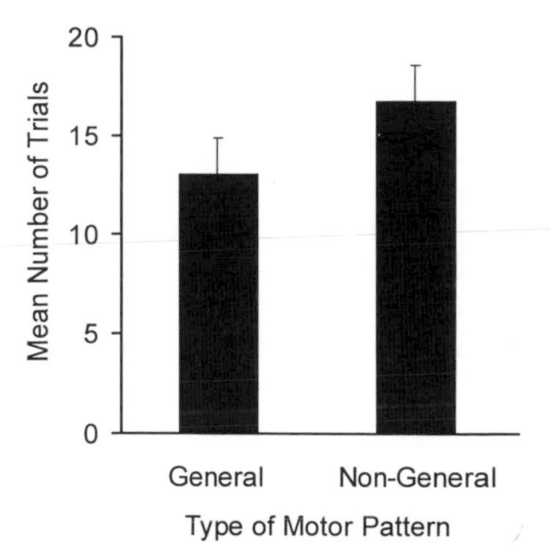

Fig. 2. Mean number of trials (plus standard error) required to perform the demonstrated actions in each of two conditions. *General*, motor patterns that were observed in the free play period; *Non-general*, motor patterns that were not observed during the free play period. [From Myowa-Yamakoshi and Matsuzawa (1999)]

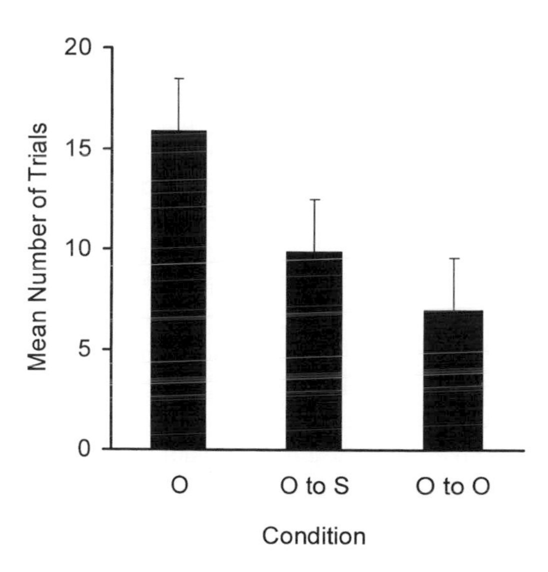

Fig. 3. Mean number of trials (plus standard error) required to perform the demonstrated actions in each of three conditions. *O*, the one-object condition; *O to S*, the one-object-to-self condition; *O to O*, the one-object-to-another condition. [From Myowa-Yamakoshi and Matsuzawa (1999)]

to manipulate each object in familiar ways. This finding indicates that their responses were somehow stimulus bound.

This study suggested that it is easier for chimpanzees to perform an action in which an object is directed toward some external location than to manipulate a single object alone. In addition, the chimpanzees were less likely to focus on the details of the demonstrator's body movements while manipulating an object; they paid more attention to the direction of the manipulated object. There were some constraints in the basic cognitive processes required to transform visual information into matching motor actions when the chimpanzees imitated human actions.

3 How Do Chimpanzees Represent Others' Actions?

It is possible that this basic difference in visual-motor information processing reflects the core differences between the social-cognitive abilities of humans and chimpanzees. The next question of importance was how the chimpanzees, with lower imitation skills, understood what the demonstrator did; in other words, how does the imitator represent the actions of others within a psychological framework (e.g., intention, desire) (Baron-Cohen 1995; Foder 1983). Nonverbal tests have shown that human infants as young as 14 to 18 months old are capable of understanding something about the others' intentions (Carpenter et al. 1998; Meltzoff 1995). Meltzoff (1995) investigated the understanding of others' intentions in 18-month-old human infants. Meltzoff took advantage of the natural tendency of human infants to reproduce the actions of others, which is referred to as the "behavioral reenactment procedure." In this experiment, the infants saw that the demonstrator attempted but failed to reach the end state of an action (e.g., he attempted to pull on the ends of a dumbbell, but his hands slipped). After the demonstration, when given the opportunity to manipulate the object themselves, the infants could spontaneously achieve the end state of the actions as often as infants who saw the successful demonstration and more often than infants in the other control conditions. Meltzoff concluded that 18-month-old infants are capable of representing others' actions in a psychological framework—from body movements to the underlying goal or intention.

Experimental research on apes' understanding of others' actions has primarily focused on the differences between intentional and accidental actions, although these studies are few (Call and Tomasello 1998; Call et al. 2005; Povinelli 1991; Povinelli et al. 1998; Premack 1986). However, the results are mixed; they display both positive (Call and Tomasello 1998; Call et al. 2005; Povinelli 1991) and negative findings (Povinelli et al. 1998; Premack 1986).

We hypothesized that chimpanzees who have limited visual-motor information processing with regard to body movement should be capable of understanding that others' intentions differ from those of humans. Using the behavioral reenactment procedure (Meltzoff 1995), Myowa-Yamakoshi and Matsuzawa (2000) investigated whether chimpanzees are capable of understanding the demonstrator's intention and focused on the structure of the demonstrated actions to determine the types of cues that would be available for chimpanzees to understand others' intentions.

The subjects were five adult chimpanzees belonging to the PRI; they were the same five chimpanzees who were tested by Myowa-Yamakoshi and Matsuzawa (1999). A human demonstrator and a chimpanzee sat face to face. Eight pairs of objects were used as test stimuli. Each pair consisted of two objects. One was called the "container," and it required different patterns of motor operation to open (pushing, pulling, or twisting). The other was called the "irrelevant tool," and it was irrelevant to the opening of the container. Each session consisted of two phases of demonstration: (a) the demonstrator attempted but failed to open the container because one of his hands slipped off the container (failure phase)

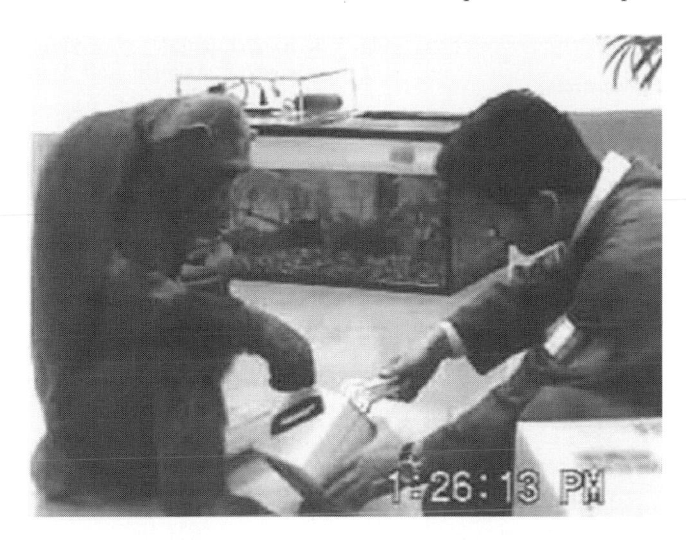

Fig. 4. A chimpanzee (Chloe), sitting face to face with a human demonstrator, is performing the demonstrated action—trying to open a box with an irrelevant tool

and (b) the demonstrator successfully opened the container (success phase). Following the demonstrations in both phases, the chimpanzees were allowed to manipulate the objects by themselves. Each chimpanzee was introduced to the two phases in the order just mentioned. Moreover, in each phase—success and failure—the demonstrator manipulated the container using one of two alternative strategies: trying to open the container (a) using the irrelevant tool or (b) by hand (Fig. 4). We counterbalanced the chimpanzees to examine the alternative within each phase. The session consisted of a 3-min free play and the following two phases of imitation test (Fig. 5).

The results revealed that the chimpanzees were able to open the containers in the failure phase, although the cases were few (less than 11% of the total performances). Overall, no clear difference was observed between the performances in the failure and success phases with regard to actually opening the container. The chimpanzees did not appear to be more successful in opening the containers even after an actual demonstration was provided. However, they manipulated the objects using the demonstrated strategy significantly more often than other strategies (Fig. 6). These findings suggest that chimpanzees anticipate others' intentions mainly by perceiving the directionality and causality of object(s) as available cues. Recently, Call et al. (2005) also revealed similar results with chimpanzees using the procedure that Meltzoff (1995) used.

It appears that chimpanzees do not understand others' intentions in the same way that humans do. Human infants understand others' intentions in a psychological framework through body movement. On the other hand, in the case of chimpanzees, the anticipation of others' forthcoming actions by perceiving the directionality and physical causality of objects is a more available cue than others' body movements performing the manipulation. Thus, compared to the

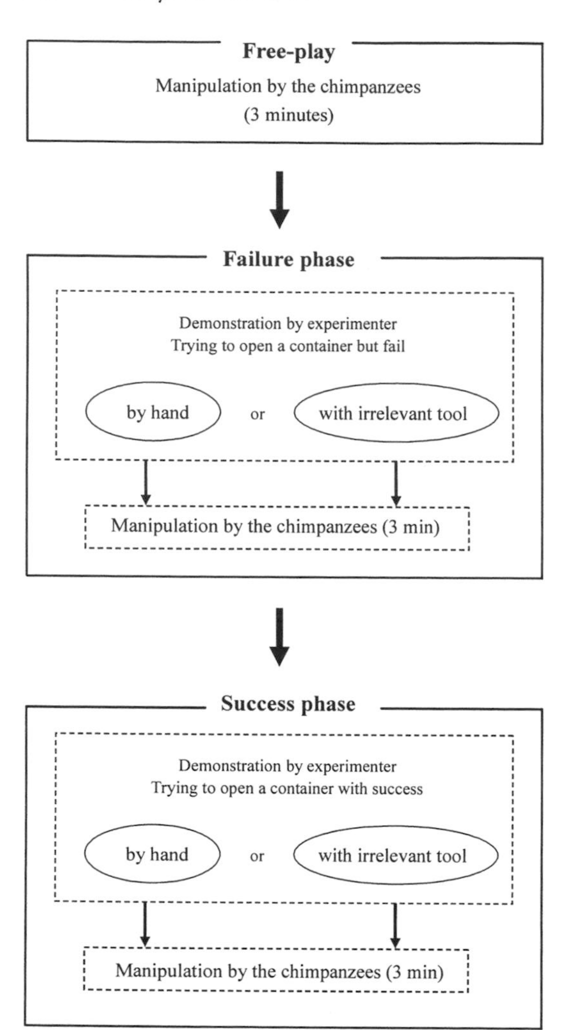

Fig. 5. Experimental procedure. [From Myowa-Yamakoshi and Matsuzawa (2000)]

current task, we may speculate that chimpanzees could find it more difficult to distinguish actions having the same outcomes or body movements to understand others' intentions.

4 Development of Imitation Abilities in Human Infants

Let us now consider the imitative abilities of human infants from a developmental perspective. The most comprehensive theory on the development of imitation is that by Piaget (1962). Piaget postulated six stages of action imitation by infants, which may be divided into three main levels. In the first level (from birth to 8 months), human imitation is restricted to imitating simple hand opening.

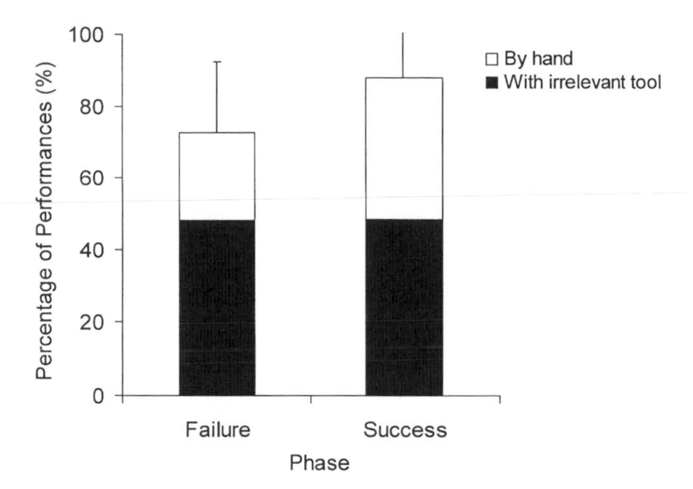

Fig. 6. Percentage of performances (plus standard error) by the chimpanzees using the identical strategies shown by the demonstrator in both the phases. [From Myowa-Yamakoshi and Matsuzawa (2000)]

This type of imitation can be accomplished through an intermodal matching process. Human infants can directly compare the demonstrator's hand movements with those of their own visible hand. In the second level (from 8 to 12 months), human infants begin to imitate facial gestures without intramodal guidance. Infants cannot see their own facial gestures, such as opening their mouths. Facial imitation depends on a cross-modal matching process. The important development in the third level is deferred imitation, and it appears at approximately 18 months. Deferred imitation is not performed during the demonstration. Piaget postulated that deferred imitation involves the infants' representational capacities and is a precursor to representing symbols such as language.

In contrast to Piaget's hypothetical framework, Meltzoff and Moore (1977) experimentally showed that human neonates can imitate some of the facial gestures [tongue protrusion (TP), lip protrusion (LP), and mouth opening (MO)] of the demonstrators. Numerous studies have been conducted on neonatal imitation including the imitation of other facial expressions (Abravanel and Sigafoos 1984; Heimann 1989; Field et al. 1982), such as eye blinking, head and finger movements, and cheek movements (Fontaine 1984; Meltzoff and Moore 1992, 1994; Reissland 1988; Vinter 1986). Meltzoff and Moore (1977, 1983) speculated that human infants can imitate motor acts performed by others through active intermodal matching (AIM), which is mediated by an innate representational system. According to the AIM hypothesis, human neonates can directly map the visible motor movements of others through their own nonvisible but felt movements.

5 The Origin of Imitative Ability: Does Imitative Ability Arise from Neonatal Imitation?

Several alternative views have been proposed on how neonatal imitation can be interpreted. For example, some researchers have insisted that this type of neonatal imitation is mediated by an "innate releasing mechanism" based on simple reflexes such as the Moro reflection (Abravanel and Sigafoos 1984; Jacobson 1979). A powerful argument supporting this view is that neonatal imitation either disappears or declines at approximately 2 to 3 months of age and reappears later (Abravanel and Sigafoos 1984; Fontaine 1984; Maratos 1982). These data are assimilated into the reflexive view by proposing that the initial drop in imitation corresponds with the inhibition of other reflexive responses.

However, Meltzoff and Moore (1992) insisted that facial imitation does not necessarily disappear at 2 to 3 months of age. They emphasized that neonatal imitation facilitates social communication, that is, it helps infants to understand the concept of "people" as opposed to "things" and aids their identification of specific people. Infants may use imitation to verify identities when engaging in face-to-face social interactions with adults. In addition, Meltzoff and Moore argued that, contrary to the reflexive view, the apparent disappearance of imitative responses observed in 2- to 3-month-olds is actually because older infants respond to people by engaging in social interaction games more frequently than neonates.

It remains unclear whether we should interpret neonatal imitation as imitative behavior. However, both these views are very similar with respect to the role of neonatal imitation. Researchers espousing these views emphasize that neonatal imitation may serve a social communicative function. Neonatal imitation may play a crucial role in attracting the attention of adults and increasing interaction opportunities for receiving care.

6 How Do Chimpanzees Acquire the Ability to Imitate?

Little is known about the existence and development of neonatal imitation in nonhuman primates (Bard and Russell 1999; Myowa 1996). It was only in 2000 that we were able to systematically investigate the imitation of facial expressions in chimpanzees immediately after birth by following a testing procedure identical to that used for human neonates (Meltzoff and Moore 1977; Myowa-Yamakoshi et al. 2004).

The subjects, two neonatal chimpanzees named Ayumu and Pal, were both born after a complete gestation period. They were reared by their biological mothers, who had participated in several cognitive experiments in PRI. They had also participated in a variety of tests related to the development of cognitive abilities (Matsuzawa 2003; Tomonaga et al. 2004).

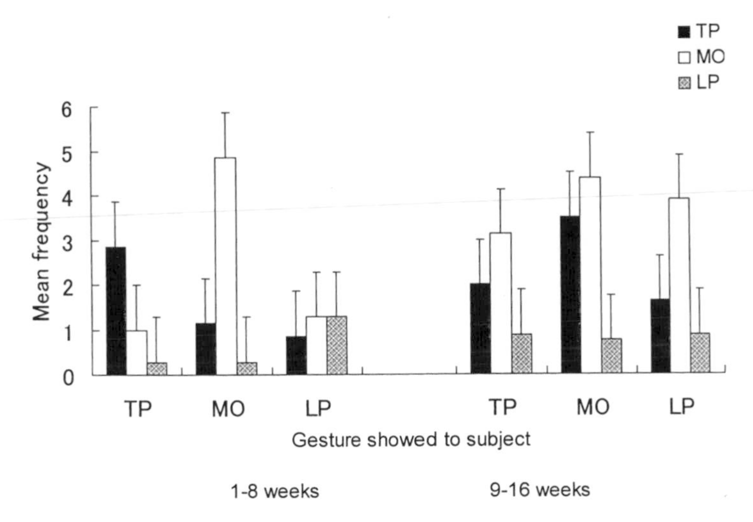

Fig. 7. The frequency of the three gestures (tongue protrusion, mouth opening, and lip protrusion) during 1 to 8 and 9 to 16 weeks of age (plus standard error). The *x*-axis plots the facial gestures shown to each chimpanzee for each period. *TP*, tongue protrusion; *MO*, mouth opening; *LP*, lip protrusion (Ayumu). [From Myowa-Yamakoshi et al. (2004)]

The test was conducted once a week for chimpanzees that were 1 to 16 weeks of age. A human tester and the chimpanzee, held by its mother, sat face to face. Before the experiment began, the tester presented the chimpanzee with an unresponsive passive face (lips closed, neutral facial expression). To sustain alertness and ensure the neonate's visual fixation on the tester's face, auditory stimulation (calling out the neonate's name once or twice) was provided before each trial. The neonate was then shown one of the following three gestures in a random order: TP, MO, or LP. In each trial, the tester demonstrated each gesture four times over a 15-s stimulus-presentation period. A 20-s response period followed immediately after. In the response period, the tester stopped making the facial gestures and displayed a passive face.

Figure 7 presents the mean frequency of each of the three gestures (TP, MO, and LP) performed by Ayumu over the two periods. These results show that at 1 to 8 weeks of age both infants were successful in producing a greater number of TP and MO responses when the TP and MO were demonstrated. However, their imitative responses of TP and MO disappeared after 9 weeks of age. Figure 8 shows the imitative responses of the three demonstrated facial gestures by Pal.

Taking these facts into consideration, we provided positive evidence for neonatal imitation in chimpanzees. At less than 7 days of age, the chimpanzees could discriminate between and imitate several human facial gestures. However, by the time they were 2 months old, the chimpanzees no longer imitated the gestures. They began to perform the MO gesture frequently in response to any of the three facial gestures presented to them. This response could be considered as "social

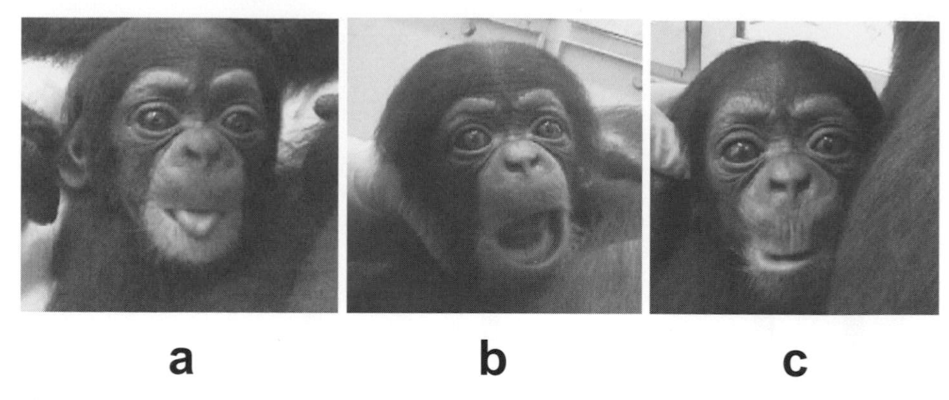

Fig. 8. Imitative responses of the three demonstrated facial gestures, tongue protrusion (**a**), mouth opening (**b**), and lip protrusion (**c**), by Pal at 2 weeks of age. [From Myowa-Yamakoshi et al. (2004)]

smiling" (i.e., play face) directed at the human experimenter. Although the reason for the disappearance of facial imitation in chimpanzees is still unknown, it is possible that the social-interactive responses toward the experimenter might have influenced the disappearance (Metlzoff and Moore 1992).

We also conducted experiments on neonatal imitation by squirrel monkeys, Japanese monkeys, and a lesser ape, a gibbon, from the time of their births. However, no clear evidence of neonatal imitation was observed (Fig. 9) (Tomonaga et al. 2003). These results suggest that they might not have the same early imitative abilities as humans and chimpanzees.

To reveal whether facial and bodily imitation in chimpanzees reappears with age, we continued to examine the imitative ability of the two chimpanzees over a period of 3 years. It was interesting to note that at around 9 months the chimpanzee infants again began to produce imitation-like responses for several facial gestures. They differentially produced the three demonstrated actions—TP, MO, and LP. In addition to the facial gestures, they also produced imitation-like responses in the form of simple bodily movements such as hitting. We may refer to their responses as "imitation-like" because their imitative responses were somewhat different from human imitation. That is, the chimpanzees' reproduction of the observed actions was always accompanied by body contact with the experimenter (Fig. 10) (Myowa-Yamakoshi 2004).

7 Body Mapping or Teleological Stance?

As suggested earlier, adult chimpanzees appear to be less sensitive to body movements than to the manipulated objects involved in the demonstrated actions. Our findings are consistent with several experimental studies of observational learning in chimpanzees. For example, Tomasello et al. (1987) and Nagell et al. (1993) investigated observational learning on tool-using behaviors in humans

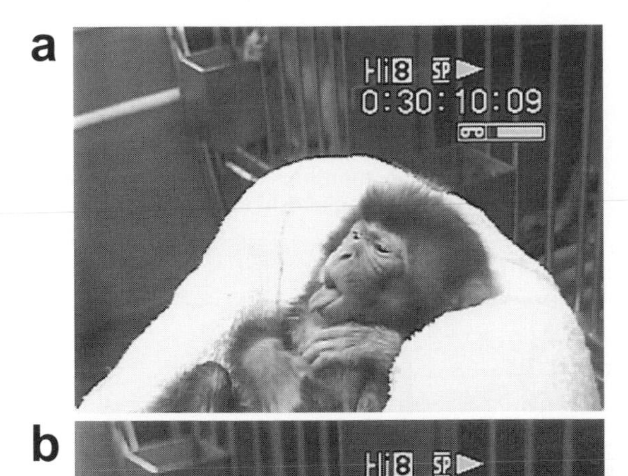

Fig. 9. Facial responses for the demonstrated facial gestures in a Japanese monkey: tongue protrusion (**a**) and mouth opening (**b**)

and captive chimpanzees. They suggested that although the human children faithfully copied the demonstrated methods of tool use, the chimpanzees did not pay attention to the precise methods involved in accomplishing the task: they tried to achieve only the results by performing the motor actions that were already in their repertoire in a trial-and-error manner. To distinguish this type of observational learning from human-specific imitative learning, Tomasello and his colleagues (Tomasello et al. 1993b; Tomasello 1999) have termed it as learning as "emulation."

Again, let us consider the imitation-like responses that the infant chimpanzees produced around 9 to 10 months of age. Did they reproduce the observed actions in a chimpanzee-specific manner, that is, emulation? In what way were their imitative outputs constructed from perceived actions performed by the model? Theoretical models proposed by recent human cognitive developmental researches appear to be useful in exploring these questions.

According to the AIM mechanism by Meltzoff and Moore, humans have an innate ability to automatically match motor programs by direct perceptual-motor mapping. This mechanism is considered to enable humans to imitate bodily gestures immediately after birth. The direct mapping view of imitative behaviors has been supported by recent neurophysiological discoveries such as mirror neurons (Fadiga et al. 1995; Rizzolatti et al. 1996).

Fig. 10. Imitative responses of the three demonstrated facial gestures, tongue protrusion (**a**), mouth opening (**b**), and lip protrusion (**c**), by Ayumu at 9 months of age. [From Myowa-Yamakoshi (2004)]

However, recent empirical evidence indicates another view—the development of imitation may be mediated by extracting the represented goals, rather than matching constituent kinematic primitives in the observed actions. For example, Bekkering et al. (2000) investigated the imitation of goal-directed actions in children 3 to 6 years of age. The demonstrated goal-directed actions consisted of touching either their right or their left ear with either an ipsilateral or a contralateral hand movement. When the contralateral hand movements were demonstrated (e.g., the demonstrator touched his left ear with his right hand reaching across his body), the children often touched their corresponding ear with an ipsilateral rather than a contralateral hand movement. Bekkering et al. insisted that the young children did not necessarily and automatically copy the demonstrated actions. In conclusion, they proposed that children may begin to learn how to build the necessary motor structures based on a goal-directed perspective instead of an active matching-to-target perspective, as postulated in the AIM mechanism. Similarly, Gergely et al. (2002) suggested that human infants modulate their imitative behavior according to the justifiability of the goal-directed actions performed by the demonstrator. They demonstrated that if the action could be rationalized by the situational constraints of the model, and if the situational constraints of the infant differed, it would prevent 14-month-olds from imitating the observed actions. Rather, they attempted to achieve the same goal by the most rational action available within their own situational constraints. Gergely and Csibra (2003) proposed that even 1-year-olds can interpret others' actions as goal directed to represent observed actions. They have termed this teleological interpretational system as the "teleological stance." Moreover, they mentioned the possibility that this early teleological stance is associated with the ability to attribute mental states to others (i.e., beliefs, desires, and intentions), an ability that would emerge only later.

We demonstrated that chimpanzee neonates also displayed the ability to imitate several facial gestures. This evidence suggests the possibility that humans and chimpanzees share the innate AIM mechanism. The phenomenon of neonatal imitation found in both the species could be mediated by a direct mapping process, based on a supramodal representational system that matches the perceptual information of the observed act with the proprioceptive information of the produced act. On the other hand, the imitation-like behaviors of older chimpanzees do not appear to be captured by the direct perceptual-motor mapping view. In contrast to human infants, the young chimpanzees only rarely match the visually perceived and motor outputs in the absence of bodily contact with the demonstrator.

Let us consider their imitation-like responses on the basis of another view—the teleological stance. It is possible that the chimpanzees can interpret the demonstrated actions by extracting their goal states. For example, they might interpret the demonstrated facial gestures, such as kissing and smiling, as the communicative signals directed toward them. It is possible that they interpret the demonstrated action 'hitting' as an invitation to a social game. Unfortunately, we were able to demonstrate only simple arbitrary bodily movements in young chimpanzees. These actions did not have the obvious nature of goals as in the case of manipulations of functional objects. Therefore, it would have been difficult to discuss their imitation-like responses by distinguishing the two views, active matching-to-target and active matching-to-goal perspectives.

It is still unclear whether and how young chimpanzees can construct motor output based on their perception of actions performed by a demonstrator. Further research is required to investigate the way in which chimpanzees develop their imitative abilities based on direct perceptual-motor mapping or teleological stance. This point is important to reveal how early imitation in chimpanzees is developmentally linked not with human-specific imitation but with a chimpanzee-specific observational learning ability, namely, emulation. Further, if the teleological stance is present in young chimpanzees, we might discuss whether or not this early ability is related to the later emergence of the mentalistic stance from an evolutionary perspective (theory of mind; Premack and Woodruff 1978).

8 Others as a "Mirror"

Finally, I must point out an external essential factor in the development of imitation, particularly in the case of humans. In the day-to-day interaction between human caregivers and infants, we notice social turn-taking behaviors that are considered to play an important role in the development of imitation. For example, caregivers pay a considerable amount of attention to infants' behaviors and react to infants' responses promptly. They introduce the infants to several facial expressions in the "exaggerated" mode, such as raising their eyebrows, opening their mouths wider, and smiling to attract the complete attention of the infants. In addition, they often imitate the responses of the infants. Similarly, infants are attracted to the caregivers' changeable and attractive gestures and respond to them. Such social imitative turn-taking games might increasingly reinforce infants' imitative abilities. It is probable that through turn-taking games with caretakers, human infants might begin to interpret others' actions as intentional or goal directed (teleological stance) by the end of the first year (Gergely and Csibra 2003).

Mothers and infants among both wild and captive chimpanzees also engage in face-to-face interactions (Fig. 11). Bard et al. (2005) investigated the number of mutual gazes between mother and infant chimpanzees belonging to PRI during the first 3 months. When the infants reached 6 to 8 weeks of age, the frequencies reached a peak; the number of mutual gazes was 27 per hour. However, it is reasonable to say that the frequency of mutual gazes in chimpanzees was not as high as that observed in humans. Moreover, in the species-typical environment, both in the wild and in captivity, not much research has been conducted on the imitative games between mother and infant chimpanzees, in contrast to humans.

Interestingly, Tomasello et al. (1993b) suggest that the chimpanzees reared in a human-like social environment (enculturated) might develop a more imitative ability than the mother-reared chimpanzees. It is possible that the imitative ability develops flexibly, depending on extended exposure to the surrounding rearing environments after birth. Is it possible that infant chimpanzees reared by enculturated chimpanzees can develop higher imitative abilities? Are there any differences in the representation that mediates the perceptions and actions in imitation between the enculturated and mother-reared chimpanzees? Is it possible that the enculturated monkeys would also develop imitative abilities? Further longitudinal developmental and comparative studies will help in reveal-

Fig. 11. Mutual gazing between a mother (Chloe) and her infant (Cleo, 1 year 11 months). (Photo by Tomomi Ochiai)

ing the relationship between species-specific biological foundations and the effect of the postnatal social experience in the development of imitation.

9 Conclusion

Our findings suggest a discontinuity between neonatal imitation and imitation that develops later in life. The capacity for neonatal imitation could be a characteristic that is common to humans and chimpanzees and has resulted from natural selection. On the other hand, the ability to imitate a broad range of whole-body actions, particularly those that do not involve objects (e.g., sign language, pantomime), appears to be an ability that is unique to humans. There may be some constraints in the cognitive processes required to transform visual information into matching motor acts in chimpanzee imitation. It is possible that this basic difference in visual-motor information processing reflects the differences in the early ability to interpret others' actions as being intentional and goal directed; it also reflects the difference in the ability to attribute mental states to others, such as beliefs, desires, and intentions in both chimpanzees and humans. It can thus be said that the later imitation might have evolved after the human lineage separated from that of chimpanzees (Myowa-Yamakoshi 2001). This more complex imitative ability in humans might have played a key role in producing unique human cultures.

Acknowledgments

The author thanks T. Matsuzawa, M. Tomonaga, M. Tanaka, H. Takeshita, T. Sato, M. Inaba, Y. Kuniyoshi, S. Hirata, N. Nakashima, A. Ueno, Y. Mizuno, S. Okamoto,

T. Imura, M. Hayashi, K. Kumazaki, N. Maeda, A. Kato, K. Matsubayashi, G. Hatano, O. Takenaka, K.A. Bard, K. Yuri, and M. Sakai for their help throughout the projects. The researches reported here and the preparation of the manuscript were financially supported by Grants-in-Aid for Scientific Research from the Japan Society for the Promotion of Science (JSPS), and the Ministry of Education, Culture, Sports, Science and Technology (MEXT) (nos. 12002009 and 16002001 to T. Matsuzawa, 13610086 and 16300084 to M. Tomonaga, 09207105 to G. Hatano, 10CE2005 to O. Takenaka, 16203034 to H. Takeshita, 16683003 to M. Myowa-Yamakoshi), the MEXT Grant-in-Aid for the 21st Century COE Programs (A2 and D2 to Kyoto University), the research fellowship to M. Myowa-Yamakoshi from JSPS for Young Scientists (no. 3642), the JSPS Core-to-Core Program HOPE, the JSPS Grant-in-Aid for Creative Scientific Research, "Synthetic Study of Imitation in Humans and Robots" to T. Sato, and the Cooperative Research Program of the Primate Research Institute, Kyoto University.

References

Abravanel E, Gingold H (1985) Learning via observation during the second year of life. Dev Psychol 21:614–623

Abravanel E, Sigafoos AD (1984) Exploring the presence of imitation during early infancy. Child Dev 55:381–392

Bard KA, Russell CL (1999) Evolutionary foundations of imitation: social cognitive and developmental aspects of imitative processes in non-human primates. In: Nadel J, Butterworth G (eds) Imitation in infancy. Cambridge University Press, Cambridge, pp 89–123

Bard KA, Myowa-Yamakoshi M, Tomonaga M, Tanaka M, Quinn J, Costall A, Matsuzawa T (2005) Group differences in the mutual gaze of chimpanzees (*Pan troglodytes*). Dev Psychol 41:616–624

Baron-Cohen S (1995) Mindblindness: an essay on autism and theory of mind. MIT Press, Cambridge

Barresi J, Moore C (1996) Intentional relations and social understanding. Behav Brain Sci 19:107–122

Bekkering H, Wohlsläger A, Gattis M (2000) Imitation of gestures in children is goal-directed. Q J Exp Psychol 53A:153–164

Call J, Tomasello M (1998) Distinguishing intentional from accidental actions in orangutans (*Pongo pygmaeus*), chimpanzees (*Pan troglodytes*), and human children (*Homo sapiens*). J Comp Psychol 112:192–206

Call J, Carpenter M, Tomasello M (2005) Copying results and copying actions in the process of social learning: chimpanzees (*Pan troglodytes*) and human children (*Homo sapiens).* Anim Cogn 8:151–163

Carpenter M, Akhtar N, Tomasello M (1998) Fourteen- through 18-month-old infants differentially imitate intentional and accidental actions. Infant Behav Dev 21:315–330

Custance DM, Whiten A, Bard KA (1995) Can young chimpanzees *(Pan troglodytes)* imitate arbitrary actions? Hayes and Hayes (1952) revisited. Behaviour 132:839–858

Fadiga L, Fogassi L, Pavesi G, Rizzolatti G (1995) Motor facilitation during action observation: a magnetic study. J Neurophysiol 73:2608–2611

Field TM, Woodson R, Greenberg R, Cohen D (1982) Discrimination and imitation of facial expressions by neonates. Science 218:179–181

Foder J (1983) The modularity of mind. MIT Press, Cambridge

Fontaine R (1984) Imitative skills between birth and six months. Infant Behav Dev 7:323–333

Gergely G, Csibra G (2003) Telelogical reasoning about actions: the naïve theory of rational action. Trends Cogn Sci 7:287–292

Gergely G, Bekkering H, Király I (2002) Rational imitation in preverbal infants. Nature (Lond) 415:755
Hayes KJ, Hayes C (1952) Imitation in a home-raised chimpanzee. J Comp Physiol Psychol 45:450–459
Heimann M (1989) Neonatal imitation, gaze aversion, and mother-infant interaction. Infant Behav Dev 12:495–505
Jacobson SW (1979) Matching behavior in the young infant. Child Dev 50:425–430
Maratos O (1982) Trends in the development of imitation in early infancy. In: Bever TG (ed) Regressions in mental development: basic phenomena and theories. Erlbaum, Hillsdale, NJ, pp 81–101
Matsuzawa T (2003) The Ai project: historical and ecological contexts. Anim Cogn 6:199–211
Matsuzawa T, Biro D, Humle T, Inoue-Nakamura N, Tonooka R, Yamakoshi G (2001) Emergence of culture in wild chimpanzees: education by master apprenticeship. In: Matsuzawa T (ed) Primate origins of human cognition and behavior. Springer, Tokyo, pp 557–574
Meltzoff AN (1988) Infant imitation after a 1-week delay: long-term memory for novel acts and multiple stimuli. Dev Psychol 24:470–476
Meltzoff AN (1990) Foundations for developing a concept of self: the role of imitation in relating self to other and the value of social mirroring, social modeling, and self practice in infancy. In: Cicchetti D, Beeghly M (eds) The self in transition: infancy to childhood. University of Chicago Press, Chicago, pp 139–164
Meltzoff AN (1995) Understanding the intentions of others: re-enactment of intended acts by 18-month-old children. Dev Psychol 31:838–850
Meltzoff AN, Gopnik A (1993) The role of imitation in understanding persons and developing a theory of mind. In: Baron-Cohen S, Tager-Flusberg H, Cohen D (eds) Understanding other minds: perspectives from autism. Oxford University Press, New York, pp 335–336
Meltzoff AN, Moore MK (1977) Imitation of facial and manual gestures by human neonates. Science 198:75–78
Meltzoff AN, Moore MK (1983) Newborn infants imitate adult facial gestures. Child Dev 54:702–709
Meltzoff AN, Moore MK (1992) Early imitation within a functional framework: the importance of person identity, movement, and development. Infant Behav Dev 15:479–505
Meltzoff AN, Moore MK (1994) Imitation, memory, and the representation of persons. Infant Behav Dev 17:83–99
Myowa M (1996) Imitation of facial gestures by an infant chimpanzee. Primates 37:207–213
Myowa-Yamakoshi M (2001) Evolutionary foundation and development in imitation. In: Matsuzawa T (ed) Primate origins of human cognition and behavior. Springer, Tokyo, pp 349–367
Myowa-Yamakoshi M (2004) Why do humans imitate? Kawade-Shobo Sinsya, Tokyo (in Japanese).
Myowa-Yamakoshi M, Matsuzawa T (1999) Factors influencing imitation of manipulatory actions in chimpanzees (*Pan troglodytes).* J Comp Psychol 113:128–136
Myowa-Yamakoshi M, Matsuzawa T (2000) Imitation of intentional manipulatory actions in chimpanzees (*Pan troglodytes*). J Comp Psychol 114:381–391
Myowa-Yamakoshi M, Tomonaga M, Tanaka M, Matsuzawa T (2004) Imitation in neonatal chimpanzees (*Pan troglodytes*). Dev Sci 7:437–442
Nagell K, Olguin R, Tomasello M (1993) Processes of social learning in the imitative learning of chimpanzees and human children. J Comp Psychol 107:174–186
Piaget J (1962) Play, dreams and imitation in childhood. Norton, New York
Povinelli DJ (1991) Social intelligence in monkeys and apes. Ph.D. thesis, Yale University, New Haven, CT
Povinelli DJ, Perilloux HK, Reaux JE, Bierschwale DT (1998) Young and juvenile chimpanzees' *(Pan troglodytes)* reactions to intentional versus accidental and inadvertent actions. Behav Process 42:205–218
Premack D (1986) Gavagai! MIT Press, Cambridge
Premack D, Woodruff G (1978) Does the chimpanzee have a theory of mind? Behav Brain Sci 1:515–526

Reissland N (1988) Neonatal imitation in the first hour of life: observations in rural Nepal. Dev Psychol 24:464–469

Rizzolatti G, Fadiga L, Gallese V, Fogassi L (1996) Premotor cortex and the recognition of motor actions. Cogn Brain Response 3:131–141

Rogers S, Pennington B (1991) A theoretical approach to the deficit in infantile autism. Dev Psychopathol 3:137–162

Tomasello M (1999) The cultural origins of human cognition. Harvard University Press, Cambridge

Tomasello M, Davis-Dasilva M, Camak L, Bard KA (1987) Observational learning of tool-use by young chimpanzees. J Hum Evol 2:175–183

Tomasello M, Kruger AC, Ratner HH (1993a) Cultural learning. Behav Brain Sci 16:495–552

Tomasello M, Savage-Rumbaugh, S, Kruger AC (1993b) Imitative learning of actions on objects by children, chimpanzees, and enculturated chimpanzees. Child Dev 64:1688–1705

Tomonaga M, Tanaka M, Matsuzawa T (2003) Development of cognition and behaviors in chimpanzees. Kyoto University Press, Kyoto (in Japanese)

Tomonaga M, Tanaka M, Matsuzawa T, Myowa-Yamakoshi M, Kosugi D, Mizuno Y, Okamoto S, Yamaguchi MK, Bard KA (2004) Development of social cognition in chimpanzees (*Pan troglodytes*): face recognition, smiling, mutual gaze, gaze following and the lack of triadic interactions. Jpn Psychol Res 46:227–235

Vinter A (1986) The role of movement in eliciting early imitations. Child Dev 57:66–71

Visalberghi E, Fragaszy DM (1990) Do monkeys ape? In: Parker S, Gibson K (eds) Language and intelligence in monkeys and apes: comparative developmental perspectives. Cambridge University Press, Cambridge, pp 247–273

Werner H, Kaplan B (1963) Symbol formation. Wiley, New York

Whiten A, Custance DM (1996) Studies of imitation in chimpanzees and children. In: Galef BG Jr, Heyes CM (eds) Social learning in animals: the roots of culture. Academic Press, London, pp 291–318

Whiten A, Custance DM, Gómez J-C, Teixidor P, Bard KA (1996) Imitative learning of artificial fruit processing in children (*Homo sapiens*) and chimpanzees (*Pan troglodytes*). J Comp Psychol 110:3–14

15
Yawning: An Opening into Empathy?

James R. Anderson[1] and Tetsuro Matsuzawa[2]

1 On Empathy in Great Apes

Comparative developmental psychologists are interested in the emergence and development of empathy in both humans and nonhumans. By empathy, we mean the ability to identify with another individual's emotions and cognitive states; it is characteristic of all normal humans from early childhood. Even today there is debate among primatologists and psychologists about whether and to what extent our nearest evolutionary neighbors, the great apes, share the capacity for empathy that we humans take for granted (Gallup 1998; Preston and de Waal 2002; Povinelli 1998; Povinelli and Vonk 2003; Tomasello et al. 2003). Although many people who work closely with these primates are convinced that they are capable of reflecting about what other individuals might be thinking, others express doubts about the extent and level at which they do this. Much of the controversy stems from the variable quality of the evidence presented in support of empathic abilities. The evidence comes from a range of observational studies and controlled experiments, and as we will see from our brief review of the literature as it concerns chimpanzees, neither source of data is problem free.

Observational studies of chimpanzees in the wild and in naturalistic groups in captivity have led to identification of a range of phenomena suggestive of the capacity for empathy in these apes (O'Connell 1995). For example, some advanced forms of deception may involve attributing intentions to other individuals and deliberately altering others' beliefs, emotions, or attentional states. The evidence for intentional deception of this type is stronger for chimpanzees than for any other species of nonhuman primates (Byrne 1995; Byrne and Whiten 1992). One of the best known examples took place in a captive group with access to a large outdoor enclosure. Menzel (1974) described how one chimpanzee, Belle, would try to feign disinterest or misdirect another chimpanzee, Rock, away from some hidden food of which only she knew the location. Rock in turn began to feign disinterest in Belle's activities, only to suddenly wheel round to detect unintentional cues from Belle about where the food might be.

[1]Department of Psychology, University of Stirling, Stirling FK 9 4LA, Scotland, UK
[2]Primate Research Institute, Kyoto University, 41 Kanrin, Inuyama, Aichi 484-8506, Japan

Not only does this example suggest that the chimpanzees understood each other's intentions, it also involves counterdeception, a phenomenon that has never been reported in any species of monkey. Goodall (1986) describes another noteworthy case of counterdeception in chimpanzees, witnessed at a feeding box at Gombe, Tanzania. One chimpanzee knew that food was available in the box but feigned disinterest because a more-dominant chimpanzee was in the vicinity. The latter made as if to leave the area, but in fact he hid behind a tree. From there he looked back to see the first chimpanzee going for the food in the box, the latter apparently believing that the dominant had left the scene. The dominant chimpanzee came back and got the food.

Although impressive, examples such as those just cited are open to criticism as evidence for cognitive empathy. The Gombe incident is an anecdote, a one-off occurrence that cannot be reinstated, whereas Menzel's procedure, although systematic and controlled (showing hidden food to one chimpanzee and then releasing the entire group), is not guaranteed to produce an outcome similar to the scenes acted out by Rock and Belle. However, a recent study with some similarities to Menzel's procedure has also described the emergence of deception by withholding information and misleading (Hirata and Matsuzawa 2001). It is to be expected, however, that tactical deception will occur only occasionally in any group, otherwise it would lose its effectiveness. Also, there is always the possibility that the deceivers have learned behavioral tactics simply for changing the behavior of others, with no regard for what the others might actually be thinking. In short, although highly suggestive, instances of possible deception cannot be accepted as the strongest evidence on which to base claims about empathy.

Fortunately, systematic observational studies of social behavior in groups of chimpanzees have provided other types of evidence for empathic abilities. For example, chimpanzee bystanders often approach to offer reassuring friendly contact to the victim of an aggressive act; this behavior is much less frequently observed in macaques (de Waal and Aureli 1997). Those authors cautiously hypothesize that the macaque–chimpanzee difference in consolation behavior may reflect species differences in empathy and thus call for further observational studies and experiments to test the hypothesis.

Some elaborate and complex experiments have been designed to explore aspects of the capacity for empathy. In one of their pioneering experiments, Premack and Woodruff (1978) showed an adult female chimpanzee, Sarah, video clips of a human faced with a problem, for example, a man trying to exit a room through a locked door. The chimpanzee was then shown several photographs, one of which depicted the solution to the problem, which in this case would be a key. Even without specific training on the task, Sarah typically selected the appropriate photograph, which the authors considered as evidence that she was able to understand the actors' intentions. One limitation of this work is that the correct response might have been based on simple association between the problem and the solution, rather than on an understanding of the actor's predicament.

Other experiments have focused on chimpanzees' attribution of knowledge to other individuals, an ability that qualifies as empathic as it implies identifying with another's cognitive (nonemotional) state. Most of these studies have assessed chimpanzees' processing of others' gaze direction as giving rise to knowledge. One study showed that chimpanzees selected the container indicated to them by a human who had watched while that one of several containers was baited with food, in preference to a container indicated by someone who did not see the baiting procedure (Povinelli et al. 1990). Other experiments have shown that chimpanzees can find hidden food through monitoring and acting on gaze cues from humans or from other chimpanzees (Itakura and Tanaka 1998, Itakura et al. 1999). If one chimpanzee can see that a more-dominant individual has looked at a piece of food, the first is less likely to make a move for that food, but may instead go for another piece that is out of view of the dominant (Hare et al. 2000).

Similar to observations from the field, evidence from experimental studies of empathy may be open to alternative interpretations. For some skeptics, in many laboratory experiments ecological validity has been too compromised. It is true that some experiments involve complicated procedures and humans acting in quite unnatural ways, such as placing a bucket over their head and then removing it just before interacting with the chimpanzee, or pointing to one object while staring at another object. Also, as with some of the more naturalistic studies, replication of results is not guaranteed; there are examples of very similar experiments leading to quite different results. Finally, there is the possibility that over repeated trials subjects might simply learn to perform the correct response, without engaging in mental attribution at all (Heyes 1998). What is clear is that the debate over empathic abilities in nonhuman species is in need of new types of data, preferably from a variety of approaches and perspectives.

2 Yawning and Empathy

Recently, we approached the debate on empathy in chimpanzees from a new angle. Our basic question was simple: Do chimpanzees show contagious yawning, as humans do? Although the link between contagious yawning and empathy might not be immediately obvious, it has been made (Lehmann 1979) and tested (Platek et al. 2003). The latter study examined the relationship between empathy and contagious yawning in humans, first by asking participants (university students) to complete the Schizotypal Personality Questionnaire (SPQ), part of which measures empathic tendencies, next asking them to interpret stories designed to measure mental state attribution, and then exposing the participants to a videotape showing a sequence of yawns. The use of video stimuli to study contagious yawning in humans was pioneered by Provine (1986), who found that more than 50% of adults would yawn within a few minutes of starting to watch a video of someone yawning repeatedly. Platek et al. reported that 40% of their subjects yawned in response to the yawn stimulus videotape, and that people who scored higher on measures of empathy and mental state attribution skills were more likely to show the effect. Furthermore, they found that contagious yawners were faster than others at recognizing their own face when it appeared on a computer screen. Platek et al. (2003) made a case for contagious yawning being related both to self-awareness, as measured by self-recognition, and to empathic tendencies, as measured by the SPQ and mental attribution tests.

Thus, the scene was set for our investigation into contagious yawning by chimpanzees. Chimpanzees are one of the few species (along with humans, other great apes, and cetaceans) known to show self-recognition, for example, in a mirror or on video (Gallup 1970; Anderson and Gallup 1999). Furthermore, as we have already indicated, there is evidence for empathic abilities in chimpanzees beyond those seen in monkeys (de Waal 1996; Preston and de Waal 2002). Therefore, the following hypothesis emerged: If self-awareness underlies the capacity for empathy, as proposed explicitly by Gallup (1982), and empathy is linked to contagious yawning (in humans; Platek et al. 2003), then chimpanzees might also be susceptible to contagious yawning. Until our interest was kindled, this possibility had never been considered in the literature. In fact, there appears to have been a widespread assumption that, although yawning is ubiquitous among vertebrates, the phenomenon of contagious yawning is uniquely human (Baenninger 1987; Lehmann 1979; Smith 1999).

3 Yawning in Chimpanzees

Chimpanzees appear to yawn in much the same way as humans do, although there are no comparative studies of physiological aspects of the act in the two species. As do humans, chimpanzees may yawn when they are tired, bored, or in a mildly disturbing situation (Goodall 1968). It is striking, however, that in all the books and journal articles describing hundreds of thousands of hours of observations of chimpanzees in captivity and in the wild, there do not appear to be any descriptions of contagious yawning, by which we mean one chimpanzee yawning in response to seeing another chimpanzee yawning.

It is important to note that chimpanzees, like humans, differ from Old World monkeys in that yawning by adult males is not a form of ritualized display expressed in situations of male–male confrontation. This type of display was noted by Darwin (1965/1872) and has been studied in macaques in some detail (Adams and Schoel 1982; Deputte 1994). Similar to human yawning, chimpanzee yawning appears to be devoid of any agonistic signaling function. But, if chimpanzee yawning is more like that of humans than that of Old World monkeys, why are there no accounts of contagious yawning in our great ape relatives? At this stage two possibilities may be considered: (1) contagious yawning in chimpanzees does not exist, or (2) contagious yawning does exist but it has been overlooked by chimpanzee researchers. This latter possibility may be true because yawning occurs at times of the day when it is unlikely to be observed, or else it occurs less frequently than in humans. It is clear that only field studies and careful observational studies of captive groups can confirm or contradict these two possibilities.

In the absence of relevant data, we designed our study to address a third possibility, which was that if chimpanzees are susceptible to contagious yawning, the effect should be observable under experimental conditions involving exposure to repeated yawn stimuli, similar to those used to study the phenomenon in humans.

4 Experimental Demonstration of Contagious Yawning in Chimpanzees

Inspired by reports that between 40% and 55% of human adults will yawn shortly after starting to watching sequences of yawns on a television monitor, we prepared two "yawn" videotapes, each showing ten short clips (6–8 s each) of chimpanzees yawning, with each clip separated by a blank (blue) screen for 6 to 10 s. One videotape featured chimpanzees from the Primate Research Institute (PRI) group, therefore highly familiar to the subjects, while the other videotape featured unfamiliar chimpanzees from the Mahale Mountains, Tanzania. Although every clip was centered around a chimpanzee yawn, the clips were otherwise varied, and included young infants yawning, close-up and middle-distance views of yawning, and chimpanzees either sitting or lying down while yawning (Fig. 1). Two "control" videotapes were also prepared; these were similar in structure to the yawn videotapes but showed chimpanzees displaying a variety of facial expressions such as grinning or threatening, but not yawning. Each videotape lasted approximately 3 min.

The main subjects were six adult female chimpanzees, ranging in age from 19 to 27 years; they were all members of a social group housed at the Primate Research Institute (PRI) of Kyoto University. Three of the females had juveniles, aged 3 years, and the latter were also included as subjects. In fact, the juveniles were considered important in this research because the only published study to date of the development of contagious yawning in human children failed to find any evidence of this in children below the age of 5 years (Anderson and Meno 2003). The chimpanzee adults and juveniles had all been subjects in a wide range of noninvasive behavioral research (Matsuzawa 2000, 2001, 2003, 2005), and all were familiar with the experimental booth. Three sides of the booth had glass walls, which allowed the chimpanzees to be filmed from different angles throughout the session. Access to the booth from the chimpanzees' outdoor enclosure was via a series of hatches and tunnels.

Each adult was tested individually except for the three mothers, who came to the booth accompanied by their juveniles. Participation in the sessions was voluntary, in that the chimpanzees could opt to come in from their outside enclosure in response to coaxing by one of the experimenters (T.M.), or they could decline. The order in which the chimpanzees were tested therefore depended on

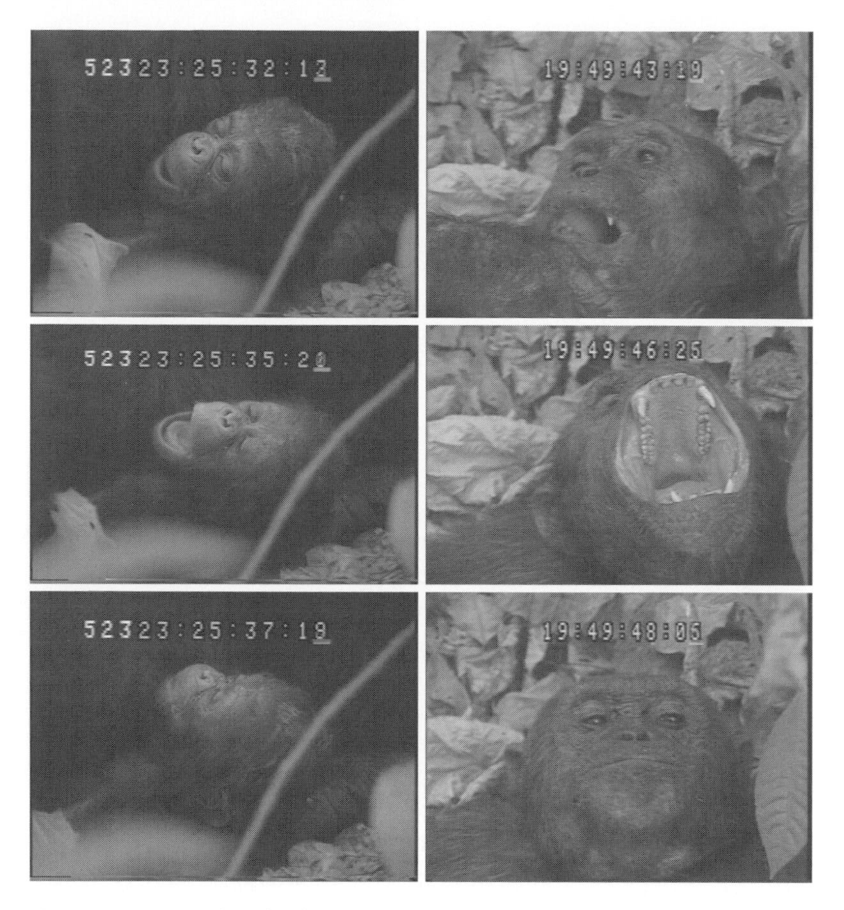

Fig. 1. Two examples of video clips of chimpanzee yawning

their willingness to participate, but in the end they were all tested in four sessions, and they saw one yawn and one control videotape in each session. Thus, each chimpanzee was exposed twice to each of the four videotapes (PRI yawn, Mahale yawn, PRI control, Mahale control).

As soon as a chimpanzee arrived in the booth, a 5-min habituation period started, during which nothing in particular happened. Immediately thereafter, a 35-cm video monitor was switched on and the first videotape was run. The monitor sat on a small table 30 cm above floor level and positioned 8 cm from the front glass wall of the booth so that it was clearly visible to the chimpanzee. When the videotape ended, the monitor was switched off and a 3-min postvideo observation period started. After that, there was a 5-min distraction period during which the experimenter (T.M.) interacted with the chimpanzee through the glass walls, talking, showing her objects, moving around the outside of the booth, etc. At the end of the distraction period, the second videotape was shown, and this was followed by another 3-min observation period. After this final period, the chimpanzee was allowed to rejoin the group in the outside enclosure. Throughout the session, apart from the distraction period, the humans in the room remained motionless and neutral, except for occasional prompts by T.M. to watch the video if the chimpanzee seemed to lose attention (although they are generally attentive to short movie clips; Morimura and Matsuzawa 2001). Any occurrences of yawning by the chimpanzees were noted in real time and verified by later analysis of videotapes of the sessions, with 100% agreement between the authors.

The results of the experiment were striking (Anderson et al. 2004). Overall, the adult chimpanzees yawned more than twice as frequently during the yawn video trials (exposure and postvideo periods combined) than during control trials (totals: 67 and 30, respectively). Individual binomial tests showed that for two of the females the difference between the yawn and control trials was extremely significant, with many more yawns during the yawn condition. Figure 2 illustrates the data for the individual adult females. The equivalent figure in the original publication erroneously showed chimpanzee Mari as yawning nine times during the control videos, whereas in fact she only did so four times, as illustrated here. These data indicate that in these adult female chimpanzees, the video-induced contagious yawning effect was shown in 33% of the sample. There were no clear differences in response to the PRI and Mahale videotapes.

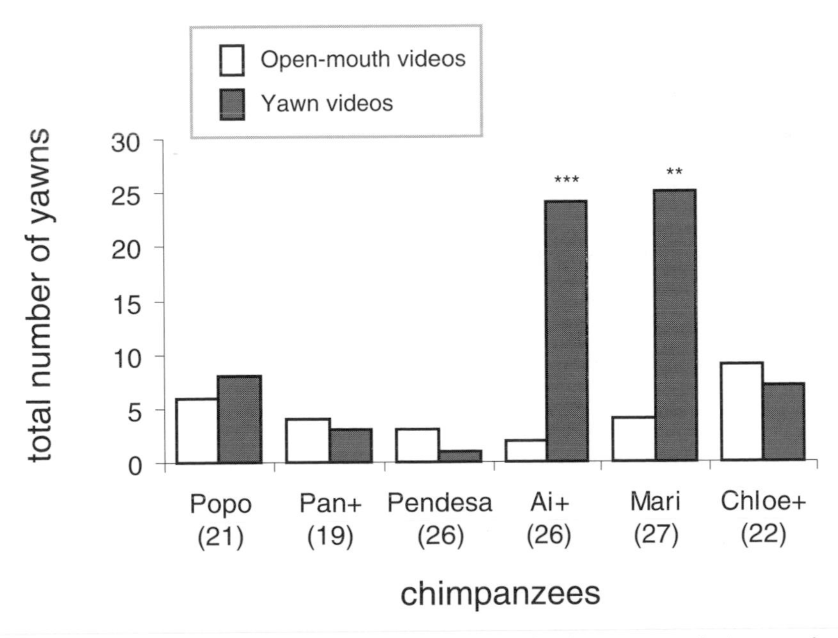

Fig. 2. Total frequency of yawns by six adult female chimpanzees during or after exposure to open-mouth videos (*open bars*) and yawn videos (*shaded bars*). Ages in years are given in *parentheses*. A *plus sign* indicates females that were tested with their infants. $^{**}P < 0.01$; $^{***}P < 0.001$

Since the publication of the original report, we have carried out some additional analyses on the data. There were no significant correlations between age of the adults and frequency of yawning in video trials, control trials, or both kinds combined. However, it is noteworthy that the contagious yawning effect was found in two of the three oldest females, all three being over 25 years old. Binomial tests indicated that for both yawn and control trials, significantly more yawns occurred in the 3-min postvideo period than during the 3-min exposure period ($P < 0.05$ and $P < 0.01$ for yawn and control trials, respectively). When the same analysis was run separately for the tapes showing the PRI and Mahale chimpanzees, the difference remained significant only for the PRI tapes (both at $P < 0.05$). Figure 3 illustrates a yawn in response to a yawn by one of the females.

A final salient outcome of the experiment concerns the three 3-year-olds. In spite of seeing the same videotapes and also seeing their mothers yawning during the tests, none of the juvenile chimpanzees ever yawned during the sessions.

5 Contagious Yawning in Chimpanzees: Discussion and Implications

Although the experiment just described must be seen to be preliminary, we hope that it has paved the way for a new approach to the thorny issue of how to assess empathic abilities through experiments with nonhumans. In contrast to many previous experiments, there is no complicated procedure in which the chimpanzees observe humans behaving in unusual ways, nor is there any experimenter-deemed "correct" response that the ape might figure out over many repetitions of the task. Indeed, a case could be made for our procedure having greater ecological validity than some of the other studies that are presented as evidence in the empathy debate. Although it is true that wild chimpanzees will not encounter videotapes showing sequences of clips depicting yawns, neither do humans under normal circumstances, and exposing subjects to video stimuli is simply a way of ensuring some control over the situation.

To discuss the experiment, we think it is enlightening to consider two comments offered by the editor of an esteemed psychological journal to which we submitted the original paper for publication. In his letter of rejection (the manuscript was not sent out for review), he wrote that because only two of the six adult female chimpanzees showed a positive response, we could not conclude that there was a group-level effect. Actually, we have never made a claim for a group-level effect. It is noteworthy that in studies in which human adults are exposed to video yawn stimuli, the percentage of participants that report yawning in response to the stimuli varies from 40% to 55% (Provine 1986; Platek et al. 2003). This result would not be considered a group-level effect either, but nobody doubts the robustness of contagious yawning in humans.

Indeed, in one respect we consider our data to be even more compelling than the human data: *Unlike humans, the chimpanzees had no idea what was expected*

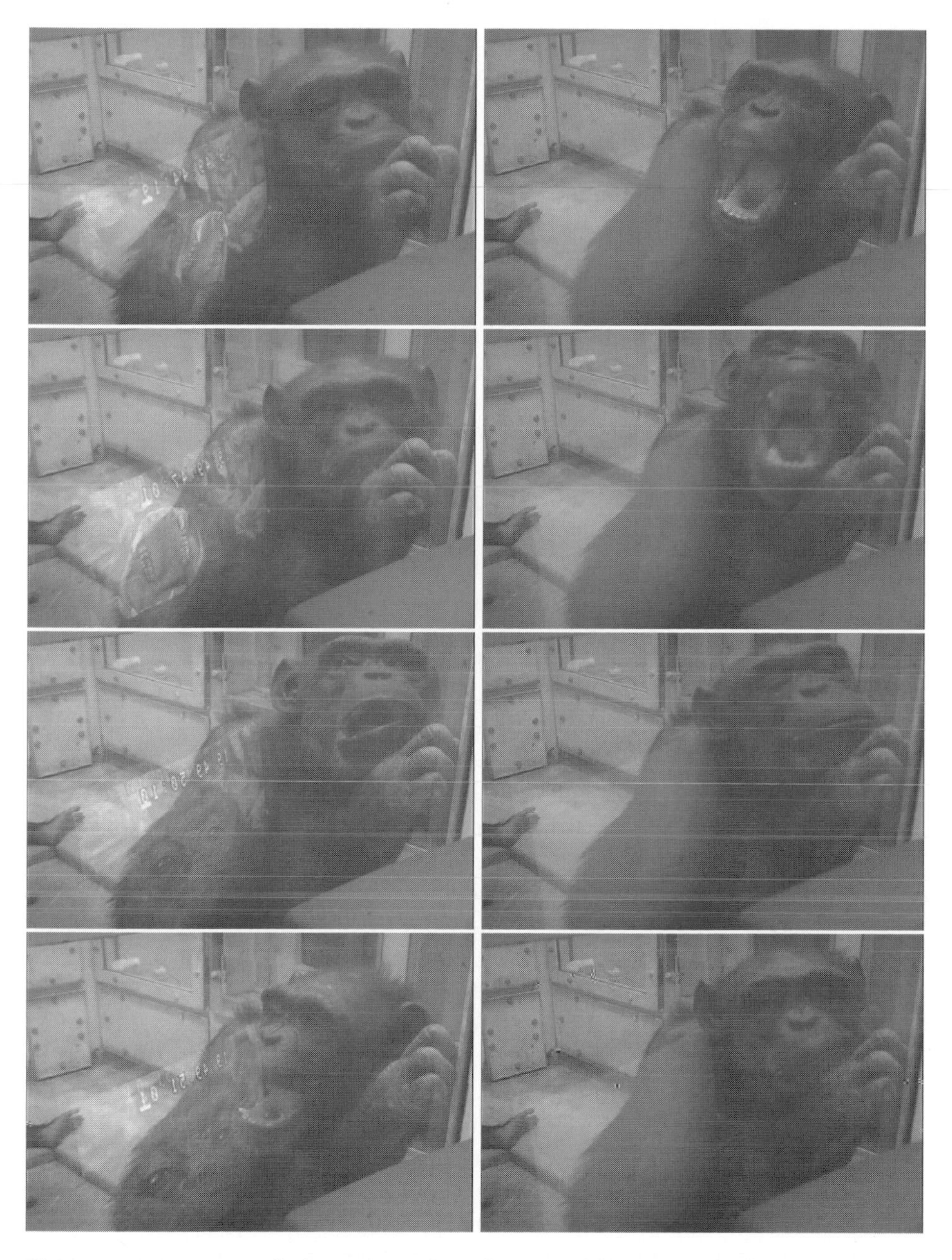

Fig. 3. A yawn response during presentation of a yawn videotape. Ai watches a yawn on the screen (*top left*), starts to yawn as the stimulus yawn ends (*low left*), continues to yawn (*top right*), and completes the yawn while the screen is blank. Visit the following site for the video scene: http://www.pri.kyoto-u.ac.jp/hope/pub/hope2004-4.html

of them. In experiments with humans, the participants are usually instructed to report their own yawns, for example, by pressing a counter. This procedure is used to avoid the inhibiting effect of being observed, but it almost certainly leads to an inflated occurrence of yawning, because by the very act of focusing on their yawning, and being likely to guess the purpose of being exposed to repeated yawn stimuli, the participants themselves are likely to induce the response. Viewed in this light, a 33% occurrence of contagious yawning in naïve chimpanzees, with no such inhibitions and no hypotheses about the experimenters' aims, is particularly impressive.

The second comment by the journal editor was that while on a safari trip he had seen several lions lying down together and yawning, and that this challenged the view that chimpanzees' contagious yawning was related to empathy. Here, he is confusing two things, namely, synchronization of activity and the contagion effect. It is true that after feeding and other essential activities have been taken care of, members of a pride of lions may start yawning during a period of rest (McKenzie 1994), but this is probably based on shared behavioral and physiological rhythms, rather than any one lion's yawns inducing the same behavior in others. Admittedly, there is no direct evidence on the issue, but we think it unlikely that lions are susceptible to contagious yawning. In fact, it is worth pointing out that the term "contagious yawning" is something of a misnomer, as yawns do not necessarily spread throughout an entire group of humans (or chimpanzees), and the likelihood of any one individual yawning in response to seeing someone else yawn probably varies as a function of a host of variables that have yet to be clarified. The research by Platek et al. (2003) has identified differences in self-recognition ability and empathic tendencies as a source of variability in humans; whether the same applies to chimpanzees is certainly worthy of investigation. Preston and de Waal (2002) offer an insightful discussion of different mechanisms underlying what we might call empathic behavior.

Among the feedback that came from colleagues upon publication of the article by Anderson et al. (2004) was the suggestion that somehow the two adult females who showed the contagious yawning effect might have done so because they had a history of particularly close contact with humans and were highly test experienced. This argument has little validity. Although there may be cases of "enculturated" apes showing better performance in tests in which humans try to get them to perform in particular ways, this kind of situation is far removed from our setup. First, all the chimpanzees live in a group, and they are much more chimpanzee oriented than people oriented. Second, the chimpanzees were spontaneously responding to images of chimpanzees, not images of humans. Finally, as we have already indicated, there was no "task" that the chimpanzee had to figure out, and no kind of reward involved (other than any possible intrinsic satisfaction from yawning).

The discovery of contagious yawning in chimpanzees has raised many other questions, a few of which we present here to close our discussion. When in ontogeny does contagious yawning start to occur, and what psychological changes cause it to occur? The fact that none of the three juveniles ever yawned during the sessions strengthens the view that the mechanisms underlying the phenomenon are similar in chimpanzees and humans, as human children below the age of 5 years also appear to show no effect (Anderson and Meno 2003). What stimuli are effective for eliciting contagious yawning? From personal experience, we can say that humans may be induced to yawn by seeing another species yawning; in fact, we can easily (and sometimes erroneously) empathize with other animals. Would chimpanzees yawn in response to seeing another species yawn? We are currently addressing this precise question experimentally.

What neural substrates are implicated in contagious yawning? Recent brain imaging studies of humans exposed to yawn stimuli have started to provide some answers to this question. Platek et al. (2005) reported different patterns of neural activity in response to watching video clips of yawning and laughing. In particular, along with brain regions associated with general face perception, there was specific activation in posterior midline cortical regions that have been associated with other aspects of self-processing, such as autobiographical memory and self-monitoring. Schürmann et al. (2005) reported greater activation in parts of the superior temporal sulcus in response to watching yawn sequences than non-yawn mouth movements. This region is known to be associated with the perception of biological motion.

Importantly, neither of these recent studies found evidence of specific activation of the so-called mirror neuron system during exposure to yawns, or of prefrontal cortical regions thought to be important in theory of mind. These results indicate that whatever underlies contagious yawning, it does not appear to be based either on conscious imitation or higher-level, "conscious" empathy. More studies on a range of nonhuman species and experimental conditions should be useful for identifying which links in the perception–action chain (see Preston and de Waal 2002) are involved in this curious and fascinating behavior.

Acknowledgment

The present study was supported by the grants (#12002009, #16002001) from MEXT and JSPS core-to-core-program of HOPE. Thanks are due to Masako Myowa-Yamakoshi, Yuu Mizuno, and Misato Hayashi for help with carrying out the experiments in chimpanzees, to Kiyonori Kumazaki and Norihiko Maeda for the care of chimpanzees, and to Kazuo Fujita for useful discussion.

References

Adams DB, Schoel WM (1982) A statistical analysis of the social behavior of the male stump-tail macaque (*Macaca arctoides*). Am J Primatol 2:249–273

Anderson JR, Gallup GG Jr (1999) Self-recognition in primates: past and future challenges. In: Haug M, Whalen RE (eds) Animal models of human emotion and cognition. American Psychological Association, Washington, DC, pp 175–194

Anderson JR, Meno P (2003) Psychological influences on yawning in children. Curr Psychol Lett: Behav Brain Cogn 11. See http://cpl.revues.org/document390.html.

Anderson JR, Myowa-Yamakoshi M, Matsuzawa T (2004) Contagious yawning in chimpanzees. Proc R Soc Lond B (Suppl) 271:S468–S470
Baenninger R (1987) Some comparative aspects of yawning in *Betta splendens, Homo sapiens, Panthera leo,* and *Papio sphinx*. J Comp Psychol 101:349–354
Byrne R (1995) The thinking ape: evolutionary origins of intelligence. Oxford University Press, Oxford
Byrne RW, Whiten A (1992) Cognitive evolution in primates: evidence from tactical deception. Man 27:609–627
Darwin C (1965/1872) The expression of the emotions in man and animals. University of Chicago Press, Chicago
Deputte BL (1994) Ethological study of yawning in primates. I. Quantitative analysis and study of causation in two species of Old World monkeys (*Cercocebus albigena* and *Macaca fascicularis*). Ethology 98:221–245
de Waal F (1996) Good natured: the origins of right and wrong in humans. Harvard University Press, Cambridge, MA
de Waal FBM, Aureli F (1997) Conflict resolution and distress alleviation in monkeys and apes. Ann N Y Acad Sci 807:317–328
Gallup GG Jr (1970) Chimpanzees: self-recognition. Science 167:86–87
Gallup GG Jr (1982) Self-awareness and the emergence of mind in primates. Am J Primatol 2:237–248
Gallup GG Jr (1998) Can animals empathize? Yes. Sci Am 9:65–75
Goodall J (1968) The behaviour of free-living chimpanzees in the Gombe Stream Reserve. Anim Behav Monogr 1:161–311
Goodall J (1986) The chimpanzees of Gombe: patterns of behavior. Harvard University Press, Cambridge, MA
Hare B, Call J, Agnetta B, Tomasello M (2000) Chimpanzees know what conspecifics do and do not see. Anim Behav 59:771–786
Heyes C (1998) Theory of mind in nonhuman primates. Behav Brain Sci 21:101–134
Hirata S, Matsuzawa T (2001) Tactics to obtain a hidden food item in chimpanzee pairs (*Pan troglodytes*). Anim Cogn 4:285–295
Itakura S, Tanaka M (1998) Use of experimenter-given cues during object-choice tasks by chimpanzees (*Pan troglodytes*), an orangutan (*Pongo pygmaeus*), and human infants (*Homo sapiens*). J Comp Psychol 112:119–126
Itakura S, Agnetta B, Hare B, Tomasello M (1999) Chimpanzees use human and conspecific social cues to locate hidden food. Dev Sci 2:448–456
Lehmann HE (1979) Yawning: a homeostatic reflex and its psychological significance. Bull Menninger Clin 43:123–136
Matsuzawa T (2000) Chimpanzee mind 1995–2000: collection of articles. Research report by the grant-in-aid from the Ministry of Education, Culture, Sports, Science and Technology, Japan
Matsuzawa T (2001) Primate origins of human cognition and behavior. Springer, Tokyo
Matsuzawa T (2003) The Ai project: historical and ecological contexts. Anim Cogn 6:199–211
Matsuzawa T (2005) Chimpanzee mind 2000–2005: collection of articles. Research report by the grant-in-aid from the Ministry of Education, Culture, Sports, Science and Technology, Japan
McKenzie AA (1994) The tonsillar evacuation hypothesis of yawning behaviour. S Afr J Sci 90:64–66
Menzel EW (1974) A group of chimpanzees in a 1-acre field: leadership and communication. In: Schrier AM, Stollnitz F (eds) Behavior of nonhuman primates. Academic Press, New York, pp 83–153
Morimura N, Matsuzawa T (2001) Memory of movies by chimpanzees (*Pan troglodytes*). J Comp Psychol 115:152–158
O'Connell SM (1995) Empathy in chimpanzees: evidence for theory of mind? Primates 36:397–410
Platek SM, Critton SR, Myers TE, Gallup GG Jr (2003) Contagious yawning: the role of self-awareness and mental state attribution. Cogn Brain Res 17:223–237

Platek SM, Mohamed FB, Gallup GG Jr (2005) Contagious yawning and the brain. Cogn Brain Res 23:448–452
Povinelli DJ (1998) Can animals empathize? Maybe not. Sci Am 9:65–75
Povinelli DJ, Vonk J (2003) Chimpanzees minds: suspiciously human? Trends Cogn Sci 7:157–160
Povinelli DJ, Nelson KE, Boysen ST (1990) Inferences about guessing and knowing by chimpanzees (*Pan troglodytes*). J Comp Psychol 104:203–210
Premack D, Woodruff G (1978) Does the chimpanzee have a theory of mind? Behav Brain Sci 1:515–526
Preston SD, de Waal FBM (2002) Empathy: Its ultimate and proximate bases. Behav Brain Sci 25:1–72
Provine RR (1986) Yawning as a stereotyped action pattern and releasing stimulus. Ethology 72:448–455
Schürmann M, Hesse MD, Stephan KE, Saarela M, Zilles K, Hari R, Fink GR (2005) Yearning to yawn: the neural basis of contagious yawning. Neuroimage 24:1260–1264
Smith EO (1999) Yawning: an evolutionary perspective. Hum Evol 14:191–198
Tomasello M, Call J, Hare B (2003) Chimpanzees understand psychological states: the question is which ones and to what extent. Trends Cogn Sci 7:153–156

16
How Social Influences Affect Food Neophobia in Captive Chimpanzees: A Comparative Approach

Elsa Addessi and Elisabetta Visalberghi

1 Introduction

Humans, chimpanzees, and capuchin monkeys are all species facing the "omnivore's dilemma" (Rozin 1977), that is, their success depends both on the propensity to eat novel foods and on the caution to explore and sample them. On the one hand, an omnivorous species should look for novel foods to enlarge its diet, to adapt to different environments, and to overcome shortage; on the other hand, it should detect and avoid the risk of ingesting poisonous substances (Freeland and Janzen 1974; Glander 1982; Milton 1993).

Food neophobia, that is, the hesitancy to eat novel foods so that only a small amount is tasted or ingested, is often viewed as a strategy to reduce the risk of being poisoned by ingesting too-large quantities of a novel food (Barnett 1963; Freeland and Janzen 1974; Glander 1982) while learning about its toxicity by ingesting a small amount and experiencing the consequences (Garcia et al. 1955; Garcia and Koelling 1966; Matsuzawa and Hasegawa 1983; Matsuzawa et al. 1983). This behavioral trait is common to many animal species, from birds to humans (warblers, *Dendroica castanea* and *D. pensylvanica*: Greenberg 1990; rats, *Rattus norvegicus:* Barnett 1958; Galef 1970; lambs, *Ovis aries*: Burritt and Provenza 1989; Provenza et al. 1995; capuchin monkeys, *Cebus apella:* Visalberghi and Fragaszy 1995; Visalberghi, et al. 2003a; rhesus macaques, *Macaca mulatta*: Weiskrantz and Cowey 1963; Johnson 1997, 2000; bonobos, *Pan paniscus*: Kano 1992; humans, Rozin 1976). In nonhuman primates, food neophobia is usually low in infants and juveniles (tufted capuchin monkeys: Fragaszy et al. 1997; Visalberghi et al. 2003a; chimpanzees: Ueno and Matsuzawa 2005) and such age differences are also evident in humans, where 4- to 7-month-old infants are less neophobic than 2- to 5-year-old children (Sullivan and Birch 1994; Birch et al. 1998).

However, because omnivores increase their chances of survival through a varied diet, food neophobia in the long run might be maladaptive and should be overcome. In fact, in primates food neophobia is influenced by several factors, such as the type of food (tufted capuchin monkeys: Visalberghi and Fragaszy

Unit of Cognitive Primatology, Institute of Cognitive Sciences and Technologies, CNR, Via Ulisse Aldrovandi 16/b, 00197 Rome, Italy

1995; chimpanzees: Visalberghi et al. 2002), the number of exposures to a novel food (tufted capuchin monkeys: Visalberghi et al. 1998; children: Birch and Marlin 1982; Sullivan and Birch 1990; Wardle et al. 2003), and the postingestion consequences (Japanese macaques: Matsuzawa and Hasegawa 1983; Matsuzawa 1983). Moreover, there is a large body of evidence that in nonhuman primates social influences increase the acceptance of novel foods (chacma baboons, *Papio ursinus*: Cambefort 1981; tufted capuchin monkeys: Visalberghi and Fragaszy 1995; Visalberghi et al. 1998; marmosets, *Callithrix jacchus*: Vitale and Queyras 1997; Yamamoto and Lopes 2004; Schrauf et al. 2004). Similarly, when children observe a familiar adult eating an unfamiliar food, they tend to eat more food compared to the situation in which the food is simply offered to them (Harper and Sanders 1975). Moreover, in adult humans the observation of the behavior of a model reduces food neophobia (Hobden and Pilner 1995).

Group living can be of great advantage in learning when, how, and on what to feed (Giraldeau 1997), and most research concerning social influences on feeding behavior has focused on the widely expressed view that social animals gain knowledge about food from conspecifics (Galef and Giraldeau 2001). However, it should be noted that if relying on others' eating behavior helps the individual to deal safely with novel foods, then it is mandatory that learning takes into account food specificity. For example, if artichokes are a novel food for individual A, then watching group members eating artichokes should affect A's likelihood of eating artichokes and not A's likelihood of eating beets; in other words, watching group members eating artichokes is a congruent source of information only about the safety of eating that same food, and not another one. Given this, to support the view that social influences foster a safe diet, we should reject two alternative explanations: the acceptance of a novel food increases (1) because of the mere presence of group members or (2) because group members are eating, regardless of what they are eating.

Surprisingly, only recently has the question of what exactly is learned from others about novel foods been experimentally addressed. We recently demonstrated that capuchins are more likely to eat a novel food when seeing group members eating (but not when group members are merely present) than when alone (Visalberghi and Addessi 2000, 2001; see also Addessi and Visalberghi 2001). However, this increase in acceptance occurs regardless of the color[1] of the food eaten by the group members. Therefore, at least in capuchins, the influence of group members allows an individual to overcome neophobia but not to learn that a food is safe. In contrast, children tested with a similar experimental paradigm are selectively oriented toward specific food targets, that is, they eat more of a novel food only when their food and the models' food are identical (Addessi et al. 2005).

The present study investigates how social influences affect neophobia and food choice in young and adult captive chimpanzees, a species more cognitively

[1]To make the observer's and the demonstrators' foods look very different from one another, they were dyed two strikingly different colors.

advanced than capuchin monkeys but less so than children (Tomasello and Call 1997). Similarly to capuchin monkeys, chimpanzees adapt to a wide range of different habitats and to food seasonality (Goodall 1986; Teleki 1989; Sugiyama and Koman 1992) by exploiting a great variety of food sources. However, very little is known about how chimpanzees respond to novel foods. According to Nishida et al. (1983), wild chimpanzees are conservative and unwilling to taste novel foods, although young individuals seem more likely to do so. When Matsuzawa and Yamakoshi (1996; see also Matsuzawa 1999; Chapter 28) provided Bossou chimpanzees with coula nuts (*Coula edulis*) in the area where they usually spend their time cracking oil-palm nuts (*Elaeis guineensis*), some chimpanzees sniffed the coula nuts, picked them up, and tried to bite them, whereas others simply ignored them. Coula nuts (whose appearance is rather different from that of oil-palm nuts) are not available in the home range of the Bossou chimpanzees but they are present in a nearby area, where the Nimba chimpanzee community lives. Only one female, probably born in the Nimba community, and therefore probably familiar with coula nuts, cracked the coula nuts open with a tool. Her behavior elicited great interest from a group of juvenile chimpanzees, some of which successfully cracked open the coula nuts in the following days.

Captive chimpanzees as well are cautious toward novel food, although in infants novel foods elicit more interest than familiar ones (Ueno and Matsuzawa 2005; see also Chapter 12; for similar findings in capuchins, see Fragaszy et al. 1997; Drapier et al. 2003). In a recent experiment, eight adult chimpanzees were offered 16 novel foods (Visalberghi et al. 2002). Marked interindividual differences in food acceptance and consumption emerged, and chimpanzees ranged from being almost completely neophobic to accepting almost all foods. Moreover, to assess whether seeing the novel food eaten by a human demonstrator affects its consumption, each novel food was presented twice with the experimenter eating the food in the presence of the chimpanzee (demonstration). Although chimpanzees were always attentive to the human eating the novel food, the demonstration did not affect the acceptance of the novel foods. The negative outcome of this study could be attributed to the type of demonstrator used, namely, a human subject. In fact, a human demonstrator may lack the salience of a conspecific demonstrator (see also Whiten and Ham 1992), and a chimpanzee–human dyad is not the setting in which information about food is expected to be transferred (although for captive chimpanzees, interactions with humans are important). Therefore, in the present study we aimed to investigate whether chimpanzees learn about food safety from the behavior of their group members (as in the capuchin monkey studies; for a review see Visalberghi and Addessi 2003; Addessi and Visalberghi, 2006). By increasing the salience of the visual input provided by the conspecific, we expected chimpanzees to pay attention not only to the behavior of the demonstrators but also to what they were eating and to behave accordingly. We tested this hypothesis both when the observer encountered only one food (experiment 1) and when the subject was presented with a choice between two novel foods (experiment 2), of which only one matched the color of the food eaten by the demonstrator. These two experiments simulate rather common situations in nature (Goodall 1986). A wild chimpanzee may well encounter a novel food while her or his group members are eating the same food, a different food, or when she or he is alone. Similarly, the chimpanzee can encounter two or more novel foods while her or his group members are selectively eating only one food.

2 Experiment 1: Does the Observation of Group Members Eating Affect the Observer's Acceptance of a Novel Food?

2.1 Methods

2.1.1 Subjects

As observers, we tested nine captive-born chimpanzees (one adult male, two adult females, two juvenile males, and four juvenile females; Table 1) living in the same group at the Wolfgang Köhler Primate Research Center/Leipzig Zoo, Max Planck Institute for Evolutionary Anthropology (Leipzig, Germany).

They were housed in indoor–outdoor areas. The indoor area consists of an indoor enclosure (430 m^2) and sleeping and observation booths (47 and 25 m^2, respectively). The outdoor area (4000 m^2) and the indoor enclosure contain natural vegetation, climbing structures, trees, water streams, and various other natural features.

Chimpanzees were fed three times a day, and browse was provided on a regular basis. In addition, they received other foods according to seasonal availability (e.g., chestnuts); other opportunities for special foraging activities (e.g., at artificial termite mounds) were also made available on a regular basis.

Table 1. Experiment 1. Assignment of foods to the experimental conditions within each block of subjects

	Experimental conditions	Subjects	Sex	Age (years)
Experimental group 1				
yellow food	Alone	Robert	M	26
green food	Different color	Trudi	F	8
red food	Same color	Fifi	F	8
Experimental group 2				
yellow food	Same color	Ulla	F	24
green food	Alone	Frodo	M	8
red food	Different color	Jahaga	F	8
Experimental group 3				
yellow food	Different color	Patrick	M	4
green food	Same color	Sandra	F	8
red food	Alone	Fraukje	F	25

In the "different-color" and "same-color" conditions (see following), the observer was tested with two demonstrators. All the subjects acted both as observer and as demonstrator, except one adult female who acted as observer only.

2.1.2 Apparatus

Chimpanzees were tested in two adjacent booths. Each booth had a Plexiglas window (73 cm × 60.5 cm × 2 cm) on its side through which the observer and the demonstrators could see each other. Between these two windows there was a space of 94 cm, in which a wooden table (93 cm × 78.5 cm × 44 cm) was inserted. The foods (see following) were placed on the wooden table. The observer and the demonstrators could reach the food through three holes (diameter, 6 cm; the holes were 23 cm apart from each other) made in their Plexiglas window.

2.1.3 Foods

We used three foods never previously tasted by the observers. They were colored green, red or yellow with Brauns Heitmann's Crazy Colors food coloring (2 ml/100 g): mashed canned green peas, colored green, mashed boiled lentils, colored red, and mashed boiled beans, colored yellow; the caloric content was 285 kj/100 g for green peas, 386 kj/100 g for lentils, and 380 kj/100 g for beans. Both sexes can discriminate these colors (Jacobs et al. 1996). Novel foods were blended in a food processor. They were all of similar texture and not particularly attractive to the apes.

The familiar food for the demonstrators was a mixture of peeled, boiled, and mashed potatoes and yogurt (in ratio 4:1; total, 500 g). In the same-color condition (see following) the potatoes were colored yellow, red, or green with Brauns Heitmann's Crazy Colors food coloring (3 ml/100 g), whereas in the different-color condition the potatoes were not colored. All the subjects liked this food very much, regardless of its color.

2.1.4 Procedure

As shown in Fig. 1, there were three experimental conditions, as follows. (a) Alone: the observer was in one booth and received the novel food, while the demonstrators were not in the adjacent booth. (b) Different color: the observer was in one booth and received the novel food, while two demonstrators were in the adjacent booth and received the not-colored familiar food. The colors of the familiar and the novel food were clearly different from one another. (c) Same color: the observer was in one booth and received the novel food, while two demonstrators were in the adjacent booth and received the familiar food of the same color as the novel food. The colors of the familiar and the novel food were identical. In each condition, the observer was given 250 g novel food; in the different-color and same-color conditions, the demonstrators were given 500 g familiar food (Fig. 2).

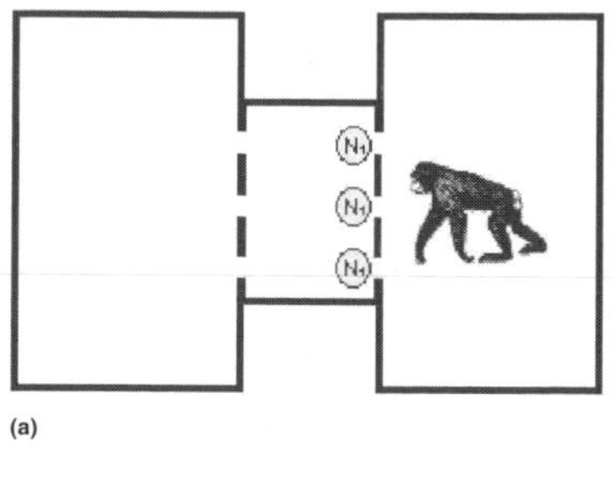

(a)

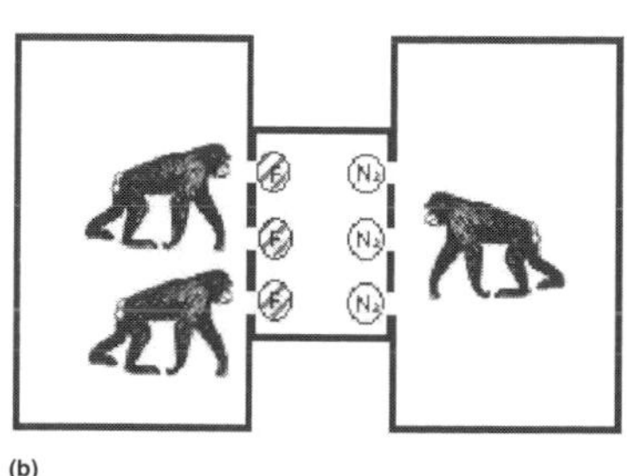

(b)

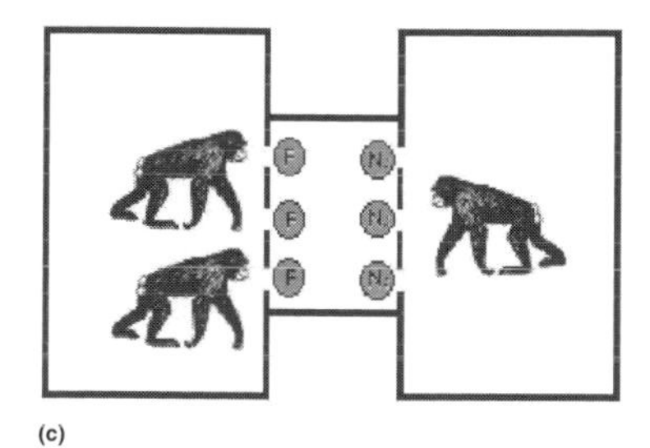

(c)

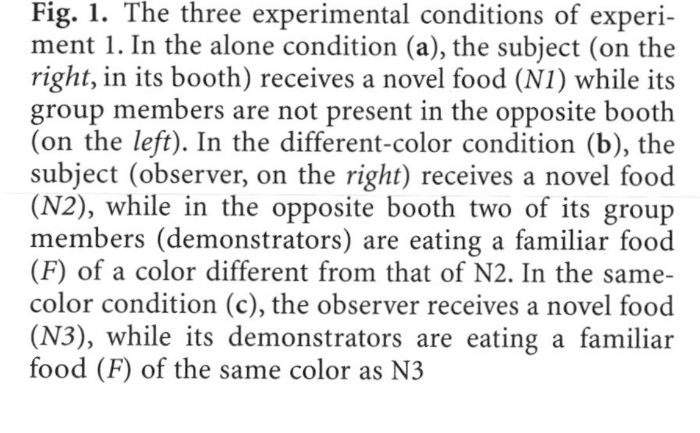

Fig. 1. The three experimental conditions of experiment 1. In the alone condition (**a**), the subject (on the *right*, in its booth) receives a novel food (*N1*) while its group members are not present in the opposite booth (on the *left*). In the different-color condition (**b**), the subject (observer, on the *right*) receives a novel food (*N2*), while in the opposite booth two of its group members (demonstrators) are eating a familiar food (*F*) of a color different from that of N2. In the same-color condition (**c**), the observer receives a novel food (*N3*), while its demonstrators are eating a familiar food (*F*) of the same color as N3

Trials were conducted at 9 a.m., 11 a.m., or 4 p.m. Chimpanzees were not food deprived; they routinely received a light meal in the early morning and in the early afternoon, so thus received this food before testing.

Because a food can be novel only once, we assigned each of the three foods to one of the three conditions within each block of three observers. Block 1 received the beans in the alone condition, the green peas in the different-color condition, and the lentils in the same-color condition; block 2 received the lentils in the alone condition, the beans in the different-color condition, and the green peas in the same-color condition; and block 3 received the green peas in the alone condition, the lentils in the different-color condition and the beans in the same-color condition (see Table 1). Observers belonging to each block were balanced for age and sex and these assignments were counterbalanced across observers.

We tested every observer in each of the three conditions once; presentations occurred on different test days. The order in which the three conditions were presented was counterbalanced across observers and blocks. Each session lasted 5 min and started as soon as the observer entered its booth and, in the different-color and same-color conditions, could see the demonstrators, which were already in their booth, eating the familiar food. We carried out the experiment between December 2001 and January 2002.

2.1.5 Behaviors Scored

An experimenter scored the behavior of the observer with instantaneous sampling every 10 s, while a fixed video camera was set to record the area by the panel in the demonstrators' booth. The following observer behaviors were scored: (1) eating behavior, that is, putting food in the mouth and chewing it or chewing food already in mouth; (2) exploration, that is, sniffing the food; (3) and visual attention to the demonstrators, both when they were eating and not eating. In addition, the latency to ingestion was scored. The novel food given to the observer was weighed before and after the session to measure the amount of food eaten.

To assess the input the observer received from its demonstrators, we scored their behavior from videotapes every 10 s (at the same sample point as the observer's behavior was scored). We counted the number of demonstrators eating in the area in front of the panel.

2.1.6 Analysis

For each behavior scored, for the latency to ingestion and for the amount of food eaten, we carried out the Friedman analysis of variance (ANOVA) for comparisons across conditions and the Wilcoxon signed-ranks test for comparisons between conditions. Moreover, we used the Mann–Whitney U test to assess whether age and sex affected observers' behavior toward novel food. Given our small sample size, we used the exact variant of each statistical test.

For the different-color and the same-color conditions, we ran Spearman correlations to evaluate (1) the relationship between the average number of observers showing visual attention to the eating demonstrators at each sample point and the average number of demonstrators eating at the same sample point, and (2) the relationship between the average number of observers eating at each sample point and the average number of demonstrators eating at the same sample point.

2.2 Results

2.2.1 Behavior of the Demonstrators

Demonstrators spent most of the session eating in the area in front of the observer. In the different-color condition there was at least one demonstrator present and eating the familiar food in 96.3% of the samples. In the same-color condition, the corresponding value was 99.0%. In fact, at the end of each session, no potato leftovers were found.

2.2.2 Behavior of the Observers

Observers showed more visual attention for the demonstrators when the demonstrators were eating than when they were not eating (different-color condition: T (the value of the Wilcoxon test) = 0, $P < 0.05$; same-color condition: $T = 0$, $P < 0.05$). In both conditions, the average number of observers showing visual attention to the eating demonstrators at each sample point was correlated with the average number of demonstrators eating at the same sample point (different-color condition: $r_s = 0.42$, $n = 30$, $P < 0.05$; same-color condition: $r_s = 0.5$, $n = 30$, $P < 0.01$). Observers' visual attention to the eating demonstrators did not differ between the different-color and same-color conditions ($T = 9.5$, $n = 9$; NS).

Observers' behavior toward the novel food did not significantly differ across conditions (latency to ingestion: $\chi^2 = 0.97$, $n = 9$; NS; exploration: $\chi^2 = 1.0$, $n = 9$; NS; eating: $\chi^2 = 0.62$, $n = 9$; NS; food eaten: $\chi^2 = 0.76$, $n = 9$; NS). Moreover, the average number of observers eating at each sample point was not significantly correlated with the average number of demonstrators eating at the same sample point (different-color condition: $r_s = -0.32$, $N = 30$; NS; same-color condition: $r_s = -0.29$, $n = 30$; NS).

Similarly, observers' behavior did not differ across trials (exploration: $\chi^2 = 5.29$, $n = 9$; NS; eating: $\chi^2 = 0.83$, $n = 9$; NS; food eaten: $\chi^2 = 0.06$, $n = 9$; NS), the only exception being latency to ingestion ($\chi^2 = 6.61$, $n = 9$; $P < 0.05$). In particular, it was significantly shorter in trial 3 than in trial 2 ($T = 0$, $P < 0.05$). Further analysis showed that this was the result of an age difference; in particular, in the third trial, juveniles showed a shorter latency to ingestion than adults ($U = 1.0$, $n_1 = 3$, $n_2 = 6$; $P < 0.05$). In the alone condition, age significantly affected also eating behavior and food eaten, with the juveniles eating and ingesting more novel food than the adults (eating behavior: $U = 1.5$, $n_1 = 3$, $n_2 = 6$; $P < 0.05$; food eaten: $U = 1.0$, $n_1 = 3$, $n_2 = 6$; $P < 0.05$). Sex did not significantly affect observers' behavior.

3 Experiment 2: Do Social Influences Affect the Observer's Choice Between Differently Colored Novel Food?

3.1 Methods

3.1.1 Subjects

As observers, we tested eight captive-born chimpanzees, the same subjects as in experiment 1, except for the alpha male. Each observer was tested with one demonstrator, and all the subjects acted both as observer and as demonstrator.

3.1.2 Apparatus

The experiment took place in the same booths as in experiment 1. All features of the apparatus were as described in experiment 1.

3.1.3 Foods

As novel food for the observer, we used mashed boiled chickpeas, a food never previously tasted by the subjects. The caloric content was 417 kj/100 g. According to the experimental group (Table 2), it was colored brown and violet or orange and blue with Brauns Heitmann's Crazy Colors food coloring (4 ml/100 g). Both sexes discriminate these colors (Jacobs et al. 1996). Novel foods were blended in a food processor. They were all of similar texture and not particularly attractive to the apes.

The familiar food for the demonstrators was a mixture of peeled, boiled, and mashed potatoes and yogurt (in ratio 4 : 1; total, 500 g). According to the experimental group (see following), it was colored brown, violet, orange, or blue with Brauns Heitmann's Crazy Colors food coloring (4 ml/100 g).

3.1.4 Procedure

The observer was in one booth and received chickpeas colored brown and violet or orange and blue, according to the experimental group. The demonstrator was in the adjacent booth and received potatoes colored brown, violet, orange, or blue, according to the experimental group (see Table 2). The observer was given 125 g each novel food; the demonstrator was given 250 g familiar food. In a session, the observer had simultaneous access to chickpeas whose color matched the color of the food eaten by the demonstrator (hereafter called "matching-color food") and to chickpeas whose color did not match the color of the food eaten by the demonstrator (hereafter called "nonmatching-color food") (Fig. 3).

We divided the eight subjects into four experimental groups ($n = 2$ each) and assigned each group to one matching color. Experimental group 1 had access to brown and violet chickpeas while the demonstrator was given brown potatoes; experimental group 2 had access to brown and violet chickpeas while the demonstrator was given violet potatoes; experimental group 3 had access to orange and blue chickpeas while the demonstrator was given orange potatoes;

Table 2. Experiment 2. Assignment of the observers to the experimental groups

Observers	Colors of the observer's novel foods	Experimental groups (defined by the color of the food eaten by the demonstrator)
Fraukje	Brown-Violet	Brown
Fifi	Brown-Violet	Brown
Sandra	Brown-Violet	Violet
Patrick	Brown-Violet	Violet
Ulla	Orange-Blue	Orange
Jahaga	Orange-Blue	Orange
Trudi	Orange-Blue	Blue
Frodo	Orange-Blue	Blue

Fig. 2. Experiment 1. Demonstrators are eating the familiar food, positioned on the table in front of the holes present in their window

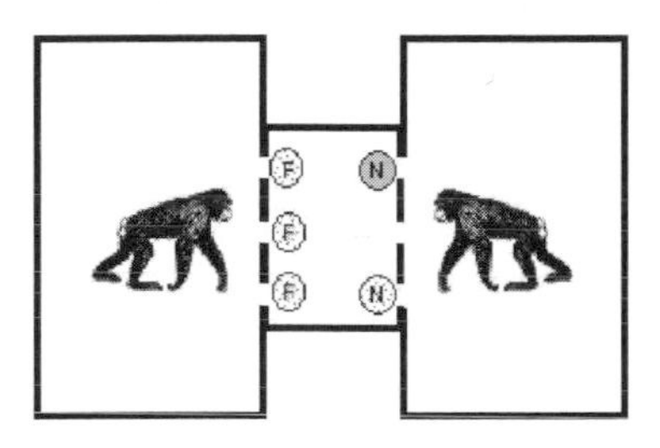

Fig. 3. Experiment 2. The observer (on the *right*, in its booth) receives a novel food of two different colors, while its demonstrator (on the *left*, in its booth) is eating a familiar food whose color matches the color of one of the two observer's foods

experimental group 4 had access to orange and blue chickpeas while the demonstrator was given blue potatoes (see Table 2). The position of the colored chickpeas was counterbalanced across subjects.

We tested every subject once. The session lasted 5 min and started as soon as the observer entered its booth and could see the demonstrator, already in her/his booth eating the familiar food. We carried out the experiment in January 2002.

3.1.5 Behaviors Scored

Behaviors scored were the same as in experiment 1. In addition, we scored the first choice of the observer, that is, which of the two foods the observer chose first.

3.1.6 Analysis

For each behavior scored, for the latency to ingestion and for the amount of food eaten, we carried out the Wilcoxon signed-ranks test for comparing the behavior between the matching-color food and the nonmatching-color food. As in experiment 1, we used the Mann–Whitney U test to assess whether age and sex affected observers' behavior toward novel food. Moreover, we ran Spearman correlations to evaluate (1) the relationship between the average number of observers showing visual attention to the eating demonstrators at each sample point and the average number of demonstrators eating at the same sample point, and (2) the relationship between the average number of observers eating the matching-color food or the nonmatching-color food at each sample point and the average number of demonstrators eating at the same sample point. Given our small sample size, we used the exact variant of each statistical test.

3.2 Results

3.2.1 Behavior of the Demonstrators

As in experiment 1, demonstrators gave to the observers the input requested by the experimental design and they ate in front of the observer in 99.6% of the sample points, on average.

3.2.2 Behavior of the Observers

Observers showed more visual attention for the demonstrators when the demonstrators were eating than when they were not eating ($T = 2.0$, $n = 8$; $P < 0.05$). However, the average number of observers showing visual attention to the eating demonstrators at each sample point was not significantly correlated with the average number of demonstrators eating at the same sample point ($r_s = 0.12$, $n = 30$; NS).

Observers' behavior toward the novel food did not significantly differ between the matching-color food and the nonmatching-color food (latency to ingestion: $T = 16.0$, $n = 8$; NS; exploration: $T = 0$, $n = 8$; NS; eating: $T = 10.0$, $n = 8$; NS; food eaten: $T = 13.0$, $n = 8$; NS) and the number of times in which the observers chose to eat first the matching-color food rather than the nonmatching-color food first was equal. The average number of observers eating at each sample point was not correlated with the average number of demonstrators eating at the same sample point (matching-color food: $r_s = 0.25$, $n = 30$; NS; nonmatching-color food: $r_s = 0.10$, $n = 30$; NS). Age and sex did not significantly affect observers' eating behavior.

4 Discussion

Chimpanzees eagerly ate the familiar food given to them when they played the role of demonstrators. In addition, we never witnessed a case in which they did not accept and eat a familiar food. Therefore, although we lack a control condition in which each observer is individually presented with familiar food, we can argue that captive chimpanzees are cautious toward novel food (see also Visalberghi et al. 2002). Moreover, in the absence of group members young chimpanzees are less neophobic than adults, which is in agreement with the observation that infant chimpanzees are more interested in novel foods than in familiar ones (Ueno and Matsuzawa 2005; see also Chapter 11). Therefore, age seems to affect food neophobia in chimpanzees, as it does in capuchin monkeys both in captivity (Fragaszy et al. 1997) and in the wild (Visalberghi et al. 2003a). According to Janson and van Schaik (1993), age differences in neophobia are the result of the youngsters' low foraging efficiency and great risk of starvation, which makes them willing to try new food. Many years ago, Itani (1958) had argued that by being more explorative young individuals might discover food sources or feeding strategies not yet exploited by their group members.

Regardless of age, chimpanzees having group members as demonstrators were not more inclined to accept novel foods or to choose a food matching in color the demonstrator's food. Surprisingly, this lack of social influence occurs despite the demonstrators eating and the observers monitoring with attention the demonstrators' behavior and food choice. Chimpanzees behaved differently from both children and capuchins. Preschool children tested with a similar paradigm were socially facilitated only when their food matched the food of the demonstrator; that is, they were sensitive to what the demonstrator was eating (Addessi et al., 2005). Capuchin monkeys accepted and ate novel foods more when facing group members eating than when alone, even if social facilitation of eating (i.e., the increased likelihood to eat when somebody else is eating; see also Clayton 1978) occurred regardless of what the demonstrators were eating (for a review, see Visalberghi and Addessi 2003; Addessi and Visalberghi, 2006). In other words, in children social influences might serve to learn about a safe diet, whereas in capuchin they serve to reduce neophobia without necessarily leading to a safe diet.

Overall our experiments with chimpanzees led to negative results, which by itself is unfortunate. However, we can rule out the possibilities that a poor experimental design and/or a small sample size accounted for the lack of the significance of most results. In fact, in other primate species the same paradigm has been successful (Visalberghi and Addessi 2003; Addessi and Visalberghi, 2006; Addessi et al., 2005). Then, how can we explain our results? First, it is possible that social facilitation of eating behavior necessitates the subjects to be hungry; this was certainly not the case in our present experiment, as well as in our previous ones, because our subjects were never food deprived before testing. However, individuals of other species did not need to be hungry for social facilitation of eating to occur. Therefore, this hypothesis is unlikely as well as impossible to test for ethical reasons. Second, captive chimpanzees might be prone to accept any food either because they are rarely, if ever, faced with potentially poisonous substances or because they do not distinguish novel from familiar foods. The behavior of our subjects did not support this second claim. Chimpanzees were cautious toward novel foods despite being in captivity (see also Visalberghi et al. 2002) and clearly treated them very differently from the familiar food they received as demonstrators. A third possibility is that eating behavior was socially facilitated and that chimpanzees inhibit such response, regardless of the type of food eaten by the demonstrator. This hypothesis is very speculative but worthwhile discussing because it allows us to interpret our data within a comparative framework.

In macaques, the observation of actions related to eating behavior activates motor representations similar to those observed. The basic mechanism underlying this phenomenon has to do with a class of visuomotor neurons, named mirror neurons, found in the macaque's premotor and parietal cortex (Gallese et al. 1996, 2002; Fogassi et al. 2005). The mirror neurons become active both when the monkey makes a specific action with its hand (or mouth) and when the monkey observes similar hand (or mouth) actions performed by another individual. Several brain imaging studies support the existence of a mirror system in humans involving the frontal and parietal areas, which are homologous to those in which mirror neurons have been found in macaques (see Rizzolatti et al. 2001). Other evidence in support of the existence of this system in humans derives from transcranial magnetic stimulation (TMS). By using this technique it has been shown that the observation of others' actions increases motor corticospinal excitability measured from various arm, hand, and mouth muscles (see Fadiga et al. 2005 for a review).

Based on the properties of mirror neurons, it was suggested that they are part of a neural system implicated in the process of action recognition, in which the visual description of an action is matched with its motor outcome (Gallese et al. 1996; Rizzolatti et al. 2001). More recently, Ferrari et al. (2005) argued that mirror neurons could be also involved in the process of response facilitation, that is, the repetition of an observed action already part of the observer's motor repertoire (as, for example, eating behavior). The property of mirror neurons to couple observed and executed actions would be suitable to allow them to participate in the response facilitation phenomenon by means of a "resonance" mechanism in which the motor system of the observer is activated specifically by observing others' actions. Evidence that this may occur derives from the TMS studied already mentioned in which action observation may increase the excitability of the corticospinal tract as a consequence of the enhancement of the motor cortex excitability.

However, with the exception of very few behaviors recently described (e.g., eating behavior; see Ferrari et al. 2005), the observation of others' behavior in primates, although this activates motor representation in the premotor cortex, does not lead to overt movement (see Chapter 13). This apparent discrepancy between neurophysiological data and behavioral data can be partly explained by the fact that during action observation the modulation pattern of corticospinal excitability can be suppressed at the spinal level, as recently demonstrated in humans using TMS (Baldissera et al. 2001; Fadiga et al. 2005). This effect has been interpreted as the expression of a mechanism serving to block overt execution of seen actions. In humans it is likely that inhibition mechanisms suppressing overt movement during action observation could involve cortical frontal areas. Support for this point comes from patients suffering from lesions in cortical frontal regions of the brain. Patients with this lesions (echopraxia) repeat automatically any action seen made by another individual, without any possibility of inhibiting repetition (Dromard 1905; Stengel et al. 1947; Lhermitte et al. 1986); it has been proposed that this uncontrolled repetition probably occurs because of a lack of inhibition at the level of the mirror system (see Rizzolatti et al. 1999). It is also possible that during development this inhibitory system require months to be fully developed. Thus, some behavioral phenomena in infants, such as the repetition of mouth movements after having observed them in adults (i.e., newborn imitation; Meltzoff and Moore 1977) could reflect this phenomenon.

Given the foregoing findings, it can be hypothesized that capuchins, chimpanzees, and humans differ in their capability to inhibit responses, with capuchins being unable to inhibit facilitated eating responses, chimpanzees being capable of inhibiting facilitated eating responses but incapable of selective inhibition, and, finally, young children being capable of selective inhibition. Future research should better investigate whether selective inhibition is affected by neuroanatomical–neurophysiological complexity and the extent to which different primate species (and individuals of different age) are able to selectively inhibit facilitated eating responses.

Finally, the following three questions still remain to be answered. (i) How can nonhuman primates learn so little by watching the behavior of others but still learn what to eat and what not to eat? (ii) How can they detect and avoid noxious foods? (iii) Why is there a strong convergence on what the individuals in a group feed on? The flavor and nutrient content of foods play an important role in determining individual choices and preferences (Simmen and Hladik 1998; Laska et al. 2000; Hladik et al. 2002; Visalberghi et al. 2003b). Taste perception provides an immediate and powerful feedback, allowing assessment of food quality (Dominy et al. 2001). Primates dislike bitter flavors, which are associated with the presence of secondary compounds (such as alkaloids and glycosides) and that can cause severe illness or even have lethal effects (Freeland and Janzen 1974; Ueno 2001). Conversely, all primate species tested so far readily accept sugars, a very important energy source (Glaser 1993). Both in human and in nonhuman primates, acceptance and rejection responses are evident already in newborns, before experiencing any consequences from the ingestion of sweet or bitter substances (gustofacial reflex; Steiner and Glaser 1984; Ueno et al. 2004).

However, taste perception is not always a reliable cue for selecting what to eat and what to avoid; in fact, some substances, as for example the lethal alkaloid dioscine, are almost tasteless (Hladik and Simmen 1996). Therefore, it is not surprising that an additional system comes into play. Nutrients and/or toxic compounds provide a feedback that animals are able to associate with the sensory properties of the food, and the metabolic consequences of eating a food efficiently directs animals' future food selection (Forbes 2001). A recent study demonstrated that after a few encounters with novel foods capuchins preferred those with a high sugar content (which is readily perceived through taste). Nevertheless, if capuchins keep encountering these same foods, their preferences become correlated with the foods' energy content. Therefore, after experiencing the consequences of ingesting the novel foods, capuchins responded to the feedback coming from the foods' energy content, and by doing so they maximize the net gain of energy (Visalberghi et al. 2003b). Similarly, preschool children learn to prefer food with a high caloric content over food with a low caloric content and use different flavors as immediate cues to distinguish foods (Birch et al. 1990).

Nevertheless, although the individual's physiology and behavior have an important role in determining diet acquisition, proximity to more knowledgeable individuals, interest in their activities, and/or opportunity to take food remains from them may foster occasions for experiencing what they eat (for a discussion, see Fragaszy and Visalberghi 2004; for a similar view in chimpanzee mother–infant pairs, Ueno and Matsuzawa 2004; see also Matsuzawa et al. 2001). Moreover, behavioral coordination, that is, the tendency to coordinate activities in space and time with those performed by group members (Coussi-Korbel and Fragaszy 1995), increases the individual's chance to engage in the same activities as its group members and thus serves as a simple and powerful social bias on individual learning.

In conclusion, these findings warn us to believe "by default" that the social context is crucial for the acquisition of a safe diet. Taste perception, a neophobic tendency toward unknown foods, and conditioned aversions and preferences seem very helpful in enabling the individual to select an adequate diet and efficiently reduce the risk of making fatal mistakes. However, group members' presence and behavior bias the individual's learning opportunities, and the combination of social facilitation, local enhancement, and stimulus enhancement does increase the chances that the food choices of a naïve individual will be canalized toward those of its group members.

Acknowledgments

We are grateful to Mike Tomasello and Josep Call for permission to carry out this study at the Wolfgang Köhler Primate Research Center of the Max Planck Institute for Evolutionary Anthropology, Leipzig. We especially thank Mike Seres for technical support, Esther Herrmann and Chikako Suda for help with data collection, and the keepers of the Wolfgang Köhler Primate Research Center/Leipzig Zoo for assistance with the chimpanzees. We further thank Gustl Anzenberger, Judith Burkart, Simone Macrì, Gloria Sabbatini, and especially Pier Francesco Ferrari for comments on an earlier version of this manuscript. This study was funded by a Ph.D. fellowship from CNR to E. Addessi and supported by grant RBNE01SZB4 from FIRB/MIUR to E. Visalberghi.

References

Addessi E, Visalberghi E (2001) Social facilitation of eating novel food in tufted capuchin monkeys (*Cebus apella*): input provided, responses affected, and cognitive implications. Anim Cogn 4:297–303

Addessi E, Visalberghi E (2006) Rationality in capuchin monkeys feeding behaviour? In: Nudds M, Hurley S (eds) Rationality in animals. Oxford University Press, Oxford, pp 313–328

Addessi E, Galloway AT, Visalberghi E, Birch LL (2005) Specific social influences on the acceptance of novel foods in 2–5-year-old children. Appetite 45:264–271

Baldissera F, Cavallari P, Craighero L, Fadiga L (2001) Modulation of spinal excitability during observation of hand actions in humans. Eur J Neurosci 13:190–194
Barnett SA (1958) Experiments on "neophobia" in wild and laboratory rats. Br J Psychol 49:195–201
Barnett SA (1963) The rat. A study in behaviour. Aldine, Chicago
Birch LL, Marlin DW (1982) I don't like it; I never tried it: effect of exposure on two-year-old children's food preferences. Appetite 3:353–360
Birch LL, McPhee L, Steinberg L, Sullivan S (1990) Conditioned flavor preferences in young children. Physiol Behav 47:501–505
Birch LL, Gunder L, Grimm-Thomas K (1998) Infants' consumption of a new food enhances acceptance of similar foods. Appetite 30:283–295
Burritt EA, Provenza FD (1989) Food aversion learning: ability of lambs to distinguish safe from harmful foods. J Anim Sci 67:1732–1739
Cambefort JP (1981) A comparative study of culturally transmitted patterns of feeding habits in the chacma baboon *Papio ursinus* and the vervet monkey *Cercopithecus aethiops*. Folia Primatol 36:243–263
Clayton DA (1978) Socially facilitated behavior. Q Rev Biol 53:373–392
Coussi-Korbel S, Fragaszy D (1995) On the relation between social dynamics and social learning. Anim Behav 50:1441–1453
Dominy NJ, Lucas PW, Osorio D, Yamashita N (2001) The sensory ecology of primate food perception. Evol Anthropol 10:171–186
Drapier M, Addessi E, Visalberghi E (2003) The response of tufted capuchin monkeys (*Cebus apella*) to foods flavored with familiar and novel odor. Int J Primatol 24:295–315
Dromard G (1905) Etude psychologique et clinique sur l'échopraxie. J Psychol 2:385–403
Fadiga L, Craighero L, Olivier E (2005) Human motor cortex excitability during the perception of others' action. Curr Opin Neurobiol 15:213–218
Ferrari PF, Maiolini C, Addessi E, Fogassi L, Visalberghi E (2005) Social facilitation of eating behavior by action observation and action hearing in macaque monkeys. Behav Brain Res 161:95–101
Fogassi L, Ferrari PF, Gesierich B, Rozzi S, Chersi F, Rizzolatti G (2005) Parietal lobe: from action organization to intention understanding. Science 308:662–667
Forbes JM (2001) Consequences of feeding for future feeding. Comp Biochem Physiol A 128:463–470
Fragaszy D, Visalberghi E (2004) Socially biased learning in monkeys. Learn Behav 32:24–35
Fragaszy DM, Visalberghi E, Galloway AT (1997) Infant tufted capuchin monkeys' behaviour with novel foods: opportunism, not selectivity. Anim Behav 53:1337–1343
Freeland WJ, Janzen DH (1974) Strategies in herbivory by mammals: the role of plant secondary compounds. Am Nat 108:269–289
Galef BG Jr (1970) Aggression and timidity: responses to novelty in feral Norway rats. J Comp Physiol Psychol 70:370–381
Galef BG Jr, Giraldeau LA (2001) Social influences on foraging in vertebrates: causal mechanisms and adaptive functions. Anim Behav 61:3–15
Gallese V, Fadiga L, Fogassi L, Rizzolatti G (1996) Action recognition in the premotor cortex. Brain 119:593–609
Gallese V, Fadiga L, Fogassi L, Rizzolatti G (2002) Action representation and the inferior parietal lobule. In: Prinz W, Hommel B (eds) Common mechanisms in perception and action: attention and performance, vol XIX. Oxford University Press, Oxford, pp 247–266
Garcia J, Koelling RA (1966) Relation of cue to consequence in avoidance learning. Psychon Sci 4:123–124
Garcia J, Kimeldorf DJ, Koelling RA (1955) A conditioned aversion towards saccharin resulting from gamma radiation. Science 122:157–158
Giraldeau LA (1997) The ecology of information use. In: Krebs J, Davies N (eds) Behavioral ecology, 4th edn. Blackwell, Oxford, pp 42–68
Glander KE (1982) The impact of plant secondary compounds on primate feeding behavior. Yearb Phys Anthropol 25:1–18

Glaser D (1993) The effects of sweeteners in primates. In: Mathlouti M, Kanters JA, Birch GG (eds) Sweet-taste chemoreception. Elsevier, London, pp 353–363

Goodall J (1986) The chimpanzees of Gombe. Harvard University Press, Cambridge, MA

Greenberg R (1990) Ecological plasticity, neophobia, and resource use in birds. Stud Avian Biol 13:431–437

Harper LV, Sanders KM (1975) The effect of adults' eating on young children's acceptance of unfamiliar foods. J Exp Child Psychol 20:206–214

Hladik CM, Simmen B (1996) Taste perception and feeding behavior in non-human primates and human populations. Evol Anthropol 5:58–72

Hladik CM, Pasquet P, Simmen B. (2002) New perspectives on taste and primate evolution: the dichotomy in gustatory coding for perception of beneficent versus noxious substances as supported by correlations among human thresholds. Am J Phys Anthropol 117:342–348

Hobden K, Pliner P (1995) Effects of a model on food neophobia in humans. Appetite 25:101–114

Itani J (1958) On the acquisition and propagation of a new food habit in the natural group of the Japanese monkey at Takasakiyama. Primates 1:84–98

Jacobs GH, Deegan JF II, Moran JL (1996) ERG measurements of the spectral sensitivity of common chimpanzee (*Pan troglodytes*). Vision Res 36:2587–2594

Janson CH, van Schaik CP (1993) Ecological risk aversion in juvenile primates: slow and steady wins the race. In: Pereira ME, Fairbanks LA (eds) Juvenile primates. Life history, development and behavior. Oxford University Press, New York, pp 57–74

Johnson EC (1997) Rhesus macaques (*Macaca mulatta*) continue to exhibit caution toward novel foods when food stressed. Am J Primatol 42:119–120

Johnson E (2000) Food-neophobia in semi-free ranging rhesus macaques: effects of food limitation and food sources. Am J Primatol 50:25–35

Kano T (1992) The last ape. Pygmy chimpanzee behavior and ecology. Stanford University Press, Palo Alto, pp 107–108

Laska M, Hernandez Salazar LT, Rodriguez Luna E (2000) Food preferences and nutrient composition in captive spider monkeys, *Ateles geoffroyi*. Int J Primatol 21:671–683

Lhermitte F, Pillon B, Serdaru M (1986) Human autonomy and the frontal lobes. Part I. Imitation and utilization behavior: a neuropsychological study of 75 patients. Ann Neurol 19:324–326

Matsuzawa T (1999) Communication and tool use in chimpanzees: cultural and social contexts. In: Hauser MD, Konishi M (eds) Design of animal communication. MIT Press, Cambridge, pp 645–671

Matsuzawa T, Hasegawa Y (1983) Food aversion learning in Japanese monkeys (*Macaca fuscata*). Folia Primatol 40:247–255

Matsuzawa T, Yamakoshi G (1996) Comparison of chimpanzee material culture between Bossou and Nimba, West Africa. In: Russon AE, Bard KA, Parker ST (eds) Reaching into thought: the minds of the great apes. Cambridge University Press, New York, pp 211–232

Matsuzawa T, Hasegawa Y, Gotoh S, Wada K (1983) One-trial long-lasting food-aversion learning in wild Japanese monkeys (*Macaca fuscata*). Behav Neural Biol 39:155–159

Matsuzawa T, Biro D, Humle T, Inoue-Nakamura N, Tonooka R, Yamakoshi G (2001) Emergence of culture in wild chimpanzees: education by master-apprenticeship. In: Matsuzawa T (ed) Primate origins of human cognition and behavior. Springer, Tokyo, pp 557–574

Meltzoff AN, Moore MK (1977) Imitation of facial and manual gestures by human neonates. Science 198:74–78

Milton K (1993) Diet and primate evolution. Sci Am 269:70–77

Nishida T, Wrangham RW, Goodall J, Uheara S (1983) Local differences in plant-feeding habits of chimpanzees between the Mahale Mountains and Gombe National Park, Tanzania. J Hum Evol 12:467–480

Provenza FD, Lynch JJ, Cheney CD (1995) Effects of a flavor and food restriction on the intake of novel foods by sheep. Appl Anim Behav Sci 43:83–93

Rizzolatti G, Fadiga L, Fogassi L, Gallese V (1999) Resonance behaviors and mirror neurons. Arch Ital Biol 137:85–100

Rizzolatti G, Fogassi L, Gallese V (2001) Neurophysiological mechanisms underlying the understanding and imitation of action. Nat Rev Neurosci 2:661–670
Rozin P (1976) The selection of food by rats, humans and other animals. In: Rosenblatt JS, Hinde RA, Shaw E, Beers C (eds) Advances in the study of behavior Vol. 6. Academic Press, New York, pp 21–76
Rozin P (1977) The use of characteristic flavorings in human culinary practice. In: Apt CM (ed) Flavor: its chemical, behavioural, and commercial aspects. Westview Press, Boulder, CO
Schrauf C, Voelkl B, Huber L (2004) The response of infant marmosets towards novel food. Folia Primatol 75(S1):410–411
Simmen B, Hladik CM (1998) Seasonal variation of taste threshold for sucrose in a prosimian species, *Microcebus murinus*. Folia Primatol 51:152–157
Steiner JE, Glaser D (1984) Differential behavioral responses to taste stimuli in non-human primates. J Hum Evol 13:709–723
Stengel E, Vienna MD, Edin LRCP (1947) A clinical and psychological study of echo-reactions. J Ment Sci 93:598–612
Sugiyama Y, Koman J (1992) The flora of Bossou: its utilization by chimpanzees and humans. Afr Study Monogr 13:127–169
Sullivan SA, Birch LL (1990) Pass the sugar pass the salt: experience dictates preference. Dev Psychol 26:546–551
Sullivan SA, Birch LL (1994) Infant dietary experience and acceptance of solid foods. Pediatrics 93:271–277
Teleki G (1989) Population status of wild chimpanzees (*Pan troglodytes*) and threats to survival. In: Heltne PG, Marquardt LA (eds) Understanding chimpanzees. Harvard University Press, Cambridge, pp 312–353
Tomasello M, Call J (1997) Primate cognition. Oxford University Press, New York
Ueno Y (2001) How do we eat? Hypothesis of foraging strategy from the viewpoint of gustation in primates. In: Matsuzawa T (ed) Primate origins of human cognition and behavior. Springer, Tokyo, pp 104–111
Ueno A, Matsuzawa T (2004) Food transfer between chimpanzee mothers and their infants. Primates 45:231–239
Ueno A, Matsuzawa T (2005) Response to novel food in infant chimpanzees. Do infants refer to mothers before ingesting food on their own? Behav Processes 68:85–90
Ueno A, Ueno Y, Tomonaga M (2004) Facial responses to four basic tastes in newborn rhesus macaques (*Macaca mulatta*) and chimpanzees (*Pan troglodytes*). Behav Brain Res 154:261–271
Visalberghi E, Addessi E (2000) Seeing group members eating a familiar food affects the acceptance of novel foods in capuchin monkeys, *Cebus apella*. Anim Behav 60:69–76
Visalberghi E, Addessi E (2001) Acceptance of novel foods in *Cebus apella*: do specific social facilitation and visual stimulus enhancement play a role? Anim Behav 62:567–576
Visalberghi E, Addessi E (2003) Food for thoughts: social learning and the feeding behavior in capuchin monkeys. Insights from the laboratory. In: Fragaszy D, Perry S (eds) Traditions in non-human animals: models and evidence. Cambridge University Press, Cambridge, pp 187–212
Visalberghi E, Fragaszy D (1995) The behaviour of capuchin monkeys, *Cebus apella*, with novel foods: the role of social context. Anim Behav 49:1089–1095
Visalberghi E, Valente M, Fragaszy D (1998) Social context and consumption of unfamiliar foods by capuchin monkeys (*Cebus apella*) over repeated encounters. Am J Primatol 45:367–380
Visalberghi E, Myowa-Yamakoshi M, Hirata S, Matsuzawa T (2002) Responses to novel foods in captive chimpanzees. Zoo Biol 21:539–548
Visalberghi E, Janson CH, Agostini I (2003a) Response towards novel foods and novel objects in wild tufted capuchins (*Cebus apella*) Int J Primatol 24:653–675
Visalberghi E, Sabbatini G, Stammati M, Addessi E (2003b) Preferences towards novel foods in *Cebus apella*: the role of nutrients and social influences. Physiol Behav 80:341–349
Vitale A, Queyras A (1997) The response to novel foods in common marmoset (*Callithrix jacchus*): the effects of different social contexts. Ethology 103:395–403

Wardle J, Cooke LJ, Gibson EL, Sapochnik M, Sheiman A, Lawson M (2003) Increasing children's acceptance of vegetables: a randomized trial of parent-led exposure. Appetite 40:155–162

Weiskrantz L, Cowey A (1963) The etiology of food reward. Anim Behav 11:225–234

Whiten A, Ham R (1992) On the nature and evolution of imitation in the animal kingdom: reappraisal of a century of research. Adv Stud Behav 21:239–283

Yamamoto ME, Lopes FA (2004) Effect of removal from the family group on feeding behavior by captive *Callithrix jacchus*. Int J Primatol 25:489–500

17
Tactical Deception and Understanding of Others in Chimpanzees

Satoshi Hirata

1 Primate Deception

Many primate species live in social groups and experience daily interactions with other group members throughout their lives. Various types of behaviors are included in these daily interactions: friendly behaviors such as grooming, kissing, and embracing; aggressive behaviors such as biting, hitting, and charging displays; and playing behaviors such as chasing and wrestling. These behaviors may convey honest, or, occasionally, deceptive signals. Whiten and Byrne (1988) focused on deceptive behaviors by primates to elucidate social intelligence.

Studying deception is difficult, as deception occurs rarely. That is, deceptive acts are recognized by others as such and ultimately fail in accomplishing their goals, as the story of the boy who cried "wolf" demonstrates. Byrne and Whiten (1990) employed a method that involved asking numerous researchers to collect as many episodes of deception as possible. The authors succeeded in gathering 253 episodes of tactical deception in primates and categorized them into 5 classes and 13 subclasses. The 5 classes included deception by concealment, distraction, creating an image, manipulation of target using social tool, and deflection of target to fall guy. They further identified 18 possible examples of intentional deception, that is, deception based on understanding the mental state of a target individual. These examples of intentional deception were found much more often among great apes (i.e., chimpanzees, bonobos, gorillas, and orangutans), suggesting that they are able to understand the mental states of others.

A landmark paper by Premack and Woodruff (1978) underlies the discussion of Whiten and Byrne (1988). Premack and Woodruff (1978) asked if a theory of mind can be applied to chimpanzees, that is, if a chimpanzee imputes mental states to himself and to others. The main purpose of Whiten and Byrne's (1988) approach to primate deception was therefore to examine the ability of primates to deceive others by recognizing the mental state of others. The present chapter describes deceptive episodes of chimpanzees and discusses their understanding of others' mental states.

Great Ape Research Institute, Hayashibara Biochemical Laboratories, 952-2 Nu, Tamano, Okayama, 706-0316 Japan

1.1 Deception by Misleading

Menzel (1971, 1974, 1975) conducted a series of studies on a group of chimpanzees living in a 1-acre outdoor enclosure. One of the most interesting tests was a naturalistic study of leadership and communication occurring in experimentally manipulated situations. First, the entire group of six chimpanzees (Shadow, Bandit, Belle, Libi, Bido, and Polly) was locked into a cage adjacent to the enclosure. Then, a human experimenter hid pieces of food, for example, under leaves or grass or behind a tree within the enclosure. One of the chimpanzees (Bandit or Belle) was taken from the group and went together with a human experimenter to the food. The chimpanzee was shown the food without being allowed to touch it, and then was returned to the group in the waiting cage. This individual was operationally called the "leader." About 2 min later, the entire group was released from the cage and allowed into the enclosure. At this time, the entire group headed straight for the food. Because the chimpanzees in this study were all young, about 3 years of age, they preferred to remain together. They often traveled together via "tandem walking" (with an arm of an individual around the waist of another individual) and by clinging (ventro-ventral contact). During testing, they also approached their food and ate it together. Thus, the leader that knew the location of hidden food successfully led its companions to the food. If individuals failed to follow, the leader then became very upset, going from one follower to the next, grimacing, tapping each on the shoulder, starting off tentatively, and then stopping to glance backward. In an extreme case, the leader screamed, grabbed a preferred companion, and dragged it in the direction of the food.

The result was different, however, if the most dominant chimpanzee Rock was present. As soon as Belle uncovered the food, Rock raced over, kicked or bit her, and took all the food. The following description by Menzel (1974) explains what happened between Belle and Rock: "Belle accordingly stopped uncovering the food if Rock was close. She sat on it until Rock left. Rock, however, soon learned this, and when she sat in one place for more than a few seconds, he came over, shoved her aside, searched her sitting place, and got the food. Belle next stopped going all the way. Rock, however, countered by steadily expanding the area of his search through the grass near where Belle had sat. Eventually Belle sat farther and farther away, waiting until Rock looked in the opposite direction before she moved toward the food at all—and Rock in turn seemed to look away until Belle started to move somewhere. On some occasions Rock started to wander off, only to wheel around suddenly precisely as Belle was about to uncover the food. Often Rock found even carefully hidden food that was 30 ft or more from Belle, and he oriented repeatedly at Belle and adjusted his place of search appropriately if she showed any signs of moving or orienting in a given direction. If Rock got very close to the food, Belle invariably gave the game away by a nervous increase in movement. However, on a few trials she actually started off a trial by leading the group in the opposite direction from the food, and then, while Rock was engaged in his search, she doubled back rapidly and got some food. In other trials when we hid an extra piece of food about 10 ft away from the large pile, Belle led Rock to the single piece, and while he took it she raced for the pile. When Rock started to ignore the single piece of food to keep his watch on Belle, Belle had temper 'tantrums'" (Menzel 1974).

Thus, there were tactics and countertactics developed during interactions between Belle and Rock. The case in which Belle led the group in a direction opposite to that of the food can be clearly considered an example of deception. From these cases, Menzel (1974) inferred that chimpanzees know what effect their own behavior is having on others.

1.2 Deception by Concealment

Matsuzawa (1991) conducted a similar study with four 4-year-old chimpanzees (Whiskey, Freida, Liza, and Opal) living in a 2,000-m^2 outdoor enclosure and in indoor areas at the University of Pennsylvania. First, the four chimpanzees were locked in an indoor area. A human experimenter went to the outdoor enclosure with one of the chimpanzees and hid a banana in view of the chimpanzee. The chimpanzee was not allowed to take the banana and was returned to the group in the indoor area. This chimpanzee was called the "witness." After several minutes, the group was released into the enclosure. On the 1st day, Opal was the witness. As soon as the group was released, Opal rushed to the banana. The remaining three chimpanzees did not seem to understand what had happened. On the 2nd day, Liza was the witness. She rushed to the banana, and the remaining three still did not understand the situation. On the 3rd day, Whiskey was the witness. He rushed to the banana, and once again the other three did not seem to comprehend the situation. Freida was too timid to be alone in the enclosure and did not take on the role of witness. On the 4th day, Opal was the witness again and she rushed to the banana, with similar results.

A change was observed on the 5th day. The banana was hidden in the northern part of the enclosure, and Liza was the witness. A short time after the group was released, Opal started to run east. Then Liza started to run north, retrieved the banana, and continued to run north. Opal noticed that Liza had the banana and chased Liza. Whiskey also began to run toward Liza and took a part of the skin of the banana from Liza. On the 6th day, the strongest individual, Whiskey, was the witness. He ran for the banana and continued running after obtaining it. The other three ran after Whiskey, but he quickly consumed the entire banana. On the 7th day, Liza was the witness. Liza headed for the banana, and Whiskey chased her. Whiskey caught up and stole the banana. On the 8th day, Whiskey was the witness. He headed for the banana, and then Liza chased him. However, Liza was weaker than Whiskey and was unable to steal the banana from him.

On the 9th day, Opal was the witness. The banana was hidden under a pole in the northwest part of the enclosure. Immediately after the group was released, Whiskey ran for an earthen pipe at the western part of the enclosure. A banana had been hidden there once before, and Whiskey searched this area carefully.

Liza followed Whiskey, and searched the same area. While the two chimpanzees were searching, Opal wandered around the exit door of the enclosure. She was performing solo play at the water tank. Freida was also playing alone. After a while, Whiskey and Liza moved to a fallen tree and Freida joined them. When the three chimpanzees went to the fallen tree, Opal slowly began to return to the exit door, then slowly headed in a westerly direction along the wall. She glanced at the three chimpanzees several times, and walked slowly toward the pole where the banana was hidden. When she was about 3 m away from the pole, she stared at the three chimpanzees. Suddenly she rushed to the pole, obtained the hidden banana, and ate it before the other three chimpanzees noticed. Clearly, this was an example of deception by concealment, that is, inhibition of attending.

1.3 Deception and Counterdeception

Hirata and Matsuzawa (2001) adopted basically the same procedure to study interactions between pairs of adult chimpanzees at the Primate Research Institute, Kyoto University. Five containers used to hide a banana were set up at the outdoor enclosure, which measured about 700 m^2. While a pair of female chimpanzees was kept inside, a human experimenter entered the outdoor enclosure and hid a banana in one of the five containers. There were two conditions: the role-divided condition and the control condition. Under the role-divided condition, one of the two chimpanzees (witness) could see where the experimenter hid the banana, while the other (witness-of-witness) could not see it directly but was allowed a view of the witness observing the outside. While the experimenter hid a banana, the witness remained in a waiting room adjacent to the enclosure and saw where the banana was hidden through a half-open door. The witness-of-witness stayed in a room adjacent to the waiting room. She could not see where the banana was hidden, but she could see the witness looking outside from a half-open door to the waiting room. Under the control condition, the two chimpanzees were brought to the waiting room and the door was closed during baiting; thus, neither could see where the banana was hidden. Under both conditions, the two chimpanzees were released into the enclosure after baiting.

The results for a pair of chimpanzees, Chloe and Pendesa, follow. Chloe served as the witness and Pendesa as the witness-of-witness for the first 8 days. Pendesa did not seek the banana for the first 3 days, except on the 2nd day when she happened to come across the banana in one of the containers after the experimenter inadvertently failed to hide the reward completely. After the 4th day, Pendesa began to search the containers by herself but did not display any action toward Chloe. Thus, the witness Chloe easily obtained the banana during this period. Role reversal was introduced on the 9th day; Pendesa served as the witness and Chloe as the witness-of-witness from the 9th to the 11th day. During this period, Pendesa headed straight to the banana and obtained it. Chloe wandered around but did not do anything in particular. Another role reversal was introduced. Chloe served as the witness and Pendesa as the witness-of-witness from the 12th to the 14th day. Pendesa began to threaten Chloe from the 11th day. Threats were followed each time by Pendesa seeking the banana alone; she found it along the way before Chloe could on the 13th day. Pendesa was thus dominant over Chloe. Then, role reversal was introduced again, and Pendesa served as the witness and Chloe as the witness-of-witness from the 15th to the 17th day. When Pendesa was the witness, she always obtained the banana, and Chloe could do nothing to prevent it.

The fourth role reversal was introduced, and Chloe served as the witness and Pendesa as the witness-of-witness from the 18th to the 31st day. Pendesa obtained the banana on the 18th and 19th days, when she first threatened Chloe and then searched for the banana alone. From the 20th day, Pendesa began to adjust her direction of movement to that of Chloe's. More precisely, after entering the enclosure, Pendesa first attempted to seek the banana by herself, and then, after Chloe had emerged, Pendesa began to approach Chloe's route from some distance away. At the same time, Pendesa began to look at Chloe more and more frequently. These strategies did not allow Pendesa to obtain the banana, however, because Chloe always arrived at the baited container before Pendesa had a chance to catch up with her. Chloe obtained the banana during the 20th to 26th days. After the 24th day, Pendesa began to run ahead of Chloe, and Chloe's initial response was to mislead Pendesa by taking an indirect route.

On the 24th day, Pendesa first entered the enclosure, and went to the right (Fig. 1.1). Then Chloe entered the enclosure and went straight ahead toward the baited container (Fig. 1.2). Pendesa looked back in the direction of Chloe, changed her route, and began to run toward Chloe. Chloe looked to the right, saw Pendesa coming, and stopped there, at a distance of about 6 m from the

Fig. 1. Behavior sequence (*1–6*) of the two chimpanzees on the 24th day. *Ch*, Chloe; *Pe*, Pendesa; *B*, baited container; *E*, empty container

baited container. Chloe turned to the left, and went toward another empty container at the left side of the enclosure (Fig. 1.3). Pendesa caught up with Chloe about 2 m from the empty container for which Chloe was heading (Fig. 1.4). Pendesa looked at Chloe, jumped up in an overbearing fashion (Fig. 1.5), and Chloe retreated diagonally from Pendesa. Then Pendesa proceeded to the empty container where Chloe had been heading, while Chloe began to approach the baited container (Fig. 1.6). Pendesa looked into the empty container, and by this time Chloe had found the banana in the baited container. Chloe also succeeded in "deceiving" in a similar manner on the 25th, 27th, and 30th days, that is, Pendesa was misled to an empty container while following Chloe to a nontarget. While Pendesa was looking in the incorrect container, Chloe returned to the target and successfully obtained the reward on those days. However, Pendesa developed a counterdeception tactic and gained access to the reward on the 26th, 28th, 29th, and 31st days. That is, Pendesa remained close by and frequently adjusted her direction to that of Chloe.

On the 28th day, Pendesa entered the enclosure. She stopped about 2 m ahead of the door and stayed there. Then Chloe put her head out through the door. Pendesa looked back at Chloe, and swung her hand threateningly toward Chloe, whereupon Chloe pouted. Pendesa advanced, and Chloe entered the enclosure. Pendesa turned back at Chloe, stood up bipedally, and swung her arms threateningly toward Chloe. Two seconds later, Pendesa advanced for 2 s but then retreated again, changing her route to match the direction of Chloe's course (Fig. 2.1). Two seconds later, Pendesa faced Chloe and Chloe stood up. Pendesa stretched both her arms around Chloe and they embraced (Fig. 2.2). One second later, Pendesa began to turn forward and withdrew her arms from Chloe.

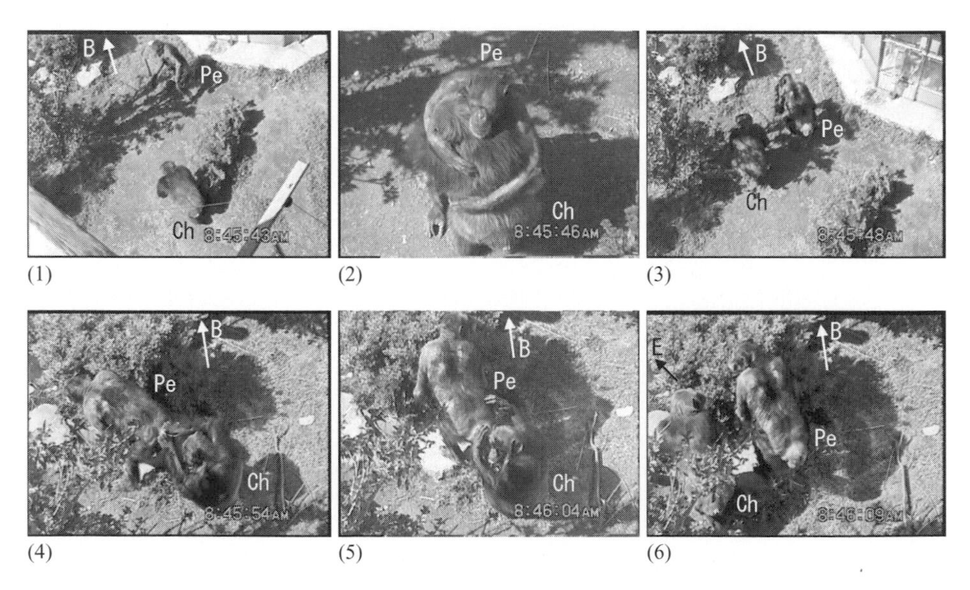

Fig. 2. Behavior sequence (*1–6*) of the two chimpanzees on the 28th day. *Ch*, Chloe; *Pe*, Pendesa; *B*, baited container; *E*, empty container

Pendesa and Chloe moved apart, headed in a forward direction, and Pendesa then went ahead of Chloe (Fig. 2.3). Three seconds later, Pendesa stopped, and Chloe came up just behind her. Pendesa then looked back at Chloe and oriented her posterior toward Chloe. Chloe embraced Pendesa from behind and inspected Pendesa's genital area with both hands (Fig. 2.4). After 11 s, Pendesa moved her body slightly forward. Chloe responded to this and touched Pendesa's waist, patting and stroking Pendesa's back rapidly with one hand and then the other (Fig. 2.5). Immediately after this, Chloe advanced toward one of the empty containers while looking at Pendesa twice, and they stared at each other (Fig. 2.6). Pendesa followed Chloe closely and appeared in front of Chloe, remaining there. Pendesa presented her rear to Chloe, who touched, stroked, and rubbed Pendesa's left instep. Nine seconds later, Chloe looked at Pendesa and moved toward the empty container. Pendesa went after Chloe and soon overtook her, about 2 m from the empty container. Then Chloe turned to the baited container while Pendesa looked into the empty container. Soon Pendesa turned back and followed Chloe, who looked back at Pendesa coming up just behind her. After 2 s, both Chloe and Pendesa arrived almost simultaneously at the baited container, and Pendesa, the witness-of-witness, obtained the banana.

From the 32nd to the 34th day of testing this pair, we introduced the control condition. Pendesa exhibited no actions but threatening once toward Chloe on the first day of the control condition. Pendesa went alone to seek the banana and obtained it on all 3 days. However, Chloe gradually lost her motivation to seek the banana toward the end of this experiment because she was repeatedly threatened and subsequently lost the reward. Chloe chose instead to stay in a neutral area of the compound during the final stage of the test, not paying any attention to Pendesa. Therefore, we decided to discontinue tests on this pair.

1.4 Implications for Understanding Others

Under these three test scenarios, there was a witness of a hidden reward. In the beginning, the situation was simple; the witness went to the hiding place that she already knew, and obtained the reward. After repeated trials, other ignorant chimpanzees or the witness-of-witness began trying to steal the reward. It was at this point that social maneuvering emerged as tactics. Rock in Menzel's (1974) study, Whiskey in Matsuzawa's (1991) study, and Pendesa in Hirata and Matsuzawa's (2001) study attempted to rob the hidden food by following and chasing the witness. The witness opposed this action by opting not to go to the hiding place in Menzel's (1974) and Matsuzawa's (1991) experiments, exhibiting deception by inhibition of attending. The witness succeeded in obtaining the reward during an unguarded moment of the ignorant chimpanzee. Furthermore, in Menzel's (1974) and Hirata and Matsuzawa's (2001) studies, the witness went in the opposite direction. After the ignorant chimpanzee or the witness-of-witness followed the witness and searched this empty area, the witness returned to the correct place and obtained the reward, exhibiting deception by misleading. The witness-of-witness in Hirata and Matsuzawa (2001) again acted against this deception by remaining very close to the witness, that is, the two chimpanzees groomed and embraced during the course of interactions, exhibiting counterdeception. Deceptive episodes tend to be generally anecdotal, as discussed by Whiten and Byrne (1988), but these three cases show that chimpanzee deceptive ability emerges quite reliably under certain experimentally created situations.

In Hirata and Matsuzawa's (2001) study, a control condition was introduced at the end of the study in which neither chimpanzee witnessed the hiding. Under this condition the dominant chimpanzee seemed to care less about the other chimpanzee, suggesting that she changed her tactic depending on whether the other chimpanzee saw the hiding procedure or not. Although a systematic comparison between conditions was not possible because of the flawed nature of introducing the two conditions, the study indicated the possibility of chimpanzee understanding regarding what others have or have not seen.

2 Understanding of Others' Visual Perception

To investigate chimpanzees' understanding of what others are seeing, Tomasello, Call, Hare, and colleagues conducted a series of experiments, including confronting chimpanzees with competition over food (Hare et al. 2000, 2001; Call 2001; Call and Carpenter 2001; Hare 2001; Tomasello et al. 2003a,b). A pair of dominant and subordinate chimpanzees was brought into two rooms on opposite sides of a third room. For example, a dominant chimpanzee stayed in the left room and a subordinate individual stayed in the right room, both of them facing a third room in the middle.

In the first set of experiments, two pieces of food, along with an opaque or transparent barrier, were placed in the center of the third room. These two pieces of food were placed apart from one another. There were two conditions: the occluder condition and the transparent barrier condition. Under the occluder condition, an opaque barrier was placed on the dominant chimpanzee's side of one of the pieces of food. Viewed by the dominant chimpanzee, only one piece of food could be seen, because the second piece of food was behind the opaque barrier. The subordinate chimpanzee could see both pieces of food: one piece of food outside the barrier and the other piece of food in front of the barrier. Under the transparent barrier condition, a barrier was placed in a way similar to the occluder condition, but the barrier was transparent. Thus, the dominant chimpanzee could also see two pieces of food: one piece of food outside the barrier and the other piece of food behind the transparent barrier. The two chimpanzees were then released into the middle room, with the subordinate individual given a brief head start to allow her time to choose. A premise in this experiment was that subordinates avoid competition over food with dominant individuals. The result of the occluder condition was that the subordinate chimpanzees went to the food in front of the barrier much more often than the food outside the barrier. Under the transparent barrier condition, the subordinate chimpanzees did not show a preference for either piece of food. These results indicated that the subordinate chimpanzees knew what the dominant chimpanzees could and could not see.

In the second set of experiments, the basic procedure was the same except that there were two barriers and only one piece of food. Under the experimental condition, the subordinate chimpanzee saw a human experimenter place a piece of food inside one of the two barriers, on the subordinate individual's side. The dominant chimpanzee was not allowed to see the hiding process because the door was closed. Under the control condition, both the dominant and subordinate chimpanzees saw the hiding process. The result was that the subordinate chimpanzees went to the food more often under the experimental condition than the control condition. Therefore, the subordinate chimpanzees knew what the dominant individuals saw and did not see. The authors also introduced other conditions to rule out alternative interpretations, for example, that the subordinate chimpanzees preferred food next to the barrier or that they were intimidated when dominant individuals observed the hiding process. The results of these variations consistently support the idea that the chimpanzees know what others can and cannot see and also what others have and have not seen. The chimpanzees understand unobservable mental states of others, at least the visual perception of others.

Povinelli and colleagues criticized Tomasello et al. (2003a,b), however, and proposed the "behavioral abstraction hypothesis" (Povinelli et al. 2000; Povinelli and Giambrone 2001; Povinelli and Vonk 2003, 2004). In this hypothesis, the chimpanzees construct abstract categories of behavior, make predictions about future behaviors that follow from past behaviors, and adjust their own behavior accordingly. Povinelli and Vonk (2003, 2004) assert that humans engage in both behavioral abstraction and mental state attribution, but there is no evidence suggesting that chimpanzees engage in mental state attribution. If an experiment relies upon behavioral invariants such as looking or gazing, it cannot clarify whether chimpanzees engage in behavioral abstraction alone or behavioral abstraction plus mental state attribution, because the chimpanzees have the chance to formulate statistical regulations of the behavior of others and make predictions of future behavior from past experience. The subordinate chimpanzees can predict from their experience of observing others what dominant chimpanzees will do if they orient toward food; the dominant will go to the food and threaten the subordinate. The subordinate chimpanzees can predict from their past experience of observing others what the dominant individuals will do if they are not present when the food is placed; the dominant will neither go to the food nor threaten the subordinate.

Povinelli and Vonk (2003, 2004) proposed an experiment in which the cue toward the inference to the mental state is arbitrary, and the subject has no exposure to others behaving in association with that cue, to test if chimpanzees or other species are capable of mental state attribution. That is, a test should be conducted that requires subjects to make an extrapolation from their own experience to the mental state of others. For example, "we let a chimpanzee interact with two buckets, one red, one blue. When the red one is placed over her head total darkness is experienced; when the blue one is similarly placed, she can still see. Now have her, for the first time, confront with others (in this case the experimenters) with these buckets over their heads. If she selectively gestures to the person wearing the blue bucket we could be highly confident that the nature of her coding was, in part, mentalistic—that is, that she represented the others as 'seeing' her" (Povinelli and Vonk 2003).

3 Naturalistic Observations and Experimental Tests

As Povinelli and colleagues discussed (Povinelli et al. 2000; Povinelli and Vonk 2003, 2004), it is true that naturalistic observations are insufficient to clarify whether behaviors result from mental attribution or behavior learning because it is almost impossible to record a complete history of an animal's interactions with others. Experimental tests are needed to address this issue, but experimental manipulations also face difficulties as irrelevant factors may influence test results.

I performed matching-to-sample tasks with two juvenile chimpanzees at the Hayashibara Great Ape Research Institute in Japan. Before my study, individuals had already learned to some extent how to solve a matching-to-sample task using a touch monitor. A sample stimulus, for example, a red circle, appeared at the bottom of a touch monitor. If the chimpanzee touched the sample stimulus, then two choice stimuli appeared at the upper part of the monitor, one of which was correct and identical to the sample stimulus, or red circle, and the other, which was incorrect and different from the sample stimulus, such as a green circle. If the chimpanzee touched the correct choice on the monitor, then a food reward was given. The two chimpanzees were able to solve the matching-to-sample task of color circles and several pictures using both colors and shapes as cues to accomplish the task.

A matching-to-sample task was then introduced in which a human–chimpanzee interactive situation was used. Two plates were placed in front of a chimpanzee, one on the left and the other on the right. Then I placed a red wooden circle in one of the plates and a green wooden circle in the other plate. I then gave a third wooden circle, red or green, to the chimpanzee, and asked it to put it into the plate having the same-colored wooden circle. If the chimpanzee succeeded, then a food reward was given. The two chimpanzees selected the correct plate at an above-chance level after 100 to 200 trials. Wooden yellow, blue, and gray circles were added, and the chimpanzees solved this matching-to-sample task after the first 20 trials. Thus, they seemed to grasp the objective of the task. When nine pairs of new objects with various shapes were introduced, the chimpanzees also solved this task after the first 20 trials. Subsequently, I used square boards on which either a black double circle or a black star was painted. The procedure was the same: a board with a double circle was placed on one of the plates and a board with a star on the other. I gave a third board with either a double circle or a star to the chimpanzee, and asked it to put it on the plate in which the board with the same mark had been placed. The two chimpanzees could not solve this task. After 3,000 trials, their performance was still at chance level.

Following this failure, the same matching-to-sample task was given using a touch monitor. A black double circle and a black star were used as stimuli, with the same size and shape as those painted on the boards. One of the two chimpanzees performed at above-chance levels after 50 trials, and the other performed above chance after 20 trials. Thus, they could solve the matching-to-sample task of a double circle and star using a touch monitor; however, when I reintroduced the task with plates and boards, the chimpanzees could not solve the task.

From the results of the touch monitor task, the chimpanzees could discriminate between a double circle and a star, although the same assertion cannot be made from the results of the plate task. The chimpanzees understood the rules of the matching-to-sample task using the plates and the touch monitor. The color, shape, and size of the double circle and the star were the same in the touch monitor task and the plate task. Although they could not pay attention to the difference in the patterns on the square boards, they could do so with the touch monitor. Thus, a change in the experimental procedure may generate different results, even if the basic structure of the task remains the same.

It is necessary to investigate chimpanzees' social cognition using rigorous experimental situations, such as a choice task with human experimenters using a red bucket and blue bucket. However, poor performance does not necessarily mean that chimpanzees lack the ability essential to successfully complete the task in ways that human investigators presume. Social cognition has evolved to solve problems that animals experience in group living. Thus, social cognition should appear in natural situations when animals interact with other group members. There is no guarantee, however, that same abilities are utilized when chimpanzees are faced with artificially created situations that are unrelated to their natural lives. Both naturalistic observations and rigorous experiments have advantages and disadvantages. The word "social" originally meant "allied" or "united." Naturalistic observations and rigorous experiments should be allied and united to further our understanding of the evolution of social cognition.

Acknowledgments

The studies were financially supported by the Japan Society for the Promotion of Science (JSPS) for Young Scientists (No. 2926) to S. Hirata, the Core-to-Core Program HOPE by JSPS, and Grants-in-Aid for Special Promotion Research (#12002009, #16002001) of the Ministry of Education, Culture, Sports, Science and Technology to T. Matsuzawa. I greatly acknowledge T. Matsuzawa for his generous guidance throughout the studies, M. Tomonaga and M. Tanaka for their support and suggestions, and all the collaborators for their assistance in data collection and analysis.

References

Byrne RW, Whiten A (1990) Tactical deception in primates: the 1990 database. Primate Rep 27:1–101
Call J (2001) Chimpanzee social cognition. Trends Cogn Sci 5:388–393
Call J, Carpenter M (2001) Do apes and children know what they have seen? Anim Cogn 4:207–220
Hare B (2001) Can competitive paradigms increase the validity of experiments on primate social cognition. Anim Cogn 4:269–280
Hare B, Call J, Agnetta B, Tomasello M (2000) Chimpanzees know what conspecifics do and do not see. Anim Behav 59:771–785
Hare B, Call J, Tomasello M (2001) Do chimpanzees know what conspecifics know? Animal Behav 61:139–151
Hirata S, Matsuzawa T (2001) Tactics to obtain a hidden food item in chimpanzee pairs (*Pan troglodytes*). Anim Cogn 4:285–295
Matsuzawa T (1991) Chimpanzee mind (in Japanese). Iwanami-Shoten, Tokyo
Menzel EW (1971) Communication about the environment in a group of young chimpanzees. Folia Primatol 15:220–232
Menzel EW (1974) A group of chimpanzees in a one-acre field. In: Shrier AM, Stollnitz, F (eds) Behavior of non-human primates, vol 5. Academic Press, San Diego, pp 83–153
Menzel EW (1975) Purposive behavior as a basis for objective communication between chimpanzees. Science 189:652–654
Povinelli DJ, Giambrone S (2001) Reasoning about beliefs: a human specialization? Child Dev 72:691–695
Povinelli DJ, Vonk J (2003) Chimpanzee minds: suspiciously human? Trends Cogn Sci 7:157–160
Povinelli DJ, Vonk J (2004) We don't need a microscope to explore the chimpanzee's mind. Mind Language 19:1–28
Povinelli DJ, Bering JM, Giambrone S (2000) Toward a science of other minds: escaping the argument by analogy. Cogn Sci 24:509–541
Premack D, Woodruff G (1978) Does the chimpanzee have a theory of mind? Behav Brain Sci 4:515–526
Tomasello M, Call J, Hare B (2003a) Chimpanzee understand psychological states: the question is which ones and to what extent. Trends Cogn Sci 7:153–156
Tomasello M, Call J, Hare B (2003b) Chimpanzee versus humans—it's not that simple. Trends Cogn Sci 7:239–240
Whiten A, Byrne RW (1988) Tactical deception in primates. Behav Brain Sci 11:233–273

Part 5
Conceptual Cognition

18
Early Spontaneous Categorization in Primate Infants—Chimpanzees, Humans, and Japanese Macaques—with the Familiarization-Novelty Preference Task

Chizuko Murai

1 Introduction

In this chapter, I introduce a series of comparative studies for categorization in infancy of Japanese monkeys (*Macaca fuscata*), chimpanzees (*Pan troglodytes*), and humans (*Homo sapiens*). Categorization is the cognitive activity of sorting objects (or events), that is, grouping objects possessing similar attributes and distinguishing those objects from others possessing dissimilar attributes. This is one of the most important activities for processing objects in the world to flexibly adapt to one's environment. How infant monkeys, infant chimpanzees, and human infants categorize objects was compared using similar experimental methods. Also, the differences and the similarities found between their early categorization were explored. In a series of experiments, the familiarization-novelty preference task, which is generally used for human infant study, was applied for all the species. None of the species received any special training involving reinforcements during the experiments. By this means, I attempted to examine spontaneous categorization by primate infants.

I first briefly describe the functions of categorization in our lives. Then, I review previous categorization studies of human infants and nonhuman primates and mention some issues that stem from these studies. Last, I report the results of the experiments. Then, species specificity for perceptual, conceptual, and cognitive development in the ability to categorize objects is discussed.

Brain Science Research Center, Tamagawa University Research Institute, 6-1-1 Tamagawa Gakuen, Machida, Tokyo 194-8610, Japan

2 What Is Categorization?

In the world, there are numerous objects. In other words, we must live in a chaotic world of objects. However, we adapt to the world and go through the days without much confusion because categorization, that is, the ability of the mind to sort and group objects based on their attributes, helps us to effectively process objects and lightens such confusion in our world. Object categorization is reflected in our usual behaviors, for example, classifying cutlery into forks and spoons, organizing the books in the bookshelf into novels, magazines, and comics, and so on. It would be terrible if we did not have this ability to categorize the objects around us; it is not imaginable. For example, when we are shopping in a department store, we go to the clothing department to buy a shirt and to the grocery department to buy bread. In this case, if the merchandise were set out without any order, we could only find things that we wanted with great difficulty. It is very useful for us that things are in order according to their kinds or use applications. Categorization is one of the fundamental cognitive processes and closely related to our life.

Many researchers have investigated categorization for a long time. In modern cognitive science, the term "categorization" refers to the recognition of discriminably different objects as members of the same category based on some internalized representations of the category. Such representation has been called schema, concept, or categorical representation (Quinn 2002). To assign objects to categories, it is necessary to recognize not only the attributes of the objects but also the relationship among them such as their similarities and differences, that is, categorical attributes. Categorization is also widely related to various domains. It is not restricted to concrete objects; it is also applied to things such as color, tone, number, and events. Furthermore, during the process of expanding previously recognized categorical attributes to some novel objects, it is necessary to judge whether such categorical attributes have to be generalized to the novel objects. For example, we recognize "dachshund" and "bulldog" as members of the "dog" category by attending to their appearance or behavioral attributes. And such categorical attributes of "dog" are generalized to "collie" but not to "giraffe." Thus, categorization is not a superficial perception of objects but is based on deeper processing of objects; of course, it depends on the level of categorization. Subjects' cognition of the world is reflected in the categories that they construct in their minds.

There are some advantages of categorization in processing objects, as follows. First, the amount of information about objects is effectively processed in the course of categorization by attending not only to perceptual attributes but also abstract relationships among them. This function would make the "cost" of information processing lower. For example, we would use some memory strategies when we have to store a lot of words (in our mind). In this case, we often retain the words after categorizing them based on their alphabetical order or meanings. Second, related to this notion, by storing information categorically we are able to retrieve the information smoothly. As in the example of a department store already mentioned, if we stored the objects in a disorderly fashion, we would take a long time to retrieve the needed information and also would often make mistakes. Third, categorization helps us to learn a novel object. When we encounter a novel object, we process the attributes that it possesses and, at the same time, refer to a familiar category. If the novel object possesses some attributes resembling those of a familiar category, this category would be activated. In this manner, categorization and formed categories facilitate human processing of objects.

3 The Question of Categorical Development in Human Infants

In our daily lives, we categorize objects naturally. Categorization may be a spontaneous and automatic process for human adults in a sense. When does such ability appear in the course of development? Early development of human categorization is one of the main issues in recent research on cognitive development.

For more than 30 years, many researchers focused on infants' perception and cognition, and experimental investigation involving categorization studies has increased. This trend was based on the progress of experimental methodologies for human infants, for example, tasks using infants' looking behavior on stimuli (e.g., scale models, photographs, line drawings). Such tasks contain the preferential looking task, the habituation–dishabituation task, the familiarization-novelty preference task, and so on. Also, with these looking procedures, it came to be possible to test younger infants who were not mature in motor skills such as reaching. Among these tasks, the familiarization-novelty preference task is widely used for research on infants' categorization. This task is dependent on a behavioral tendency of infants; they prefer novel or interesting objects to familiar or unattractive objects, and look longer at the former than at the latter. When this task is used to test infants' categorization of objects, infants are at first presented with a number of objects (e.g., beagle, poodle, husky) from one category (e.g., dog) repeatedly until their looking response comes down to the defined criterion; in other words, until they are familiar with the category. This is the familiarization phase. After this phase, the test phase is conducted. In the test phase, infants are simultaneously or sequentially presented with two novel objects: one is a new object (e.g., shepherd) from the familiar category and another is an object (e.g., a Persian cat) from a novel category In this task, the results are interpreted as evidence that infants formed a categorical representation during the experiment when they showed significant decrease of looking time at objects (that is, familiarization) during the familiarization phase and that they showed significant recovery of their looking time for the novel-category object when they generalized familiarization to a new object from the familiar category. It is thought that these responses of infants indicate that infants regard different objects from one category as alike—in other words, group these objects—and moreover that they exclusively distinguish the category from another category.

With the looking procedure or the task using infants' examination of the objects (e.g., the object-examination task, the generalized imitation method, and the sequential touching task; Mareschal and Quinn 2001), some previous studies suggested that human infants possess the basic ability for categorization (Mandler and McDonough 1993, 1998; Pauen 2002; Rakison and Butterworth 1998; Younger and Fearing 1999, 2000). For example, some researchers reported human infants in their early development form some categorical representations, for example, "cat," "dog," "mammal," "furniture," by attending to perceptual properties of objects (Behl-Chadha 1996; Quinn et al. 1993; Quinn and Johnson 2000).

Recently, based on this evidence of early categorization in infants, researchers have come to focus on when infants begin to categorize objects during their developmental course. To obtain empirical evidence for the emergence of categorization, recent studies tested infants shortly after birth. For example, Quinn et al. (2001) examined whether neonates in the first few days of life could categorize geometrical figures with the familiarization-novelty preference task (Quinn et al. 2001). In this study, they reported that neonates categorized open or closed figures (e.g., crosses and triangles). These interesting results suggested the possibility that the ability for categorization in humans is based not only on experiences after birth but also on innate factors. This issue is worthwhile but not simple. If humans possess an innate foundation to categorize objects, how and when we are endowed with this ability must be reconsidered. However, it is very difficult for researchers to investigate categorization in neonates or fetuses experimentally because of methodological limitations. Thus, we should provide another way to approach this interesting and difficult issue. In light of this, it seems very helpful to research the origins of human categorization not only in terms of ontogeny but also in terms of phylogeny, which is studying from the aspect of evolutionary factors to examine how human beings obtain the ability for categorization. Recently, the term "evolution" has become familiar to researchers in developmental psychology. To understand differences and similarities between humans and nonhuman primates by comparing among species seems to be a good way for us to rethink what we know about humankind.

4 Comparative Study for Categorization

It is known that humans and nonhuman primates share some behavioral attributes that are rooted in a common ancestor. For example, the primitive grasp reflex that appears in neonates is thought to be shared behavior with nonhuman primates. This is requisite behavior that is brought out when nonhuman primates cling to their mothers' back or abdomen in their childhood. Previous studies of cognitive development in human infants reported that human infants show various cognitive behaviors including categorization in their early development, e.g., memory (Meltzoff 1988; Rovee-Collier and Hayne 1987), object perception (Kellman and Spelke 1983), causal understanding (Baillargeon et al. 1992; Luo et al. 2003), and social understanding (Repacholi and Gopnik 1997). These cognitive behaviors are functional in early development because they are potentially important skills for human infants to adapt to their environment. In light of this, the matter of the emergence of such cognitive functions is important. There is the possibility of finding clues of the roots of human cognition in our neighbors such as macaques or chimpanzees, for example, the grasp reflex mentioned earlier. The present study is based on this approach to explore the emergence and the developmental process of human behaviors through comparing humans and nonhuman primates (or nonhuman animals).

4.1 Categorization in Nonhuman Primates

Because categorization is a very important skill for processing information in the world, it is not surprising that categorization is not restricted to humans but is also seen in nonhuman primates. For example, previous animal studies reported that many species of nonhuman primates respond categorically to objects: squirrel monkeys (*Saimiri sciureus*) (Phillips 1996); baboons (*Papio anubis*) (Bovet and Vauclair 1998); rhesus monkeys (*Macaca mulatta*) (Neiworth and Wright 1994); gorillas (*Gorilla gorilla gorilla*) (Vonk and MacDonald 2002); and chimpanzees (*Pan troglodytes*) (Tanaka 2001). These findings provide empirical evidence of good cognitive capacity in nonhuman primates and important cues for exploring the character of human categorization and its evolutionary origin.

However, most animal studies have some experimental limitations, as some investigators note (Brown and Boysen 2000). For example, in animal studies, the subjects are usually trained to respond categorically to objects during the experiments, whereas the spontaneous categorization of participants is examined in human studies. Training is often conducted to lead the subjects' responses to the required answer by reinforcing their correct responses. In this case, the concern is whether the subjects would be able to learn the category. Such categorization studies for nonhuman primates certainly show that they possess the ability for learning some categories. However, to compare categorization itself in nonhuman primates and humans it is necessary to investigate not only trained categorization but also spontaneous categorization without training in nonhuman primates. Moreover, there are not enough data on early categorization by nonhuman primate infants. In other words, there is a lack of information concerning the developmental process and developmental change in categorization of nonhuman primate infants. Although it has been reported that infant macaques after the first day of life discriminated objects such as simple figures with training, suggesting the possibility of the formation of categorical responses (Jolly 1972), it is not certain how infant macaques develop categorical responses later. Thus, it seems to be important to examine when and how nonhuman primates start to form categories and develop this ability, for a fuller understanding of categorization in nonhuman primates and the properties of the development of categorization in humans and nonhuman primates.

For these reasons, it can be said that study to directly compare early spontaneous categorization in humans and nonhuman primates is very important to allow us to understand categorization in both kinds of primates well.

4.2 Do Primate Infants—Macaques, Chimpanzees, and Humans—Categorize?

In the present study, early spontaneous categorization in macaque infants, chimpanzee infants, and human infants was examined. First, early categorization in primate species was examined by testing primate infants (less than 2 years after birth) as already described. Second, spontaneous categorical responses in all species were assessed using a task without specific training, including reinforcement for correct responses. Next, categorical responses in the three species were compared using similar tasks. To conduct this study, the experimental procedure generally used for human infants, that is, the so-called familiarization-novelty preference task, was used for nonhuman primate infants.

Previously, some studies comparing the cognition of human infants with that of adult primates have been conducted, but with little regard to the developmental stage of the subjects. However, to compare the developmental process of cognition between species, it is necessary to examine what kinds of behaviors are found in the course of development in each species. By doing so, it would be possible to infer whether some cognitive behaviors in each species would develop in the same way and what factors would affect their development. It should be noted that we must recognize the variety of cognitive behaviors in each species, but it is difficult to carefully consider such variety of behaviors in comparing cognitive development between species. In general, it is thought that chimpanzees would develop 1.5 times and macaques 4 times as quickly as humans do based on their life span or physical development. However, presently no criterion to compare development between species using their cognitive behaviors as a measure has been established. Thus, in the future, it is necessary to set a standard to compare cognitive development between species.

In this context, the previous study suggested that human infants and chimpanzee infants showed similar categorical responses in the same task (Spinozzi 1993; Spinozzi et al. 1998). In these studies, Spinozzi and her colleagues compared the development of spontaneous sorting behavior by chimpanzee infants between the ages of 15 and 54 months and by human infants between the ages of 6 and 24 months, using sets of simple objects that were logically related in form (cups, rings, blocks, and crosses) and color (blue, green, yellow, and red) as stimuli. The participants were presented with a set of six objects in each trial. These objects were divided into two classes that differed from each other in one of their two properties, either color (three red cups and three yellow cups) or form (three red cups and three red rings), or differed in both color and form (three blue crosses and three green blocks). They analyzed participants' constructive manipulation with these sets of objects, and concluded that chimpanzees spontaneously sorted objects based on both properties and that development of sorting behavior in chimpanzee infants seemed to be similar to that of human infants. However, for now, there are not enough data from the same aspect as these studies. It goes without saying that more data are needed to establish a standard to compare cognitive development between species.

The main aim of the present study is to examine, by comparatively investigating the categorization among three primate species, whether there were some species-specific properties and similarities in categorization among the species to allow us to have a fuller understanding of categorization in those species. If there are some similarities among species, the possibility of the primate origin of categorization may be suggested. And if there are species-specific properties, it would help to know the attributes of categorization that each species has uniquely developed during evolution. In the following, first, I introduce the experiments for chimpanzee infants and human infants (Murai et al. 2005) and then that for infant Japanese macaques (Murai et al. 2004).

5 Spontaneous Categorization in Chimpanzee Infants

5.1 Method

5.1.1 Subject

Three infant chimpanzees (*Pan troglodytes*), named Ayumu (male), Cleo (female), and Pal (female), participated in this experiment. Ayumu was tested at the age of 14 to 23 months, Cleo was tested at the age of 12 to 20 months, and Pal was tested at the age of 10 to 18 months. They were born at the Primate Research Institute of Kyoto University and were reared by their mothers from birth.

5.1.2 Stimuli

The stimuli were three-dimensional lifelike scale models from three global-level categories: mammal, furniture, and vehicle (Fig. 1). The mammals were made from rubber or plastic and had no moving parts. The furniture exemplars were made from wood or plastic, and the vehicles were made from plastic or metal. Any moving parts of furniture and vehicles (e.g., drawer, wheels, doors) were glued so that they were immobile.

5.1.3 Procedure

In the experiment, the familiarization-novelty preference task using participants' examination behavior to the objects as a measure was used. Here, "examination" was defined as the behavior of looking at the stimulus objects while

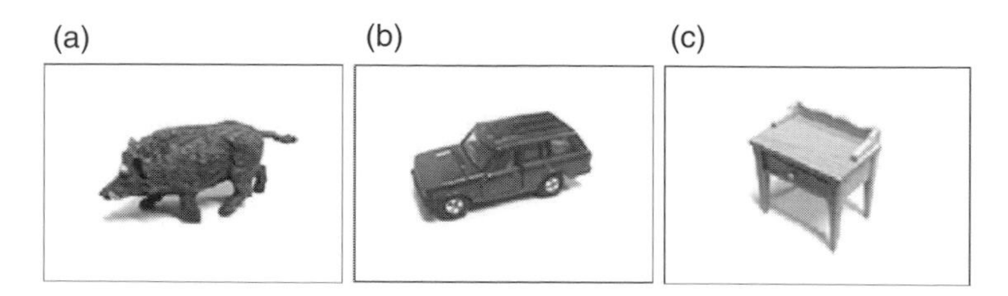

Fig. 1. Examples of stimulus objects from (**a**) mammal, (**b**) vehicle, and (**c**) furniture categories

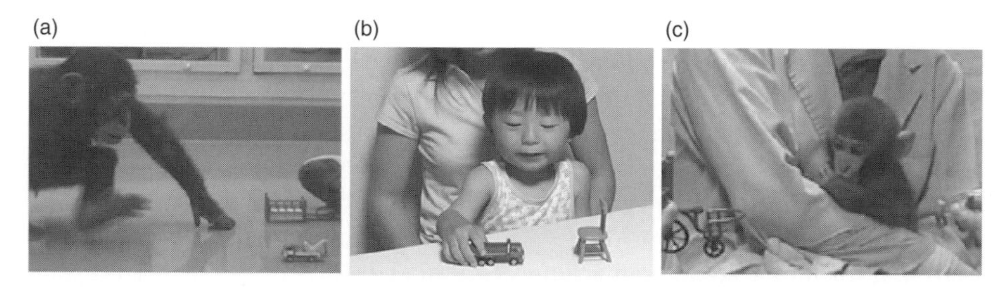

Fig. 2. Experimental situation in each species. *Left*, infant chimpanzee; *middle*, human infant; *right*, infant Japanese macaque

touching them. No participants were trained during the experiments. This task consisted of two phases: the familiarization phase and the test phase. During the familiarization phase, the participant was sequentially presented with four different familiarization objects from one category (e.g., mammal). Each trial lasted 15 s, beginning with the participant's first examination of the object; intertrial interval (ITI) was about 10 s. Then, in the test phase, the participant was presented with a pair of test objects: one was a new object from the now-familiar category (mammal) and the other was from a novel category (e.g., vehicle). The left–right positioning of the two objects was counterbalanced across sessions. Each trial lasted 15 s, beginning with the participant's first examination. Thus, one session consisted of four familiarization trials and one test trial. Figure 2a shows the experimental situations. All trials were videotaped for later scoring. To assess the amount of examination time, one of two coders calculated the number of frames (one frame $= \frac{1}{30}$ s) in which the participants examined the object. Behaviors such as simply looking and simply touching (or mouthing) without looking were not considered.

Actually, participants were tested with two kinds of test pairs for each familiarization condition. For example, when the participants were familiarized with mammals, they were tested with a mammal (novel familiar-category) and a furniture exemplar (novel category), or a mammal and a vehicle (novel category). Either kind of test pair was used in one session, the order being counterbalanced across sessions in each condition. All participants received 12 sessions (6 sessions for each of two test pairs) in each familiarization condition, that is, they received 36 sessions overall.

5.1.4 Preference Test

Preference tests were conducted to examine whether the participants had any intrinsic preference for a particular category. The procedure was almost the same as in the categorization test described previously, except the participants were presented with neutral objects during familiarization. In the familiarization phase, the participants were presented with four neutral objects (wooden bricks varying in shape and color). Then, in the test phase, they were presented with a pair of objects from two categories. Test objects were chosen randomly for each session from the set of stimuli used in the category test. The pair was presented twice, and the left–right positioning of objects was counterbalanced across trials. Three contrasts were generated from combinations of three categories: mammal versus furniture, vehicle versus furniture, and mammal versus vehicle. One type of contrast was tested in a session, with the order counterbalanced across sessions. All participants received 8 sessions for each contrast, that is, they received 24 sessions overall.

5.2 Results

5.2.1 Preference Test

For each contrast, mean examination times for each category object in the test phase were analyzed to ensure whether participants preferred one of two categories. As a result, Ayumu and Cleo did not have any intrinsic preference for a particular category. However, it was showed that Pal preferred mammals to two other categories (Table 1).

Table 1. Mean (M) and standard error (SE) of examination times (in seconds) in the preference test by chimpanzee infants

	Category contrast					
	Mammal vs. furniture		Vehicle vs. furniture		Mammal vs. vehicle	
	Mammal	Furniture	Vehicle	Furniture	Mammal	Vehicle
Ayumu						
M	1.91	2.52	1.97	3.22	1.53	2.26
SE	0.36	0.59	0.56	0.52	0.22	0.34
Cleo						
M	1.65	2.04	3.00	1.74	2.41	1.35
SE	0.52	1.00	0.57	0.66	0.30	0.54
Pal						
M	2.89	0.71	1.46	1.44	3.80	0.31
SE	0.65	0.30	0.59	0.68	0.46	0.18

5.2.2 Category Test

If participants categorized objects, their examination times should gradually decrease in the familiarization phase, that is, they should be familiarized with the category. Also, their examination times to novel-category objects should be longer than those to novel familiar-category objects in the test phase. That is, they should prefer novel-category objects to novel familiar-category objects.

In the familiarization phase, participants' examination times were averaged across the first two trials (the first block) and across the last two trials (the second block) for each familiarization condition. Then, the mean examination times of both blocks were analyzed to investigate whether participants' examinations decreased from the first block to the second block. As Table 2 shows, examination times did not significantly decrease across blocks, suggesting that participants were not familiarized with the category.

In the test phase, for each familiarization condition, the mean examination times to novel familiar-category objects and that to novel-category objects were analyzed to ensure whether participants' examinations to the novel-category objects were significantly longer than that to the novel familiar-category objects. The results indicated that participants examined the novel-category objects longer than novel familiar ones, suggesting that all participants significantly preferred novel-category objects to novel familiar-category objects as a whole (Fig. 3).

Also, additional analyses to examine participants' novelty preference in each category contrast—mammal versus furniture, vehicle versus furniture, and mammal versus vehicle—were conducted to examine whether participants' novelty preference was bidirectional in all category contrasts, regardless of which category the participants were familiarized with, because there is the pos-

Table 2. Mean (M) and standard error (SE) of examination times (in seconds) in the familiarization blocks by chimpanzee infants

	Familiar category					
	Mammal		Furniture		Vehicle	
	Block 1	Block 2	Block 1	Block 2	Block 1	Block 2
Ayumu						
M	1.04	1.83	4.01	3.28	3.47	3.32
SE	0.31	0.21	0.54	0.36	0.57	0.49
Cleo						
M	0.80	0.58	2.49	1.75	2.46	0.79
SE	0.23	0.22	0.81	0.56	0.85	0.22
Pal						
M	2.14	2.55	3.67	3.35	1.29	1.73
SE	0.48	0.87	0.87	0.72	0.52	0.59

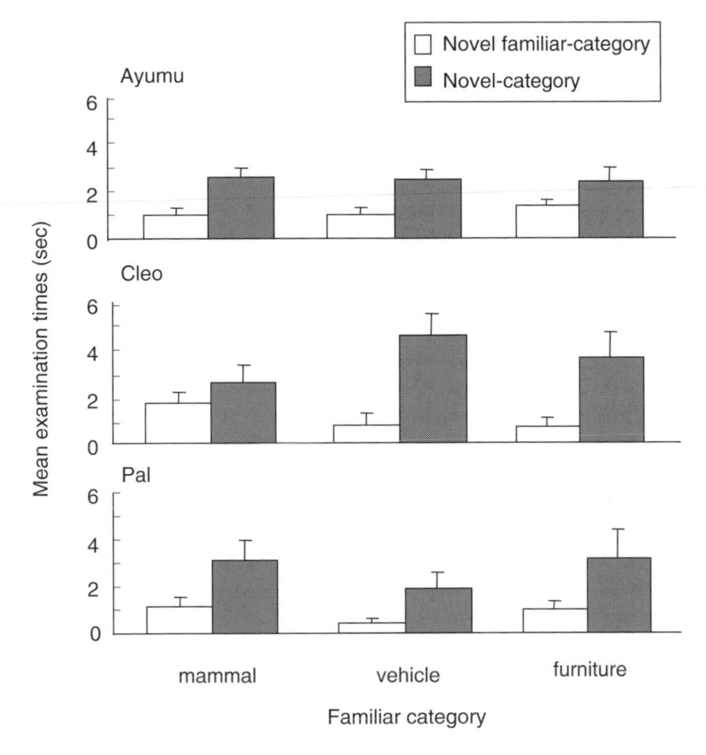

Fig. 3. Mean examination times (with *SE error bars*) for each test object in each familiarization condition by chimpanzee infants

sibility that the combination of the familiarization objects and the novel test objects would affect the participants' novelty preference. For example, it was reported that human infants who were familiarized with the cat category preferred the novel dog category in the test phase, whereas infants who were familiarized with the dog category did not prefer the novel cat category in the test phase (Quinn et al. 1993).

There were two kinds of test pairs for each category contrast. For example, in the contrast of mammal versus vehicle, there were two test pairs of mammal-vehicle (M-V) and vehicle-mammal (V-M). The M-V pair meant that the participants were familiarized with mammals, then tested with mammal (novel familiar-category) and vehicle (novel category), while the V-M pair meant that they were familiarized with vehicle, then tested with vehicle (novel familiar-category) and mammal (novel category). For each test pair, mean examination times to each test object were analyzed. The results showed that participants' examination times to novel-category objects were significantly longer than those to novel familiar-category objects in every test pair (Fig. 4). These results indicated that the significant novelty preference was bidirectional in all category contrasts, regardless of which category the participants were familiarized with.

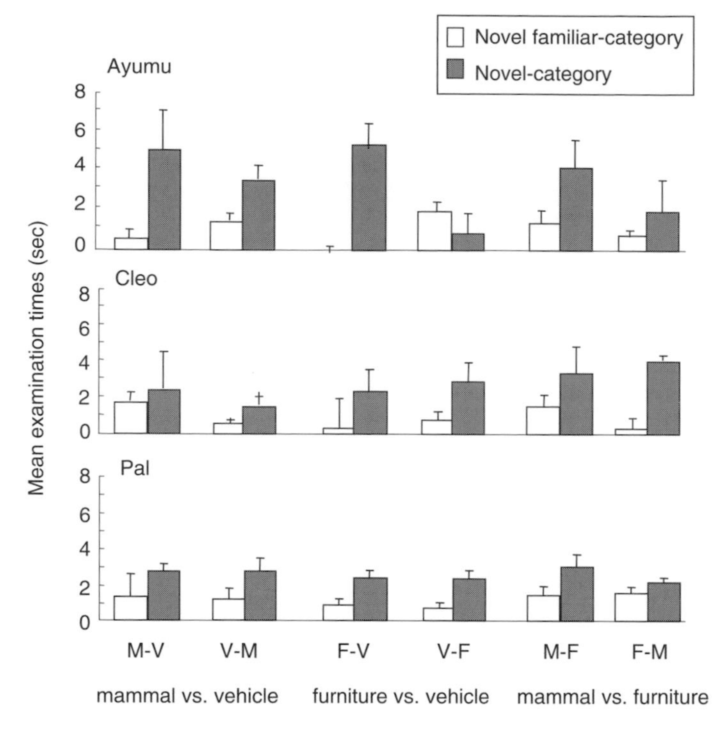

Fig. 4. Mean examination times (with *SE error bars*) for each test object in each test pair by chimpanzee infants

5.3 Brief Discussion

The important result of this experiment is that all participants showed significant novelty preference in the test phase; this suggested that chimpanzee infants would spontaneously discriminate among three categories.

However, no chimpanzee infant showed a significant decrease of examination time during the familiarization phase; that is, they were not familiarized with the category even after repeated presentations of the category objects. I discuss the possible reasons for these results in the general discussion after investigating whether such responses would be unique for chimpanzee infants. However, at least, it can be said that these results would not reflect the participants' inability to process similarities and properties among the familiarization objects, because, if the participants had not categorically processed the objects at all, they should randomly respond to the test objects in the test phase, and the consistent novelty preference found in this study would not have been appeared.

Also, it had to be noted that Pal showed an intrinsic preference for mammals in the preference test. Such a preference would accentuate her novelty preference when the mammal category was presented to her as a novel-category object. However, she similarly showed a significant novelty preference when she was familiarized with the mammal category. Thus, Pal's response in the test phase was not simply dominated by her intrinsic preference but was based on the experiences during the familiarization phase.

These results indicate that chimpanzee infants spontaneously form some categorical representations of global-level categories, which suggests that early spontaneous categorization is not restricted to human infants but is shared with chimpanzee infants.

6 Spontaneous Categorization in Human Infants

6.1 Method

6.1.1 Subjects

Sixty-four human infants participated in this experiment: 48 infants participated in the category test and the remaining 16 infants participated in the preference test. In the category test, 48 participants were divided into two age groups: the younger group (24 infants aged 10–17 months; mean age, 14 months and 7 days), and the older group (24 infants aged 18–24 months; mean age, 21 months and 11 days). In each age group there were 12 girls and 12 boys. In the preference test, 16 infants were divided into two age groups: 8 in the younger group (4 girls, 4 boys aged 12–17 months; mean age, 14 months and 29 days) and 8 in the older group (3 girls, 5 boys aged 18–22 months; mean age, 21 months and 9 days).

6.1.2 Stimuli

The same scale models as in the chimpanzee experiment were used (see Fig. 1).

6.1.3 Procedure

The procedure was essentially the same as in the experiment for chimpanzee infants, except that two test trials were done in a session (in the chimpanzee experiment, one test trial was done per session). In this experiment, four familiarization objects from one category were presented to the participants in the familiarization phase, and then the test pair was presented to the participants twice in the test phase. The left–right positioning of two objects was counterbalanced across trials. Thus, in this experiment, each session consisted of four familiarization trials and two test trials, and two sessions were conducted for each infant. Figure 2b shows the experimental situation.

Each infant was randomly assigned to one of three familiarization conditions: mammal, furniture, and vehicle. Therefore, eight infants were included in each condition for each age group. In each condition, the infant was first familiarized with four objects from one category (e.g., mammal), and tested with the first test pair (e.g., mammal versus furniture). After the first session, the second familiarization phase started, in which four new objects from the same category were presented with the infant as in the first session, and then the infant was tested with the second test pair (e.g., mammal versus vehicle). The order of presentation of the two kinds of test pairs was counterbalanced across infants in each condition.

6.1.4 Preference Test

The preference test procedure was essentially the same as in the chimpanzee experiment. Each infant received three sessions. In the first session, the infant was presented with four neutral objects (wooden bricks), then tested with one of three test pairs (mammal versus furniture, mammal versus vehicle, and furniture versus vehicle). Each test pair was presented twice. The left–right positioning of two objects was counterbalanced across trials. In the second and the third sessions, infants were similarly familiarized with four wooden bricks, then tested with one of the remaining pairs. The order of presentation of the three kinds of test pairs was counterbalanced across infants in both age groups.

6.2 Results

6.2.1 Preference Test

For each category contrast, participants' examination times in the test phase were analyzed to ensure whether participants significantly preferred either of two categories in each age group. The results showed that there was no preference for any particular category in both age groups (Table 3).

6.2.2 Category Test

For each familiarization condition, participants' mean examination times in the familiarization phase were analyzed to examine whether their responses would decrease from the first block to the second block in each age group. The results indicated that participants' examination times significantly decreased across blocks in both age groups (Table 4).

Table 3. Mean (M) and standard error (SE) of examination times (in seconds) in the preference test by human infants

	Category contrast					
	Mammal vs. furniture		Vehicle vs. furniture		Mammal vs. vehicle	
	Mammal	Furniture	Vehicle	Furniture	Mammal	Vehicle
Younger						
M	3.96	3.16	4.42	3.92	4.25	5.45
SE	0.99	1.00	1.28	0.68	0.75	0.96
Older						
M	5.18	6.01	6.82	3.35	3.60	5.95
SE	0.97	0.67	0.92	0.79	0.98	1.20

Table 4. Mean (M) and standard error (SE) of examination times (in seconds) in the familiarization blocks by human infants

	Familiar category					
	Mammal		Furniture		Vehicle	
	Block 1	Block 2	Block 1	Block 2	Block 1	Block 2
Younger						
M	5.36	3.97	8.72	8.11	7.08	6.94
SE	0.81	0.67	0.67	0.81	0.68	0.87
Older						
M	9.18	6.70	7.33	6.41	10.08	9.26
SE	0.74	1.03	0.66	0.77	0.73	0.99

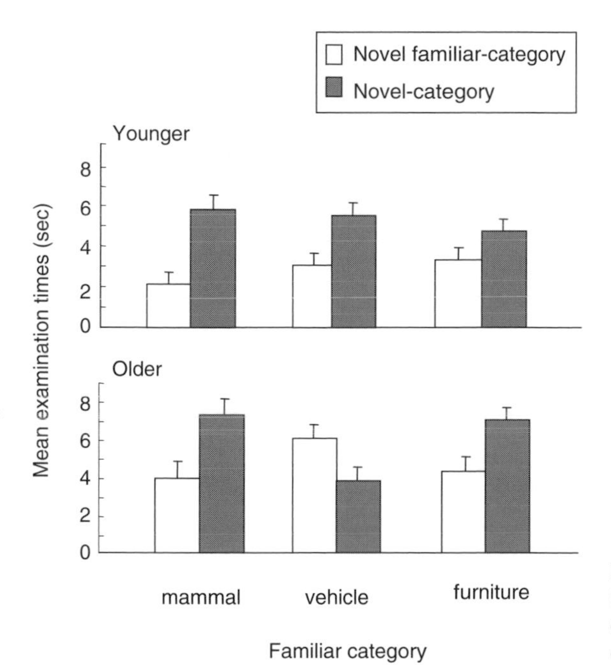

Fig. 5. Mean examination times (with *SE error bars*) for each test object in each familiarization condition by human infants

In the test phase, for each familiarization condition, the mean examination times for each test object (novel familiar-category objects and novel-category objects) were analyzed to ensure whether participants significantly preferred novel-category objects to novel familiar-category objects in each age group. The results indicated that participants showed significant novelty preference when they were familiarized with mammals and furniture. However, they did not show such preference when they were familiarized with vehicles (Fig. 5).

In additional analyses, for each kind of test pair from three category contrasts (mammal versus furniture, vehicle versus furniture, and mammal versus

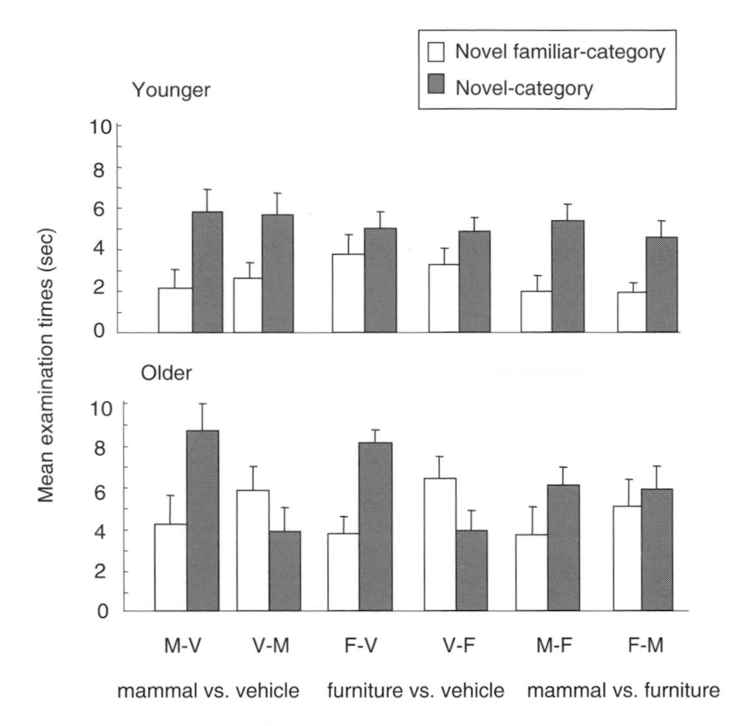

Fig. 6. Mean examination times (with *SE error bars*) for each test object in each test pair by human infants

vehicle), infants' examination times for each test object were analyzed as in the chimpanzee experiment. The results revealed that infants showed bidirectional significant novelty preference in mammal versus furniture; however, such responses were not shown in the other two contrasts. As Fig. 6 shows, infants did not express the novelty preference when they were familiarized with vehicles.

6.3 Brief Discussion

In this experiment, evidence that human infants showed significant decrement in examination time during the familiarization phase was obtained; that is, they were familiarized with the categories. This finding is the one of the differences from the results obtained with chimpanzee infants.

Human infants also showed significant novelty preference in the test phase. These results of significant familiarization and novelty preference in infants provide strong evidence that human infants categorized global-level objects. However, they showed significant novelty preference when they were familiarized with mammals and furniture but not with vehicles. These results suggested that human infants formed categorical representation of mammals and furniture, at least. These results are not inconsistent with previous results of global-level categorization in human infants (Behl-Chadha 1996). However, when infants were familiarized with vehicles, they did not show such preference. Although the reason for the results is not clear, there are some possible reasons, for example, that vehicle objects are much more variable compared to another category of objects. In addition, as Figs. 5 and 6 show, this tendency was more salient in the older infants.

6.3.1 Analyses of Human Infants' Object Manipulation

In this context, it was often observed that the older infants tended to manipulate vehicle objects functionally, that is, they moved or slid the vehicle in a straight line, even though the vehicle wheels were fixed and immovable. Such functional object manipulations were also observed in the younger infants, although not as frequently. As the experimenter did not show the manipulations of vehicle objects to the infants during the experiment, it was certain that infants did not learn to slide the vehicles through observing vehicle objects actually moving. Thus, it was expected that infants had already known how to move the vehicles. If this is the case, infants tend to examine the vehicles for a long time even after repeated presentations of them during the familiarization phase, as they might try to actually manipulate them (infants could take the objects only in the test phase; they did not take them in the familiarization phase).

To confirm this possibility, infants' specific manipulations in the test phases were assessed. Of all 64 infants (32 younger infants and 32 older infants) who participated in the preference test and the category test, two coders counted the number of infants who produced two types of object manipulations from video records: sliding vehicles and making mammals hop. "Sliding vehicle" was defined as moving the vehicle back and forth, not in one direction, more than once; "hopping mammal" was noted when the infant made mammals hop more than once.

Of 32 infants, 19 older infants performed sliding vehicle, compared to 11 younger infants. Most importantly, of 8 older infants who were familiarized with vehicles, 7 performed sliding vehicle. In stark contrast, only 1 younger infant performed this movement in the same condition. It seems likely that the sliding action of vehicles would be facilitated in vehicle-familiarized older infants, and then they would continue examining the vehicles even after repeated presentations of them. In addition, in both age groups, only 2 infants performed mammal-hopping actions. This result indicates that, for human infants, scale models of vehicles afford sliding movements; this has been reported in previous studies of human infants (Rakison and Butterworth 1998; Rakison and Cohen 1999). Such manipulations of vehicles by infants do not depend on the specific context of the present study but on the infants' previous experiences and their understanding of object movement. This finding leads us to emphasize that the lack of novelty preference does not indicate that older infants failed to form categorical representations of vehicles. Furthermore, it is important that infants were clearly familiarized with vehicle objects in the familiarization phase.

The present results suggested that human infants formed categorical representations. Moreover, some interesting unique responses such as consistent familiarization and functional object manipulations were found. In the general discussion, I mention the properties of categorical responses in each species based on these similarities and differences that appeared through the comparative studies.

7 Spontaneous Categorization in Macaque Infants

7.1 Method

7.1.1 Subjects

Two male infant Japanese macaques (*Macaca fuscata*), named Romio and Tim, participated in this experiment. Romio was tested at the age of 3 to 5 months and Tim at the age of 2 to 4 months. They were born at the Primate Research Institute of Kyoto University and reared from birth by human caregivers.

7.1.2 Stimuli

Similar scale models from the mammal, furniture, and vehicle categories were used as in the prior experiments (see Fig. 1).

7.1.3 Procedure

As in the prior experiments, the familiarization-novelty preference procedure without any training was used. However, there are three different points from the prior experiments: the measure used for the analyses, the way to present familiarization objects, and the presentation time of familiarization objects. In this experiment, participants' looking behavior, not examination, was used as a measure. In the familiarization phase, the participant was sequentially presented with two pairs of familiarization objects from one category twice (e.g., a pair of cat and cow from the mammal category was presented in the first and third trial, and a pair of pig and dog was presented in the second and fourth trial). The left–right positioning of objects was counterbalanced across trials. Each familiarization trial lasted 20 s and ITI was 10 s. In the test phase, participants were presented with a pair of test objects twice: one was a new object from the familiar category (i.e., fox), and the other was from a novel category (i.e., desk). The left–right positioning of objects was counterbalanced across trials. Each test trial lasted 15 s and ITI was 10 s. Hence, one session consisted of four familiarization trials and two test trials. Figure 2c shows the experimental situations. Both subjects received 8 sessions for each familiar category (4 sessions for each test pair), that is, they received 24 sessions overall.

7.1.4 Preference Test

Eight stimuli were randomly chosen from each of three categories. During the preference test, pairs of objects from two categories were presented to the participant. Four pairs were generated from each of three contrasts (mammal versus furniture, vehicle versus furniture, and mammal versus vehicle), and a pair was presented to the participant for 15 s in each trial. Each participant received one session of 12 trials.

7.2 Results

7.2.1 Preference Test

For each category contrast, participants' looking times at each test object were calculated and divided by the total looking time for both test objects. Looking time was then converted to a percentage score, termed a preference score. The mean preference scores for each category were compared to chance level (50%) for each participant. The results indicated that neither participant had a significant preference for a particular category (Table 5).

7.2.2 Category Test

In the familiarization phase, participants' looking times were averaged across the first two trials (the first block) and across the last two trials (the second block) for each familiarization condition. Then, the mean looking times of both blocks were analyzed to investigate whether participants' looking decreased from the first block to the second block, for each participant. As Table 6 shows, looking time did not significantly decrease across blocks in either participant, suggesting that they were not familiarized with the category.

Table 5. Mean preference scores (%) in the preference test by infant Japanese macaques

	Category contrast					
	Animal vs. vehicle		Furniture vs. vehicle		Animal vs. furniture	
	Animal	Vehicle	Furniture	Vehicle	Animal	Furniture
Romio						
Mean preference	49.9	50.1	27.5	72.5	68.3	31.7
SD	17.38		11.06		22.27	
N	4		4		4	
t (vs. chance)	0.003		2.034		0.821	
Tim						
Mean preference	47.8	52.2	47.6	52.4	58.3	41.7
SD	17.64 14.67			16.56		
N	4		4		4	
t (vs. chance)	0.128		0.167		0.501	

Table 6. Mean looking times (in seconds) in the familiarization blocks by infant Japanese macaques

	Trial block	
Familiar category	1 & 2 trials	3 & 4 trials
Romio		
Animal		
Mean	8.99	8.43
SD	0.71	1.12
Furniture		
Mean	7.65	5.46
SD	1.00	0.57
Vehicle		
Mean	9.40	9.28
SD	0.60	0.94
Tim		
Animal		
Mean	6.88	5.73
SD	0.54	0.86
Furniture		
Mean	7.45	9.43
SD	0.79	1.7
Vehicle		
Mean	9.94	9.09
SD	0.40	1.22

SD, standard deviation

In the test phase, for each familiarization condition, the mean preference scores to novel familiar-category objects and to novel-category objects were calculated as in the preference test. Then, for each participant, these scores were analyzed to ensure whether the preference score for the novel-category objects was higher than that for the novel familiar-category objects. The results indicated that the mean score for the novel-category objects was significantly higher than that for novel familiar ones in mammal and furniture conditions, suggesting that both participants showed significant novelty preference in those conditions (Fig. 7). Moreover, for Romio, overall mean preference score for novel-category objects was $M = 71.27$, and that for novel familiar-category objects was $M = 28.73$. For Tim, overall mean preference score for novel-category objects was $M = 71.23$ and that for novel familiar-category objects was $M = 28.77$.

In additional analyses, for each kind of test pair from three category contrasts, the mean preference scores for novel-category objects were calculated. Then, these scores were compared to chance level (50%) for each participant. The results revealed that both participants showed bidirectional significant novelty preference in mammal versus furniture; however, such responses were not

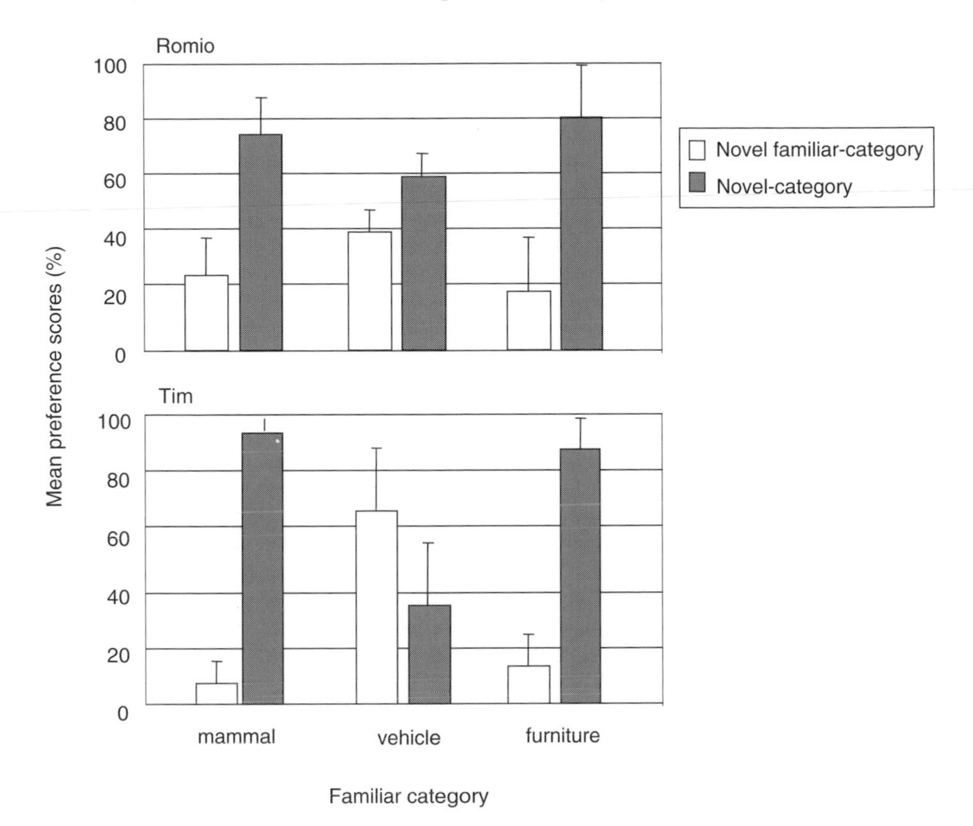

Fig. 7. Mean preference scores (with *SE error bars*) for each test object in each familiarization condition by infant Japanese macaques

shown in the other two contrasts (Fig. 8). The participants did not show novelty preference when they were familiarized with vehicles.

7.3 Brief Discussion

The main result of this experiment was that infant macaques showed spontaneous categorical discrimination in the form of novelty preference. This finding suggested that infant macaques could categorically process the objects based on their properties, that is, Japanese macaques in their early development possess the ability to form categorical representations.

However, there are some questions about the present findings. First, participants did not show significant familiarization, which seemed to be a similar response to that of chimpanzee infants. Then, what did participants do during the familiarization phase? If they saw only the familiarization stimuli, they would randomly respond to the objects in the test phase regardless of the categorical

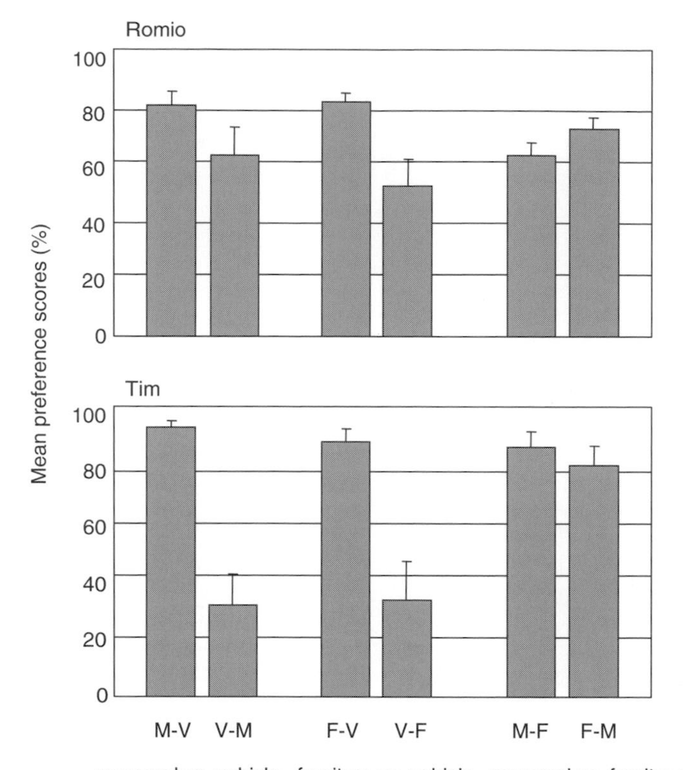

Fig. 8. Mean preference scores (with *SE error bars*) for the novel-category object in each test pair by infant Japanese macaques

difference between them. However, they significantly preferred novel-category objects as a whole. These results suggested that they would process the similarities among the familiarization objects, although this was not reflected in the explicit familiarization.

Second, participants did not express novelty preference when they were familiarized with vehicle objects. The reason could not be determined. However, as one possibility, we guessed that it was difficult for participants to recognize the categorical similarities among the objects. Nevertheless, they certainly discriminated vehicle objects from the other two category objects, because they preferred novel vehicle objects when they were familiarized with mammal and furniture categories.

The present results suggested that infant Japanese macaques would form categorical representations of global-level categories without any training. Of course, we should be cautious to interpret these results as evidence of categorization by infant macaques in general, because evidence of familiarization was lacking and the number of participants was small. However, it can be said that the present study would be meaningful in terms of providing data showing early categorization in macaques.

8 General Discussion

8.1 The Similarities Among Three Species

The present study was conducted to examine early categorization in Japanese monkeys, chimpanzees, and humans. As a result, it was indicated that there was a similarity in categorization among those three species. In all species, significant preference for novel-category objects was shown. For human infants, many studies have reported evidence of discrimination between familiar and novel categories (Behl-Chadha 1996; Younger and Fearing 1999, 2000). Thus, the present results in human infants were not surprising. However, the results provide us with another implication in terms of comparative development; that is, not only human infants but also nonhuman primate infants could form categorical representations. The present study implied that those three primates in their early development share the basic ability to form categorical representations.

Moreover, the present study indicated that the familiarization-novelty preference task using participants' looking or examination behavior as the measure was useful to test spontaneous categorization in nonhuman primate infants. However, further investigations of categorical behavior by primate infants with other tasks and categories objects are needed. I believe that such studies provide us much significant data and deeper understanding of primate categorization.

8.2 The Differences Among the Three Species

In a series of experiments it was indicated that there were some species-specific responses. First, only human infants showed consistent significant familiarization in the familiarization phase. It was suggested that human infants possess the capacity to group the familiarization objects as members of the same category by recognizing the similarities among them beyond the perceptual variety of the global-level category members. In contrast, the other two species did not show this tendency. The reasons were not clear. However, there were some possible reasons, such as inadequate time and/or numbers of presentation of familiarization objects and variety of familiarization objects. In the present study, different familiarization objects, not a single object, were presented to the participants. Possibly, such variety of familiarization objects might retain the participants' attention in the course of the familiarization phase, with the result that salient familiarization would not occur. Indeed, Behl-Chadha (1996) reported that 3- to 4-month-old human infants also did not show familiarization when they were presented with different familiarization objects (e.g., various chairs) and suggested that the diversity of familiarization objects would maintain their attention. As another reason, Japanese monkeys and chimpanzees could realize the similarities among familiarization objects only when they were presented with an object from the contrast category in the test phase. Therefore, these results suggested that nonhuman primates might subsequently form categorical representations of a familiar category that excludes novel-category objects.

These differences among the species would indicate species-specific cognitive ability underlying categorical responses. In this context, some recent studies suggested that there are differences in the level of object processing between nonhuman primates and humans. Quinn et al. (2002) suggested that human infants show some flexibility in object processing by attending not only to local but also to global properties of objects. Thus, very young infants may process object information at multiple levels even when dealing with complex objects (Quinn et al. 2002). Object processing in human infants seems to result in them easily detecting whole properties of the objects and recognizing similar properties among the objects. Such object processing also would be helpful to categorize the various objects by their appearance from the global-level category. In contrast, chimpanzees and macaques of all ages appear more likely to attend to local properties than global ones when processing objects (Fagot and Tomonaga 1999). Thus, there was the possibility that the present results would stem from species specificity in the level of object processing.

Second, functional object manipulation was also unique for human infants. In the present experiment, infants often manipulated an object functionally, especially when they were presented with vehicle objects, suggesting that human infants ontologically discriminated between the objects in terms of object movement. They moved vehicle objects (sliding) but not objects from other categories. Such object manipulations indicated that, by 2 years of age, human infants might have developed certain kinds of knowledge about object meanings such as object-appropriate movement (Mandler 1992; Rakison and Cohen 1999), in addition to the ability for categorical discrimination or grouping. Infants' functional manipulations suggested that human infants seemed to attend to object attributions such as the expected movement as well as visible information about objects, for example, their features. This tendency may play an important role in forming some conceptual categories such as "animal." On the other hand, chimpanzee infants did not show such object manipulation although they had an opportunity to manipulate the objects. Currently, so far as I know, there is no research to investigate whether nonhuman primate infants associate certain objects with certain kinds of movements. However, to examine the understanding of object movements in nonhuman primates, if not in the form of manipulation, is important because it would be helpful to know whether nonhuman primates possess conceptual understanding of objects.

8.3 The Developmental Process of Categorization in Primates

The present study provided evidence of early categorization in nonhuman primates and indicated similarities and differences in categorization among the three species. All three species seemed to share the basic ability to form categorical representation; however, issues of early categorization and the developmental process in nonhuman primates are still open. For example, the research by Spinozzi and her colleagues mentioned earlier (Spinozzi et al. 1998; Spinozzi 1993) suggested that the spontaneous sorting behavior shown in chimpanzees aged 15 to 54 months seemed to be similar that in human infants aged 6 to 24 months. The present study suggested that chimpanzee infants aged 10 to 23 months and human infants aged 10 to 24 months showed a similar ability to form categorical representations. However, performance by chimpanzee infants was found to be comparable to that by younger human infants (10–17 months), considering that younger human infants' examination times were relatively shorter and functional object manipulations did not frequently appear. Thus, it was assumed that the developmental rate and the process in categorization were different in each primate species, whereas at the same time there are actually some similarities among species. To resolve these issues, further studies for primate categorization at various developmental periods are required.

As I described earlier, in general lifespan or physical development of participants was used as the measure to compare developmental rate among species. However, as is one of the aims of this study, it is meaningful to establish a standard to compare cognitive development among species. Such work would help us to understand the properties of cognitive development in humans and nonhuman primates and also the primate origins of human cognition. Thus, it is necessary to investigate what kinds of cognitive behaviors appear in various stages of development and how they develop in each species. I hope the present study will contribute to fill in the blanks of research about the process of cognitive development in primates.

Acknowledgments

This study was supported by JSPS Grants-in-Aid for scientific research 12002009, MEXT 21st Century COE program D-10 to Kyoto University, and the Cooperative Research Program of the Primate Research Institute, Kyoto University. I am very grateful to Drs. S. Itakura, T. Matsuzawa, M. Tomonaga, and M. Tanaka for their insightful suggestion and for helping to conduct the experiments.

References

Baillargeon R, Needham A, De Vos J (1992) The development of young infants' intuitions about support. Early Dev Parent 1:69–78

Behl-Chadha G (1996) Basic-level and superordinate-like categorical representations in early infancy. Cognition 60:105–141

Bovet D, Vauclair J (1998) Functional categorization of objects and of their pictures in baboons (*Papio anubis*). Learn Motiv 29:309–322

Brown DA, Boysen ST (2000) Spontaneous discrimination of natural stimuli by chimpanzees (*Pan troglodytes*). J Comp Psychol 114:392–400

Fagot J, Tomonaga M (1999) Global-local processing in humans (*Homo sapiens*) and chimpanzees (*Pan troglodytes*): use of a visual search task with compound stimuli. J Comp Psychol 113:3–12

Jolly A (1972) The evolution of primate behavior. Macmillan, New York
Kellman PJ, Spelke ES (1983) Perception of partly occluded objects in infancy. Cogn Psychol 15:483–524
Luo Y, Baillargeon R, Brueckner L, Munakata Y (2003) Reasoning about a hidden object after a delay: evidence for robust representation in 5-month-old infants. Cognition 88:B23–B32
Mandler JM (1992) How to build a baby: II. Conceptual primitives. Psychol Rev 99:587–604
Mandler JM, McDonough L (1993) Concept formation in infancy. Cogn Dev 8:291–318
Mandler JM, McDonough L (1998) Studies in inductive inference in infancy. Cogn Psychol 37:60–96
Mareschal D, Quinn PC (2001) Categorization in infancy. Trends Cogn Sci 5:443–450
Meltzoff AN (1988) Infant imitation and memory: nine-month-olds in immediate and deferred tests. Child Dev 59:217–225
Murai C, Tomonaga M, Kamegai K, Terazawa N, Yamaguchi MK (2004) Do infant Japanese macaques (*Macaca fuscata*) categorize objects without specific trainings? Primates 45:1–6
Murai C, Kosugi D, Tomonaga M, Tanaka M, Matsuzawa T, Itakura S (2005) Can chimpanzee infants (*Pan troglodytes*) form categorical representations in the same manner as human infants (*Homo sapiens*)? Dev Sci 8(3):240–254
Neiworth JJ, Wright AA (1994) Monkeys (*Macaca mulatta*) learn category matching in a nonidentical same-different task. J Exp Psychol Anim Behav Process 20:429–435
Pauen S (2002) Evidence for knowledge-based category discrimination in infancy. Child Dev 73:1016–1033
Phillips KA (1996) Natural conceptual behavior in squirrel monkeys (*Saimiri sciureus*): an experimental investigation. Primates 37:327–332
Quinn PC (2002) Early categorization: a new synthesis. In: Goswami U (ed) Handbook of childhood cognitive development. Blackwell, Malden, pp 84–101
Quinn PC, Johnson MH (2000) Global-before-basic object categorization in connectionist networks and 2-month-old infants. Infancy 1:31–46
Quinn PC, Eimas PD, Rosenkrantz SL (1993) Evidence for representations of perceptually similar natural categories by 3-month-old and 4-month-old infants. Perception 22:463–475
Quinn PC, Slater AM, Brown E, Hayes RA (2001) Developmental change in form categorization in early infancy. Br J Dev Psychol 19:207–218
Quinn PC, Bhatt RS, Brush D, Grimes A, Sharpnack H (2002) Development of form similarity as a gestalt grouping principle in infancy. Psychol Sci 13:320–328
Rakison D, Butterworth GE (1998) Infants' use of object parts in early categorization. Dev Psychol 34:49–62
Rakison DH, Cohen LB (1999) Infants' use of functional parts in basic-like categorization. Dev Sci 2:423–432
Repacholi BM, Gopnik A (1997) Early reasoning about desires: evidences from 14-and 18-month-olds. Dev Psychol 33:12–21
Rovee-Collier C, Hayne H (1987) Reactivation of infant memory: implications for cognitive development. Adv Child Dev Behav 20:185–238
Spinozzi G (1993) Development of spontaneous classificatory behavior in chimpanzees (*Pan troglodytes*). J Comp Psychol 107:193–200
Spinozzi G, Natale F, Langer J, Schlesinger M (1998) Developing classification in action: II. Young chimpanzees (*Pan troglodytes*). Hum Evol 13:125–139
Tanaka M (2001) Discrimination and categorization of photographs of natural objects by chimpanzees (*Pan troglodytes*). Anim Cogn 41:100–115
Vonk J, MacDonald SE (2002) Natural concepts in a juvenile gorilla at three levels of abstraction. J Exp Anal Behav 78:315–332
Younger BA, Fearing DD (1999) Parsing items into separate categories: developmental change in infant categorization. Child Dev 70:291–303
Younger BA, Fearing DD (2000) A global-to-basic trend in early categorization: evidence from a dual-category habituation task. Infancy 1:47–58

19
Processing of Shadow Information in Chimpanzee (*Pan troglodytes*) and Human (*Homo sapiens*) Infants

Tomoko Imura[1], Masaki Tomonaga[2], and Akihiro Yagi[1]

1 Introduction

Shadow information, which not only is available in the real world but is also a useful cue for depth perception in two-dimensional pictures or photographs, is categorized as one of the pictorial depth cues. Shadows are classified into two types depending on how they are formed on surfaces: one is "attached" shadow and the other is "cast" shadow. Attached shadows are formed when a surface itself obstructs the light falling on it. Attached shadows provide three-dimensional shapes of objects such as convexity or concavity (Ramachandran 1988a,b; Kleffner and Ramachandran 1992). In contrast, cast shadows occur when one surface occludes another surface from the light source. The shape of a cast shadow can be used for identification of an object shape (Mamassian et al. 1998; Norman et al. 2000; Berbaum et al. 1984; but see Erens et al. 1993). Spatial relationships between cast shadows and casting objects provide effective information about spatial arrangements of objects (Yonas et al. 1978). Especially, motion of the cast shadow improves perception of the spatial layout by adults (Kersten et al. 1997).

This chapter initially reviews studies concerning pictorial depth perception from shadows for human adults and then describes findings from two different approaches: human developmental studies, and studies from a comparative and developmental standpoint. Based on the findings, we discuss issues to be explored in the future.

2 Depth Perception from Shadows in Human Adults

Previous studies for human adults reveal that human visual system processes shadow information relying on several a priori assumptions. First, there is only one light source illuminating the whole scene, and second, this light source shed the light from "above" in the retinal coordinates. These assumptions enable us

[1]Department of Integrated Psychological Science, Kwansei Gakuin University, 1-1-155 Uegahara, Nishinomiya, Hyogo 662-8501, Japan
[2]Primate Research Institute, Kyoto University, 41 Kanrin, Inuyama, Aichi 484-8506, Japan

to perceive the top bright and bottom dark circles as convex and the top dark and bottom bright circles as concave from attached shadows, as shown in Fig. 1. Also, we can also perceive the three-dimensional spatial layout of objects from cast shadows (Fig. 2).

Kleffner and Ramachandran (1992) showed that two assumptions affected the perception of shapes from attached shadows in human adults. They examined the effects of the shading direction using a visual search task. In the vertical shading condition (top bright and bottom dark and vice versa), it was easy to detect a concave shape among convex shapes. However, in the horizontal shading condition (left bright and right dark and vice versa), it was difficult to detect the oppositely shaded target among the distractors. The differences in performance based on shading direction suggest that two assumptions actually affect the visual search task. In addition, the fluency of the search speed (i.e., increment in response times per one search item) was dependent on the number of distractors in vertical shading, implying that depth by shading is processed preattentively in the early vision (that is, "pop-out"). As additional evidence for preattentive processing, search asymmetry was also found (Kleffner and Ramachandran 1992; Braun 1990, 1993; Sun and Perona 1996; cf. Treisman and Gormican 1988). It was easier to detect a concave target among convex distractors than vice versa.

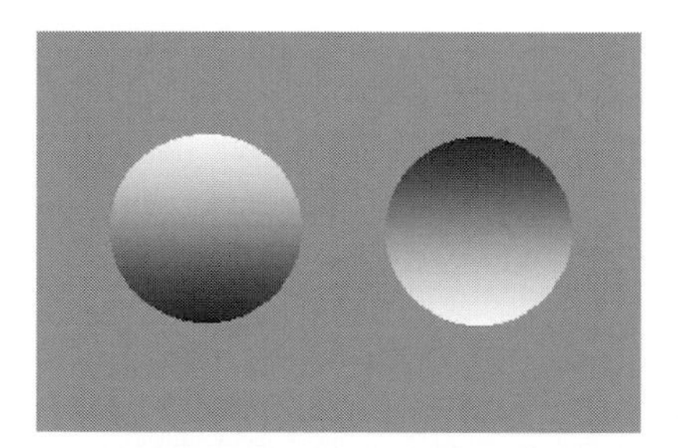

Fig. 1. We can perceive the *top bright* and *bottom dark circles* as convex (*left*) and the *top dark* and *bottom bright circles* as concave (*right*) from attached shadows

Fig. 2. A cast shadow moving diagonally in a motion to that of a ball produces the impression that the ball is receding in depth (*left*), whereas a horizontal trajectory of a cast shadow produces the impression of a ball floating above the floor (*right*)

The other, third constraint was found by Kersten et al. (1997). They revealed that motion of the cast shadow improved perception of the spatial layout, and the stationary light source assumption enables us to infer the locations of objects from moving cast shadows. They created animations called "ball-in-a-box" consisting of a ball and a cast shadow moving in a room with walls and a floor, and demonstrated that cast shadow motion could cause a strong impression of the object moving in depth. Manipulations of a cast shadow motion produced a significant change in the impressions of the three-dimensional motion of the object. As illustrated in Fig. 2, for adults, a cast shadow moving in a motion diagonal to that of a ball produces the impression that the ball is receding in depth, whereas a horizontal trajectory of a cast shadow produces the impression of a ball floating above the floor. Even when the motion of the cast shadow was actually generated by moving the light sources, the human visual system perceived motion of the ball (Mamassian et al. 1998) because the light sources were assumed to be stationary.

These three assumptions are seemed to be valid for humans who adapted to an environment where a single sun is shining from above and seems to be stationary. If the ability to perceive pictorial depth might be learned through one's own experience, infants who have no (or very little) experience of spontaneous manipulations or nonhuman animals adapted to different environments from humans might process pictorial depth information in a different way from human adults. We summarize the human developmental studies and comparative–developmental studies in the next section, including our studies, which explored the effect of experiences on a priori assumptions from a comparative–developmental standpoint.

3 Human Developmental Studies

A study explored how human children acquire a priori assumptions on pictorial depth perception from shadows. Yonas et al. (1979) investigated the use of frames of reference in interpreting attached shadow information with children from 3 to 8 years old using discrimination task. They found that 3-year-old children depended on the retinal frame of reference, whereas 8-year-olds showed almost equal dependence on both gravitational and retinal frames of reference. These results suggest that children could perceive the depth from an attached shadow based on "single light source" and "lighting from above" assumptions and come to refer to more-abstract frames of reference as they grow older.

On the other hand, the developmental origin of depth perception from shadow information has been investigated using reaching responses or looking behavior as dependent measures. Studies using reaching responses based on infants' preference for the three-dimensionally nearer of two objects showed that 7-month-old human infants discriminated shapes defined by attached shadows in two-dimensional photographs. This developmental emergence corresponds with other pictorial depth cues [familiar size (Yonas et al. 1982); occlusion (Granrud and Yonas 1984); shading (Granrud et al. 1985); linear perspective (Arterberry et al. 1989); relative size (Yonas et al. 1985)]. However, using the paired-comparison familiarity–novelty preference procedure, 3-month-old infants detected the differences defined both by the shading and by line junctions (Bhatt and Waters 1998). Bhatt and Bertin (2001) also reported that 3-month-old infants detected the differences defined by only a line junction cue. In contrast, Durand et al. (2003) showed unclear evidence of representational capacity from linear perspective cues in 3- and 4-month-olds. However, there are no attempts to clarify whether infants' perception of depth from attached or cast shadows is based on a priori assumptions such as single light source, lighting from above, and stationary light source.

To address this question, we conducted a series of experiments with human infants using looking time as a dependent measure. First, we examined the effect of shading directions on shape from attached shadow for 3- and 4-month-old infants. The stimulus displays were composed of 20 disks arranged in a 5 × 4 grid on a gray background. Vertical and horizontal shading conditions differing in the directions of shading were prepared (Fig. 3). The displays contained 4 oppositely shaded disks. This area was called the target, and we examined detection of the target using the familiarity–novelty procedure. Experiments consisted of a 15-s familiarization phase and a 15-s test phase. Infants were familiarized with two arrays consisting of homogeneous shaded disks, whereas in the test phase one side of the array was changed so that it contained an embedded target. If infants could detect a target defined by shading, as human adults do, they would look at the arrays containing a target more in the test phase, especially when the shading direction was vertical and the target was "concave" as in human adults. Table 1 shows mean target preference score during the test trials. Four-month-olds looked significantly longer at the target side in the vertical shading condition, whereas with horizontal shading they did not exhibit the target preference. In contrast, 3-month-old infants did not show any preferences to these test arrays. Furthermore, we examined asymmetry in the detection of convex versus concave shapes. Four-month-old infants failed to

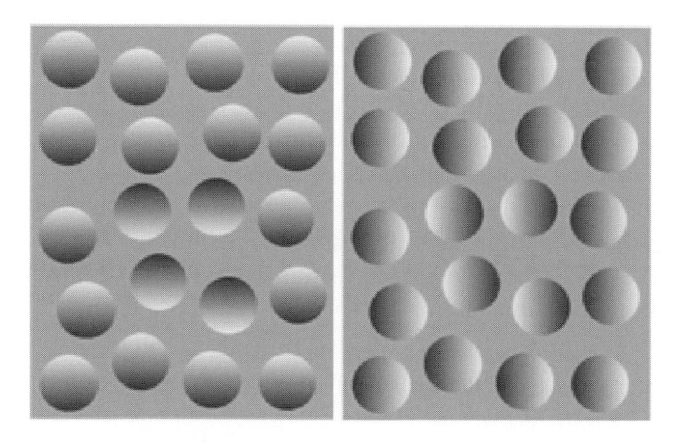

Fig. 3. Examples of test stimuli that were composed of 20 disks arranged in a 5 × 4 grid on a gray background and contained 4 oppositely shaded disks. Conditions of *vertical* and *horizontal shading* were prepared

Table 1. Mean target preference scores during the test trials

Shading condition	Mean (SEM)	*t* (M vs chance, 50%)	*P* (two-tailed)
3-month-olds (*n* = 14)			
Vertical	49.03 (5.60)	0.17	n.s.
Horizontal	49.83 (5.58)	0.03	n.s.
4-month-olds (*n* = 14)			
Vertical	62.82 (5.29)	2.43	<.05
Horizontal	42.46 (6.27)	1.20	n.s

SEM, standard error of the mean; M, mean; n.s., nonsignificant

detect the target in the vertical shading condition when the target was convex. The findings of these experiments suggest that concave shapes were much easier to detect than convex shapes for 4-month-olds, and these results are similar to those of visual search for adults. Our results suggest that 4-month-olds process shading information based on the two assumptions, single light source and lighting from above, in the same way as adults.

To assess the stationary light source assumption for infants, we examined the perception of motion trajectory of object from the moving cast shadow in infants at 4 to 7 months old using a habituation–dishabituation procedure in the other experiments (Imura et al. in press). We investigated whether infants could discriminate the motion trajectories of the ball from the motion of cast shadows using "ball-in-a-box" animations (Kersten et al. 1997). We prepared two kinds of events: one was perceived by adults as moving in "depth" and the other as flowing "up" (see Fig. 2). The events were produced by manipulating only the motion trajectories of a cast shadow. We used a habituation–dishabituation procedure in which the experimental session consisted of four habituation trials and two test trials. A trial was terminated when the infants looked away from the monitor for more than 2 s or continued to look at the monitor for up to 40 s. The duration of each test trial was fixed to 20 s. Each infant was habituated to the "depth" event in which a ball and a cast shadow moved from the bottom left to the top right. After the habituation trials, each infant was tested with the two novel events. In the "depth" event, the trajectories of a ball and a cast shadow were symmetrical to those of the habituation event. In the "up" event, the ball motion was identical to that in the habituation event, except that the cast shadow moved horizontally. If infants, similar to adults, perceived the habituation event as motion in depth then they were expected to exhibit a significant novelty response to the "up" event. The mean looking time during the test trials is shown in Fig. 4. As this figure illustrates, 6- to 7-month-olds looked longer during the "up" event than during the "depth" event, whereas 4- to 5-month-olds showed no difference in looking time between the two test events. These findings indicate that 6- and 7-month-old infants might discriminate between the "depth" and the "up" motions from the cast shadow trajectory. An additional experiment

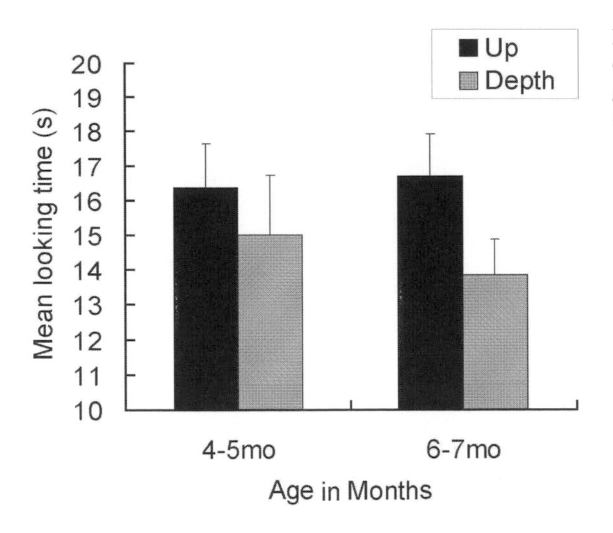

Fig. 4. The mean looking time during the test trials for each age group. *Error bars* indicate SEM across participants

revealed that this discrimination was not based on preferences for two-dimensional "approach–avoidance" motions of ball and shadow contained in an "up" event but was explained by the three-dimensional perception of object motion trajectory. Overall, it is concluded that 6- to 7-month-old-infants infer the object position from moving cast shadow using the stationary light source assumption.

When the findings from the two studies are considered together, human infants perceive, at least as early as the latter half of the first year of life, pictorial depth defined by attached and cast shadows based on a priori assumptions. These results are consistent with the previous findings that the ability of pictorial depth perception develops between 5 and 7 months of age (Granrud 1986). Furthermore, our results suggest that infants at this age might process shadow information in the same way as adults. Based on these findings, without experience or knowledge about light sources, assumptions regarding the shape from shading might be acquired in early life.

4 Comparative–Developmental Studies

There have been some previous studies on the perception of pictorial depth using nonhuman animals from comparative-cognitive (Reid and Spetch 1998) and neurophysiological perspectives (Hanazawa and Komatsu 2001).

Some of the studies tried to explore more directly whether a priori assumptions concerning attached and cast shadows were learned through experiences (Hershberger 1970; Hess 1950, 1961). In these studies, chicks were raised in artificial environments where a controlled light source provided illumination. The results were, however, inconsistent across researchers. Hess (1950, 1961) found that chicks were influenced by the position of the light source (top or bottom), but Hershberger (1970) found no effects of lighting conditions.

Tomonaga (1998) examined the perception of shape from shading in adult chimpanzees and humans using visual search and texture segregation tasks from the comparative–cognitive standpoint. Chimpanzees have evolved in three-dimensionally rich tropical rain forests that have different features from the savannah environment to which humans have adapted. Results of this study did not differ from those reported in previous experiments in the case of the human subjects, but opposite patterns were obtained for the chimpanzees. Chimpanzees showed easier detection in the horizontal than in the vertical shading condition. Tomonaga suggested two possible explanations for this discrepancy. The first is that chimpanzees had different assumptions from those of humans for the processing of shading information as a result of adaptation to relatively different environments. The second is that they relied more on local cues than on pictorial depth cues from shading, such as brightness contrast (cf. Miura and Kawabata 1999, 2000). Further exploration of these possibilities will reveal the effect of experience on a priori assumptions.

On the other hand, there are relatively few studies from the point of view of comparative–cognitive development. Gunderson et al. (1993) tested infant pig-tailed macaques on the perception of pictorial depth from linear perspective and relative size differences. They found that macaque infants at the age of 7 to 8 weeks reliably reached with their arms for the "near" object when they viewed the stimuli in the monocular condition. However, there are no reports of studies with nonhuman primate infants focusing on the perception of depth from shadows.

We examined the perception of depth from shading in three infant chimpanzees named Ayumu, Cleo, and Pal, at 4 to 10 months of age under the preferential reaching task (Imura and Tomonaga 2003). These three chimpanzees were raised by their biological mothers and lived in an outdoor compound (Matsuzawa 2003). This experiment was conducted as a part of a project of longitudinal study on chimpanzee development, and they had participated in a number of other tests during this experiment (Tomonaga et al. 2004). We recorded reaching and looking time as behavioral measures based on the study by Gunderson et al. (1993), using photographs of convex and concave as test stimuli. As in the case of human infants (Granrud et al. 1985), the aim was to examine the transfer of reaching responses from three-dimensional convex objects to two-dimensional convex patterns in the chimpanzee infants.

We prepared three sets of displays. The first one included three-dimensional toys and their photographs. Toys and photographs were attached to the left and right sides of gray panels (actual size, 30.0 cm × 12.0 cm). These displays were used in the baseline trials to assess occurrences of reaching response. The second display included three-dimensional convex and concave shapes: gray table tennis balls (3.6 cm in diameter) were cut in half and attached to the gray panels as convex and concave at a distance of 18.0 cm. The third display was composed of two-dimensional photographs of convex and concave (Fig. 5). These displays were used in the test trials to examine infants' ability of seeing shape from shading. We presented these displays to subjects, and their reaching response to

Fig. 5. An example of a test stimulus used in the two-dimensional convex (*left*)–concave (*right*) test trials. It must be noted that the two circles are flat. There is the information of shading only

Fig. 6. A chimpanzee infant (Ayumu) looks at the panel and reaches toward the three-dimensional toy (half-cut peach) in a baseline trial (photo courtesy of the *Mainichi* Newspaper, Japan)

the stimuli was measured. Based on previous studies, human infants prefer to reach for the nearer of the two objects. If chimpanzee infants also prefer to reach for three-dimensional toys rather than their photographs, and three-dimensional convex than concave shapes, reaching response would be appropriate as a dependent measure. And also, infants' reaching is based on perceived depth, reaching preference should occur also in the two-dimensional convex rather than concave.

Each display was presented at 30 cm away from the infant's face and then each trial was started. When the infant reached for and first touched one of the stimuli or 30 s passed, the trial was terminated (Fig. 6). The experiment consisted of two types of sessions. The first type was a baseline session in which only the toy set was used. A baseline session consisted of 12 trials for Ayumu and 6 trials for Cleo

Table 2. Mean relative reaching response (%) during the test sessions

	3D convexity	3D concavity	2D convexity	2D concavity
Ayumu	75	0	36.1	8.3
Cleo	70	13.3	53.3	13.3
Pal	59.3	11.1	37	18.5
Average	68.1	8.1	42.1	13.4

3D, three-dimensional; 2D, two-dimensional

and Pal. The second type was a test session in which all three sets of stimuli appeared. A test session consisted of 8 toy trials (baseline trials), 2 three-dimensional convex/concave trials, and 2 two-dimensional convex–concave trials for Ayumu; and 4 toy trials, 1 three-dimensional convex–concave trial, and 1 two-dimensional convex–concave trial for Cleo and Pal. On average, Ayumu received 2 baseline and 1 test sessions per week, and Cleo and Pal received 2 baseline and 2 test sessions per week. In total, Ayumu received 51 sessions (33 baseline and 18 test sessions), Cleo received 58 sessions (29 baseline and 29 test sessions), and Pal received 56 sessions (29 baseline and 27 sessions).

Table 2 shows the percentage of reaching response to convex or concave in the test displays. All chimpanzee infants reached significantly more often for the three-dimensional convex than for the concave shape. Furthermore, the important point is that two of the three infants reached significantly more often for the photographic convex shape than for the concave one. These results suggest that chimpanzees perceive pictorial depth defined by shading information as early as the latter half of the first year of life. These findings were seemingly consistent with the results from human infants. Granrud et al. (1985) reported that human infants began to reach for the pictorial convex between 5 and 7 months of age. The present results from chimpanzees were consistent with the results from humans, although we must also bear in mind species differences in developmental speed.

In this experiment, we used "naturalistic" (photographic) shading patterns instead of computer graphics shading patterns. These naturalistic shading patterns contained richer information than artificial patterns, including, for example, shadows cast on the background surface, highlighted areas on the surface of the objects, and relative differences in global brightness between convex and concave. These factors should be further examined in the future.

5 Conclusions

In this chapter, we reviewed the studies on the perception of shape from shadows from developmental and comparative-developmental standpoints to explore the effects of experience on a priori assumptions for processing of shadows; single light source, lighting from above, and stationary light source.

Developmental studies for human infants showed that 4-month-old infants process attached shadow information based on the two assumptions. Also, 6- to 7-month-old infants could discriminate the motion trajectory of the ball from moving cast shadow. These findings suggest that a priori assumptions are available for infants without experience of manipulation of the objects or self locomotion. Furthermore, comparative–developmental studies revealed that chimpanzee infants perceive pictorial depth defined by attached shadows at least as early as the latter half of the first year of life. Considering that there are differences in processing attached shadows between chimpanzees and humans (Tomonaga 1998), further studies are needed whether chimpanzee infants process shadow information based on a priori assumptions in the same way as humans.

The present results may suggest that extensive postnatal visual experience of two-dimensional stimuli with pictorial depth (such as TV, photographs, and paintings) may not be chiefly responsible for the emergence of the perception of pictorial depth. The perception of pictorial depth is to a large extent biologically determined and has been acquired during the course of primate evolution (and avian evolution, cf. Hershberger 1970; Hess 1950, 1961; Regolin and Vallortigara 1995; Reid and Spetch 1998). The question that remains, then, is the relationship between the perception of pictorial depth and binocular depth perception. Human infants become sensitive to most pictorial depth cues at around 5 to 7 months of age, preceded by the emergence of sensitivity to binocular disparity and motion parallax cues (Granrud 1986). Thus, sensitivity to pictorial depth cues can be considered a generalization from or association with these cues (Gunderson et al. 1993). In the future, we need to explore systematically the developmental relationship between sensitivity to monocular and binocular depth cues from a comparative perspective.

There are many pictorial depth cues other than attached and cast shadows. In humans, the developmental relationships among these cues have already been discussed (Granrud 1986). From the comparative perspective, there may exist a certain adaptive hierarchy among pictorial depth cues. For animals living in the forest, shading information, occlusions, or line junctions may be more important, whereas linear perspective or texture gradients might play a more critical role in flying animals. For example, occlusion is a difficult cue for the perception of object unity in pigeons (Fujita 2001), but not in hens (Forkman and Vallortigara 1999). In experiments exploring the Ponzo illusion, macaques have been shown to be less sensitive to linear perspective cues than humans (Fujita 1996). Species differences based on the environments to which animals have adapted thus do exist. Future explorations of the relationship among pictorial depth cues in various species of animals will contribute to our understanding of the development and evolution of the perception of pictorial depth.

Acknowledgments

We thank Dr. Tetsuro Matsuzawa, Dr. Masayuki Tanaka, and the staff of the Department of Behavioral and Brain Sciences, Primate Research Institute, Kyoto

University for research with chimpanzees. We also thank Dr. Masami K Yamaguchi, Dr. So Kanazawa, Nobu Shirai, and Yumiko Otsuka for their helpful comments during experiments for human infants. This research was financially supported by Grants-in-Aid for Scientific Research from the Ministry of Education, Culture, Sports, Science and Technology (MEXT), Japan (nos. 12002009, 15500172, 16002001, 16300084), a Grant-in-Aid for Scientific Research from Japan Society for the Promotion of Science (JSPS), Japan (to Tomoko Imura), and the Cooperative Research Program of the Primate Research Institute, Kyoto University to Tomoko Imura.

References

Arterberry M, Yonas A, Bensen AS (1989) Self-produced locomotion and the development of responsiveness to linear perspective and texture gradients. Dev Psychol 25:976–982
Berbaum K, Bever T, Chung CS (1984) Extending the perception of shape from known to unknown shading. Perception 13:479–488
Bhatt RS, Bertin E (2001) Pictorial depth cues and three-dimensional information processing in early infancy. J Exp Child Psychol 80:315–332
Bhatt RS, Waters SE (1998) Perception of three-dimensional cues in early infancy. J Exp Child Psychol 70:207–224
Braun J (1990) Focal attention and shape-from-shading. Perception 19:A112
Braun J (1993) Shape-from-shading is independent of visual attention and may be "texton". Spat Vision 74:311–322
Durand K, Lecuyer R, Feichtel M (2003) Representation of third dimension: the use of perspective cues by 3- and 4-month-old infants. Infant Behav Dev 26:151–166
Erens RG, Kappers AM, Koenderink JJ (1993) Perception of local shape from shading. Percept Psychophys 54:145–156
Forkman B, Vallortigara G (1999) Minimization of modal contours: an essential cross-species strategy in disambiguating relative depth. Anim Cogn 2:181–185
Fujita K (1996) Linear perspective and the Ponzo illusion: a comparison between rhesus monkeys and humans. Jpn Psychol Res 38:136–145
Fujita K (2001) Perceptual completion in rhesus monkeys (*Macaca mulatta*) and pigeons (*Columba livia*). Percept Psychophys 63:115–125
Granrud CE (1986) Binocular vision and spatial perception on 4- and 5-month-old infants. J Exp Psychol: Human Percept Perform 12:36–49
Granrud CE, Yonas A (1984) Infants' perception of pictorially specified interposition. J Exp Child Psychol 37:500–511
Granrud CE, Yonas A, Opland EA (1985) Infants' sensitivity to the depth cue of shading. Percept Psychophys 37:415–419
Gunderson VM, Yonas A, Sargent PL, Grant-Webster KS (1993) Infant macaque monkeys respond to pictorial depth. Psychol Sci 4:93–98
Hanazawa A, Komatsu H (2001) Influence of the direction of elemental luminance gradients on the responses of V4 cells to texture surfaces. J Neurosci 21:4490–4497
Hershberger W (1970) Attached-shadow orientation perceived as depth by chickens reared in an environment illuminated from below. J Comp Physiol Psychol 73:407–411
Hess EH (1950) Development of chick's responses to light and shade cues of depth. J Comp Physiol Psychol 43:112–122
Hess EH (1961) Shadows and depth perception. Sci Am 204:138–148
Imura T, Tomonaga M (2003) Perception of depth from shading in infant chimpanzees (*Pan troglodytes*). Anim Cogn 6:253–258
Imura T, Yamaguchi KM, Kanazawa S, Shirai N, Otsuka Y, Tomonaga M, Yagi A (2006) Perception of motion trajectory of object from the moving cast shadow in infants. Vis Res 46:652–657

Kersten D, Mamassian P, Knill DC (1997) Moving cast shadows induce apparent motion in depth. Perception 26:171–192
Kleffner DA, Ramachandran VS (1992) On the perception of shape from shading. Percept Psychophys 52:18–36
Mamassian P, Knill DC, Kersten D (1998) The perception of cast shadows. Trends Cogn Sci 2:288–295
Matsuzawa T (2003) The Ai project: historical and ecological contexts. Anim Cogn 6:199–211
Miura K, Kawabata H (1999) The relative orientation difference of shaded disks in search task. In: Proceedings of the 63rd annual convention of the Japanese Psychological Association in Aichi, p 373 (in Japanese)
Miura K, Kawabata H (2000) Effects of edge orientation and polarity of shaded disks on search task. In: Proceedings of the 64th annual convention of the Japanese Psychological Association in Kyoto, p 398 (in Japanese)
Norman JF, Dawson TE, Raines SR (2000) The perception and recognition of natural object shape from deforming and static shadows. Perception 29:135–148
Ramachandran VS (1988a) Perception of shape from shading. Nature (Lond) 331:163–166
Ramachandran VS (1988b) Perceiving shape from shading. Sci Am 259:58–65
Regolin L, Vallortigara G (1995) Perception of partly occluded objects by young chicks. Percept Psychophys 57:971–976
Reid SL, Spetch ML (1998) Perception of pictorial depth cues by pigeons. Psychoanal Bull Rev 5:698–704
Sun JY, Perona P (1996) Preattentive perception of elementary three dimensional shapes. Vision Res 36:2515–2529
Tomonaga M (1998) Perception of shape from shading in chimpanzees (*Pan troglodytes*) and humans (*Homo sapiens*). Anim Cogn 1:25–35
Tomonaga M, Tanaka M, Matsuzawa T, Myowa-Yamakoshi M, Kosugi D, Mizuno Y, Okamoto S, Yamaguchi MK, Bard KA (2004) Development of social cognition in infant chimpanzees (*Pan troglodytes*): face recognition, smiling, gaze, and the lack of triadic interactions. Jpn Psychol Res 46:227–235
Treisman A, Gormican S (1988) Feature analysis in early vision: evidence from search asymmetries. Psychol Rev 95:15–48
Yonas A, Goldsmith LT, Hallstrom JL (1978) Development of sensitivity to information provided by cast shadow in pictures. Perception 7:333–341
Yonas A, Kuskowski M, Sternfels S (1979) The role of frames of reference in the development of responsiveness to shading information. Child Dev 50:495–500
Yonas A, Pattersen L, Granrud CE (1982) Infants' sensitivity to familiar size as information for distance. Child Dev 53:1285–1290
Yonas A, Granrud CE, Pattersen L (1985) Infants' sensitivity to relative size information for distance. Dev Psychol 21:161–167

20
Color Recognition in Chimpanzees (*Pan troglodytes*)

Toyomi Matsuno[1], Nobuyuki Kawai[2], and Tetsuro Matsuzawa[1]

1 Introduction

Well-developed color perception is an important characteristic of primate vision. In contrast to other mammalian species, most primates have trichromatic vision with more than three types of photoreceptors, each of which has different light absorption characteristics. Researchers consider this trichromaticity to be an evolutionary adaptation to facilitate finding edible young leaves and ripe fruits among foliage (Dominy and Lucas 2001; Sumner and Mollon 2000a,b). The color processing mechanism of primates is so elaborate that humans are able to discriminate very slight differences (1–2 nm wavelength) in visible light.

Although humans are able to discriminate between myriad colors, in daily life we tend to recognize colors in a more general way. For example, when we see a rainbow, we recognize bands of only a few colors (i.e., usually five to seven colors), whereas in reality the rainbow is a continuous spectrum. This method of color processing is called color categorization.

Color categorization has been a controversial cross-discipline topic. One main question is whether color categorization is universal. If color categorization is universal, this would imply that the human visual system is innately hardwired to perceive specific color categories and that humans share basic color categories consisting of specific clusters of colors, regardless of cultural differences or native language (Berlin and Kay 1969). Some, however, have claimed that there are no such universal or hardwired color categories, that color categories are acquired through language and/or perceptual learning and color space can be arbitrarily divided into different color name categories (Roberson et al. 2000).

With this issue in mind, an essential question is whether nonhuman primates possess this type of color recognition. In nature, nonhuman primates do not have a developed language, but they do see their richly colored environment with trichromatic vision. Therefore, a comparative approach to test characteristics of color categorization in nonhuman primates could be of great benefit when investigating categorical color recognition and the role of language.

[1]Primate Research Institute, Kyoto University, 41 Kanrin, Inuyama, Aichi 484-8506, Japan
[2]Graduate School of Information Science, Nagoya University, Furo-cho, Chikusa-ku, Nagoya 464-8601, Japan

Several studies on color recognition have been conducted on chimpanzees, which are the closest evolutionary relative to humans (Fujiyama et al. 2002; Stone et al. 2002). They have visual recognition quite similar to that of humans (Matsuzawa 2003; Tomonaga 2001). In addition, their ability to learn symbols makes them exceptional nonhuman subjects (Gardner and Gardner 1969; Premack 1971; Rumbaugh 1977); several studies have benefited from this ability by applying symbol recognition to color names (Asano et al. 1982; Matsuzawa 1985). Therefore, chimpanzees are integral to a comparative study of categorical color recognition.

This chapter provides an overview of color recognition studies in chimpanzees. A more detailed summary of the controversial discussion on human color categorization is followed by a review of chimpanzee color perception, acquisition of color names, and color classification. A discussion of the contributions of chimpanzee studies examines steps to clarify the color categorization mechanisms in the human cognitive system, and the chapter concludes with thoughts on the future of color recognition studies in chimpanzees.

2 Color Recognition in Humans

Berlin and Kay (1969) first reported universality of color terms. They surveyed 98 languages and found a striking commonality in color classification among various human cultures. They found that well-developed languages seem to have 11 basic color terms (red, green, yellow, blue, orange, purple, brown, pink, white, black, and gray). Accordingly, the order of color term appearance in cultures with fewer than 11 basic color terms obeys a certain rule, which they call the "evolution of basic color terms." They also found that the best example within each category (focal color) is consistent among cultures and suggested that categorical foci are more important than categorical boundaries (see also Regier et al. 2005).

This universality and uniqueness of basic color terms were first reported in linguistic surveys but have also been confirmed in several experimental studies (Guest and Van Laar 2000; Sturges and Whitfield 1997; Uchikawa and Shinoda 1996; Zollinger 1988). Heider (1972) tested the Dani people of western New Guinea, who use only two achromatic basic color terms, and found that in a simple recognition task, Dani subjects were better at remembering the focal colors of eight basic chromatic color terms than they were at remembering nonfocal colors. In addition, Dani people learned associations between focal colors and arbitrary words better than they did associations between nonfocal colors and new words. These findings suggest that in humans color categories are universal and independent of language. Boynton and Olson (1990) also reported the salience of basic color terms in their color-naming experiments. In their study, the use of chromatic basic color terms yielded greater consistency within subjects, greater consensus between subjects, and shorter response times than did any other color terms. These results seem to support Berlin and Kay's (1969) theory, suggesting that humans have an innate physiological ability to perceive specific common categories of color and that these perceptual constraints underlie universal color recognition and the acquisition of basic color terms.

Developmental studies have also investigated the hypothesis that categorical color recognition is innately constrained to a universal form (Davies et al. 1994). Bornstein et al. (1976) tested 4-month-old infants using a habituation paradigm and found that infants partition the color spectrum into the same four categories (red, green, yellow, and blue) as do human adults. Franklin et al. (2005b) also showed that preverbal infants perceive color categories by assessing their eye movements when they follow color targets. Research has shown that children's knowledge of color terms does not affect such perceptual color categorization (Franklin et al. 2005a), while perceptual color categorization seems to influence the developmental process of color term learning (Mervis et al. 1975; Pitchford & Mullen, 2003).

For years, researchers have accepted the basic idea of color universality with some reconsideration about the definition and range of basic color terms (Crawford 1982; Greenfield 1986; Kay and McDaniel 1978). However, some recent studies have contradicted this "universal" view (Özgen 2004; Saunders and van Brakel 1997) by providing new findings (Özgen and Davies 2002; Roberson et al. 2000) that the conventional hypothesis cannot explain.

An investigation of color naming by the Berinmo people of Papua New Guinea contradicted the usual English classification, which researchers considered one example of "universal" classification (Davidoff et al. 1999). The Berinmo people use five different color terms, and their borders markedly deviate from the universal categories that simply combine 2 or more categories to represent 11 color categories of developed language by five color terms (as discussed in Berlin and Kay 1969). There was no agreement between the two language groups on the positions of focal colors. Further investigation of categorical color perception in the two language groups revealed that the Berinmo exhibited categorical effects that differed from English subjects, depending on the category border of their own language (Roberson et al. 2000).

Research in the field of perceptual learning has also raised questions about the universality of color categories. Özgen and Davies (2002) trained English-language speakers to categorize colors across a novel boundary and found that the acquired category had the same effects on color perception as the "natural" color category (cf. Bornstein and Korda 1984; Boynton et al. 1989). These results imply that human color categories are not rigidly constrained but are rather flexible and can be arbitrarily reorganized through intensive perceptual learning.

These recent findings have led some researchers to assert that color categorization is not universal and is based predominantly on language and associated perceptual learning (Özgen 2004; Roberson et al. 2000). The lack of findings regarding a firm neural basis for universal constraints on color categorization also encourages these discussions on the relativity of color categories.

However, there are still questions to be answered about the hypothesis claiming that color categorization is linguistically relative and unconstrained. First, it is difficult for the hypothesis to explain the reports of developmental studies in preverbal infants. Second, flexibility in categorization and innate constraints are not exclusive. It is possible that color categorization could be universally constrained but modified by strong environmental influences such as language and culture.

Thus, there are two divergent hypotheses to explain color categorization in humans. One argues that color categorization is equally constrained in all humans and therefore color categorization is universal regardless of differences in culture or language. The other denies such universal constraints and claims that categorical color partitions are relatively determined under the influence of language and perceptual learning. However, current findings about human color categorization seem to provide contradictory evidence for these theories, fully supporting neither. A more plausible option might be a compromise between these two opposing theories, embracing both the relatively loose constraints on color categorization and its flexibility. In any case, further investigation is needed.

One reason that this issue is so complex is that studies using adult humans are inevitably influenced or constrained by the subjects' own acquired language and environment, which varies among cultures; this makes it difficult to determine whether color categorization is universal or relative. Therefore, future developmental studies and also studies involving animals (which will not involve language) could add greatly to the literature on color categorization.

3 Color Perception in Chimpanzees

The visual perception of chimpanzees is similar to that of humans (Tomonaga 2001). Several investigations have indicated similarities in the underlying color perception mechanisms. Grether (1940a–d, 1942) used an elaborate series of behavioral experiments to compare color perception between chimpanzees and humans. In the experiments, young chimpanzees were trained in discrimination tasks using spectral light as stimuli to reveal hue discrimination thresholds, color equation characteristics, color and brightness contrast thresholds, and visible spectral limits. The response tendencies of chimpanzees in every test did not differ greatly from those of human subjects.

A direct measurement of spectral sensitivity using electroretinogram (ERG) flicker photometry provided further comparative data. Jacobs et al. (1996) showed that chimpanzees use triadic differentiation of ERG signal patterns depending on the adaptation light in a manner similar to humans, suggesting that they also have S-, M-, and L-cone types. The study also demonstrated that chimpanzees and humans have a similar spectral sensitivity function, except that the chimpanzees had a slightly higher sensitivity to short-wavelength light and slightly lower sensitivity to long-wavelength light.

Recent studies on the photopigment gene also support theories that the two species have similar low-level processing of color perception (Deeb et al. 1994;

Dulai et al. 1994). These studies determined certain amino acid sequences that code the light-sensitive visual pigments composed of a protein moiety (opsin). They found that chimpanzee L- and M-cone opsin sequences are highly homologous to the equivalent human pigments and that chimpanzee pigment absorption characteristics predicted by substitution of the amino acid were almost the same as in humans.

Given these reports, it is likely that chimpanzee color vision is similar to that of humans, at least in parts of the early stages of color processing. If this is the case, the next issues to be addressed are the later stages, that is, color recognition such as color categorization.

4 Acquisition of Color Terms and Color Naming in Chimpanzees

Relative to the intensive cross-disciplinary investigations in early color vision, there are far fewer studies on higher-level color recognition in chimpanzees. Among those, one well-known approach is training in color symbol acquisition (Essock 1977; Matsuzawa 1985). Researchers have used this kind of study in the wide-ranging and longitudinal "ape language" research projects (Matsuzawa 2001, 2003; Rambaugh 1977).

At the Primate Research Institute of Kyoto University, chimpanzees were trained to use arbitrary geometric figures (lexigram) and Chinese kanji characters as color names (Fig. 1). Initial training began when chimpanzees were approximately 2 years old (Asano et al. 1982; Matsuzawa 1985). As a result of training in this matching task, all three chimpanzees learned to match a symbolic color name to an object's color.

Matsuzawa (1985) used a female chimpanzee named Ai to further investigate color naming and classification. Matsuzawa tested the chimpanzee's generalization of color naming to untrained new colors and compared the resulting properties of color classification to those of humans using the Munsell color space, which consists of the three dimensions of color properties: hue, saturation, and brightness. At the time of testing, Ai was 4 years old and had learned 11 lexigrams of symbolic color names (see Fig. 1) corresponding to the 11 basic color terms (Berlin and Kay 1969) by being trained to match the symbols with 11 specific Munsell color chips. In the experiments, Ai was presented with a Munsell

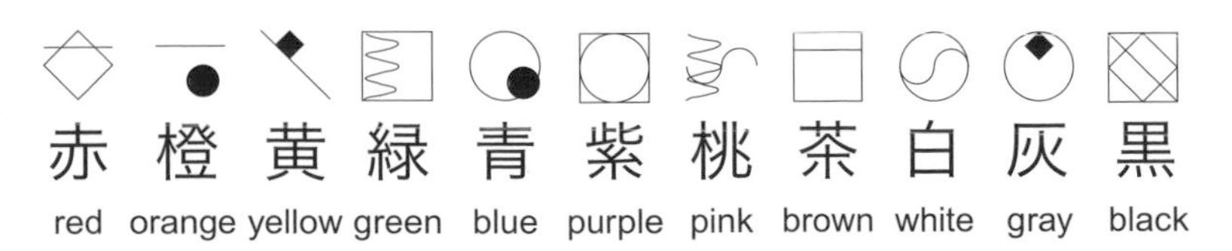

Fig. 1. Lexigrams (*top*) and kanji characters (*middle*) for 11 color names acquired by the chimpanzees

color chip at the beginning of each trial and was required to choose a corresponding key from among the 11 color-name keys. Color chips varied in the dimensions of brightness (experiment 1), hue (experiment 2), or hue and brightness (experiment 3).

In experiment 1, Ai produced three groups (black, gray, white) during the variation of brightness in achromatic color chips, suggesting the continuum of brightness that she saw was divided into three categories. In experiment 2, she was tested with 40 chromatic color chips, representing the color circle; she exhibited consistent judgments about categorical responses throughout each session. In addition, in both experiments Ai exhibited longer response times when the dimensions of chips bordered two adjacent categories.

After Matsuzawa (1985) confirmed the validity of the procedure and the categorical responses in experiments 1 and 2, a direct comparison between chimpanzees and humans was conducted. In experiment 3, Ai was required to name 215 color chips that varied in 40 hues and 7 brightness levels and had a maximum saturation for a given hue and brightness value. Figure 2 shows the results; the areas in which color chips were consistently matched to a single color name during all three trials are unshaded and the areas in which a chip was matched to more than a single color name are shaded. Ai's classification was very consistent (74% of the chips were matched with a single color name), as were results from the human subject (79%), and the areas and borders of named colors were very similar to those of the human subject. Matsuzawa (1985) also showed that dispersion of the focal color points in the 20 languages reported by Berlin and Kay (1969) fell into an area to which the chimpanzee consistently gave a color name. These results suggest that even though chimpanzees have no language in nature, they do have the ability to use symbolic color names to describe perceived colors in their environment and also that they classify or categorize perceived colors in the same manner as do humans.

5 Color Classification in Chimpanzees Skilled and Unskilled in the Use of Symbolic Color Names

Ai, the chimpanzee referred to in the previous section, was trained from a young age to use symbolic color names as communication tools to describe her cognitive world. However, such intensive and long-term training using color stimuli and the acquisition of the symbols themselves can influence color recognition. To further understand chimpanzee color recognition, we need to understand color classification in naive chimpanzees that have not been influenced by acquired color symbols.

Matsuno et al. (2004) investigated color classification of two chimpanzees under identical experimental conditions. Ai was one of the subjects; at the time of testing, she was 23 years old and had learned 11 kanji characters as symbolic color names in addition to the lexigrams. She had received continuous training and maintained her performance in symbolic matching-to-sample tasks (color-

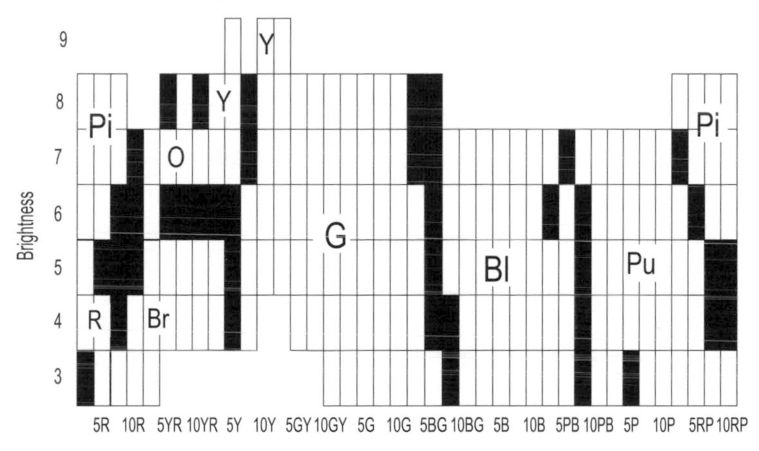

Fig. 2. Color naming of chips with maximum saturation for a given hue and brightness value for a chimpanzee (*top*) and a human observer (*bottom*). The *horizontal axis* shows hue changes; the *vertical axis* shows brightness; *rectangles* represent individual color chips; *solid circles* indicate training colors. The named colors are abbreviated as follows: *Bk*, black; *Bl*, blue; *Br*, brown; *G*, green; *O*, orange; *Pi*, pink; *Pu*, purple; *R*, red; *Y*, yellow

to-symbol, symbol-to-color, and symbol-to-symbol; Kawai and Matsuzawa 2003; Suzuki and Matsuzawa 1997). The other subject was Pendesa, who was the same age as Ai, but although she had other cognitive skills comparable to Ai, such as identical matching abilities (Sousa and Matsuzawa 2001) and line drawing (Iversen and Matsuzawa 1997), she was less experienced in symbolic color names. When Pendesa was 21 years old, she was first given a symbolic matching-to-sample task for an experiment other than color recognition (Sousa and Matsuzawa 2001). She was not as accurate in the task as Ai, and because of her limited training, her understanding of the symbols was incomplete.

Matsuno et al. (2004) adopted a "nonlinguistic" test (the color matching method) to perform a direct comparison of color classification by these two chimpanzees under the same conditions. In the experiment, each chimpanzee was presented with color matching-to-sample tasks using a color CRT monitor. One of the 124 test colors was selected from the available range of CRT colors and presented as a sample stimulus in each test trial. Two colors were selected for comparison from nine standard colors that Ai had stably matched to color symbols of red, green, yellow, blue, orange, purple, brown, and pink during symbolic matching-to-sample tasks (cf. Sousa and Matsuzawa 2001). In other words, the nine standard colors were used for the nine corresponding color lexigrams. Researchers used results from the two-alternative forced-choice task to assess the consistency of responses and the consensus of consistent responses between subjects.

The two subjects exhibited similar distributions of consistently matched test colors on the chromaticity diagrams, with some spatial clusters around standard colors (Fig. 3). The location of the centroid for each color (calculated by averaging the coordinates of all test colors to which each subject responded consistently with the standard color) was also very proximate between the two subjects (Fig. 4). However, there were some distinct differences in the two subjects' classifications: Ai produced consistent color responses significantly more often (65% of the test colors) than did Pendesa (45%). For example, Ai made consistent matches using every standard color, while Pendesa infrequently provided consistent responses for standard blue and brown. Correspondingly, the subjects did not share consistent responses to standard blue and brown. The subjects also exhibited different tendencies in response times; in addition, Ai exhibited significantly shorter response times during consistent responses than during inconsistent responses (as seen in Matsuzawa 1985), but Pendesa did not display the same tendency.

The similarity and dissimilarity between these two chimpanzees shed some light on the way that chimpanzees categorize color. The distribution of classified colors and centroids suggests that subjects perceive and group colors in a similar way, regardless of whether they are skilled in using symbolic color names. However, Pendesa's low levels of consistency and indistinguishable response times during consistent and inconsistent color choices indicate that a skilled subject has an advantage. This result implies that experience in color discrimination and/or color-naming training may have a refining influence on categorical color recognition in chimpanzees.

6 Summary of Studies on Color Recognition in Chimpanzees

As we have discussed, studies have shown that chimpanzees are not only capable of learning symbolic color names, but that they classify colors in the same way as humans. Furthermore, this common color classification is exhibited regard-

(a)

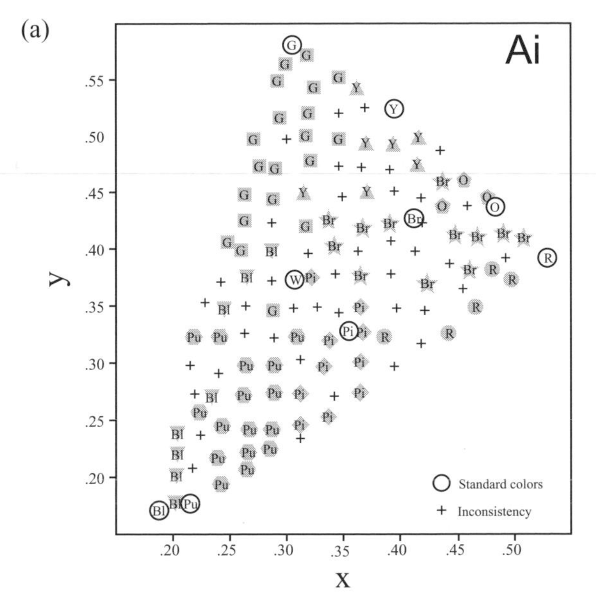

(b)

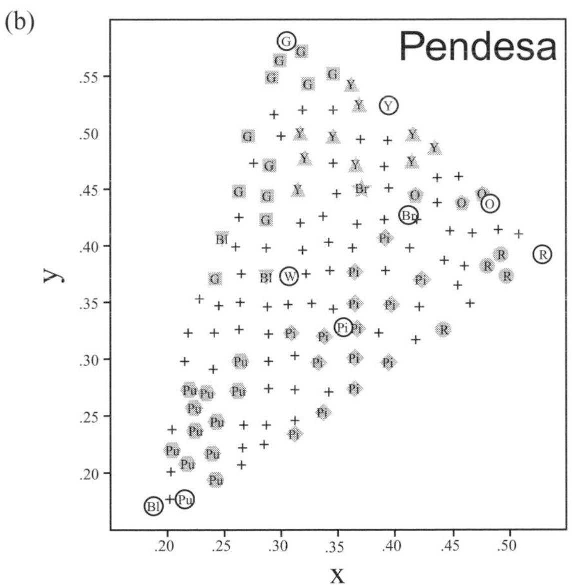

Fig. 3. Test colors plotted on CIE 1931 *x*–*y*-chromaticity diagrams, showing consistently chosen standard colors (*Bl*, blue; *Br*, brown; *G*, green; *O*, orange; *Pi*, pink; *Pu*, purple; *R*, red; *Y*, yellow) or inconsistent responses (*crosses*). Standard colors were also plotted with *open circles*. **a** Ai's consistent and inconsistent responses to test colors; **b** Pendesa's consistent and inconsistent responses to test colors

less of symbolic skills, although data suggest that experience with color symbols refines classification. These findings suggest that chimpanzees and humans share an underlying mechanism not only for color perception but also for symbolic and/or categorical color recognition.

In the context of discussions on human color categories, researchers have interpreted this interspecies commonality as support for the universality of color categorization and for the existence of underlying neural constraints. It does appear that, under similar perceptual and cognitive constraints, both species

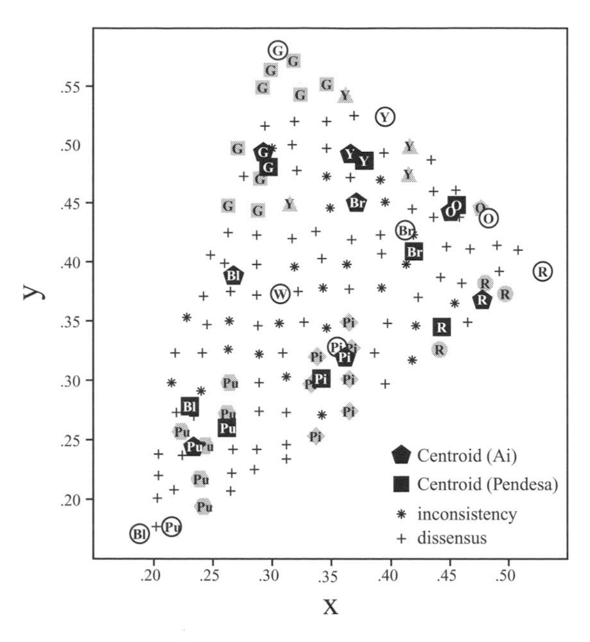

Fig. 4. Consensus between subjects and centroids (*white letters*). *Black letters* indicate test colors to which both subjects consistently responded with the same standard color; *asterisks* indicate inconsistent responses by both subjects; *crosses* indicate other patterns of responses. The centroid for a standard color was calculated by averaging the coordinates of all test colors to which each subject responded consistently with the standard color

classify colors in the same way. However, this does not assure the universality of color categorization, because the alternative explanation of perceptual learning and linguistic relativity (Özgen 2004) has not yet been refuted.

In Matsuzawa's (1985) study, Ai classified colors in the same way that humans do, using symbols representing the 11 basic color terms. A proponent of the universal view would claim that because of the universality of constraints on color categorization, a chimpanzee could learn symbols and classify colors in the same manner as humans. However, a proponent of the relative view would claim that chimpanzees do not have universal constraints just as humans are not constrained by innate universal color categories, and that acquisition training in the same 11 basic color terms that humans use constrained the chimpanzee classification to be the same as humans.

Matsuno et al.'s (2004) study of an unskilled chimpanzee had a procedural weakness on this point. Although the unskilled chimpanzee was not constrained by learned symbols, the task required that she use nine colors corresponding to the nine chromatic basic color terms. An alternative explanation might be that such constraints brought about the apparent similarity to classifications constrained by universal categories. The more stable and consistent classification by the skilled chimpanzee also implies that color categorization in chimpanzees is also somehow flexible, as in humans.

Studies on color classification in chimpanzees are not able to end the arguments about the mechanisms of color categorization, but they do provide valuable insights into the shared mechanisms for categorizing colors as well as the role that symbols play in categorical color recognition.

7 Future Research into Chimpanzee Color Recognition

The use of chimpanzees as subjects of color categorization studies has advantages over other nonhuman primates (Sandell et al. 1979) and nonprimate animals (Jones et al. 2001; Wright and Cumming 1971; von Frisch 1950); not only do they have an architecture in early color vision similar to humans because of evolutionary proximity but they are capable of being tested using linguistic as well as perceptual approaches. That ability means we can not only probe categorical color recognition in subjects without language but can also test the influence that the acquisition process of symbolic color names has on it. We can also compare two groups of chimpanzees that were reared in the same environment but differ only in skills of color symbol use. Furthermore, we can arbitrarily control the number and properties of color symbols to be acquired. By limiting the number of basic color terms or teaching symbols that divide color space in ways totally difference from the basic color terms, chimpanzees with different "knowledge" about colors may or may not exhibit color categorization totally different from those who learned the basic color terms. These types of study will contribute to an understanding of the relationships between color symbols and color categorization.

Chimpanzees are also capable of learning more complex tasks that may be too difficult for preverbal infants. Verbal instructions are as ineffective for chimpanzees as they are for human infants, but chimpanzees can participate in long-term training and are able to learn various cognitive tasks, such as matching-to-sample (Matsuno et al. 2004) and visual search tasks (Tomonaga 2001), which provide a variety of ways to test subjects. In addition, in contrast to developmental studies, chimpanzees skilled and unskilled in the use of color symbols can be tested during the same developmental stages, with well-developed early vision and a cognitive capability comparable to human adults (Kawai and Matsuzawa 2000).

These comparative approaches could provide further understanding about not only the evolutionary continuity of color recognition but also its underlying mechanisms shared by primate species. In addition, these investigations could bring new insights into more general cognitive abilities concerning the role of symbols that humans acquired during the process of hominization.

References

Asano T, Kojima T, Matsuzawa T, Kubota K, Murofushi K (1982) Object and color naming in chimpanzees (*Pan troglodytes*). Proc Jpn Acad 58(B):118–122

Berlin B, Kay P (1969) Basic color terms: their universality and evolution. University of California Press, Berkeley

Bornstein MC, Kessen W, Weiskopf S (1976) Color vision and hue categorization in young human infants. J Exp Psychol Hum Percept Perform 2:115–129

Bornstein MH, Korda NO (1984) Discrimination and matching within and between hues measured by reaction times: some implications for categorical perception and levels of information processing. Psychol Res 46:207–222
Boynton RM, Olson CX (1990) Salience of chromatic basic color terms confirmed by three measures. Vis Res 30:1311–1317
Boynton RM, Fargo L, Olson CX, Smallman HS (1989) Category effects in color memory. Color Res Appl 12107–12123
Crawford TD (1982) Defining "basic color terms". Anthropol Linguistics 24:338–343
Davies I, Corbett G, McGurk H, Jerrett D (1994) A developmental study of the acquisition of colour terms in Setswana. J Child Lang 21:693–712
Davidoff J, Davies I, Roberson D (1999) Colour categories in a stone-age tribe. Nature (Lond) 398:203–204
Deeb SS, Jorgensen AL, Battisti L, Iwasaki L, Motulsky AG (1994) Sequence divergence of the red and green visual pigments in great apes and humans. Proc Natl Acad Sci U S A 91:7262–7266
Dominy NJ, Lucas PW (2001) Ecological importance of trichromatic vision to primates. Nature (Lond) 410:363–365
Dulai KS, Bowmaker JK, Mollon JD, Hunt DM (1994) Sequence divergence, polymorphism and evolution of middle-wave and long-wave visual pigment genes of great apes and old world monkeys. Vis Res 34:2483–2491
Essock SM (1977) Color perception and color classification. In: Rumbaugh DM (ed) Language learning by a chimpanzee: the Lana project. Academic Press, New York, pp 207–224
Franklin A, Clifford A, Williamson E, Davies I (2005a) Color term knowledge does not affect categorical perception of color in toddlers. J Exp Child Psychol 90:114–141
Franklin A, Pilling M, Davies I (2005b) The nature of infant color categorization: evidence from eye movements on a target detection task. J Exp Child Psychol 91:227–248
Fujiyama A, Watanabe H, Toyoda A, Taylor TD, Itoh T, Tsai S, Park H, Yaspo M, Lehrach H, Chen Z, Fu G, Saitou N, Osoegawa K, de Jong PJ, Suto Y, Hattori M, Sakaki Y (2002) Construction and analysis of a human-chimpanzee comparative clone map. Science 295:131–134
Gardner RA, Gardner BT (1969) Teaching sign language to a chimpanzee. Science 165:664–672
Greenfield PJ (1986) What is grey, brown, pink, and sometimes purple: the range of "wild-card" color terms. Am Anthropol 88:908–916
Grether WF (1940a) Chimpanzee color vision. I. Hue discrimination at three spectral points. J Comp Psychol 29:167–177
Grether WF (1940b) Chimpanzee color vision. II. Color mixture properties. J Comp Psychol 29:179–186
Grether WF (1940c) Chimpanzee color vision. III. Spectral limits.. J Comp Psychol 29:187–192
Grether WF (1940d) A comparison of human and chimpanzee spectral hue discrimination curves. J Exp Psychol 28:419–427
Grether WF (1942) The magnitude of simultaneous color contrast and simultaneous brightness contrast for chimpanzee and man. J Exp Psychol 30:69–83
Guest S, Van Laar D (2000) The structure of colour naming space. Vis Res 40:723–734
Heider ER (1972) Universals in color naming and memory. J Exp Psychol 93:10–20
Iversen I, Matsuzawa T (1997) Model-guided line drawing in the chimpanzee (*Pan troglodytes*). Jpn Psychol Res 39:154–181
Jacobs GH, Deegan JF, Moran JL (1996) ERG measurements of the spectral sensitivity of common chimpanzee (*Pan troglodytes*). Vis Res 36:2587–2594
Jones CD, Osorio D, Baddeley RJ (2001) Colour categorization by domestic chicks. Proc R Soc Lond B 268:2077–2084
Kawai N, Matsuzawa T (2000) Numerical memory span in a chimpanzee. Nature (Lond) 403:39–40
Kawai N, Matsuzawa T (2003) Transitive law and its generalization in a chimpanzee. In: Tomonaga M, Tanaka M, Matsuzawa T (eds) Cognitive and behavioral development in chimpanzees (in Japanese). Kyoto University Press, Kyoto, pp 415–418

Kay P, McDaniel CK (1978) The linguistic significance of the meanings of basic color terms. Language 54:610–646
Matsuno T, Kawai N, Matsuzawa T (2004) Color classification by chimpanzees (*Pan troglodytes*) in a matching-to-sample task. Behav Brain Res 148:157–165
Matsuzawa T (1985) Colour naming and classification in a chimpanzee (*Pan troglodytes*). J Hum Evol 14:283–291
Matsuzawa T (2001) Primate origins of human cognition and behavior. Springer, Tokyo
Matsuzawa T (2003) The Ai project: historical and ecological contexts. Anim Cogn 6:199–211
Mervis CB, Catlin J, Rosch E (1975) Development of the structure of color categories. Dev Psychol 11:54–60
Özgen E (2004) Language, learning, and color perception. Curr Direct Psychol Sci 13:95–98
Özgen E, Davies IRL (2002) Acquisition of categorical color perception: a perceptual learning approach to the linguistic relativity hypothesis. J Exp Psychol Gen 131:477–493
Pitchford NJ, Mullen KT (2003) The development of conceptual colour categories in preschool children: influence of perceptual categorization. Vis Cogn 10:51–77
Premack D (1971) Language in chimpanzee? Science 172:808–822
Regier T, Kay P, Cook RS (2005) Focal colors are universal after all. Proc Natl Acad Sci U S A 102:8386–8391
Roberson D, Davies I, Davidoff J (2000) Color categories are not universal: replications and new evidence from a stone-age culture. J Exp Psychol Gen 129:369–398
Rumbaugh DM (1977) Language learning by a chimpanzee: the Lana project. Academic Press, New York
Sandell JH, Gross CG, Bornstein MH (1979) Color categories in Macaques. J Comp Physiol Psychol 93:623–635
Saunders BAC, van Brakel J (1997) Are there nontrivial constraints on colour categorization? Behav Brain Sci 20:167–179
Sousa C, Matsuzawa T (2001) The use of tokens as rewards and tools by chimpanzees (*Pan troglodytes*). Anim Cogn 4:213–221
Stone AC, Griffiths RC, Zegura SL, Hammer MF (2002) High levels of Y-chromosome nucleotide diversity in the genus *Pan*. Proc Natl Acad Sci U S A 99:43–48
Sturges J, Whitfield TWA (2000) Salient features of Munsell colour space as a function of monolexemic naming and response latencies. Vis Res 37:307–313
Sumner P, Mollon JD (2000a) Catarrhine photopigments are optimized for detecting targets against a foliage background. J Exp Biol 203:1963–1986
Sumner P, Mollon JD (2000b) Chromaticity as a signal of ripeness in fruits taken by primates. J Exp Biol 203:1987–2000
Suzuki S, Matsuzawa T (1997) Choice between two discrimination tasks in chimpanzees (*Pan troglodytes*). Jpn Psychol Res 39:226–235
Tomonaga M (2001) Investigating visual perception and cognition in chimpanzees (*Pan troglodytes*) through visual search and related tasks: from basic to complex processes. In: Matsuzawa T (ed) Primate origins of human cognition and behavior. Springer, Tokyo, pp 55–86
Uchikawa K, Shinoda H (1996) Influence of basic color categories on color memory discrimination. Color Res Appl 21:430–439
von Frisch K (1950) Bees. Their vision, chemical senses, and language. Cornell University Press, Ithaca
Wright AA, Cumming WW (1971) Color-naming functions for the pigeon. J Exp Anal Behav 15:7–17
Zollinger H (1988) Categorical color perception: influence of cultural factors on the differentiation of primary and derived basic color terms in color naming by Japanese children. Vis Res 28:1379–1382

21
Auditory–Visual Crossmodal Representations of Species–Specific Vocalizations

Akihiro Izumi

1 Introduction

Understanding others' status is necessary to live in complex societies. Chimpanzees seem to use various cues such as facial expressions and vocalizations to understand others' status. Vocalizations of other individuals are particularly important to understand social events that are invisible. To respond adequately to other chimpanzees' vocalizations, vocal individuality is principal information. We previously examined vocal individual recognition using an auditory-visual matching-to-sample task in a captive chimpanzee (Kojima et al. 2003). The subject chimpanzee correctly selected the picture of the vocalizer in response to species-specific vocalizations (pant hoots, pant grunts, and screams). The chimpanzee seems to use crossmodal representations of vocalizations to perform the matching task. Savage-Rumbaugh et al. (1988) demonstrated that two chimpanzees and a bonobo perform symbolic crossmodal tasks using artificial symbol systems. Although the results suggest that these apes have symbolic representations, it is still unknown how these species represent species-specific communication signals.

From early stages of development, humans demonstrate various levels of crossmodal representations of their speech. Dodd (1979) revealed that 10- to 16-week-old infants are aware of the synchronicity of auditory speech and lip movements. Walker-Andrews and her colleagues (1991) demonstrated that infants can match faces and voices based on gender before they reach 6 months of age. McGurk and MacDonald (1976) showed that adult humans integrate auditory and visual speech when identifying consonant-vowel syllables. This phenomenon, known as the McGurk effect, was observed in 5-month-old infants (Rosenblum et al. 1997). In the preferential looking paradigm study, 4.5-month-old infants gazed longer at a vocalizing face that matched the heard vowel sound than at a face vocalizing a different vowel (Kuhl and Meltzoff 1982, 1984; Patterson and Werker 1999). Walton and Bower (1993) used the operant-choice sucking procedure and demonstrated that 6- to 8-month-old infants prefer to receive possible face–voice pairs than impossible pairs. These results have

National Institute of Neuroscience, National Center of Neurology and Psychiatry, 4-1-1 Ogawa-Higashi, Kodaira, Tokyo 187-8502, Japan suggested that prelinguistic infants recognize the correspondence between auditory and visual speech at various levels.

We examined the audiovisual crossmodal representations of a chimpanzee (Izumi and Kojima 2004). The chimpanzee performed an auditory-visual matching-to-sample task. The test stimuli in the study were movie clips instead of still pictures. Movie clips provide rich information of individual status including mouth movements. In experiment 1, movie clips of silent and vocalizing chimpanzees were used to examine whether the chimpanzee understands the appearance of other chimpanzees who vocalize or do not. In experiment 2, four types of vocalizations (pant hoot, pant grunt, food grunt, and scream) and vocal faces were presented to test the ability to match the types of vocalizations and the vocalizing faces. In both experiments, the chimpanzee was required to identify vocalizing individuals to receive rewards. The effects of the status (or movements) of the stimulus chimpanzees on the matching performance of the subject chimpanzee were examined.

2 General Methods

2.1 Subject

The subject was an 18-year-old female chimpanzee (*Pan troglodytes*) named Pan who has been extensively trained to perform the auditory-visual matching-to-sample task (Hashiya and Kojima 2001a, b; Kojima et al. 2003). Pan has a daughter named Pal, and they were together during the experimental sessions.

2.2 Apparatus

The experiments were conducted in an experimental booth (2.4 m wide × 2.0 m deep × 1.8 m high). A 21-inch computer monitor with a touch panel system was placed in one corner of the booth and connected to a personal computer outside the booth. The resolution of the monitor was 1024 × 768 pixels (width × height). The computer controlled the behavioral procedure and data collection using a customized program. Auditory stimuli were generated with the computer and were presented via a speaker located outside the booth. Because the booth was not soundproofed, sounds were easily transmitted to the chimpanzee. The auditory level was calibrated with a sound level meter (model 215; Quest Electronics).

2.3 Stimuli

The auditory stimuli were vocalizations of chimpanzees who were members of Pan's group and familiar to Pan. In our previous investigation (Kojima et al. 2003), Pan showed a nearly perfect performance in identifying vocal individuals of pant hoots, pant grunts, and screams. The vocalizations were prepared with 16-bit precision and a 48-kHz sampling rate. The stimulus level was approximately 80dB sound pressure level (SPL). Each test movie depicted clearly the face of one of the chimpanzees whose vocalizations were used as sample stimuli. Original sounds of these movies were muted. Vocalizations and movies were recorded in the chimpanzee enclosures of the Primate Research Institute, Kyoto University, in 2002.

2.4 Procedure

The task was a two-choice crossmodal matching-to-sample (Fig. 1). Initially, a start key (a purple rectangle; 10 cm wide × 4 cm high) was displayed on the monitor, which Pan was required to respond to repetitively (three to six times) to initiate a trial. Multiple responses were required because we intended to make Pan attentive to the monitor. In each trial, a sample stimulus and two test stimuli were presented simultaneously. The sample stimulus was a vocalization of a chimpanzee (sample individual), and the test stimuli were two movie clips (matching and nonmatching movies) of chimpanzees. The matching movie showed the sample individual who vocalized the sample vocalization, whereas the nonmatching movie showed a different chimpanzee. These movies were presented side-by-side on the monitor. There was an intertrial interval (ITI) of 15 s after Pan selected one of the two test movies. If Pan correctly chose the matching movie, various food rewards were dispensed manually during the ITI; however, she got no reward if she incorrectly chose the nonmatching movie.

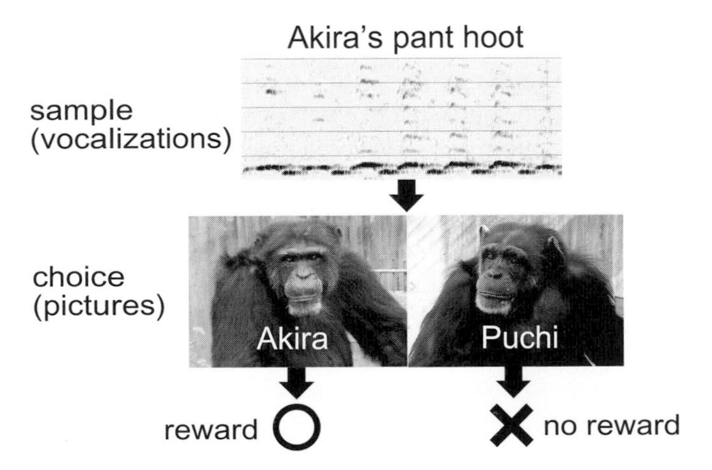

Fig. 1. Schematic representation of a trial. In each trial, Pan was rewarded when she chose a predetermined "match" picture in response to a sample vocalization. In the trial shown, the sample vocalization is Akira's pant hoot and the correct response is to choose Akira's picture

3 Experiment 1: Understanding of the Appearances of Vocalizing or Silent Individuals

This experiment was designed to test whether chimpanzees understand the appearance of other chimpanzees when vocalizing or when not. We examined the effect of the stimulus of chimpanzees' appearance (vocalizing or silent) in the movies on the performances of vocal individual recognition. The subject chimpanzee was required to choose the movie of the vocal individual in response to a pant-hoot vocalization and was not required to attend to the individual's appearance.

3.1 Methods

The vocal stimuli were composed of pant hoots of five chimpanzees, three females (named Ai, Pendesa, and Reiko) and two males (Akira and Gon), all of whom were members of Pendesa group and familiar to Pan. In our previous investigation (Kojima et al. 2003), Pan showed a nearly perfect performance in identifying vocal individuals of pant hoots, pant grunts, and screams. Four pant-hoot vocalizations were prepared for each chimpanzee, and each vocalization was used only once per session.

According to the pairing of the matching and nonmatching movies, there were four types of trials, namely vocal-vocal, vocal-silent, silent-vocal, and silent-silent conditions (Fig. 2). The name of each condition represents whether the

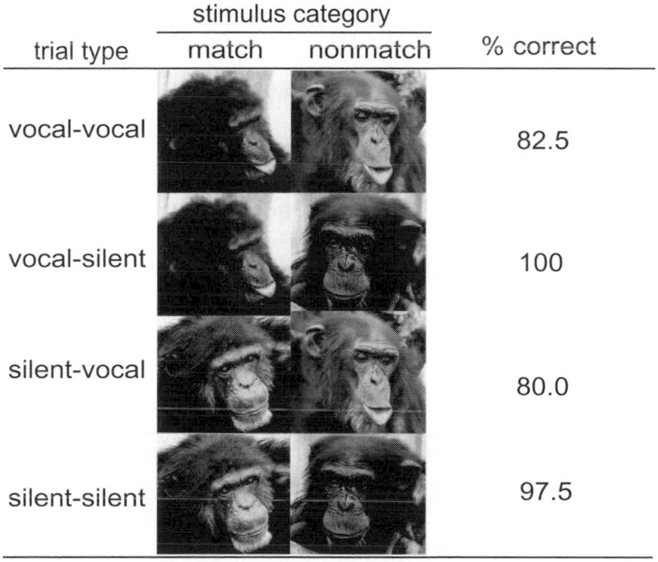

	stimulus category		
trial type	match	nonmatch	% correct
vocal-vocal			82.5
vocal-silent			100
silent-vocal			80.0
silent-silent			97.5

Fig. 2. Examples of stimulus pairs and percentages of correct responses in four types of trials in experiment 1. In the case shown, the sample stimulus is Akira's pant hoot and a baited response is to choose Akira's picture (i.e., match)

matching and nonmatching movies depicted vocalizing or silent chimpanzees. Under silent-vocal conditions, for example, the chimpanzee in the nonmatching movie vocalized while the chimpanzee in the matching movie did not. Under this condition, Pan had to choose the matching silent movie because she had to choose the match movie (sample individual) regardless of whether the movie depicted a vocalizing face or not. Even if the matching movie was a vocalizing movie (under vocal-vocal and vocal-silent conditions), the sample vocalization and the test movie were not synchronized because these stimuli came from different utterances. Each session contained 20 trials. Data from the initial 8 sessions were used for analysis.

3.2 Results and Discussion

The percentages of correct responses are shown in Fig. 2. Performance was affected by the stimulus conditions [χ^2 (3, $N = 160$) = 13.889, $P = 0.003$]. Performance under the vocal-silent condition was superior to that under the vocal-vocal (Fisher's exact test: $P = 0.046$) and silent-vocal conditions ($P = 0.031$). Performance under the silent-silent condition was superior to that under the silent-vocal condition ($P = 0.028$). In this block, Pan tended to respond to the vocalizing movies even if they were nonmatching. The fact that the matching performance was affected by the status of the stimulus chimpanzee (vocalizing or silent) suggests that Pan understood vocalizations are from vocalizing faces.

4 Experiment 2: Understanding of the Correspondence Between Vocalization and Face Types

Pan possessed crossmodal representation of vocalizations, that is, she understood to some extent the relationship between vocal sounds and vocalizing faces. The aim of the next experiment was to investigate in detail these crossmodal representations of vocalizations. We prepared four types of vocal sounds (pant hoot, pant grunt, food grunt, and scream) and corresponding vocalizing faces, and Pan was examined to determine whether she understood the correspondence between vocalization and face type.

4.1 Methods

We prepared four types of vocalizations: pant hoots (PH), pant grunts (PG), food grunts (FG), and screams (Scr). Two types of vocalizations were prepared for each of the chimpanzees in the three pairs. There were two vocal sounds and one movie of a vocalizing face for each type of vocalizations by each chimpanzee. These types of vocalizations were of different social contexts (Marler and Tenaza 1977; Goodall 1986). Chimpanzees protrude their lips when they vocalize pant hoots and pant grunts, and they vocalize screams with bared-teeth faces. No par-

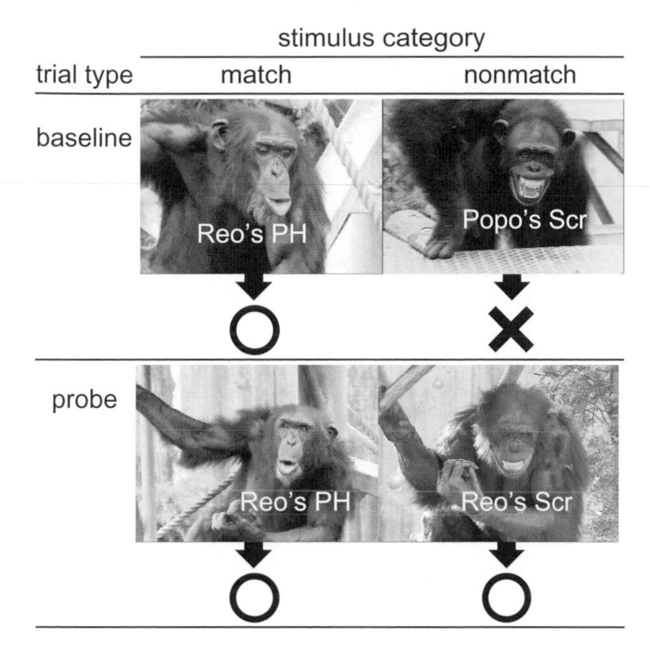

Fig. 3. Examples of the stimulus pairs in baseline and probe trials. Sample vocalization is Reo's pant hoot for both cases. In baseline trials, the vocal-individuality cues and/or the vocalization-type cues were valid for making correct responses. In probe trials, both the match and nonmatch stimuli depicted the sample individual, and responses to either were rewarded

ticular lip movements seem to accompany food grunts; however, the vocalizers were eating something in the stimulus movies.

Two types of trials were conducted, baseline and probe trials (Fig. 3). A daily session consisted of 24 trials including 12 baseline and 12 probe trials. Pan completed 8 sessions.

The baseline trials were intended to maintain Pan's matching performance based on vocal individuality, and the procedure was similar to that in experiment 1. In the matching movie, the sample chimpanzee vocalized the same type of vocalizations as the sample vocal stimuli, whereas in the nonmatching movie, a different chimpanzee vocalized another type of vocalization. As in the previous experiment, Pan was rewarded when she made a correct response (choosing the matching movie). In these baseline trials, the vocal-individuality cues and/or the vocalization-type cues were valid for making correct responses.

The probe trials were designed to test whether Pan understood the correspondence between vocalization and face type. Because we intended to examine whether Pan understood such correspondence without extensive training, Pan was not required to respond based on the vocalization types. Both the matching and nonmatching movies depicted the sample individual, and responses to either were rewarded. In the matching movie, the sample individual vocalized the same type of vocalization as the sample vocalization, while in the nonmatching movie, the same individual vocalized another type of vocalization. It was predicted that Pan would prefer to respond to the matching movies if she knew the voice–face correspondence.

Table 1. Numbers of match and nonmatch responses in four conditions in experiment 2

Conditions	Trials	Response type		Binominal test
		Match	Nonmatch	
Baseline	95	77	18	$P < 0.001$
Probe: PH-PG	32	17	15	ns
Probe: PH-FG	32	24	8	$P = 0.008$
Probe: PH-Scr	30	22	8	$P = 0.018$

4.2 Results and Discussion

The results of the experiment are shown in Table 1. Sometimes Pal, the infant, responded instead of Pan, and therefore the data from such trials were discarded. Match responses indicate a response to a matching movie instead of to a nonmatching movie and were equivalent to correct responses during the baseline trials. Except for under the pant hoot–pant grunt condition, the numbers of match and nonmatch responses were different significantly (two-tailed binominal test: baseline, $P < 0.001$; pant hoot–food grunt condition, $P = 0.008$; pant hoot–scream condition, $P = 0.018$). This result suggests that Pan used vocal-type cues both under the pant hoot–food grunt and pant hoot-scream conditions.

The reason she did not choose matching movies under the pant hoot–pant grunt condition might be the commonality of the two types of vocalizations. Faces during both pant hoots and pant grunts are characterized by protruded lips. Furthermore, the stimulus chimpanzees in these trials (Pendesa and Reiko) sometimes emitted pant hoots and pant grunts in a sequence, particularly when they approached dominant male chimpanzees.

5 General Discussion

The results of the first experiment suggest that the subject chimpanzee Pan recognized the appearances of other chimpanzees whether they vocalized or not. The second experiment examined whether Pan understood the correspondence of vocalization type and vocalizing face. The results suggested that she did understand such correspondence under limited conditions. These results extended previous findings and revealed that not only can Pan identify vocal individuals but she also understands the status of chimpanzees by listening to their vocalizations. In other words, Pan seems to possess crossmodal representations of the species-specific vocalizations.

However, only one chimpanzee who had been extensively trained to perform the crossmodal matching task (Hashiya and Kojima 2001a,b; Kojima et al. 2003) participated in the present study, and therefore we cannot conclude immediately that chimpanzees in general possess similar crossmodal representations of vocalizations. Although we trained Pan to perform the crossmodal matching-to-sample task based on vocal individuality (Kojima et al. 2003), we did not teach her the correspondence between vocal sounds and vocalizing faces. It seems reasonable therefore to consider chimpanzees are equipped to match the vocal sounds and faces during everyday life.

Crossmodal representations of vocalizations might help chimpanzees to understand complex social interactions containing various kinds of auditory and visual information. Similar to monkeys (Kojima 1985; Colombo and D'Amato 1986), auditory short-term memory in chimpanzees is fragile compared with visual memory (Hashiya and Kojima 2001). If crossmodal representation of vocalizations makes vocalization memory robust, it might assure the contribution of vocalizations in understanding invisible individuals.

To some extent, humans perceive phonetic information only by observing mouth/lip movements (speech reading: e.g., Bernstein et al. 2000). The movie clips in the present study might contain various contextual cues, such as the directions of gazes and head movements, and therefore it was not possible to conclude whether the subject chimpanzee matched corresponding vocalizations and vocalizing movies through mouth/lip movements alone. Recent investigations revealed that vocalizing faces activates the auditory cortex in both humans and monkeys (humans: Calvert 2001; monkeys: Ghazanfar et al. 2005). These results imply common neural processes that support similar abilities exist in these species.

Using the matching-to-meaning task, Parr (2001) demonstrated the ability of chimpanzees to understand the emotional states of other chimpanzees based on facial expressions. Her chimpanzees successfully matched various emotional movies and chimpanzee facial expressions based on emotional meanings. For example, the chimpanzees matched a hypodermic needle and the bared-teeth display, both of which relate to negative emotions in chimpanzees. In future studies, it will be necessary to examine whether chimpanzees match vocalizations and visual stimuli based on emotional value using methods similar to those used in the present study.

As in previous studies (Davenport et al. 1975), it was also difficult for Pan to acquire crossmodal matching of various object sounds (Hashiya and Kojima 2001a,b). Because understanding relationships between objects and arbitrary sounds seems a prerequisite to spoken language, such a difficulty in chimpanzees is impressive. In contrast to the difficulties with object sounds, conspecific vocalizations and pictures seem not to be difficult to match. After extensive training to match sounds and pictures of humans and objects (Hashiya and Kojima 2001a,b), Pan easily acquired the task to match group mates' vocalizations and pictures (i.e., vocal individual recognition; Kojima et al. 2003). In the present study, Pan matched the corresponding vocal sounds and vocalizing faces without training. Ghazanfar and Logothetis (2003) revealed that rhesus monkeys (*Macaca mulatta*) are able to recognize the correspondence between their species-specific vocalizations and vocalizing faces. The subject monkeys were presented with two side-by-side movies of a monkey articulating two distinct types of vocalizations ("coo" and "threat"); only the sound track of one of the movies was played. Without training in crossmodal relationships, the monkeys looked longer at the movie with an accompanying vocal sound, showing crossmodal recognition of vocalizations.

There might be two levels of crossmodal matching: matching of natural and arbitrary relationships. Although both levels of matching are easy for humans, chimpanzees might experience difficulty in matching sounds and objects in arbitrary relationships. Relationships between vocalizations and vocalizing faces seems to be natural, or at least something more than arbitrary, for conspecifics because animals perceive these vocalization and face pairs very frequently from the early stages of development, and they themselves produce such vocalizations. Matching of auditory and visual stimuli with natural relationships can be regarded as the process of understanding sound sources. An extreme example of natural relationships might be synchronized sounds and object movements. Although humans were revealed to perceive synchronicity of speech sounds and lip movements from the very early stages of development (Dodd 1979), whether chimpanzees practically understand these relationships has never been examined. To clarify the differences between vocal representations of humans and nonhumans, it should be important to examine how deeply nonhuman primates understand such auditory-visual relationships.

Acknowledgments

The author is grateful to Shozo Kojima for his advice and collaboration throughout this study. This work was supported by MEXT grants (13410025, 12002009) and a grant for the biodiversity research of the 21st Century COE (A14).

References

Bernstein LE, Demorest ME, Tucker PE (2000) Speech perception without hearing. Percept Psychophys 62:233–252

Calvert GA (2001) Crossmodal processing in the human brain: insights from functional neuroimaging studies. Cereb Cortex 11:1110–1123

Colombo M, D'Amato MR (1986) A comparison of visual and auditory short-term memory in monkeys (*Cebus apella*). Q J Exp Psychol 38B:425–448

Davenport RK, Rogers CM, Russell IS (1975) Cross-modal perception in apes: altered visual cues and delay. Neuropsychologia 13:229–235

Dodd B (1979) Lip reading in infants: Attention to speech presented in- and out-of-synchrony. Cogn Psychol 11:478–484

Ghazanfar AA, Logothetis NK (2003) Facial expressions linked to monkey calls. Nature (Lond) 423:937–938

Ghazanfar AA, Maier JX, Hoffman KL, Logothetis NK (2005) Multisensory integration of dynamic faces and voices in rhesus monkey auditory cortex. J Neurosci 25:5004–5012

Goodall J (1986) The chimpanzees of Gombe: patterns of behavior. Belknap, Cambridge

Hashiya K, Kojima S (2001a) Acquisition of auditory-visual intermodal matching-to-sample by a chimpanzee (*Pan troglodytes*): comparison with visual-visual intramodal matching. Anim Cogn 4:231–239

Hashiya K, Kojima S (2001b) Hearing and auditory-visual intermodal recognition in the chimpanzee. In: Matsuzawa T (ed) Primate origins of human cognition and behavior. Springer, Tokyo, pp 155–189

Izumi A, Kojima S (2004) Matching vocalizations to vocalizing faces in a chimpanzee (*Pan troglodytes*). Anim Cogn 7:179–184

Kojima S (1985) Auditory short-term memory in the Japanese monkey. Int J Neurosci 25:255–262

Kojima S, Izumi A, Ceugniet M (2003) Identification of vocalizers by pant hoots, pant grunts and screams in a chimpanzee. Primates 44:225–230

Kuhl PK, Meltzoff AN (1982) The bimodal development of speech in infancy. Science 218:1138–1141

Kuhl PK, Meltzoff AN (1984) The bimodal representation of speech in infants. Infant Behav Dev 7:361–381

Marler P, Tenaza R (1977) Signaling behavior of apes with special reference to vocalization. In: Sebeok TA (ed) How animals communicate. Indiana University Press, Bloomington

McGurk H, MacDonald J (1976) Hearing lips and seeing voices. Nature (Lond) 264:746–748

Parr LA (2001) Cognitive and physiological markers of emotional awareness in chimpanzees (*Pan troglodytes*). Anim Cogn 4:223–229

Patterson ML, Werker JF (1999) Matching phonetic information in lips and voice is robust in 4.5-month-old infants. Infant Behav Dev 22:237–247

Rosenblum LD, Schmuckler MA, Johnson JA (1997) The McGurk effect in infants. Percept Psychophys 59:347–357

Savage-Rumbaugh S, Sevcik RA, Hopkins WD (1988) Symbolic cross-modal transfer in two species of chimpanzees. Child Dev 59:617–625

Walker-Andrews AS, Bahrick LE, Raglioni SS, Diaz I (1991) Infants' bimodal perception of gender. Ecol Psychol 3:55–75

Walton GE, Bower TGR (1993) Amodal representation of speech in infants. Infant Behav Dev 16:233–243

22
Spontaneous Categorization of Natural Objects in Chimpanzees

Masayuki Tanaka

1 Introduction

Categorization is the ability to distinguish among individual objects and events in the world and recognize some of them as equivalent on one basis. Objects and events in the world are physically different from one another, and they are distributed continuously around us. Categorizing objects and events and organizing the world is one of the most important abilities of animals. Humans and nonhuman animals spontaneously construct a variety of categories from birth. Categorization does not have only one basis. It is often based on perceptual similarity, but sometimes on thematic relationships, in which the individual is aware of the objects together. In this chapter, I compare the ability of categorization between humans and chimpanzees, which are the closest relative to humans, and discuss the specialization of the categorization ability in humans.

Rosch et al. (1976) suggested that the world contains intrinsically separate things, and that the world is structured because real-world attributes do not occur independently from each other. They described basic level categories. The basic level has the most numbers of common features and is differentiated from other category members. Basic levels of categorization emerge early in infancy in humans (Behl-Chadha 1996; Mandler and Bauer 1988).

Categorization is also one of the most important abilities of nonhuman animals for survival. An animal species that eats plants must classify leaves or fruits into "food" or "nonfood." The shape or color of natural objects varies in the world, but animals seem to be able to classify them. Previous studies have revealed that nonhuman animals have a great ability to classify natural objects or artificial stimuli into two or more categories. The previous studies used the so-called concept formation paradigm. For example, Herrnstein and Loveland (1964) showed that pigeons could classify the slides on the basis of whether the slides contained human images. The subjects could transfer their responses to novel slides that were not used in training. After Herrnstein and Loveland (1964), many studies have reported that nonhuman animals can form various categories. Some categories are based on natural entities: people, trees, water, cats,

Primate Research Institute, Kyoto University, 41 Kanrin, Inuyama, Aichi 484-8506, Japan

flowers, and primate species (Austs and Huber 2001; Bhatt et al. 1988; D'Amato and Van Sant 1988; Herrnstein et al. 1976; Vonk and MacDonald 2004; Yoshikubo 1985). Others are based on artificial entities: car, chair, cartoons, pseudobutterfly, and alphabet/numerical characters (Bhatt et al. 1988; Cerella 1980; Jitsumori 1996; Vauclair and Fagot 1996).

In the concept formation paradigm, subjects are trained to respond or not to respond to category examples that human researchers have defined. Therefore, these studies revealed that nonhuman animals have the potential to form such categories, but it is still unclear what types of categories the animals spontaneously form and use in their lives. This chapter treats the abilities that nonhuman animals spontaneously use to categorize objects in their world. In the studies discussed in this chapter, subjects were not trained to respond to specific categories, and the types of categories that they formed were examined.

2 Categorization on the Basis of Perceptual Attributes

Natural objects are often categorized on the basis of some common perceptual attributes. Humans can discriminate category examples and recognize them as equal. It is still unclear whether nonhuman animals categorize in the same way. The first study treats the abilities of discrimination and categorization.

2.1 Discrimination and Categorization of Category Examples

In studies of animal categorization, subjects are trained with a small number of examples from each category before being tested using novel examples. When subjects continue to respond to novel examples in the same way that they respond to training examples, the results are said to demonstrate open-ended categorization. These studies, however, lack an important control needed to concretely infer categorization: the demonstration that examples from the same category are different.

In humans, categorization is assumed to occur when observers respond in the same manner to different stimuli (Behl-Chadha 1996). Only a few animal studies have addressed the issue of within-class discrimination (Thompson 1995; Vauclair and Fagot 1996; Wasserman et al. 1988). Wasserman et al. (1988) trained pigeons to discriminate individual examples within each of four categories. Their results showed more errors for within-category discrimination than for between-category discrimination. Vauclair and Fagot (1996) showed that baboons could categorize the alphanumeric characters B and 3 in various font styles. After the first experiment, an identity matching-to-sample task was used to assess the issue of within-class discrimination. Results showed that examples from the same category were discriminably different, suggesting that baboons developed open-ended categorical procedures. Altogether, these studies suggest that nonhuman primates and pigeons can sort perceptually different items into the same class.

In humans, the class in which objects are classified might depend on relations between objects. That is, a collie might be classified as a dog in one context, but as an animal in another context. Roberts and Mazmanian (1988) reported that pigeons and monkeys had difficulty discriminating between birds and other animals, or between animals and nonanimals, although they learned to discriminate between kingfishers and other birds. Results suggest that the pigeons and the monkeys did not form basic level categories. If so, we should assess the ability of categorization in chimpanzees, phylogenetically the closest relatives to humans, to consider the evolution of the ability of categorization. Tanaka (2001) assessed categorization abilities in four adult female chimpanzees in the following experiments.

2.1.1 Methods

Subjects

The subjects had been previously tested in various experiments on cognitive abilities (Kawai and Matsuzawa 2000; Matsuzawa 2003; Tanaka 1996, 1997; Tomonaga and Matsuzawa 2002). They lived with seven other chimpanzees in a stimulating outdoor compound with many plants (Ochiai and Matsuzawa 1998).

Stimuli

Four natural categories (flowers, trees, weeds, and ground surface) were used as experimental categories. The subjects could see the objects in the four categories in their daily lives and they would promote category formation as humans do. Digital images of familiar items, the same kinds of which existed in the circumstances of the subjects, were used in experiment 1 (Fig. 1).

Procedure

Experiment 1 consisted of a discrimination training phase and a categorization test phase.

Discrimination training: The subjects were individually trained to discriminate individual species within a class (i.e., azalea, camellia, Japanese cherry, and dandelion) and between classes. In a matching-to-sample task, the subjects were to choose images of the same item as the sample among four comparisons. The images of the samples were different images of the same type of sample item. The comparison stimuli were from the same category under one condition (S-trials) and from different categories under another condition (D-trials).

Categorization test: Probe trials were shown after the training. In the probe trials, the sample and positive comparison stimuli were different items from the same category, and the foils were selected from among the three other test categories. The stimuli used in the test trials were chosen as follows. The sample

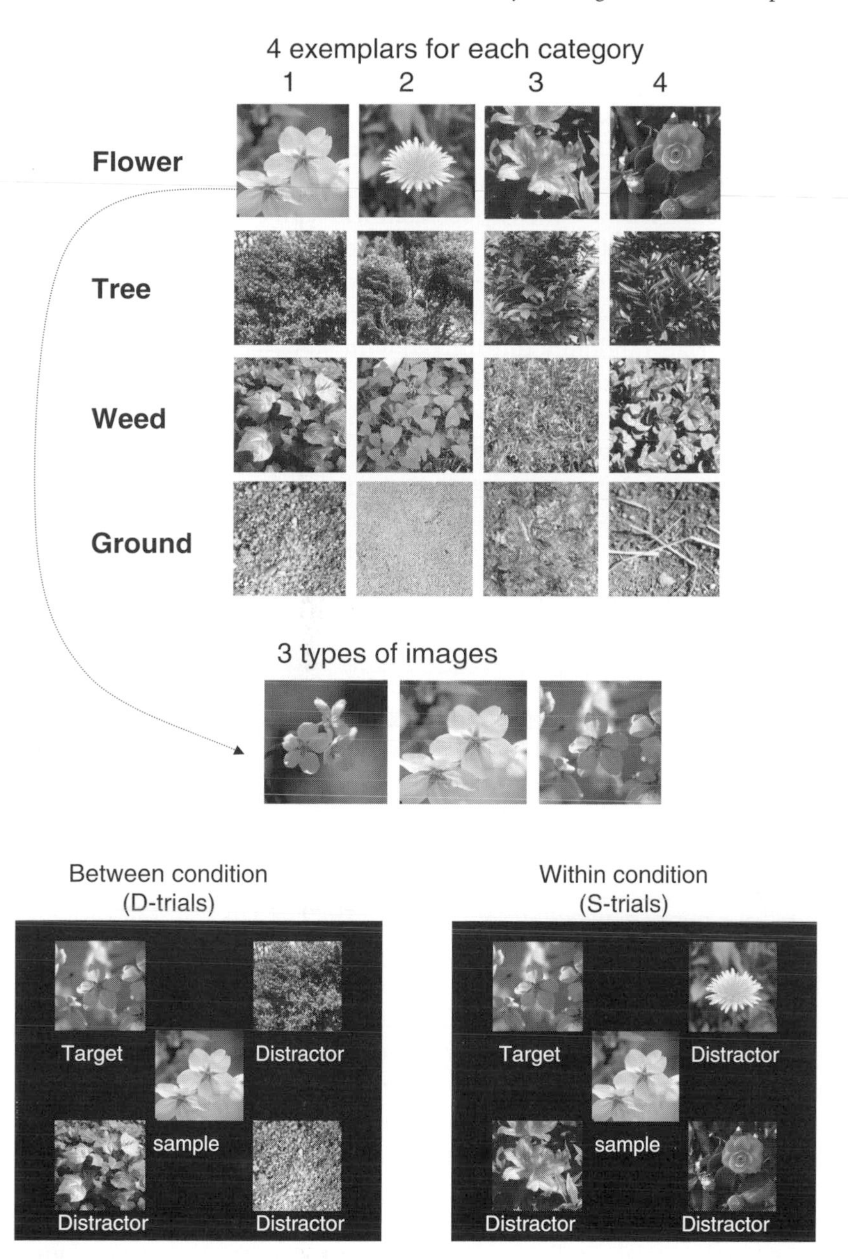

Fig. 1. *Upper.* Examples of stimuli used in experiment 1 in Tanaka (2001). Each category has four types of example. *Middle.* There were three types of digitized images of each example. *Bottom.* Two conditions of a matching-to-sample task. In both conditions, the images of a sample and a target were different, but showed the same example. Distractors were from the same category as that of the target in the *Between* condition, and each distractor was from a different category in the *Within* condition

and comparison stimuli used in the test trials were chosen considering performance achieved during the last ten training sessions. First, only two items of each category were chosen according to the best matching performance when presented as a sample in S-trials. These items were used as samples in the test trials. Second, for each item chosen as a test sample, two comparison items were chosen, which were the least frequently selected when an error was made in the S-training-trials. These items were used for positive comparison in the test trials. Note that this procedure for stimulus selection ensured that the test sample and positive comparison stimuli were discriminably different. The test sessions consisted of 16 probe trials randomly intermixed with 96 baseline trials (48 S- and 48 D-trials) similar to those of the training phase. Each test sample stimulus was presented twice during a session, once with each of two positive comparison stimuli with which it was paired. In each probe trial, three distractors were selected from the three categories of items that were different from the sample category.

2.1.2 Results and Discussion

Discrimination training: All subjects could choose the correct images of the same item as the sample either in S-trials or D-trials. The subjects consistently showed better performance in the D-trials (mean correct = 92.9) than in the S-trials (mean correct = 60.7) in the last ten training sessions. All the chimpanzees exceeded 80% correct performance or higher in the D-trials after the 11th training session. Performance after the 11th session became lower than 80% in the S-trials but still exceeded the chance level (25%) for each subject (binominal test; all $P < 0.05$). A category by test condition analysis of variance (ANOVA) was computed for performance data obtained in the last ten training sessions. This ANOVA revealed a significant main effect of test conditions [$F(1, 24) = 36.6$, $P < 0.001$] but no significant effect of category [$F(3, 24) = 1.13$, $P = 0.359$] and no significant condition by category interaction [$F(3, 24) = 0.16$, $P = 0.922$], suggesting similar response behaviors for the four categories of items. The results of the training revealed that the chimpanzees could match different images of the same item and suggested that the items of different categories were easier for the chimpanzees to discriminate than those within the same category.

Categorization test: The subjects achieved 83.8% correct performance on average in the baseline trials. Individual baseline performance was above chance for both the D- (96.5% correct) and S-trials (71.1% correct) (binomial tests, all $P < 0.001$). Figure 2 shows the total percentage of correct performance in the D- and S-trials (baseline). Each subject showed the same results as those in the training. Categories (e.g., flower, tree, weed, and ground) by conditions (S- and D-trials) ANOVA performed on the number of correct trials revealed that the main effect of conditions was significant [$F(1, 24) = 26.8$, $P < 0.0001$], showing reduced performance in the S-trials compared with that in the D-trials. In the test trials, not all the subjects chose different items of the same category as

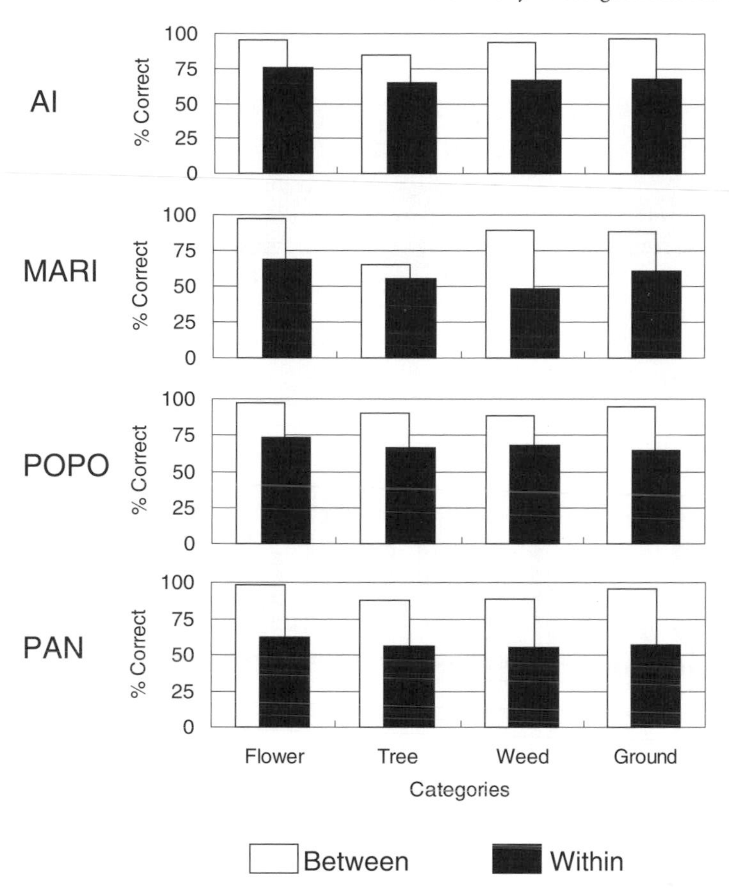

Fig. 2. Percentage of correct responses in *Between* and *Within* conditions for each subject in experiment 1 in Tanaka (2001)

samples in the S-trials, but they chose significantly more numbers of different items from the same category as the sample in more than four pairs of samples and comparisons in which there were no images of the same items as the sample (Fig. 3). The results suggest that the chimpanzees could recognize perceptually discriminable items from the same category as the same and that they could change the level of categorization according to the composition of the stimuli. That is, the chimpanzees are able to categorize not only different photographs of the same item but also different items from the same category.

Such categorization with an embedded structure is fundamental in humans.

2.1.3 Transfer to Unfamiliar Items

In experiment 2, Tanaka (2001) used unfamiliar items from the same four categories. The items were those that the chimpanzees had not seen before. The chimpanzees showed almost the same performance as that in experiment 1. That

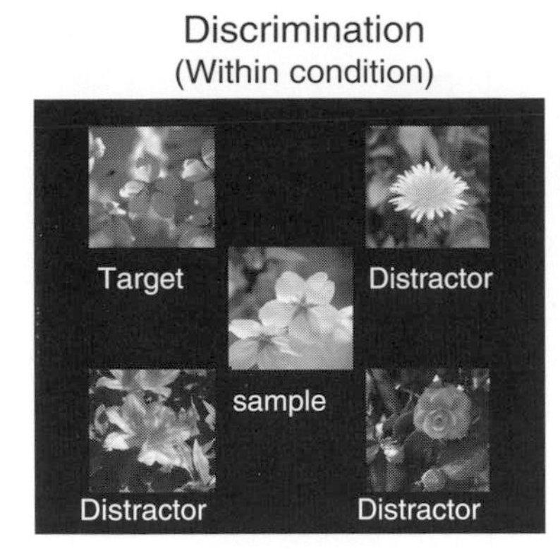

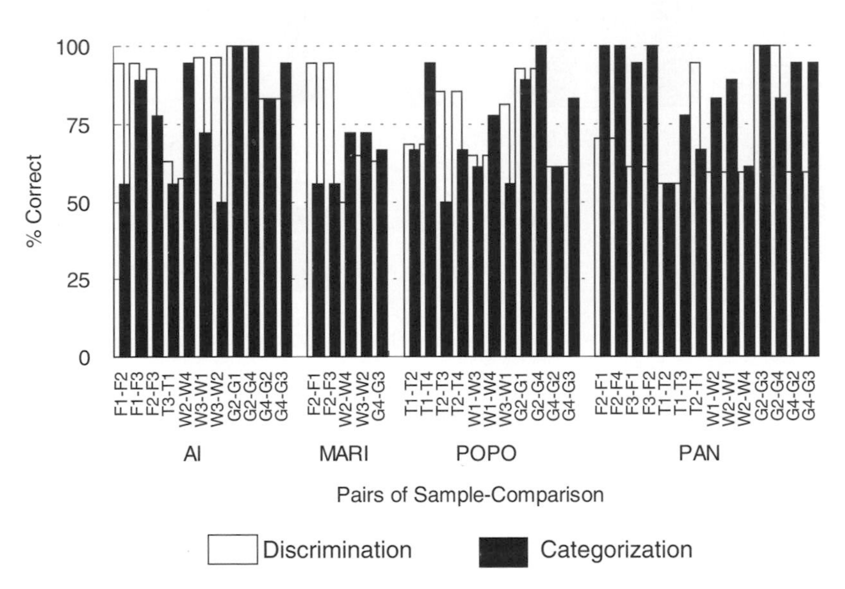

Fig. 3. *Upper*. Examples in discrimination trials (*left*, Within condition) and categorization trials (*right*, Test condition). The sample and target was from the same category but were different species. The target in the categorization trial was presented as a distractor in the discrimination trials. *Bottom*. Percentage of correct responses in the discrimination and categorization trials. The *white bars* indicate performance in the discrimination trials; the *black bars* indicate performance in the categorization trials. There are only combinations of the sample and the target in the categorization trials, in which the subjects made significantly more correct choices in either discrimination or categorization trials

is, the chimpanzees could discriminate the items from the same categories when the same items as the sample were used as comparison stimuli, and the subjects could match the discriminable item from the same category as the sample when the same items as the sample were not in the comparison stimuli. Altogether, the present experiments demonstrated that chimpanzees spontaneously categorize perceptually discriminable items from the same category, as humans do, and that the categories are applicable to novel items as well as to familiar items.

2.1.4 General Discussion

The results of experiments 1 and 2 suggest the following. (1) Chimpanzees were able to match images of natural objects according to the category to which they belonged. (2) Within-class discrimination was more difficult for the chimpanzees than between-class discrimination. (3) The experimental examples that the animals could categorize were discriminably different. (4) Selection of the response stimulus could be made considering either the types of objects or their category, depending on the type of problem that had to be solved. In particular, the chimpanzees could to some extent select a photograph of an item from the same category as the sample when there was no photograph of the same item as the sample.

Thompson (1995) reported that the notion that different objects have common class attributes, which permit them to be distinguished from each other, is the core of conceptual categorization. So far, only a few studies have addressed this issue experimentally (Vauclair and Fagot 1996; Wasserman et al. 1988), in contrast to the many studies that have assessed concept formation in nonhuman animals. The main purpose of the study was to verify whether the subjects could effectively discriminate items from the same category, before testing whether the subjects could categorize the items. The present study suggests that chimpanzees categorize real objects in the same manner as humans do.

Of course, abilities of visual perception and cognition are fundamental to that of categorization. Tomonaga (2001) summarized that discrimination ability in chimpanzees was comparable to that in humans. He revealed that chimpanzees could detect an item with or without a feature in a stimulus array. It is not surprising that chimpanzees discriminated each category of examples. That is, it is expected that chimpanzees could form example-level categories. This study also revealed that chimpanzees could spontaneously form a higher level of categories that contained four different examples (i.e., flower, tree, weed, and ground). In addition, this study suggests that the chimpanzees could select which level of categorization should be used according to the combination of stimuli trial by trial.

There seemed to be no necessity for the chimpanzees to discriminate between trees and weeds, flowers and trees, or between flowers and weeds. However, the subjects quickly learned to discriminate items in D-trials. The results suggest that the chimpanzees had already formed such categories before the present study and that the categories might have some significance for the chimpanzees. The results of the present study are in contrast with those of Roberts and Mazmanian (1988). These authors suggest that pigeons and monkeys have difficulties in sorting stimuli at a level that is considered basic for humans (i.e., bird vs other animals). Moreover, the nonhuman subjects failed to transfer to novel examples. Of course, humans were able to discriminate bird from nonbird slides, and animal from nonanimal slides, and to transfer to novel examples. Humans organize categories at different levels. That is, humans could classify an object into categories of basic level (e.g., dog), superordinate level (e.g., animal), or subordinate level (e.g., collie). In particular, humans are likely to classify at the basic level, where there is a higher within-category similarity and a higher between-category dissimilarity (Rosch et al. 1976). The study revealed that chimpanzees could indeed match the examples in both subordinate (e.g., dandelion) and basic level categories (e.g., flowers), apparently in the same way that humans do. The study did not make it clear whether chimpanzees could match at the superordinate level. From a human perspective, the flower, tree, and weed categories might be considered more natural than the ground category because they correspond to real objects existing in nature. Moreover, the flower, tree, and weed categories belong to the superordinate plant category. Interestingly, no difference emerged in terms of the discrimination performance between the ground category and the more natural flower, tree, and weed categories. The results suggest that chimpanzees are not sensitive to the naturalistic character of these latter categories, at least in the experiments (Fig. 4).

Thus, chimpanzees spontaneously form categories of natural objects in their circumstances. It is possible for the chimpanzees to apply the categories to

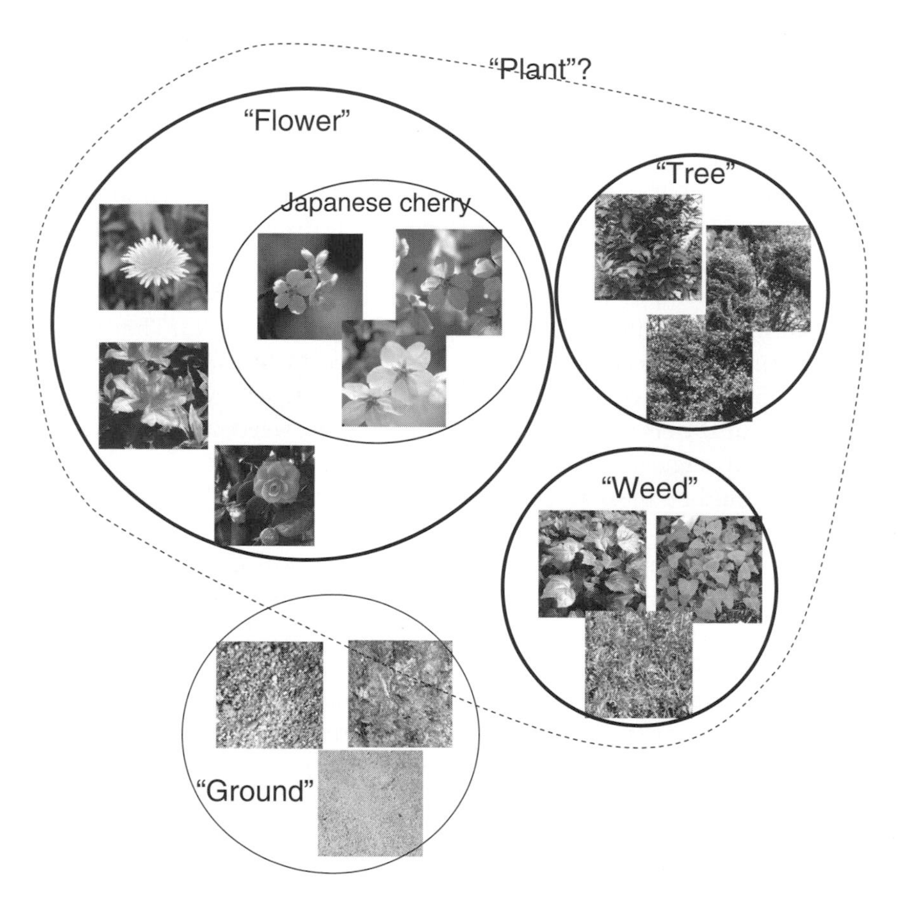

Fig. 4. Schema of relationship of categories suggested by Tanaka (2001). Each category has subcategories of examples similar to the subcategory of the Japanese cherry. The examples are shown in Fig. 1

not only familiar examples but also to novel ones. The level of categorization might correspond to the basic level in human categorization. Categorization at the superordinate level in nonhuman animals is discussed later in this chapter.

2.2 Spontaneous Categorization as Preference

We have other ways of investigating the ability of categorization than the paradigm of the so-called concept formation or the above matching-to-sample paradigm. We often have preference for the specific category level as well as specific individual items. Nonhuman animals would also have preference for specific categories spontaneously. Previous studies revealed that macaque monkeys showed differential preference for visual stimuli (Humphrey 1972, 1974; Humphrey and Keeble 1974). In particular, macaque monkeys showed preference for the visual stimuli of their own species (Swartz and Rosenblum 1980).

Fujita and Matsuzawa (1986) developed an automated procedure for assessing the preference of nonhuman animals through a sensory reinforcement procedure. In their study, a female chimpanzee was trained to press a button to see a variety of color slides. Slides were presented as long as the subject kept pressing the button. Repeated pressings within 10 s after a previous release produced the same slides again. The slide was changed if 10 s had passed after releasing the button. Analysis of response duration and response interval revealed a clear difference between slides with humans and those without humans, with the chimpanzee preferring to view the former. Using this procedure, Fujita and his colleagues (Fujita 1987, 1990, 1993a; Fujita and Watanabe 1995; Fujita et al. 1997) demonstrated that macaque species tend to show greater interest in slides of their own species. For example, Japanese macaques (*Macaca fuscata*) preferred to observe slides of Japanese macaques over those of other macaque species, such as rhesus macaques (*Macaca mulatta*). In contrast, rhesus macaques preferred to observe slides of rhesus macaques over those of other macaques. Fujita pointed out that such differential interest might help prevent interbreeding among closely related species.

Fujita (1990, 1993b) also revealed that social experience in infanthood might influence an animal's preference for a particular species. He used subjects with variously restricted social experience (i.e., either reared by humans, or with conspecific heterospecific peers). In this study, rhesus macaques tended to prefer seeing rhesus macaques regardless of their age or social experience. However, Japanese macaques with restricted experience tended to prefer to see rhesus macaques over Japanese macaques. Although there are many studies that used macaque species, there are few studies that used great apes. The studies using great apes suggested that these apes have genetically programmed preference to images of conspecifics rather than those of other great ape species or humans.

Tanaka (2003) developed a new method of evaluating visual preference using a touch-sensitive screen. This method, which is based on a form of sensory reinforcement, is called the free-choice task. All that the subjects have to do is to touch stimuli on a touch-sensitive screen. The stimuli that the subject touched were moved inside a frame at the top of the display. In each trial of the task, the subjects are giving the chance to choose three times to estimate their order of preference. A food reward is often delivered irrespective of which stimulus the subject chose, but choices are not always reinforced to extinguish superstitious behaviors (e.g., a choice on the basis of position of the stimuli).

2.2.1 Methods

Subjects

The study aimed to investigate species preference in five adult chimpanzees reared by humans, but with access to social interaction with fellow chimpanzees (see Table 1). Two subjects were born in Africa and received at the laboratory at 1 year of age. The other subjects were born in captivity and reared by humans immediately after birth. They lived with other chimpanzees in a captive community of the Primate Research Institute, Kyoto University (Table 1).

Stimuli

Stimuli were 5.6-cm digitized color images (198 × 198 pixels, 24-bit color bitmap file) generated from color photographs. In test 1, there were three stimulus sets,

Table 1. Subjects and their profiles

Name	Age[a] at test	Birth	Age[b] to PRI	Notes
Ai	23	Africa (wild)	1:01	Reared by humans together with Mari and another male infant
Mari	23	Africa (wild)	1:06	Moved to JMC at the age of 9 and joined as part of the chimpanzee group of JMC; she returned to PRI at the age of 19
Pendesa	22	JMC	2:09	Reared by humans immediately after birth; she lived with Ai, Mari, and the other infants moved to PRI
Popo	17	PRI	0:0	Reared by humans and lived with her brother and sister (Pan) during her infancy; she joined the adult group at the age of 10
Pan	16	PRI	0:0	Reared by humans and lived with her brother and sister (Popo) during her infancy; she joined the adult group at the age of 9

PRI, Primate Research Institute, Kyoto University; JMC, Japan Monkey Center, Inuyama, Aichi, Japan

[a]Age when the subject was tested

[b]Age when the subject arrived at the Primate Research Institute, Kyoto University

each of which consisted of four categories: (1) human and great apes (four genera: *Homo*, *Pan*, *Gorilla*, *Pongo*); (2) Haplorhine (four families: Hominidae, Pongidae, Hylobatidae, Cercopithecidae); and (3) primate sets (five superfamilies: Hominoidea, Cercopithecoidea, Ceboidea, Lemuroidea, and Lorisoidea). Each category in each set consisted of 10 different examples. Each example was used in only one of the stimulus sets. Only the Hominoidea category in set 3 consisted of three subcategories (Hominidae, Pongidae, Hylobatidae) of 10 different examples each (i.e., 30 examples). That is, sets 1 and 2 consisted of 40 examples and set 3 consisted of 60 examples. The images of humans did not include Japanese people whom the subjects met every day. Instead, the images of humans included a wide variety in terms of race, age, and sex, as the focus of the study was preference based not on familiarity but on biological category. In test 2, the stimuli were a subset of those used as stimulus in set 3 (Primates) in test 1. Ten images were used from Pongidae, Hominidae, Hylobatidae, and Cercopithecidae categories to create a new set: (1) Haplorhine—each image was processed to monochrome format to produce another stimulus set; (2) Haplorhine (black and white)—the background of each image from stimulus set 1 was erased to produce the third stimulus set; (3) Haplorhine (without background)—that is, each image from set 3 appeared against a white background while the images of the individual subjects remained color.

Procedure

The subjects were presented with digitized color images of various species of primates on a CRT screen. Their touch responses to the images were reinforced by food reward irrespective of which image they touched. The images of humans did not include the Japanese people whom the subjects met every day. Instead, the photographs of humans included a wide variety in terms of race, age, and sex, as the focus of the study was preference based not on familiarity, but on biological category.

Data Analysis

Each choice was scored according to the order of choice in a trial. The first choice scored 3 points, the second choice scored 2 points, and the third choice scored 1 point. That is, the stimuli that the subjects chose early on in a trial were taken to be those that the subjects preferred. The score for each category within each stimulus set was summed up separately for the five subjects.

2.2.2 Results

Test 1. Figure 5 shows the score in each subject in test 1. In set 1, every subject except Mari chose the images of *Homo* (i.e., human) much more often than those in the other categories. Every subject chose images of gorillas far below

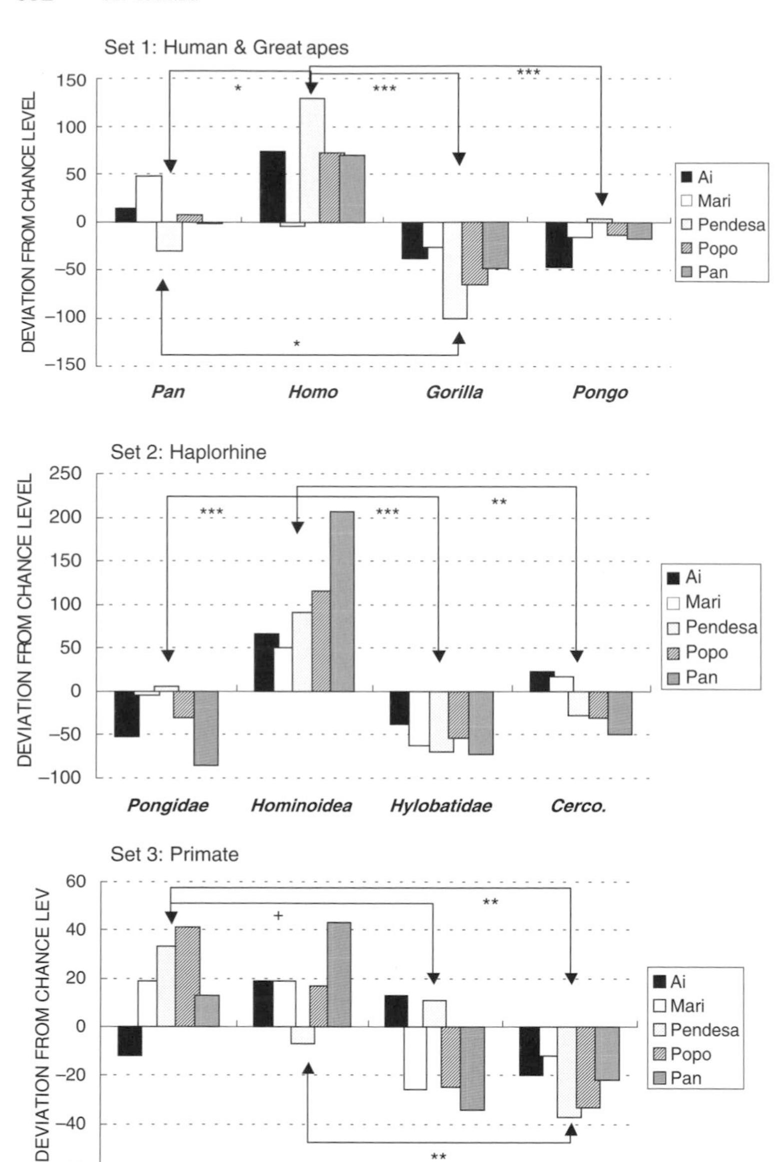

Fig. 5. Standardized score of each category in three stimulus sets. Each *bar* indicates the score of the category from which the chance level score was subtracted. Chance level score was calculated as follows: if the subject touched the stimuli randomly, each category should have been chosen 30 times as the first choice, the second choice, and the third choice over 12 sessions. Consequently, the chance level score was 180 (i.e., $30 \times 3 + 30 \times 2 + 30 \times 1$). *Cerco.* on the *x*-axis indicates the Cercopithecidae/Cercopithecoidea category. Asterisks indicate significant difference between the scores of two categories: *** $P < 0.001$, ** $P < 0.01$, * $P < 0.05$, $^{+} P < 0.08$

the chance level. A one-way ANOVA of stimulus category was conducted and revealed that the main effect of stimulus category was significant [$F(3,16) = 12.7$, $P < 0.001$]. Tukey's HSD (honestly significant difference) test revealed that the score of *Homo* (mean, 247.6) was significantly higher than that of *Pan* (mean, 187.0; $P < 0.05$), *Gorilla* (mean, 123.8; $P < 0.001$), and *Pongo* (mean, 161.6, $P < 0.01$). Tukey's test also revealed that the score of *Pan* was significantly higher than that of *Gorilla* ($P < 0.04$). In set 2, every subject chose the photographs of Hominidae (i.e., human) much more often than those of the other categories. A one-way ANOVA of stimulus category was conducted and revealed that the main effect of stimulus category was significant [$F\ (3,16) = 16.9$, $P < 0.0001$]. Tukey's HSD test revealed that the score of Hominidae (mean, 286.0) was significantly higher than that of Pongidae (mean, 146.6; $P < 0.001$), Hylobatidae (mean, 120.8; $P < 0.001$), and Cercopithecidae (mean, 166.6; $P < 0.01$). There was no difference among the scores of the other three categories. In set 3, the difference among the categories was the smallest among the three sets. A one-way ANOVA revealed that the main effect of stimulus category was significant [$F(3,16) = 7.23, P < 0.01$]. Tukey's HSD test revealed that the scores of Hominoidea (mean, 198.8) and Cercopithecoidea (mean, 198.2) were significantly higher than that of prosimian (Lemuroidea and Lorisoidea: mean, 155.2; $P < 0.01$). The scores of Hominoidea and Cercopithecoidea were nearly significantly higher than Ceboidea (mean, 167.8; $P < 0.08$).

Test 2. Figure 6 shows the score of each subject in test 2. The score of Hominidae was the highest in most of the stimulus sets and most of the subjects. In set 1, the score of the Hominidae category (i.e., human) was much higher than those of the other categories for each subject. A one-way ANOVA of stimulus category was conducted and revealed that the main effect of stimulus category was significant [$F(3,16) = 8.09$, $P < 0.01$]. Tukey's HSD test revealed that the score of Hominidae (mean, 336.2) was significantly higher than those of Pongidae (mean, 138.4; $P < 0.01$), Hylobatidae (mean, 91.4; $P < 0.001$), and Cercopithecidae (mean, 154.0; $P < 0.01$). There was no difference between the scores of the other three categories. The results revealed that the subjects tended to choose images of humans (i.e., Hominidae). In the case of stimulus set 2, only one subject, Pan, showed a very clear tendency to choose images of Hominidae. Four of the five subjects did not show a clear tendency to choose images from a specific category. However, three chimpanzees, Mari, Pendesa, and Popo, chose images in the Hominidae category more often than those in the other categories. A one-way ANOVA of stimulus category was conducted and revealed that the main effect of stimulus category was significant [$F(3,16) = 4.20, P < 0.05$]. Tukey's HSD test revealed that the score of Hominidae (mean, 246.2) was higher than those of Pongidae (mean, 153.2; $P < 0.05$) and Hylobatidae (mean, 143.0; $P < 0.03$), but not significantly different from that of Cercopithecidae (mean, 177.6). In set 3, the results were very similar to those in set 1; that is, four of the five subjects chose the images of Hominidae more often than those of the other categories. A one-way ANOVA of stimulus category revealed that the main effect of stimulus category was significant [$F(3,16) = 8.09$, $P < 0.01$]. Tukey's HSD test

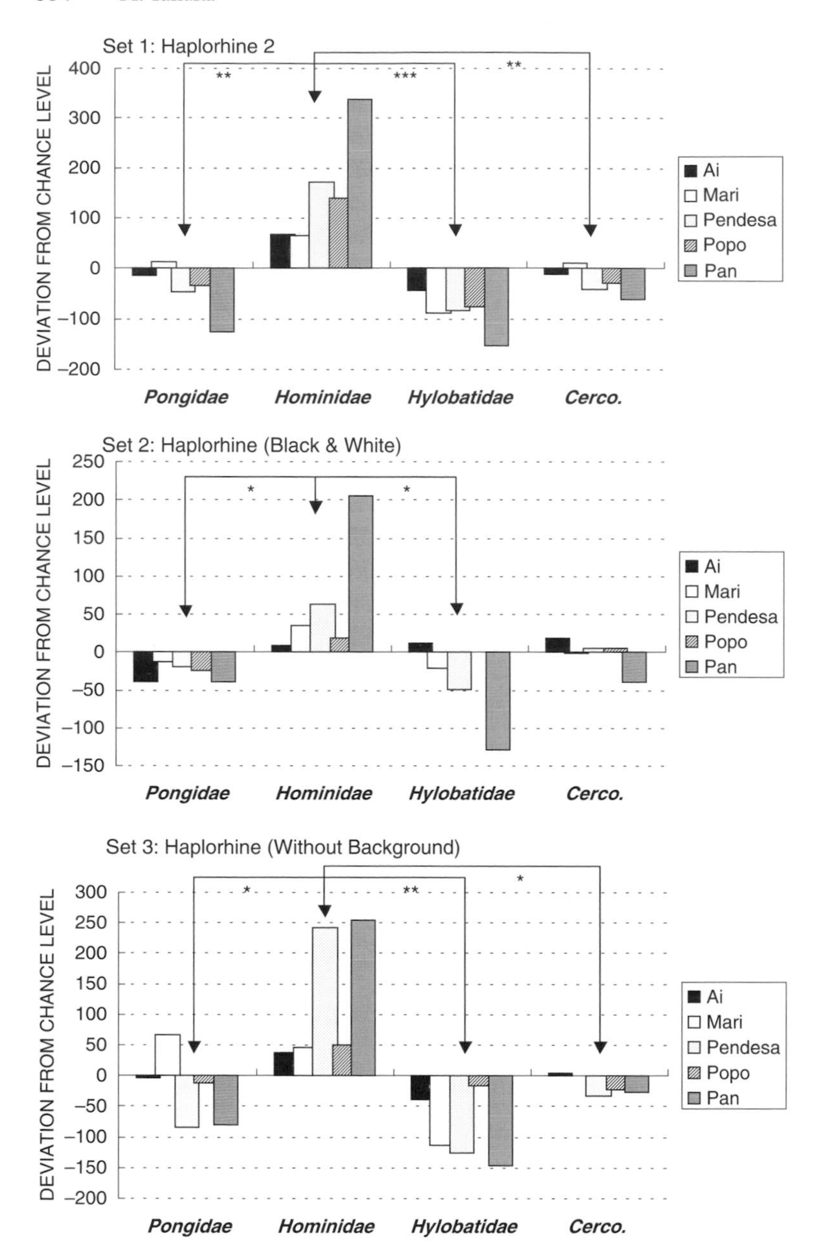

Fig. 6. Standardized score of each category in three stimulus sets. Each *bar* indicates the score after the subtraction of the chance level score (180) from the raw score. *Cerco.* on the *x*-axis indicates the Cercopithecidae/Cercopithecoidea category. *Asterisks* indicate significant differences between the scores of two categories: $^{***}P < 0.001$, $^{**}P < 0.01$, $^{*}P < 0.05$

revealed that the score of Hominidae (mean, 306.0) was significantly higher than those of Pongidae (mean, 57.2; $P < 0\ 05$), Hylobatidae (mean, 92.2; $P < 0.01$), and Cercopithecidae (mean, 164.6; $P < 0.05$). There was no difference among the scores of the other three categories.

2.2.3 Discussion

The study revealed the following three points. (1) Every subject tended to choose photographs of humans more often than any other category of primates. The preference was consistent across different stimulus sets varying in color and in background. (2) The degree of visual preference was not in accordance with phylogenetic distance from chimpanzees. (3) The subjects' preference for the photographs of humans was reduced in the case of monochromatic photographs in comparison with color ones.

The study revealed that there was a difference in preference within the Pongidae category, to which the subjects belonged. In particular, the chimpanzees chose the images of chimpanzees significantly more often than those of gorillas, their closest phylogenetic relatives. The score of orangutans was intermediate between chimpanzees and gorillas. Thus, the subjects' preference did not correspond to phylogenetic classification on the basis of morphological similarity to humans (i.e., taxonomy). However, the subjects tended to choose the images of Hominoidea and Cercopithecoidea more often than those of Ceboidea, or prosimians in test 1. The results suggest that visual preference in the chimpanzees might be contrastive among phylogenetically close species but indifferent among phylogenetically distant species. Figure 8 shows the supposed categorization in the subjects that summarized the results.

Fujita (1990, 1993b) suggested that social experience in infanthood might influence preference. All five chimpanzees had been in captivity for at least 16 years. They were reared by humans immediately after birth, or at least from 1.5 years of age. Their preference might have developed through social experience, especially that during infancy.

Recent studies support the effects of social experience on visual preference. Tanaka et al. (2004) used methods similar to those of Tanaka (2003) and showed that one infant chimpanzee, who was reared by her mother and lived in a captive chimpanzee community, showed preference for the images of chimpanzees over those of humans. Now we are going to investigate visual preference in infant chimpanzees to assess the effects of social experience during infanthood. Tanaka and Uchikoshi (2004) assessed visual preference in a juvenile agile gibbon that was reared by humans from birth. The gibbon also showed a preference for images of humans over those of gibbons. These studies suggest that apes are influenced in their visual preference to specific species by their social experience, particularly in the early stage of their lives. These results also suggest that the apes would form a specific category on the basis of their life histories.

3 Categorization on the Basis of Nonperceptual Attributes

Concerning the ability to categorize on the basis of nonperceptual information in nonhuman animals, there are studies of stimulus equivalence. Stimulus equivalence consists of reflexivity, symmetry, transitivity, and reversed transitivity, which are relationships between stimuli established in a matching-to-sample task (Sidman and Tailby 1982). Reflexivity is identified by the formation of the same-different concept: If the subject is trained to match X to X, then it should be able to match Y (a novel stimulus) to Y without explicit training. Symmetry requires functional interchangeability between sample and comparison stimuli: If the subject is trained to match X to Y, then it should be able to match Y to X. Transitivity is a derived association of stimuli that have had no direct association: If the subject matches X to Y and Y to Z, then it should be able to match X to Z. Reversed transitivity is a reverse association of transitivity: If the subject matches X to Y and Y to Z, then it should be able to match Z to X.

Some studies have revealed that monkeys and chimpanzees showed some evidence of reflexivity, symmetry, and transitivity of relationships (Tomonaga et al. 1991; Yamamoto and Asano 1995), but it is difficult in general to establish stimulus equivalence in nonhuman animals. Most of the previous studies dealt with arbitrary relations involving artificial stimuli (e.g., geometric figures). There are very few studies that have considered relationships among concrete, real objects (e.g., complementary relationships: bottle and cap). Such relationships established during actual handling might facilitate the integration of information.

3.1 Categorization on the Basis of Thematic Relations

Humans can learn various types of relationship among objects and use the information about such relationships in new situations. Moreover, humans can also integrate such information. For example, a person puts money in a safe and locks it with a key. Then the person puts the key in a desk drawer. When the person needs money, he or she will probably go to the desk rather than to the safe: The person uses information about the new relationship between the money and the desk. This is an obvious example of an ability of integration of relationships among the objects. Such ability to use information may be found in nonhuman animals at some level, because nonhumans as well as humans live in complex environments in which objects and organisms are related in various ways.

Many studies have revealed that nonhuman animals, particularly nonhuman primates, are able to learn various types of abstract relationship. For example, there are studies concerning the same–different concept in monkeys (D'Amato et al. 1986; Fujita 1983) and chimpanzees (Oden et al. 1988). There are also studies concerning arbitrary relations such as the so-called language training in great apes (Gardner and Gardner 1969; Matsuzawa 1985; Premack 1976; Rumbaugh 1977; Savage-Rumbaugh 1986). Moreover, Premack and Premack (1983) claimed that language-trained chimpanzees could use the relationship between relationships to solve an analogy task. For example, in a matching-to-sample format the subject was presented with a pair of oranges as the sample, a pair of apples as the correct alternative, and a banana and an apple as the incorrect alternative. Conversely, the subject was presented with an orange and an apple as the sample, a pineapple and a pear as the correct alternative, and a pair of pears as the incorrect alternative.

Some studies suggest that chimpanzees are able to learn relationships in one situation and use information about such relationships in a new situation (Gillan et al. 1981; Itakura 1994; Premack 1976; Savage-Rumbaugh et al. 1978). Most previous studies dealt with arbitrary relations involving artificial stimuli (e.g., geometric figures). There are very few studies that have dealt with relationships among concrete, real objects (e.g., complementary relationships: bottle and cap). Such relationships established during actual handling might facilitate the integration of information about object–object relationships.

Tanaka (1995) provided evidence showing that five chimpanzees, including the subjects of this study, could use complementary relationships between objects (e.g., a bottle and a cap) to sort objects in an object-sorting task that was a modification of one used by Matsuzawa (1990). In the object-sorting task, the subjects had been trained to place three objects on two trays on the basis of identity. Then the subjects were tested on their ability to sort novel triads that consisted of a complementary pair (e.g., a bottle and a cap) and a neutral object. After the subjects had been trained to assemble complementary pairs, the frequency with which the subjects placed a complementary pair together on the same tray increased significantly. These results suggest that the chimpanzees learned complementary relationships during the assembling training and used the information on the relationships to solve the sorting task. In Tanaka's (1995) study, the chimpanzees distinguished between familiar and unfamiliar objects. The chimpanzees were presented with one novel object and two different familiar objects that were used in sorting training on the basis of identity. The subjects spontaneously put the familiar objects on one tray and one novel object on another tray, although the subjects had been trained to put different objects on different trays. These results suggest that the chimpanzees could find some concrete relationships among objects to classify the objects that the subjects had used in object-manipulating situations.

Moreover, Tanaka (1996) demonstrated that one chimpanzee recognized more than one relationship among different objects, which the chimpanzee learned through manipulating experiences. In the first experiment, one female chimpanzee learned to match one part of an assembled object to its other part, match a tool for the assembled object, match a container to its tool, and match a tool to its container. In the experiment, the subjects assembled the parts, using the tool to open the assembled object, or put the tool into the corresponding container (Fig. 7). After the subject could choose the corresponding item according to the sample item, the subject participated in other experiments. In the following experiment, the subjects learned to choose digitized images of the sample

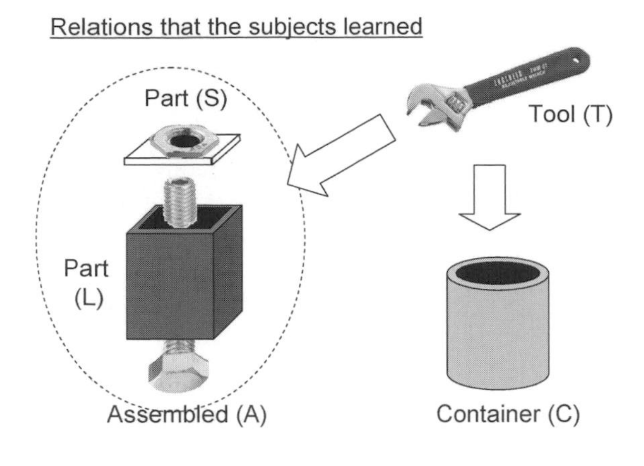

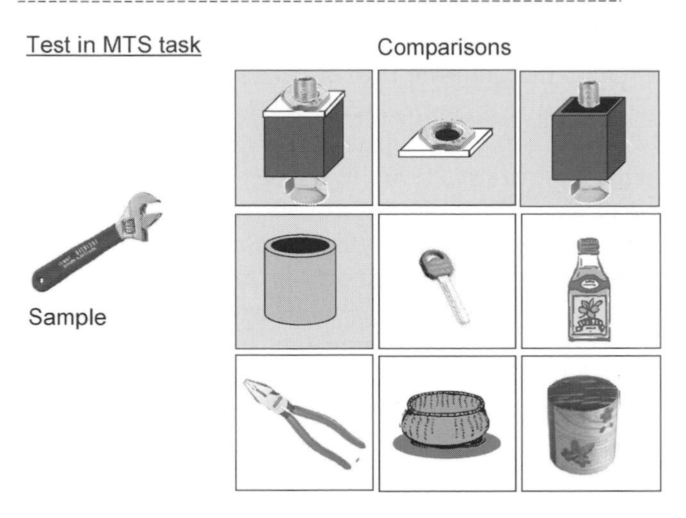

Fig. 7. *Upper*. Schema of relationships that the subjects learned in the training in Tanaka (1996). In training, the subjects actually handled the objects. *Bottom*. Schema of the condition in the test trials. One object was presented as a sample, and then nine images were presented in the CRT. To indicate which sample was related to the sample object, the background of the images of the objects related to the sample is shown in *gray* in this schema, but the background was green in all images in the test

objects used in the first experiment. Then the subjects were tested whether they could use the information learned in the first experiment to choose the images of related objects when an image of the sample was not among the comparison stimuli but images of the objects that the subject had learned to match to the sample object were (see Fig. 7). Although the subject was reinforced irrespective of her choices, she chose images of items related to the sample when there were no images of the sample object (Fig. 8). The third experiment tested whether the subjects could discriminate the pictures of related objects and showed that the subject was able to match an image of the object among the objects that were

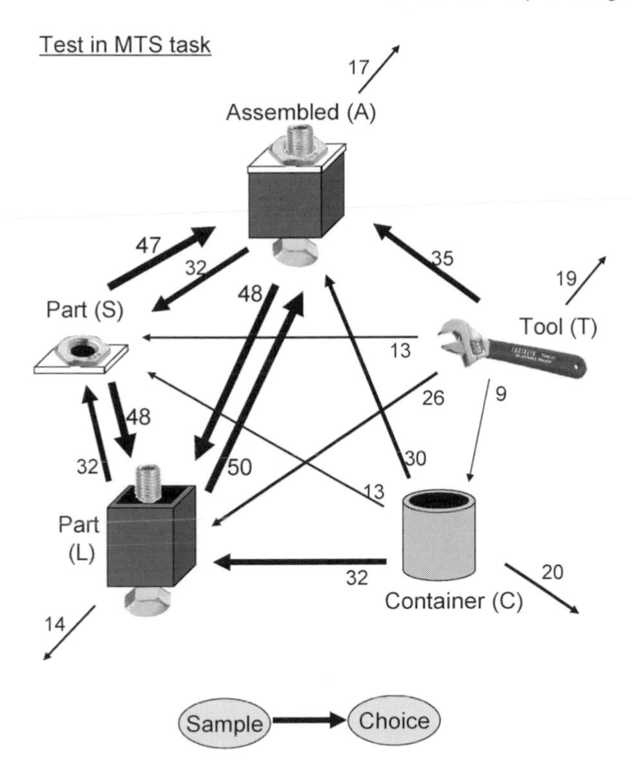

Fig. 8. Schema of subject's choices in test. The *arrowhead* indicates which item the subject chose when the item at the origin of an *arrow* was a sample (i.e., sample to choice). *Arrows that point away from ellipses* indicate the choices of images of unrelated objects. The *number beside an arrow* indicates the percentage of frequency of choices of 90 choices. *Wider arrows* indicate that the subject more often chose arrowed objects. If the percentage was less than 5%, the arrows were omitted in this schema

related in the first experiment and could distinguish each object from the related objects.

The results in Tanaka (1996) revealed the following three points. (1) The subject acquired the relationships among the real objects that had different shape and color. (2) The subject could discriminate the two-dimensional image of the item that she had manipulated before and match the two-dimensional image of the sample object on the monitor when the three-dimensional object was presented as the sample. (3) The subject could choose the image of the object that had some kind of relationship to the sample object in her previous experience when the image of the sample object was not on the monitor.

The results suggested that the chimpanzee could use their knowledge about the relationships among objects to classify the objects in another situation. It was difficult to interpret whether the subject found a perceptual similarity even among the assembled parts and the tool. The objects were easily discriminable from one another, and the subject could match the image of the sample with the other images from the test trials. It is also surprising that when the container was the sample, the subject very often chose the assembled object or the parts, which were disassembled by the tool that was to be put in the sample container. The results suggest that the chimpanzee not only determined the relationship between two objects in the experiment (i.e., assembling the parts, using the tool to open the assembled object, and putting the tool in the corresponding container), but also integrated these relationships to organize semantic networks of the objects.

Thematic relationships are said to be often used by young children in classification, but older children and adults use taxonomic relationships more often than thematic relationships. However, even human adults often used thematic relations under some conditions (Lin and Murphy 2001). This type of relationship still has a role in conceptual representations in humans as well as in chimpanzees.

3.2 Category Formation on the Basis of Functions

Humans make a variety of categories, which lead to a hierarchically organized taxonomic system. Categories may be classified generally into two types, taxonomic and thematic categories. Taxonomic categories appear in a hierarchical system based on asymmetrical inclusion relationships, wherein a superordinate category (e.g., animal) includes and references subordinate categories (e.g., dog). Most members of categories in general use (e.g., dog, chair) resemble one another or share some features. Such categories are called basic level categories (Rosch et al. 1976). Superordinate categories (e.g., animal, furniture), however, are usually based on general functional features rather than perceptual resemblance. Many studies of human adults and children have dealt with these categories, and developmental research has shown that taxonomic categories, particularly superordinate ones, appear later than thematic categories (Fenson et al. 1989; Lucariello et al. 1992; Markman and Hutchinson 1984).

Some studies showed that nonhuman animals can use information on categories learned in one situation in other situations (Itakura 1994; Savage-Rumbaugh et al. 1980; Tanaka 1995, 1996). Savage-Rumbaugh et al. (1980) showed that three chimpanzees who had been given artificial language training were able to use information about categories based on whether the objects were edible or inedible. The task was to respond to a lexigram (complex geographical figure) representing a function (i.e., food or tool) when one of six training objects was presented as the sample. After reaching a learning criterion, the authors tested the generalization of this skill by presenting five additional foods and five additional tools. Two chimpanzees correctly categorized almost all ten novel items in trial 1, but one chimpanzee correctly categorized only three items. However, she sorted all ten novel items correctly in trial 1. Thus, the authors suggested that the three chimpanzees were able to conceptualize food and tools and concluded that the ability to organize this information is similar at a symbolic level.

Tanaka (1997) examined the ability of one female chimpanzee to form categories of objects on the basis of their function (i.e., tool, container, and food). In my previous study (Tanaka 1996), the subject learned the relationship between two objects by manipulating and in matching-to-sample tasks. That is, the subject learned which tool was used for the object, and which container was used for the tool. The food items were used as rewards in the training. In experiments 1, 2, and

3, a matching-to-sample task presented real objects as samples and digitized images of the objects for comparisons. Experiment 1 tested whether she could match images of objects related to the sample when there was no image of the sample among the comparisons. The subject showed a notable tendency to choose images of objects that were complementary to the sample in the preliminary training, but she chose few images of objects from the same functional category as the sample. In experiment 2, there were four comparisons, none of which were images of the objects that were complementary to the sample; only one comparison showed an image of an object from the same functional category as the sample. The subject chose images of objects from the same functional category as the sample at only the chance level. In experiment 3, the subject was trained to match an image of an object from the same functional category as the sample, but showed no improvement in her performance. In experiment 4, the subject was trained to choose a lexigram corresponding to the functional categories. After the subject matched the lexigrams correctly on the basis of functions, she correctly matched some of the untrained objects on the basis of their functions. These results suggest that matching a common lexigram to more than one sample facilitates the development of functional categories.

In the study, a pattern of results very similar to those of Tanaka (1996) was reproduced. The chimpanzee matched the objects that were related to one another in the previous manipulation phase. The chimpanzee, however, did not use information about relationships based on function in matching-to-sample test (see Fig. 9). The chimpanzee had difficulty in learning to match images of items from the same functional category as the sample, which was probably because of the influence of her previous strategy in identity matching training. However, after the chimpanzee learned to match a lexigram to two different

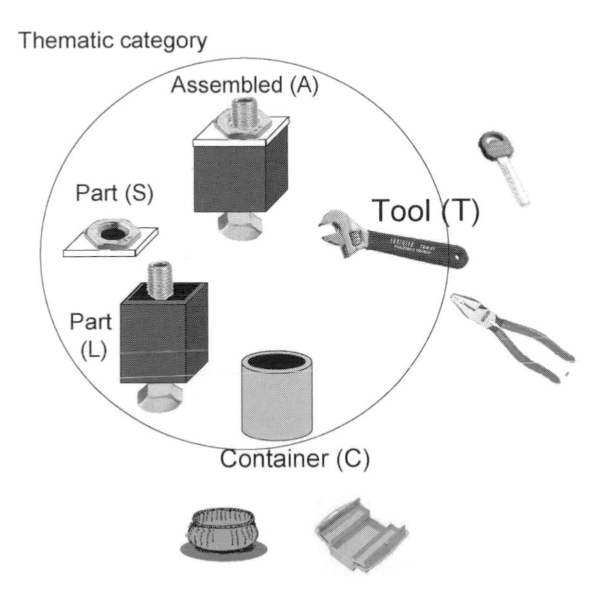

Fig. 9. Schema of the thematic category that the chimpanzee formed through her experience, as suggested by Tanaka (1996, 1997). The schema also suggest that the categories based on function (e.g., tool, container) are not formed spontaneously in chimpanzees

items on the basis of functional categories, she was able to choose correct responses for some untrained items. The results suggest that the categories of objects were based on concrete experience in chimpanzees. For example, the objects belonging to the tool category were not related to each other but were categorized on the basis of abstract function (e.g., to disassemble the box). It may be difficult for chimpanzees to categorize the objects on the basis of abstract attributes.

Premack and his colleagues (Oden et al. 1988; Premack 1976, 1983a,b; Thompson and Oden 1993) suggested the effects of language training on cognitive processes in chimpanzees. Premack and Premack (1983) claimed that language training appeared to convert an animal with a strong bias for responding to appearances into one that can respond on an abstract basis. Namely, a language-trained chimpanzee, Sarah, was able to use the relationship between objects (i.e., same or different) to solve analogy problems, but juvenile chimpanzees, who were untrained in language, were not able to do so (Oden et al. 1988; Premack 1983a). The results found by Tanaka (1997) appeared to be consistent with the hypothesis of Premack and Premack (1983). The chimpanzee used in Tanaka (1997) had not learned training for a language-like system, but training for responding to the lexigrams on the basis of abstract, functional properties may reduce a tendency to respond on the basis of perceptual resemblance.

Some studies suggest that language training is not necessary for nonhuman animals to classify objects on the basis of function (Savage-Rumbaugh et al. 1978; Bovet and Vauclair 1998). However, the subjects were tested to classify between food and nonfood in both studies. Tanaka (1997) reported that it is not necessary for the subject to use a lexigram to classify food. The subject was able to choose the image of different food items from the sample in experiment 3. Food (i.e., edible or not) is a special category, particularly for nonhuman animals. In addition, the subject was very much aware of foods. The animals are not aware of the other objects (i.e., tool, or toy) except during the experiment time. A rich experience would facilitate the formation of a category of food.

4 Conclusion

This chapter considered the abilities of categorization in nonhuman animals in comparison with that in humans. We are able to categorize objects on the basis of various attributes. From another point of view, we have much information about objects in the world. We often acquire such information through individual experience. The information consists of perceptual and nonperceptual attributes. Figure 10 shows various attributes of an object. Perceptual attributes contain common features that are often used to form natural categories, which contain the attributes that are individually significant (i.e., factors of preference). Nonperceptual attributes contain relationships to a specific event or situation. Function is also a nonperceptual attribute that is necessarily related to a specific event.

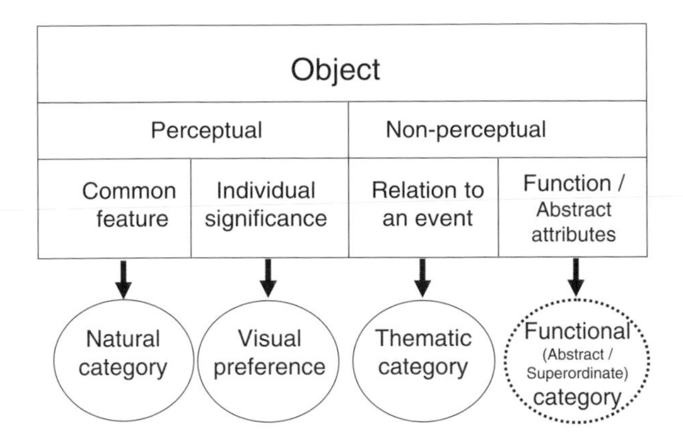

Fig. 10. Schema of an object and its perceptual and nonperceptual attributes. *Arrows* indicate which types of categories were formed by the attribute

The studies in this chapter treated the abilities of categorization in chimpanzees. The studies revealed that chimpanzees spontaneously use either perceptual or nonperceptual attributes to form categories. First, chimpanzees could form categories of objects in their environment. The study showed that chimpanzees discriminated category examples and categorized them in the same session. These results suggest that chimpanzees used different perceptual attributes and change the level of categories according to the combination of objects. Second, nonhuman animals form categories on the basis of individual preference, which may be influenced by the individuals' experience. The study in chimpanzees revealed that visual preference was not genetically programmed.

Third, chimpanzees could form categories on the basis of thematic relationships as humans do. Chimpanzees had acquired the relationships among objects through handling experience, and they used the knowledge in different task situations. Such relationships are based on object assembly, familiarity, tool-manipulated objects, or spatial closeness. All relationships can be actually experienced by the individuals themselves. Chimpanzees as well as humans are able to integrate information about relationships among objects and form thematic categories. Fourth, humans form categories on the basis of function regardless of a specific event, or situation. It seems difficult for chimpanzees to form categories on the basis of functions (e.g., tool). In such a category, each member did not actually relate in the same situation, and the individual has no experience in handling with the category members at the same time. The results suggest that the categories of objects were based on concrete experience in chimpanzees. It may be difficult for chimpanzees to categorize the objects on the basis of abstract attributes. The formation of categories on the basis of function may require the ability to operate on an abstract property of objects, such as function. Some chimpanzees that had extensive training in symbol use showed an ability to operate on abstract properties for classifying objects (Gillan et al. 1981; Premack 1983a,b). Such training might influence the ability of categorization in chimpanzees and perhaps in other species of great apes.

5 Summary

Categorizing objects and events and organizing the world is one of the most important abilities of animals. Categorization in humans is defined as an ability to distinguish among individual objects and events in the world and recognize some of them as equivalent on one basis. Animals who have the ability of categorization could categorize unfamiliar or novel items as well as familiar items in their environment. The abilities of categorization in nonhuman animals were studied in the "concept formation" paradigm. That is, subjects were tested on whether they could respond on the basis of a concept that was human defined, regardless of its significance to the animals. The studies in chimpanzees revealed that chimpanzees and humans had the ability to categorize various objects without specific training. Categorization is often based on intrinsic perceptual attributes, or natural correlations of features, which are called basic level, but sometimes on relationships related to a specific situation (i.e., thematic relationships). That is, individuals who are aware of some objects in one situation can classify those objects into one group. It is reported that young children often categorize objects in such a manner, but studies with chimpanzees showed that chimpanzees also categorize objects on the basis of thematic relationships. However, previous studies also suggested the difficulty of chimpanzees in forming categories on the basis of functions, or nonperceptual abstract attributes. Humans often form such categories and organize a hierarchical system of taxonomy. This difficulty of chimpanzees may be linked to the abilities necessary for humans to acquire a language system.

Acknowledgments

This study was partly supported by the MEXT Grants-in-Aid for Scientific Research, nos. 12002009, 12710037, 15730334, 16002001, 10CE2005, and for the 21st Century COE Program (A-14). I thank Mr. S. Nagumo for his technical assistance in programming and interfacing. I also thank the staff members of the Center for Human Evolution Modeling Research, Primate Research Institute, Kyoto University.

References

Aust U, Huber L (2001) The role of item- and category-specific information in the discrimination of people versus nonpeople images by pigeons. Anim Learn Behav 29:107–119

Behl-Chadha G (1996) Basic-level and superordinate-like categorical representations in early infancy. Cognition 60:105–141

Bhatt RS, Wasserman EA, Reynolds WF, Knauss KS (1988) Conceptual behavior in pigeons: categorization of both familiar and novel examples from four classes of natural and artificial stimuli. J Exp Psychol Anim Behav Proc 14:219–234

Bovet D, Vauclair J (1998) Functional categorization of objects and of their pictures in baboons (*Papio anubis*). Learn Motiv 29:309–322

Cerella J (1980) The pigeon's analysis of pictures. Pattern Recogn 12:1–6

D'Amato MR, Van Sant P (1988) The person concept in monkeys (*Cebus apella*). J Exp Psychol Anim Behav Proc 14:43–55
D'Amato MR, Salmon DP, Loukas E, Tomie A (1986) Processing of identity and conditional relations in monkeys (*Cebus apella*) and pigeons (*Columba livia*). Anim Learn Behav 14: 363–373
Fenson L, Vella D, Kennedy M (1989) Children's knowledge of thematic and taxonomic relations at two years of age. Child Dev 60:911–919
Fujita K (1983) Formation of the sameness-difference concept by Japanese monkeys from a small number of color stimuli. J Exp Anal Behav 40:289–300
Fujita K (1987) Species recognition by five macaque monkeys. Primates 28:353–366
Fujita K (1990) Species preference by infant macaques with controlled social experience. Int J Primatol 11:553–573
Fujita K (1993a) Role of some physical characteristics in species recognition by pigtail monkeys. Primates 34:133–140
Fujita K (1993b) Development of visual preference for closely related species by infant and juvenile macaques with restricted social experience. Primates 34:141–150
Fujita K, Matsuzawa T (1986) A new procedure to study the perceptual world of animals with sensory reinforcement: recognition of humans by a chimpanzee. Primates 27:283–291
Fujita K, Watanabe K (1995) Visual preference for closely related species by Sulawesi macaques. Am J Primatol 37:253–261
Fujita K, Watanabe K, Widarto TH, Suryobroto B (1997) Discrimination of macaques: the case of Sulawesi species. Primates 38:233–245
Gardner RA, Gardner BT (1969) Teaching sign language to a chimpanzee. Science 165:664–672
Gillan DG, Premack D, Woodruff G (1981) Reasoning in the chimpanzee. I: Analogical reasoning. J Exp Psychol Anim Behav Proc 7:1–17
Herrnstein RJ, Loveland DH (1964) Complex visual concept in the pigeon. Science 146:549–550
Herrnstein RJ, Loveland DH, Cable C (1976) Natural concepts in pigeons. J Exp Psychol Anim Behav Proc 2:285–311
Humphrey NK (1972) "Interest" and "pleasure": two determinants of a monkey visual preference. Perception 1:395–416
Humphrey NK (1974) Species and individuals in the perceptual world of monkeys. Perception 3:105–114
Humphrey NK, Keeble GR (1974) The reaction of monkeys to "fearsome" pictures. Nature (Lond) 251:500–502
Itakura S (1994) Symbolic representation of possession in a chimpanzee. In: Parker S, Mitchell R, Boccia M (eds) Self-awareness in animals and humans: developmental perspectives. Cambridge University Press, New York, pp 241–247
Jitsumori M (1996) A prototype effect and categorization of artificial polymorphous stimuli. J Exp Psychol Anim Behav Proc 22:405–419
Kawai N, Matsuzawa T (2000) Numerical memory span in a chimpanzee. Nature (Lond) 403:39–40
Lin EL, Murphy GL (2001) Thematic relations in adults' concepts. J Exp Psychol Gen 130:3–28
Lucariello J, Kyratzis A, Nelson K (1992) Taxonomic knowledge: what kind and when? Child Dev 63:978–998
Mandler JM, Bauer PJ (1988) The cradle of categorization: is the basic level basic? Cogn Dev 3:247–264
Markman EM, Hutchinson JE (1984) Children's sensitivity to constraints on word meaning: taxonomic versus thematic relations. Cogn Psychol 16:1–27
Matsuzawa T (1985) Use of numbers by a chimpanzee. Nature (Lond) 315:57–059
Matsuzawa T (1990) Spontaneous sorting in human and chimpanzee. In: Parker ST, Gibson KR (eds) "Language" and intelligence in monkeys and apes. Cambridge University Press, New York, pp 451–468
Matsuzawa T (2003) The Ai project: historical and ecological contexts. Anim Cogn 6:199–211
Ochiai T, Matsuzawa T (1998) Planting trees in an outdoor compound of chimpanzees for an enrichment. In: Hare VJ, Worley E (eds) Proceedings of the 3rd international conference on environmental enrichment. The shape of enrichment Inc., San Diego, CA, pp 355–364

Oden DL, Thompson RKR, Premack D (1988). Spontaneous transfer of matching by infant chimpanzees (*Pan troglodytes*). J Exp Psychol Anim Behav Proc 14:140–145
Premack D (1976) Intelligence in ape and man. Erlbaum, Hillsdale, NJ
Premack D (1983a) Animal cognition. Annu Rev Psychol 34:351–362
Premack D (1983b) The codes of man and beasts. Behav Brain Sci 6:125–167
Premack D, Premack AJ (1983) The minds of an ape. Norton, New York
Roberts WD, Mazmanian DS (1988) Concept learning at different levels of abstraction by pigeons, monkeys, and people. J Exp Psychol Anim Behav Proc 14:247–260
Rosch E, Mervis CB, Gray WD, Johnson DM, Boyes-Braem P (1976) Basic objects in natural categories. Cogn Psychol 8:382–439
Rumbaugh DM (1977) Language learning by a chimpanzee. Academic Press, New York
Savage-Rumbaugh ES (1986) Ape language: from conditioned response to symbol. Columbia University Press, New York
Savage-Rumbaugh ES, Rumbaugh DM, Smith ST, Lawson J (1980) Reference: The Linguistic essential. Science 210:922–924
Sidman M, Tailby W (1982) Conditional discrimination vs. matching to sample: An expansion of testing paradigm. J Exp Analysis Behav 31:23–44
Swartz KB, Rosenblum LA (1980) Operant responding by bonnet macaques for color video-taped recordings of social stimuli. Anim Learn Behav 8:311–321
Tanaka M (1995) Object sorting in chimpanzees (*Pan troglodytes*): classification based on identity, complementarity, and familiarity. J Comp Psychol 109:151–161
Tanaka M (1996) Information integration about object-object relationships in chimpanzees. J Comp Psychol 110:323–335
Tanaka M (1997) Formation of categories based on functions in chimpanzees (*Pan troglodytes*). Jpn Psychol Res 39:212–225
Tanaka M (2001) Discrimination and categorization of photographs of natural objects by chimpanzees (*Pan troglodytes*). Anim Cogn 4:201–211
Tanaka M (2003) Visual preference by chimpanzees (*Pan troglodytes*) for photos of primates measured by a free choice-order task: implication for influence of social experience. Primates 44:157–165
Tanaka M, Mizuno Y, Yamamoto S (2004) Visual preference for photos of primates by mother-reared chimpanzee infants. Poster presented at the 2nd international workshop for young psychologists on evolution and development of cognition, Kyoto, Japan, November 13–14, 2004. Abstract retrieved from http://www.psy.bun.kyoto-u.ac.jp/COE21/record/2IWYPabst.pdf
Tanaka M, Uchikoshi M (2004) Visual preference for photos of primates in a human-rared gibbon. Primate Res 20 Supplement: 23. (in Japanese)
Thompson RK (1995) Natural and relational concepts in animals. In: Roitblatt H, Meyer JA (eds) Comparative approaches to cognitive science. MIT Press, Cambridge, pp 175–224
Thompson RKR, Oden DL (1993) "Language training" and its role in the expression of tacit propositional knowledge by chimpanzees (*Pan troglodytes*). In: Roitblat LM, Herman LM, Nachtgall PE (eds) Language and communication: Comparative perspectives. LEA, Hillsdale, NJ
Tomonaga M (2001) Investigating visual perception and cognition in chimpanzees (*Pan troglodytes*) through visual search and related tasks: from basic to complex processes. In: Matsuzawa T (ed) Primate origins of human cognition and behavior. Springer, Tokyo, pp 55–86
Tomonaga M, Matsuzawa T (2002) Enumeration of briefly presented items by the chimpanzee (*Pan troglodytes*) and humans (*Homo sapiens*). Anim Learn Behav 30:143–157
Tomonaga M, Matsuzawa T, Fujita K, Yamamoto J (1991) Emergence of symmetry in a visual conditional discrimination by chimpanzees (*Pan troglodytes*). Psychol Rep 68:51–60
Vauclair J, Fagot J (1996) Categorization of alphanumeric characters by Guinea baboons: within-and between-class stimulus comparison. Curr Psychol Cogn 15:449–462
Vonk J, MacDonald SE (2004) Levels of abstraction in orangutan (*Pongo abelii*) categorization. J Comp Psychol 118:3–13

Wasserman EA, Kiedinger RE, Bhatt RS (1988) Conceptual behavior in pigeons: categories, subcategories, and pseudocategories. J Exp Psychol Anim Behav Proc 14:235–246
Yamamoto J, Asano T (1995) Stimulus equivalence in a chimpanzee (*Pan troglodytes*). Psychol Rec 45:3–21
Yoshikubo S (1985) Species discrimination and concept formation by rhesus monkeys (*Macaca mulatta*). Primates 26:285–299

23
Cognitive Enrichment in Chimpanzees: An Approach of Welfare Entailing an Animal's Entire Resources

Naruki Morimura

1 Introduction

1.1 The Stimulating Life of Wild Chimpanzees

Wild chimpanzees (*Pan troglodytes*) live in a surrounding environment that includes mountains, rivers, forests, grasslands, and an enormous variety of plants and animals that inhabit it. Through coexistence and competition with those living things, wild chimpanzees learn numerous strategies to survive throughout their lives. They learn, for example, to identify poisonous and edible foods, the locations and times for getting food, and how to access a food patch, find a food at the patch, and process the food for consumption. Along with that learning, they acquire knowledge of materials that are available as tools. Moreover, through identification of other members of a party to which they themselves belong, chimpanzees form social groups. Members of a group sometimes cooperate to hunt and share prey.

In various contexts, a chimpanzee receives stimuli from the environment by identifying a stimulus according to their circumstances and experiences. They perceive a sensory stimulus by processing its characteristics, then represent and integrate them through more-complex processing such as memory, learning, and reasoning for understanding a situation. Laboratory work has illuminated a variety of cognitive competence of chimpanzees. For example, chimpanzees learn a lexigram and use it for naming of things (Matsuno et al. 2004). They have a numerical competence (Matuzawa 1981; Boysen and Hallberg 2000) and short-term memory comparable to that of adult human beings (Kawai and Matsuzawa 2000). They manufacture a tool and use it for problem-solving tasks (Tonooka et al. 1997). In a social context, they obtain foods by adjusting their own behavior to a situation, misleading others, and deception (Hirata and Matsuzawa 2001). Based on understanding a situation, chimpanzees undertake decision making and then act. These courses of processing sensory stimuli from their surroundings are essential for a wild chimpanzee to get what they want, solve problems, and, ultimately, to allow their survival and reproduction. For those reasons,

Great Ape Research Institute, Hayashibara Biomedical Laboratories, 952-2 Nu, Tamano, Okayama 706-0316, Japan

what they perceive, understand, and do can be considered as the bases of their lives.

1.2 Cognition and Animal Welfare

In contrast, captive chimpanzees live in an environment where human caretakers assure sufficient nutrition and safety throughout their lives. They are assured of their survival for long periods and reproduction of more offspring than their wild counterparts. However, such a captive condition differs vastly from and is much poorer than that in the wild in terms of exertion of cognitive ability. For captive chimpanzees, daily life requires no perception, understanding, or decision making for their survival and reproduction. For example, they need not move to explore a food environment because a human caretaker provides food for them on time each day. A human caretaker never serves things that cannot be eaten. Therefore, captive chimpanzees have no knowledge of, nor do they need to learn, what is edible; nor do they risk their survival on the need to pay attention to what might be poison. All foods served by a human caretaker are immediately edible without any other processing before consumption. They also need not cooperate with other group members for something to do. Although captive chimpanzees are not always completely safe, they face no risk of predation. On the other hand, the beginning of feeding in the wild condition starts with exploration of foods in their habitat. Even if a chimpanzee locates and accesses what it wants, it would be required to process a food as preparation for consumption. Sometimes chimpanzees might use a material as a tool and cooperate with other conspecifics to do something. Wild chimpanzees are targeted by predators and must remain alert to movements by neighboring chimpanzee groups. A captive condition, therefore, not only implies that sensory stimuli surrounding chimpanzees in their enclosure are poorer than those in the wild but that captive chimpanzees have no necessity to put sensory stimuli to some use for their survival and reproduction. This fact indicates that the processing of sensory stimuli surrounding captive chimpanzees is not essential for their existence.

Such an environment of captive chimpanzees seems to distort their behavior. Numerous studies have indicated that captive chimpanzees develop abnormal behavior with various behavioral types depending on their facilities (Hook et al. 2002). Coprophagy, urophagy, regurgitation/reingestion, unusual posturing, stereotypical behaviors such as rocking, and self-orality are attributable to a poor environment (Davenport and Menzel 1963; Berkson and Mason 1964; Davenport and Rogers 1968; Walsh et al. 1982; Capitanio 1986). Those behaviors are considered to be self-stimulation that chimpanzees adopt to cope with boredom and stress in a restricted environment. Capitanio (1986) pointed out that body rocking helps an infant compensate for the lack of motion stimulation normally provided by a mother.

From the viewpoint of animal welfare, promoting the psychological well-being of captive primates was addressed in the 1985 amendments to the Animal

Welfare Act (Animal Welfare Act 1985). For wild-derived species such as chimpanzees, a major goal of psychological well-being is to make the behavior of a captive individual comparable to that of the wild counterpart. This goal requires satisfying criteria such as the following: (1) coping with its physical and social environment, (2) engaging in species-typical behavior, (3) eliminating maladaptive and pathological behavior, and (4) maintaining a balance of temperament and the absence of chronic signs of distress (National Research Council 1998).

For these purposes, enormous efforts have been devoted to the practice of environmental enrichment (Segal 1989; Lutz and Novak 2005). Environmental enrichment is classifiable into several subcategories: feeding enrichment, physical structural enrichment, social enrichment, and sensory enrichment. For example, feeding enrichment includes hiding foods in an enclosure (Anderson and Chamove 1984) and introducing a puzzle feeder and an artificial termite mound (Nash 1982). Social enrichment is done for an individual to learn social skills for social interaction, copulation, and rearing of offspring (Bloomsmith and Baker 2001). In terms of animal cognition, some environmental enrichment programs provide an opportunity for animals to exert and train various cognitive abilities. Through the use of artificial termite mounds, for example, chimpanzees can learn how to use materials as tools. Social enrichment such as group living with a parous female helps breeding by nonparous females. Both tool use and care of offspring are behaviors that are based on the physical and social intelligence which a chimpanzee acquires throughout their life. Thereby, environmental enrichment enhances a condition that captive chimpanzees perceive, understand, and—make decisions on the environment surrounding them. Environmental enrichment implies the importance of a life that enhances the exertion of chimpanzees' cognitive abilities.

1.3 Cognitive Enrichment

Here, one conjecture arises. Chimpanzees have needs in their lives that stimulate their cognitive competence. In other words, they need an environment in which a chimpanzee fully expresses their cognitive ability through their life. Its enrichment can be called "cognitive enrichment." Such environmental enrichment is intended to maintain a captive environment for the exertion of their cognitive competence in various contexts of daily life. Few cognitive enrichment studies have been reported to date (Brooks 2004; Citrynell 1998), but the importance of an environment in the development of an animal's cognitive competence was pointed out long ago (Davenport et al. 1973). Nevertheless, whether a chimpanzee is seeking such an opportunity to behave based on cognitive competence remains unknown. The behavior of an animal's cognitive competence is roughly classifiable into two processes. One is related to a process to input information from an environment surrounding an animal. The other is a process to output an action onto an environment by an animal. This chapter specifically addresses chimpanzees' preferences and cognitive background on both input and output processes over their environment. We then discuss the implications for cognitive enrichment.

2 Preference and Recognition of Movies in Chimpanzees

2.1 Sensory Enrichment and Sensory Reinforcement

To compensate for the lack of visual, auditory, olfactory, and tactical stimuli from a captive environment, sensory enrichment has been widely implemented, especially for individual caged chimpanzees in laboratory facilities (Brent 2001). Chimpanzees in laboratories, for example, have few objects or scenes to view from their cages because a view from a cage is constant and less stimulating in itself. As an attempt at visual sensory enrichment, TV programs and movie clips are presented to caged chimpanzees. Bloomsmith and Lambeth (2000) reported that chimpanzees spend 38.4% of their time watching television when TV programs are presented. Brent and Stone (1996) described long-term effects of TV presentation as enrichment. They presented TV commercial video clips for about 23 months. Results indicated that chimpanzees actually watched television for 1.5% of the presented time. Chimpanzees' interest in watching television continued for a long time, even though their level of interest was not high.

Moreover, the preference for visual stimuli has been demonstrated as a result of enormous laboratory work, especially in the fields of operant conditioning. After the 1950s, sensory stimuli such as light and sound, deemed irrelevant to physiological needs such as water and food, were considered to function as a reinforcing stimulus and modify the frequency of certain emergent behaviors for various species. That is, a sensory stimulus derived from an animal's surrounding environment functions as a primary reinforcer in an operant response. The phenomenon is called sensory reinforcement (Matsuzawa 1981). Using the paradigm of sensory reinforcement, Fujita and Matsuzawa (1986) presented colored slides of pictures to a chimpanzee, Ai, without any food rewards. Ai was able to choose a picture to see continuously if she repeatedly touched a button within 10 s after the previous picture was released. The result indicated that Ai demonstrably viewed pictures that showed humans longer and discriminated pictures portraying humans from those which showed no humans. The chimpanzee chose to look at a particular picture by preference, not for a food reward.

2.2 Cognitive Studies of Chimpanzees Using Movie Stimuli

Examples of sensory enrichment and sensory reinforcement imply that chimpanzees preferred to view some visual stimuli. The way of stimulus presentation in the examples is apparently beyond the natural context in their life such as feeding, interacting with conspecifics, and so on. Why did chimpanzees show their interest in watching pictures and movie clips? Does a TV presentation of sensory enrichment function identically for cognition as if chimpanzees see a scene in a daily life? Knowing how chimpanzees perceive visual stimuli can present some implications for further investigation.

To determine how animals perceive and retain what they see in daily life, one method used to investigate memory in animals is the serial probe recognition task (Sands and Wright 1980). Wright et al. (1985) demonstrated that pigeons, monkeys, and humans all showed serial position effect in tests of list memory. Lists consist of several discrete items. Thus, it might be suggested that a sight in daily life comprises a series of scenes. The processes involved in perceiving, retaining, and retrieving sights in daily life might be expected to be similar to processes of perceiving, retaining, and retrieving moving images on a monitor. Considering, for example, digital video recording, which consists of 30 still frames per second, the memory of a movie might approximate the memory of a list comprising vastly numerous still frame pictures. However, qualitative differences exist between memory processes of a list and those for a movie. We perceive several distinct parts to a movie based on our subjective feeling. Additionally, spatial and temporal changes in the composition of a movie occur continuously over a series of scenes. These characteristics might engender different processes in the memory of movies and lists. The study of movie memory provides an opportunity to investigate how animals perceive and retain sights in their daily lives from the standpoint of comparative cognitive science.

Although simple presentations of TV programs and movie clips have been done in large environmental enrichment programs (Rumbaugh et al. 1989), cognitive studies using movies are few. Premack and Woodruff (1978) presented 30-min-long movie clips of a 14-year-old chimpanzee, Sara. The movies in the experiment depicted several scenes in which a human encounters a problem. After showing a movie clip, several photographs of the scenes of problem solving appeared. In the task, Sara was able to choose a correct photograph. Menzel et al. (1985) showed that chimpanzees were able to obtain a food reward that was placed at an opposite side of a wall by using a hall, along with visual cues from a mirror and live video images shown on a television. Although a video image was shown as flipped vertically and/or horizontally, chimpanzees were able to access a piece of fruit.

Itakura and Matsuzawa (1993) examined a chimpanzee's ability of acquiring personal pronouns. In the training trial of the experiment, a chimpanzee, Ai, was presented a movie clip that human A approach to human B. After training, Ai became able to describe subjects (human A), actions, and objects (human B) after viewing a video clip. Eddy et al. (1996) investigated self-recognition of chimpanzees. In the experiment, a mirror image and a videotape of chimpanzees were presented for 20 min to two groups of chimpanzees, one aged 3 years and the other aged 7 to 10 years. Both groups of chimpanzees responded to a mirror and a video image similarly: they glowered and moved their hands and feet. To the contrary, body exploration was apparent only in chimpanzees of 7 to 10 years of age toward a mirror image. Moreover, O'Connell and Dunbar (2005) examined the ability of understanding causality. They presented movie clips of food, humans, and wild chimpanzees in both normal and reverse film sequences using habituation–dishabituation paradigms. The viewing time of movies by dishabituation was longer in the transition from a normal to a reverse sequence than in the transition from reverse to normal. Those results suggest that chimpanzees responded to causality in a movie sequence rather than just the change in the movie's composition.

Cognitive study of movies on chimpanzees consistently demonstrates that chimpanzees understand the movies. They do not see a movie as colors flickering on a screen. They perceive and represent an image from the movie's contents. Again, do chimpanzees watch a movie just as they see a scene in daily life? Therefore, Morimura and Matsuzawa (2001) investigated the process in the memory of a movie clip for chimpanzees.

2.3 Memory of Movies by Chimpanzees

This study was intended to investigate memory processes of movies by chimpanzees. First, using a movie-to-movie matching-to-sample task, the ability of chimpanzees to discriminate movies was tested in experiment 1. Second, using a movie-to-still matching-to-sample task, the movie recognition task was tested in experiment 2.

Four adult chimpanzees at the Primate Research Institute of Kyoto University were the subjects. All four individuals were female (Ai, 21 years; Pendesa, 21 years; Chloe, 17 years; Pan, 14 years). Before this study, all had participated in various experiments, including matching-to-sample tasks (Matsuzawa 2003). Only Ai had prior experience with tasks involving movie stimuli (Itakura and Matsuzawa 1993). Maintenance and experimental conditions conformed to the Guide for the Care and Use of Laboratory Primates in the Primate Research Institute of Kyoto University.

2.4 Experiment 1: Movie-to-Movie Matching-to-Sample Task

The movie task premises that chimpanzees can perceive a movie. It is assumed that if chimpanzees perceive a movie accurately, they should be able to discern one movie from another. To date, few studies have directly demonstrated that capability. Therefore, this study was intended to investigate chimpanzees' ability to perform the movie-to-movie matching-to-sample task. Furthermore, to examine the process by which chimpanzees accomplish the task on their very first encounter with movie images, this experiment was conducted without any prior training of subjects.

Fig. 1. A subject performs the movie matching-to-sample task

Experiments were conducted in an experimental booth (2.5 m × 2.0 m × 2.0 m) for chimpanzees (Fig. 1). The booth had a 21-inch color monitor with a touch-sensitive panel (SMT2; MicroTouch) on one wall. A universal feeder (BFU-310; Biomedica) was placed above the monitor. This device delivered a piece of apple or a raisin as a reward to the food tray placed under the monitor. A personal computer (HP-808; A ONE) was used to control the experimental events and to record experimental data. Visual Basic language (Microsoft) was used for programming the experimental tasks.

The source for the experimental stimuli was a recording of wild chimpanzee behavior in the Mahale Mountains National Park, Tanzania (Nishida 1990). Stimuli comprised ten color movie clips. The duration of each stimulus was 5 s. None of the ten movie clips contained overlapping scenes. Each was a continuous scene with few compositional changes. These clips were translated separately into digital movie files in MPEG format. They were presented against a dark background as stimuli 6.5 cm wide and 5.0 cm high on the monitor. The positions of sample and comparison stimuli on the monitor were constant in every trial.

A trial proceeded as illustrated in Fig. 2. At the start of a trial, a white circle, approximately 4.0 cm in diameter, appeared in the lower right area of the monitor. After a subject touched the white circle on the monitor, the circle disappeared and a sample stimulus appeared in the center of the lower half of the monitor: a 5-s movie clip then began to play immediately. Touching the sample had no effect during this playback, thereby allowing the full-length presentation of the clip to the subject. After the movie finished, if a subject touched the sample stimulus, the sample stimulus disappeared and two movies as comparison stimuli immediately appeared in the upper right and left corners and played simultaneously for 5 s. Touching the comparisons also had no effect during the playback. In each trial, if a subject touched the correct stimulus, a chime sounded for 1 s and food reward (a piece of apple or a raisin) was delivered. A buzzer sounded for 1 s and was followed by the next trial if a subject touched the incorrect stimulus. To examine all possible combinations of two comparison stimuli

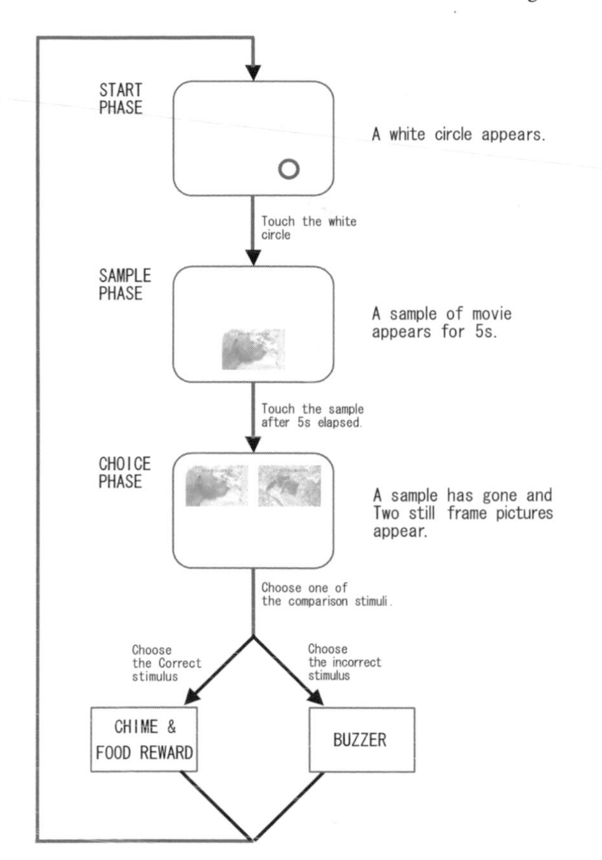

Fig. 2. Schematic representation of the procedure used in the movie matching-to-sample task

($10 \times 9 = 90$ patterns) and to counterbalance the position effects (right and left), all subjects received two sessions of 90 trials each.

Figure 3 shows the cumulative percentage of correct responses scored by each of four subjects. Cumulative accuracy fluctuated strongly in the first half but was relatively steady in the latter half. This fact suggests that chimpanzees learned to discriminate among the movies. Ai began to respond correctly and reliably after the first seven trials of the first session. In other words, Ai was able to discriminate between stimuli in the movie-to-movie matching-to-sample task from the beginning. We set a statistically significant criterion of $P < 0.05$ in which animals completely learned the task when the cumulative accuracy was under 5% of the level in the binomial test. According to that definition, Ai continued to make correct choices and satisfied the criterion at the 5th trial: Chloe was at the 16th, Pan at the 87th, and Pendesa at the 98th trial. Of the four chimpanzees, Ai learned the task most quickly, followed by Chloe, Pan, and Pendesa. The overall final proportions of correct responses in the movie-to-movie matching-to-sample task were 83.3% (Ai), 78.9% (Chloe), 65.0% (Pan), and 66.7% (Pendesa), all of which were significantly higher than the chance level (Ai, $\chi^2 = 45.000$, $P < 0.001$; Chloe, $\chi^2 = 32.780$, $P < 0.001$; Pan, $\chi^2 = 11.749$, $P < 0.001$; Pendesa, $\chi^2 = 10.286$, $P = 0.001$).

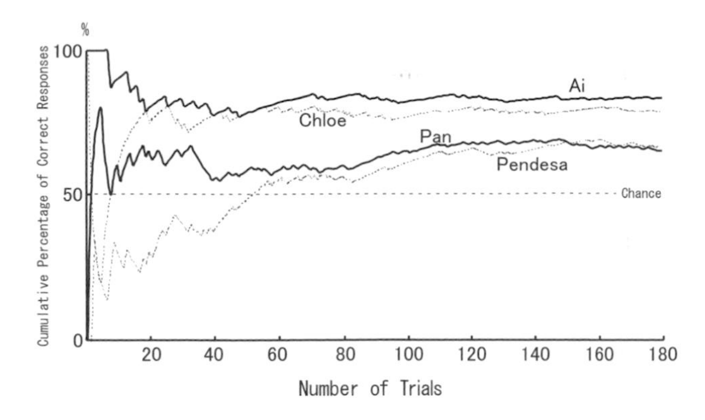

Fig. 3. Cumulative percentage of correct responses for all chimpanzees for the movie-to-movie matching-to-sample task of experiment 1. The percentage at each trial indicates the average accuracy, as calculated from the first to the targeted trial

All chimpanzees demonstrated the ability to discriminate movies from the very first session onward. These results demonstrate that naive chimpanzees soon learn to discriminate among movie clips even when novel movies are presented.

2.5 Experiment 2: Movie-to-Still Matching-to-Sample Task

Experiment 2 investigated the chimpanzees' ability to recall a movie. Using a movie-to-still matching-to-sample task, the movie recognition task was tested using all four subjects. In this recognition task, following the presentation of a movie clip as the sample stimulus, two still frame pictures were presented as comparison stimuli. One was a still frame picture taken from the sample stimulus and the other was a still frame picture taken from another movie stimulus. We assumed that the more accurately chimpanzees retained movie memory, the more frequently they would select the correct comparison stimulus. Based on the chimpanzees' responses in this task, we examined the extent to which chimpanzees retained the contents of a movie.

Furthermore, if scenes within a movie change drastically and/or frequently within a short period, the load to retain it would be expected to increase. Therefore, the composition is one possible characteristic that is responsible for increasing the load on memory processing. To investigate compositional effects on the memory of a movie, Experiment 2 examined the animals' performance under two conditions: the continuous-movie condition (continuous condition) and the discrete-movie condition (discrete condition). In the continuous condition, stimuli were movies in which composition changed gradually over time. On the other hand, in the discrete condition, stimuli were movies in which composition included sudden changes from scene to scene. By comparing the respective performances under these two conditions, the effects of the movies' characteristics on memory processes were examined.

The subjects and the apparatus were identical to those in experiment 1. The sample stimuli were color movie clips in MPEG format. In the continuous condition, the sample stimuli were the same ten movies as those used in experiment 1 because the movies had no sudden change in composition. All movie stimuli consisted of one continuous scene each. In the discrete condition, ten new movie stimuli were created using the same wild chimpanzee footage as that in experiment 1 as the source. All these novel stimuli involved sudden changes in scenes, of which the time of occurrence and the frequency varied depending on the movie (Fig. 4). In all other aspects they resembled the movie stimuli of experiment 1.

The comparison stimuli were color still frame pictures in JPEG format (Fig. 4; 10 movies × 6 still frame pictures). These color pictures were taken from all ten movie stimuli in each condition. Each movie stimulus file was sampled repeatedly at 1-s intervals from 0 s to 5 s, creating the selected six still frame images (0-s, 1 s, 2 s, 3 s, 4 s, and 5 s). These still frame images were translated into color photographs in JPEG format. The same procedure was applied to all movie stimuli in both conditions. Consequently, 60 color still frame pictures for each condition were available as comparison stimuli.

The basic procedure was identical to that in experiment 1, except for the following points. In experiment 2, sample stimuli were movies and comparison stimuli were still-frame pictures. In all trials, after showing a 5-s movie as the sample stimulus, two still frame pictures were presented as comparison stimuli. One was the correct comparison stimulus: it depicted a scene that was included in the sample movie clip. The other was an incorrect comparison stimulus, a scene taken from one of the other nine movies.

Of the two conditions, the continuous condition was tested first, followed by the discrete condition. Within any given session, all comparison stimuli for recognition were of a fixed frame position. For example, in one session, still pictures showing the 5th s frames of all ten movie samples were used as comparison stimuli throughout. To examine all possible combinations of two comparison stimuli ($10 \times 9 = 90$ patterns) and to counterbalance the position effects (right and left), testing for one frame position consisted of two sessions of 90 trials each. These procedures were applied to all six frame positions from the 0 s to the 5 s. Consequently, each subject received 12 sessions in all for each condition. The order of testing different frame positions changed randomly according to subjects and conditions.

In the discrete condition, we noted a serial position effect in the memory of movie clips by chimpanzees. The results are plotted in Fig. 5 according to the performance at each frame position under both conditions for all the chimpanzees. Comparing the two conditions, performances differ markedly in terms of the shape of the serial position functions. In the continuous condition, accuracy was uniformly high in all frame positions [analysis of variance (ANOVA); $F\,(5,18) = 0.096$, $P = 0.996$]. In the discrete condition, the serial position functions showed good performance at frame positions nearer the end—a recency effect [ANOVA; $F\,(5,18) = 44.460$, $P < 0.0001$]. Repeated-measures ANOVA on the two conditions was applied at six frame positions each. Results of analyses showed a significant difference between continuous and discrete conditions

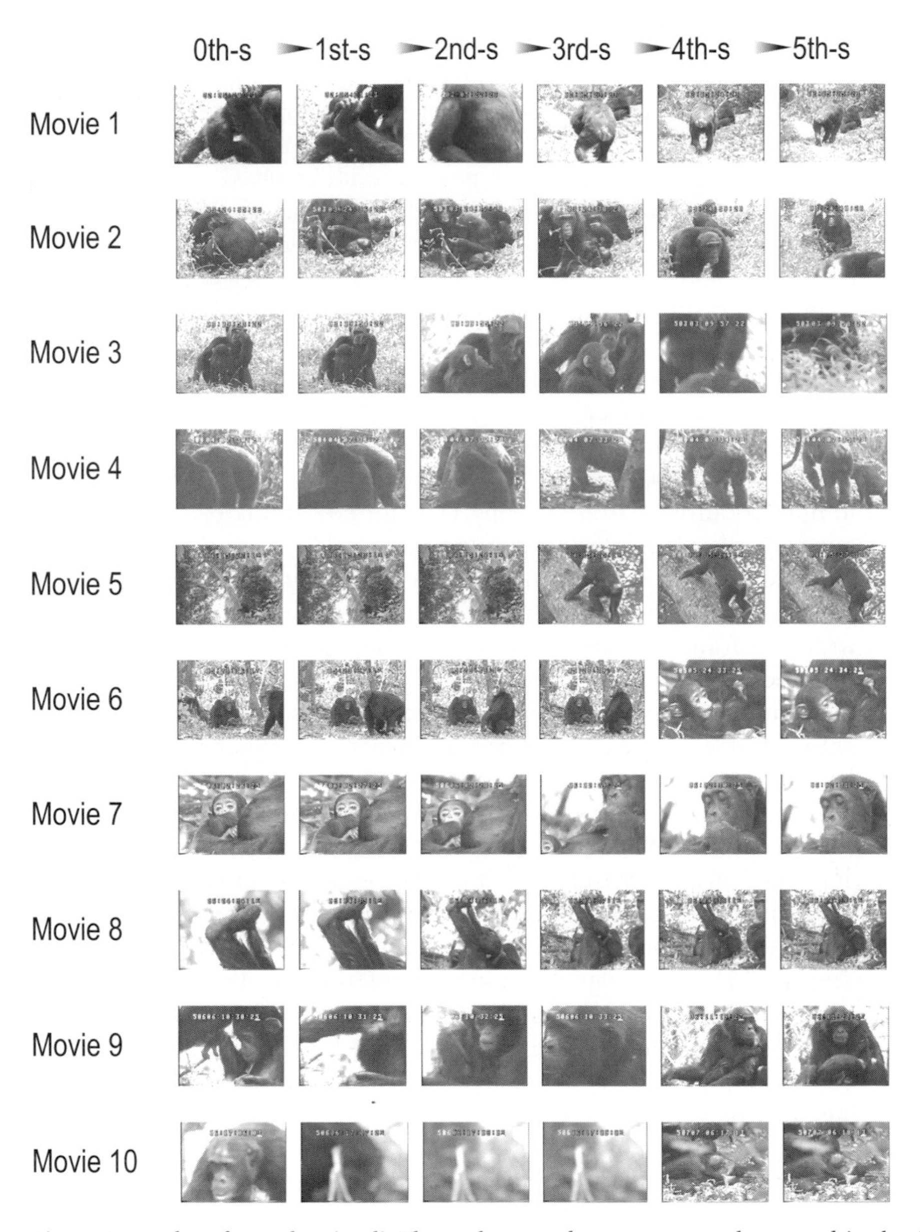

Fig. 4. Examples of sample stimuli. These photographs were among those used in the discrete-movie condition of experiment 2. Video footage taken from The Wild Chimpanzees at Mahale Mountains, by Miho Nakamura, 1997. Copyright 1997 by the ANC Corporation, Japan

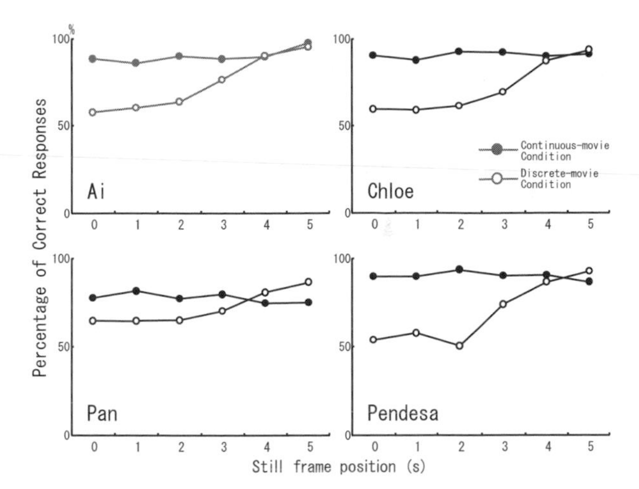

Fig. 5. Serial position functions showing memory performance at six frame positions under both conditions in experiment 2

[$F(1, 30) = 21.62$, $P = 0.004$]. Statistical support for serial position effects in the memory of movies was provided only in the discrete-movie condition. This tendency was observed consistently in all four subjects.

The differences in the subjects' performance in the two conditions suggested that the characteristics of composition in the movies might affect their memory processing. The occurrence of sudden changes of movies was a major difference between the continuous condition and the discrete condition. In the continuous condition, no sudden change was apparent in the composition within a movie. On the other hand, in the discrete condition, sudden changes occurred one or two times within each movie. The decline in accuracy in first-half frame positions (0 s, 1 s, and 2 s) in the discrete condition might have resulted from the drastic change in scene compositions within a movie. On the other hand, no sudden change of composition was apparent in the movies of the continuous condition, and the accuracy at each frame position was correspondingly similar among the six frame positions. Furthermore, even in the discrete condition, little difference existed in the compositions of the 4-s and 5-s frame positions. Consequently, the compositional similarity is likely to be a major reason why all four subjects' accuracy on the task was similarly high in all six frame positions of the continuous condition and in the last two frame positions of the discrete condition.

2.6 Movie Perception in Chimpanzees

This study demonstrated the following two points. (1) Chimpanzees can discriminate between movie clips in their very first encounter with such stimuli. (2) Characteristics of composition in movies affect the chimpanzees' memory processes. The recency effect appeared only in the discrete condition. This result suggests that when the scene composition in movies includes sudden drastic changes, chimpanzees retained the movie clips under a similar memory process as that used for items of a list. Wright et al. (1985) demonstrated that pigeons, monkeys, and humans consistently showed the recency effect when required to respond with 0-s delay, that is, immediately after the end of the list. In the present study, the recognition of a still frame picture was conducted with 0-s delay after presentation of a movie clip as the sample stimulus. The recency effect in the discrete condition was similar to that observed in studies of memory for list items. These results suggest that chimpanzees might retain movies along with the temporal order of the movies' constituent scenes.

Moreover, the highly accurate performance in the continuous condition of experiment 2 supports the notion that chimpanzees recognize a still frame picture of movie based on the compositional characteristics of movies. In the continuous condition, stimuli were movies in which scene compositions changed gradually. Even if chimpanzees retain only the second half of a movie (3 s, 4 s, and 5 s) in memory, they can respond correctly in the recognition test about the first half of a movie clip (0 s, 1 s, and 2 s) based on compositional characteristics. In fact, all four subjects showed consistent high accuracy in the continuous condition. Overall, chimpanzees probably responded on the basis of the temporal order as well as compositional characteristics of our movie stimuli.

In conclusion, these movie experiments demonstrated that chimpanzees retain movie stimuli based on their characteristics such as the composition and the temporal order of its constituent scenes. Results suggest the possibility that chimpanzees understand a movie clip as a film sequence similarly to humans. Therefore, we conclude that chimpanzees prefer to watch television and movie clips with some understanding of what they see.

3 Voluntary Work in Chimpanzees

3.1 Voluntary Work in the Laboratory

In the wild, when, where, and how a chimpanzee acts are determined by individuals' choices. However, under captive conditions, human caretakers largely control the animals' behavior. To permit the behavior of captive animals to be "natural" through environmental enrichment programs, not only behavioral repertoires and activity budgets but also the processes through which an animal decides to take action, that is, the voluntary nature of the behavior, should be considered to simulate the wild state.

In a preference test, some animals choose to work to get food rather than to obtain it without work. This phenomena is known as contra-freeloading in the field of operant conditioning. It was found for the first time in pigeons and rats that pressed a disk or a lever to get food rewards when identical food was freely available (Neuringer 1969). Menzel (1991) reported that chimpanzees went out of their way to solve a discrimination task for food rewards, even though they were simultaneously able to obtain the same food freely. Bonnet macaques also preferred to solve computerized video tasks whether receiving food or not

(Washburn and Hopkins 1994). Caged pigtailed macaques and rhesus macaques tried to get a food from a puzzle feeder rather than a freely accessible food box (O'Connor and Reinhardt 1994; Reinhardt 1994). These animals were allowed an alternative to performing a task to get food. Nevertheless, they actively chose to participate in the task itself rather than to acquire a food primarily.

These findings indicate that chimpanzees and other primate species prefer to do something before eating, even though solving a computer task and a puzzle can be considered to be a load in terms of cognition and motor skill. This finding might be another example that chimpanzees are seeking a life that stimulates their cognitive competence—a process to output an action onto their environment. However, a computer task and a puzzle feeder are devices that are artificially set for the experiments. Whether a chimpanzee prefers such a load before eating in a more natural context is not clear. A feeding enrichment program is available to investigate the chimpanzee's choice of voluntary work in the natural context.

A number of feeding enrichment programs have been used. Among these, unpredictable feeding schedules (Bloomsmith and Lambeth 1995), increasing the number of food provisions (Morimura and Ueno 1999), dispersion of food distribution (Grief et al. 1992), and the hiding of food in the animals' enclosures (Anderson and Chamove 1984) all resulted in animals choosing when and/or where they fed. Moreover, an increased variety of foods (Glick-Bauer 1997) and introduction of an artificial termite mound and a puzzle feeder (Nash 1982; Gilloux et al. 1992) might be seen as feeding enrichment, ensuring the voluntary nature of behavior in the processing aspect of feeding, in other words, how an animal feeds. However, most usual feeding enrichments permit an animal to use only one type of processing for each food. In brief, interaction with the environment is always limited to one particular procedure, whereas animals in the wild have several choices to obtain a food, and freely select their behavior. Tool-using behavior of chimpanzees in an outdoor enclosure was investigated under experimental settings to determine whether it was applicable in a more naturalistic context (Morimura 2003).

The subjects were four infant chimpanzees at the Great Ape Research Institute (GARI) of Hayashibara Biomedical Laboratories. Two were males (Loi, 5 years; Zamba, 5 years), and the others were females (Tsubaki, 4 years; Mizuki, 3 years). Although all had participated in experiments before this study, including matching-to-sample tasks, the present experiment represents their first experience at a tool-using task. The care and use of the chimpanzees adhered to the Guide for the Care and Use of Great Apes of the Great Ape Research Institute, Hayashibara Biomedical Laboratories.

3.2 Experiment 3: Juice Drinking Task in Naturalistic Context

This study permitted voluntary tool-using behavior in chimpanzees by experimentally enhancing freedom of choice in the processing aspect of feeding. Wild chimpanzees at Bossou, Guinea use leaves for drinking water inside the natural hollow of a tree (Sugiyama 1995; Tonooka 2001) As a simulation of this water-drinking behavior, tube feeders filled with orange juice were presented to chimpanzees. In this setup, chimpanzees were able to access juice by dipping their hands directly into the liquid or by using various objects as tools. Using similar tube feeders, Tonooka et al. (1997) precisely investigated the process of acquiring tool-using behavior for individuals and its transmission within a group of chimpanzees. They found that this type of tube feeder induced tool-using behavior of captive chimpanzees for accessing juice in a similar context in the wild. Under this condition, when chimpanzees use tools actively, tool-using behavior can be regarded as a spontaneous behavior in a "natural" context.

Figure 6 shows a representative transparent acrylic tube (30 cm long by 10 cm wide) that served as the juice-holding device. The tube's top and bottom were covered with acrylic boards. A hole halfway along the length of the tube provided chimpanzees with access to the juice. Tubes were attached to the concrete wall with stainless steel parts. The feeders were positioned such that the top of the device was 50 cm from the ground. Four separate feeders were set up over a distance of 3 m along the wall to deter monopolization by dominant individuals.

Before the experiment, a preliminary test was carried out to assess tool use by the chimpanzees. In this test, a variant type of tube feeder, which was filled with orange juice, was presented to the same chimpanzees. The mouth of this tube feeder was too narrow for chimpanzees to reach through with their hands. During the 60-min pretest sessions, chimpanzees were allowed to behave freely. Observations were carried out from a location that was hidden from the chimpanzees to prevent any observer influence on the subjects. Pretest results showed that all chimpanzees used tools to obtain juice. As sessions progressed, chimpanzees came to favor a straw as the principal tool because of its ready availability.

The experiment was conducted in the outdoor enclosure shown in Fig. 7. One 60-min experimental session was carried out per day; it was repeated for 30

Fig. 6. The juice delivery apparatus used in experiment 3

Fig. 7. Outdoor enclosure showing the activities of chimpanzees

sessions. Each feeder was filled with 1 l of commercially available orange juice. The orange juice was diluted with an equivalent volume of water to prevent excess intake of sugar by the chimpanzees. The start of the session was defined as the moment the chimpanzees entered the outdoor enclosure. During the course of an experimental session, chimpanzees were allowed to behave freely. Because the feeders were removed at the end of each session, chimpanzees had no opportunity to manipulate the feeders at times other than during the experimental period. The total weight of the remaining liquid was measured to determine if any juice was left over in any of the four feeders. Chimpanzee behavior was also recorded using digital video recording devices (DCR-TRV8; Sony). The experimenter remained invisible to the chimpanzees during the trials to minimize observer influence on the subjects' behavior. Data were gathered by reviewing the videotapes and were analyzed quantitatively in terms of the methods used by chimpanzees to obtain juice from the device, the frequency of each method, and temporal patterns in its occurrence.

Criteria for tool-using behavior were the following. In line with Inoue-Nakamura and Matsuzawa's (1997) classification of behaviors necessary for nut cracking, tool-using behavior to access juice from the tube feeders was divided into four processes: (1) picking up an object by hand to be used as a tool, (2) inserting the chosen object into a feeder, (3) withdrawing juice using the object, and (4) transferring the object into the mouth. Tool-using behavior by the chimpanzees was regarded as successful when they performed these four behaviors in the appropriate order.

As a result, the methods used by the chimpanzees to extract juice from the feeders were classified into one of two categories: (1) obtaining juice directly using body parts, and (2) obtaining juice with the aid of tools. The first type can be further subdivided into two categories: (a) drinking juice or licking drops splattered on the feeders' exterior and (b) accessing juice by dipping the hand into the feeder. Among these observed methods, chimpanzees most commonly employed tools to obtain juice from the feeders (Fig. 8). All four subjects showed

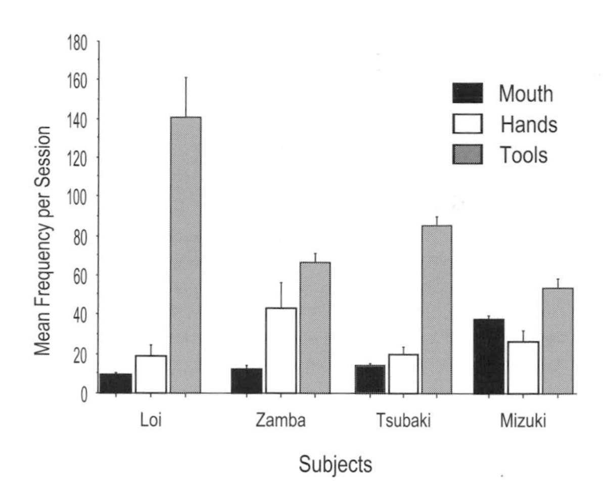

Fig. 8. Mean frequency of different methods of juice access by chimpanzees

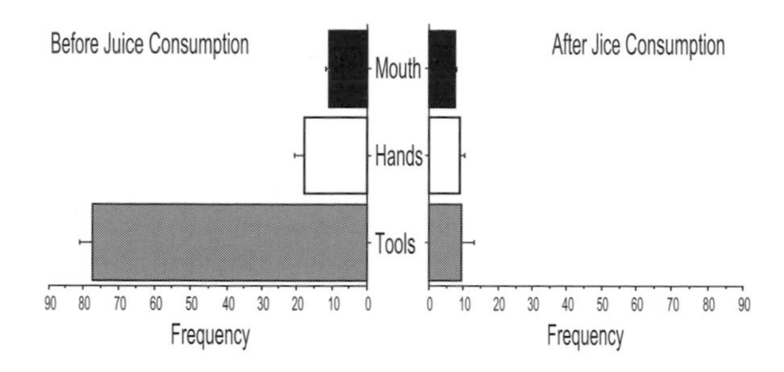

Fig. 9. Comparison of the mean frequency of the use of mouth, hands, and tools by four chimpanzees between the periods of before juice consumption (initial 20 min) and after juice consumption (final 40 min) of trials

this tendency. Chimpanzees most frequently used a straw as the tool, whereas other materials such as a piece of cloth and timber were also used on occasion. Among these objects, a straw was used in 98.7% of all successful cases of tool-using behavior. The use of feeders peaked immediately after the beginning of trials; that pattern decreased rapidly with time. In all experimental sessions, chimpanzees consumed the juice supplied in the containers within 20 min of the beginning of the trial. This fact reflected that chimpanzees actively used the feeders as long as juice remained in the feeders. However, the use of feeders tended to lessen as the volume of remaining juice in the tubes decreased. However, even after 20 min had elapsed, subjects intermittently visited the feeders and did not cease using them completely. Chimpanzees were observed to visit the feeders in turn and place straws into the tubes to wipe and lick the small amounts that they were still able to retrieve. The use of feeders by chimpanzees differed markedly between the periods of before juice consumption (initial 20 min) and after juice consumption (final 40 min) of an experimental session, in terms of the use of mouths, hands, and tools (Fig. 9). Results of repeated-measures ANOVA to compare the frequency of the use of mouth, hands, and tools before juice consumption showed that tools were used significantly more often than were other methods [F (3, 2) = 22.051, $P < 0.0001$].

Nevertheless, after juice consumption, no statistical difference was apparent among the frequencies of the three methods [F (3, 2) = 1.988, P = 0.0667]. This fact indicates that chimpanzees used tools selectively for accessing juice.

3.3 Spontaneous Tool Use of Chimpanzees

This study demonstrated that when chimpanzees are presented with various methods to obtain orange juice, such as mouths, hands, and tools, they employ all available choices. Chimpanzees might concentrate on a preferred method when multiple methods are available as options. However, whereas chimpanzees in this study mainly accessed juice using tools, they also used other procedures. Furthermore, the use of mouth and hands continued to occur intermittently until the end of the experiment, regardless of the presence or absence of juice. Consequently, chimpanzees opted for one among several choices in all situations.

It can be inferred that this experiment succeeded in giving chimpanzees a chance to feed while providing them an opportunity for free choice. As a result, it stimulated the chimpanzees to exhibit tool-using behavior, which is a behavior that is shown by chimpanzees in the wild. The results of this study also support the idea that environmental enrichment with highly positive results can simultaneously enhance freedom of choice in animal behavior (Menzel 1991; O'Connor and Reinhardt 1994).

In conclusion, chimpanzees preferentially use a tool for juice consumption. Once they consumed almost all the juice, they intermittently attempted to access juice by every method including mouths, hands, and tools equally. Although chimpanzees could drink juice in several ways in the experimental condition, they concentrated on tool using. Therefore, we conclude that chimpanzees prefer to exert their cognitive competence in the context of output process interacting with their environment.

4 Discussion and Conclusions

4.1 What Is Cognitive Enrichment?

Results of these experiments indicated that chimpanzees actively choose to exert their cognitive ability when they see a movie and when they work for food using tools. Chimpanzees do not watch a movie clip just because they are interested in colors flickering on the screen. Chimpanzees see it with the understanding of it as a movie, as do humans. They recognize the movie's contents, its positional relationships of left, right, top, and bottom, and causal relationships. They also retain memories of a movie based on its contents and composition. They pay attention to the movie and view it longer than usual when a scene of an unusual causal relationship is presented in the movie clip. Moreover, chimpanzees do not use tools out of a necessity to get food. They prefer to drink juice using wads of straw as a tool rather than by hand. They manufactured wads of straw with an understanding of how it works and how to use it effectively for drinking juice. Therefore, chimpanzees prefer to see visual stimuli and to use their motor skills to affect their surroundings with perception, understanding, and decision making. The experiments with movies, sensory enrichment, and sensory reinforcement are examples of input processes from which a chimpanzee receives sensory stimuli from a surrounding environment. Experiments to assess spontaneous tool use, voluntary work, and contra-freeloading are examples of output processes that a chimpanzee uses to operate on an environment. The two examples show that chimpanzees prefer the exertion of their cognitive competence in both input and output processes interacting with their environment.

The daily life of wild chimpanzees is deeply connected to the environment surrounding them based on their own cognitive ability. Their exertion of cognitive competence can be considered their own strategy to construct a connection with their surroundings in various contexts of daily life. Their respective preferences, as revealed using visual stimuli and motor skills, can be regarded as playing a role of starter for the construction of both input and output connections with their environment. This explanation is extremely similar to the example of contra-freeloading. Inglis et al. (1997) pointed out that contra-freeloading could be adaptive for animals living in changing environment to gather information from their surroundings. Therefore, environmental enrichment should incorporate chimpanzees' needs for exerting their cognition. This approach implies the importance of cognitive enrichment, which primarily focuses on ensuring the exertion of chimpanzee's cognitive competence onto its surroundings in both input and output processes. That is, a captive condition should not only be an environment that stimulates an animal in various contexts of daily life but also an environment that enhances operation on surroundings by animals there. When a captive environment possesses both functions, at least, it is expected that a chimpanzee can construct a functional relationship with an environment. Natural habitats of chimpanzees cannot be fully reconstructed in a captive condition. Alternatively, the functional relationship between chimpanzees and their environment in the wild might be transformed into maintaining the condition of captive chimpanzees. An approach of cognitive enrichment would provide substantial progress along this course.

Moreover, cognitive enrichment provides a central feature of welfare that is applicable to all captive chimpanzee individuals. Cognitive competence differs from individual to individual. Individual differences in their competence arise from the different developmental stages, ages, disease, and disability. Among captive chimpanzees, some individuals exist, for example, who have difficulty using tools and interacting with other conspecifics in their own group as a complication of their rearing history (Brent et al. 1995). Compared to mother-reared individuals, nursery-reared chimpanzees display some behavioral deficits including inadequate copulation skill (King and Mellen 1994) and maternal incompetence (Rogers and Davenport 1970). Maintaining such individuals in conditions that are comparable to wild conditions puts them at risk of injury and disease. Individual differences in cognitive competence and its deficits should also be considered from the viewpoint of animal welfare. The approach of cognitive enrichment might provide a standard that is applicable to different individuals. Assuring the exertion of cognitive competence is comparable to satisfying the individual's desire for something to do on its own way in various contexts of daily life. Consequently, the concept of cognitive enrichment allows that functional relationships between chimpanzee individuals and their environments might differ among individuals. For this reason, cognitive enrichment provides an approach to welfare that involves an animal's entire resources.

4.2 Cognitive Enrichment for Satisfying Psychological Well-Being of Captive Chimpanzees

Chimpanzees are likely to utter loud food calls when they locate a tree that is loaded with fruit (Goodall 1986). A chimpanzee, Loi in GARI, found food and ran around repeatedly while eating it. In the same situation, Zamba gathered up almost all the food at one place and made a bed-like circle of food around him before eating (Fig. 10). Their frolicking behavior with laughter related to food seems to show that they were satisfied mentally before they were satisfied physically.

Similarly to the example given above, some chimpanzees' behavior is not linked directly to their fitness through survival and reproduction. Watching movies and using tools in experimental situations are not necessary activities. Rather, the chimpanzees do it by choice, implying that their need for behavior seems not to be goal oriented but process oriented. In feeding, resting, interacting with conspecifics, cooperating with others, and so on, chimpanzees behave in their own way according to individual preferences during the course of accomplishing their ends. Another importance of cognitive enrichment implies

Fig. 10. When a chimpanzee, Zamba, found an enormous amount of food in an experiment, he made a bed-like circle around himself at the first setout. Thereafter, he devoted his attention to eating food of his own choice. (Photo by GARI)

a need for ensuring freedom on their way of behavior in captive chimpanzees. The findings of these experiments, especially the experiment of spontaneous tool use, indicated that ensuring their behavioral freedom might engender comparable behavior of captive chimpanzees to their wild counterparts. Consequently, pathological behavior such as abnormal behavior and chronic distress are expected to be diminished or eliminated. Promoting the psychological well-being of captive chimpanzees might be comparable with the consideration of individuals' way of behavior based on their cognitive ability and their own preferences. Therefore, cognitive enrichment is essential to accomplish the psychological well-being of captive chimpanzees.

Moreover, cognitive competence of various animals in captivity increasingly draws a high degree of attention from the viewpoint of animal welfare (Milgram 2003; Toates 2004). The knowledge of cognitive enrichment is expected to expand to other captive animals such as zoo animals, laboratory animals, domestic animals, and companion animals for improvement of their welfare. However, studies of cognitive enrichment have just been launched. What kind of cognitive competence they have and when and how they use it in daily life remain unknown. Further study of cognitive enrichment is needed to provide a linkage among various disciplines, especially comparative cognitive science. It is not an exaggeration to say that the beginning of studies of animal welfare starts with an understanding of animals themselves. The cognitive enrichment approach would provide concrete steps to accomplish animal welfare entailing an animal's entire resources.

Acknowledgments

The present study was financed by grants from the Ministry of Education, Science, Sports, and Culture of Japan (nos. 07102010 and 12002009) to Tetsuro Matsuzawa and a Toyota Foundation Grant (99-A-283 and D00-A-207) to Naruki Morimura. I wish to thank Drs. T. Matsuzawa and Y. Ueno of Kyoto University for their support and helpful advice offered during this research project and valuable suggestions on the manuscript. I also thank the staff members both of the section of Language and Intelligence of Primate Research Institute, Kyoto University and the Great Ape Research Institute, Hayashibara Biomedical Laboratories, for their kind help during the study.

References

Anderson JR, Chamove AS (1984) Allowing captive primates to forage. In: Standards in laboratory animal management. Symposium proceedings, vol 2. The Universities Federation for Animal Welfare, Potters Bar, pp 253–256

Animal Welfare Act 1985. Animal Welfare Act as amended (7 USC, 2131–2156)

Berkson G, Mason WA (1964) Stereotyped behaviors of chimpanzees: relation to general arousal and alternative activities. Percept Motor Skills 19:635–652

Bloomsmith MA, Baker KC (2001) Social management of captive chimpanzees. In: Brent L (ed) Care and management of captive chimpanzees. American Society of Primatologists. San Antonio, pp 205–241

Bloomsmith MA, Lambeth SP (1995) Effects of predictable versus unpredictable feeding schedules on chimpanzee behavior. Appl Anim Behav Sci 44:65–74

Bloomsmith MA, Lambeth SP (2000) Videotapes as enrichment for captive chimpanzees (*Pan troglodytes*). Zoo Biol 19:541–551

Boysen ST, Hallberg KI (2000) Primate numerical competence: contributions toward understanding nonhuman cognition. Cogn Sci 24:423–443

Brent L (2001) Behavior and environmental enrichment of individually housed chimpanzees. In: Brent L (ed) Care and management of captive chimpanzees. American Society of Primatologists, San Antonio, pp 147–171

Brent L, Stone AM (1996) Long-term use of televisions, balls, and mirrors as enrichment for paired and singly caged chimpanzees. Am J Primatol 39:139–145

Brent L, Bloomsmith MA, Fishe SD (1995) Factors determining tool-using ability in two captive chimpanzee (*Pan troglodytes*) populations. Primates 36:265–274

Brooks LR (2004) The effects of cognitive enrichment on the stereotypic behaviors of a male western lowland gorilla (*Gorilla gorilla gorilla*). Abstract, Rocky Mountain Gorilla Workshop, Calgary, June 25–28

Capitanio JP (1986) Behavioral pathology. In: Mitchell G, Erwin J (eds) Comparative primate biology: vol 2. Part A: Behavior, conservation, and ecology. Liss. New York, pp 411–454

Citrynell P (1998) Cognitive enrichment: problem solving abilities of captive white-bellied spider monkeys. Primate Eye 66:16–17 (abstract)

Davenport RK, Menzel EW Jr (1963) Stereotyped behavior of the infant chimpanzee. Arch Gen Psychol 8:99–104

Davenport RK, Rogers CM (1968) Intellectual performance of differentially reared chimpanzees. I. Delayed response. Am J Ment Defic 72:674–680

Davenport RK, Rogers CM, Rumbaugh DM (1973) Long-term cognitive deficits in chimpanzees associated with early impoverished rearing. Dev Psychol 9:343–347

Eddy TJ, Gallup GG Jr, Povinelli DJ (1996) Age differences in the ability of chimpanzees to distinguish mirror-images of self from video images of others. J Comp Psychol 110:38–44

Fujita K, Matsuzawa T (1986) A new procedure to study the perceptual world of animals with sensory reinforcement: recognition of humans by a chimpanzee. Primates 27:283–291

Gilloux I, Gurnell J, Shepherdson D (1992) An enrichment device for great apes. Anim Welf 1:279–289

Glick-Bauer M (1997) Behavioral enrichment for captive cotton-top tamarins (*Saguinus oedipus*) through novel presentation of diet. Lab Primat Newsl 36:1–3

Goodall J (1986) The chimpanzees of Gombe: patterns of behavior. Harvard University Press, Cambridge

Grief L, Fritz J, Maki S (1992) Alternative forage types for captive chimpanzees. Lab Primat Newsl 31:11–13

Hirata S, Matsuzawa T (2001) Tactics to obtain a hidden food item in chimpanzee pairs (*Pan troglodytes*). Anim Cogn 4:285–295

Hook MA, Lambeth SP, Perlman JE, Stavisky R, Bloomsmith MA, Schapiro SJ (2002) Intergroup variation in abnormal behavior in chimpanzees (*Pan troglodytes*) and rhesus macaques (*Macaca mulatta*). Appl Anim Behav Sci 76:165–176

Inglis IR, Forkman B, Lazarus J (1997) Free food or earned food? A review and fuzzy model of contrafreeloading. Anim Behav 53:1171–1191

Inoue-Nakamura N, Matsuzawa T (1997) Development of stone tool use by wild chimpanzees (*Pan troglodytes*). J Comp Psychol 111:159–173

Itakura S, Matsuzawa T (1993) Acquisition of personal pronouns by a chimpanzee. In: Roitblat H, Herman L, Nachtigall P (eds) Language and communication: comparative perspectives. Erlbaum, Hillsdale, NJ, pp 347–363

Kawai N, Matsuzaw T (2000) Numerical memory span in a chimpanzee. Nature (Lond) 403:39–40

King NE, Mellen JD (1994) The effects of early experience on adult copulatory behavior in zoo-born chimpanzees (*Pan troglodytes*). Zoo Biol 13:51–59

Lutz CK, Novak MA (2005) Environmental enrichment for nonhuman primates: theory and application. ILAR J 46:178–191
Matsuno T, Kawai N, Matsuzawa T (2004) Color classification by chimpanzees (*Pan troglodytes*) in a matching-to-sample task. Behav Brain Res 148:157–165
Matuzawa T (1981) Sensory reinforcement: the variety of reinforcers. Jpn Psychol Rev 24:220–251 (in Japanese with English summary)
Matsuzawa T (2003). The Ai project: historical and ecological contexts. Anim Cogn 6:199–211
Menzel EW Jr (1991) Chimpanzees (*Pan troglodytes*): problem seeking versus the bird-in-hand, least-effort strategy. Primates 32:497–508
Menzel EW Jr, Savage-Rumbaugh ES, Lawson J (1985) Chimpanzee (*Pan troglodytes*) spatial problem solving with the use of mirrors and televised equivalents of mirrors. J Comp Psychol 99:211–217
Milgram NW (2003) Cognitive experience and its effect on age-dependent cognitive decline in beagle dogs. Neurochem Res 28:1677–1682
Morimura N (2003) A note on enrichment for spontaneous tool use by chimpanzees (*Pan troglodytes*). Appl Anim Behav Sci 82:241–247
Morimura N, Matsuzawa T (2001) Memory of movies by chimpanzees (*Pan troglodytes*). J Comp Psychol 115:152–158
Morimura N, Ueno Y (1999) Influences on the feeding behavior of three mammals in the Maruyama Zoo: bears, elephants and chimpanzees. J Appl Anim Welf Sci 2:169–186
Nash, VJ (1982) Tool use by captive chimpanzees at an artificial termite mound. Zoo boil 1:211–221
National Research Council (1998) Psychological well-being of nonhuman primates. National Academy Press, Washington, DC
Nishida T (1990) The chimpanzees of the Mahale Mountains: sexual and life history strategies. University of Tokyo Press, Tokyo
Neuringer AJ (1969) Animals respond for food in the presence of free food. Science 166:399–401
O'Connell S, Dunbar RIM (2005) The perception of causality in chimpanzees (*Pan* spp.). Anim Cogn 8:60–66
O'Connor E, Reinhardt V (1994) Caged stumptailed macaques voluntarily work for ordinary food. In Touch 1:10–11
Premack D, Woodruff G (1978) Does the chimpanzee have a theory of mind? Behav Brain Sci 1:515–526
Reinhardt V (1994) Caged rhesus macaques voluntarily work for ordinary food. Primates 35:95–98
Rogers CM, Davenport RK (1970) Chimpanzee maternal behavior. In: Bourne G (ed) The chimpanzee, vol 3. Immunology, infections, hormones, anatomy, and behavior of chimpanzees. Karger, Basel, pp 361–368
Rumbaugh DM, Washburn D, Savage-Rumbaugh ES (1989) On the care of captive chimpanzees: methods of enrichment. In: Segal EF (ed) Housing, care and psychological well being of captive and laboratory primates. Noyes, Park Ridge, NJ, pp 357–375
Sands SF, Wright AA (1980) Serial probe recognition performance by a rhesus monkey and a human with 10- and 20-item lists. J Exp Psychol Anim Behav Proc 6:386–396
Segal EF (1989) Housing, care and psychological well being of captive and laboratory primates. Noyes, Park Ridge, NJ
Sugiyama Y (1995) Drinking tools of wild chimpanzees at Bossou. Am J Primatol 37:263–269
Toates F (2004) Cognition, motivation, emotion and action: a dynamic and vulnerable interdependence. Appl Anim Behav Sci 86:173–204
Tonooka R (2001) Leaf-folding behavior for drinking water by wild chimpanzees (*Pan troglodytes verus*) at Bossou, Guinea. Anim Cogn 4:325–334
Tonooka R, Tomonaga M, Matsuzawa T (1997) Acquisition and transmission of tool making and use for drinking juice in a group of captive chimpanzees (*Pan troglodytes*). Jpn Psychol Res 39:253–265

Walsh S, Bramblett C, Alford P (1982) A vocabulary of abnormal behavior in restrictively reared chimpanzees. Am J Primatol 3:313–319

Washburn DA, Hopkins WD (1994) Videotape- versus pellet-reward preferences in joystick tasks by macaques. Percept Motor Skills 78:48–50

Wright AA, Santiago HC, Sands SF, Kendrick DF, Cook RG (1985) Memory processing of serial lists by pigeons, monkeys, and people. Science 229:287–289

Part 6
Tools and Culture

24
Cognitive Development in Apes and Humans Assessed by Object Manipulation

MISATO HAYASHI[1], HIDEKO TAKESHITA[2], and TETSURO MATSUZAWA[1]

1 Object Manipulation Studies

Primate species are characterized by hands that enable them to manipulate objects in various ways. Primates use their hands not only for the processing of food items but also for manipulating nonedible objects during their daily lives. Object manipulation is a direct way of interacting with and changing the surrounding environment. Thus, object manipulation skills may reflect cognitive capabilities in primate species that contribute to survival.

One long-term focus of research in the area of object manipulation has been the study of tool-using ability in animals. There have been a number of reports on tool use from a variety of perspectives, from species that perform tool use and tool manufacture among nonprimates (crows: Hunt 1996; Weir et al. 2002) and primates (capuchins, macaques, baboons, and apes; see review by van Schaik et al. 1999), examining the variety and flexibility of tool use (McGrew 1994; Parker and Gibson 1977) or the causal understanding involved (Povinelli 2000; Visalberghi and Limongelli 1994). Although primates are equipped with the skill to manipulate objects in various ways, only a limited number of primate species use tools in their natural habitat. This realization might suggest that tool use is not only a question of manual skill or ecological factors but that certain cognitive processes are also involved in producing the flexible tool-using and tool-manufacturing behaviors observed.

Several authors have provided definitions of tool use. Matsuzawa (2001) describes tool use as a set of behaviors utilizing a detached object to obtain a goal that is adaptive in the biological sense. Takeshita and van Hooff (1996) define tool use as behaviors in which an individual uses a single or multiple detached environmental object(s) as an intermediary to efficiently change the environment and thus obtain a goal. Westergaard (1993) proposes that the use of tools can be seen as a combinatorial action that results in the attainment of an immediate goal.

[1]Primate Research Institute, Kyoto University, 41 Kanrin, Inuyama, Aichi 484-8506, Japan
[2]School of Human Cultures, The University of Shiga Prefecture, 2500 Hassaka-cho, Hikone, Shiga 522-8533, Japan

One precursor of tool-using behavior is the ability to combine objects through manipulation. Researchers have focused on manipulation patterns involving such combining of objects using different terminology. Torigoe (1985) used the term "secondary manipulation," defined as object manipulation in relation to some specific feature of the environment. This term was pitted against "primary manipulation," referring to object manipulation with no relation to another object or in global relation to substrates. According to this definition, rolling a manipulandum on the floor was classified as primary whereas rolling it in a water vessel was classified as secondary. Fragaszy and Adams-Curtis (1991) used the term "relational and combinatorial acts," which include the placement of an object in relation to another and also the placement of an object in relation to a substrate (including the body as a type of substrate). The term "combinatorial manipulation" has been used by Westergaard (1992, 1993) and Westergaard and Suomi (1994), defined as the placement of an object in relation (or in contact) with another object. Matsuzawa (1994) used "object-association manipulation" to refer to an action simultaneously involving at least two objects, such as putting a nut on a stone. The term "orienting manipulation," which was used by Takeshita (1994) and Takeshita and Walraven (1996), distinguished three types of orienting manipulation: orientation to a substrate, orientation to the subject's own body, and orientation to another detached object. Takeshita (2001) used "combinatory manipulation" instead of orienting manipulation, as this term was thought to be more comprehensive. Hayashi and Matsuzawa (2003) coined "object–object combination" to describe combinatory manipulations on detached objects, excluding combinations toward substrates or the individual's own body.

In this chapter, the term combinatory manipulation is used to describe relating a detached object to something else (including substrates or objects; see details in the following section) and object–object combination is used to specify when the targets of combinatory manipulation are detached objects only. Figure 1 illustrates a way of categorizing different types of object manipulation. Object manipulation starts out from manipulating a single object using a single action and develops into manipulating multiple objects with multiple actions. Combinatory manipulation forms a subset of object manipulation and can be divided into four types. The first is "combination to self," such as placing an object against a part of one's own body. The second is "combination to other individual," such as handing an object to someone else. The third is "combination to substrate," such as hitting the floor with an object. The fourth is "combination to object" (what we call object–object combination), such as putting an object into a container.

Tool use is based on these four types of combinatory manipulation: cleaning the body with a leaf, throwing a branch to threaten another individual, putting leaves into water and retrieving them for drinking, or using a pair of stones to crack open nuts. Tool use is a means to achieving an immediate goal and it has a functional importance in problem solving. Object manipulation can be linked with play or communicative contexts as well. In object play, an individual does not pursue a direct goal with the manipulation, but he may be stimulated by the manipulation itself or by producing some effect on the environment. Objects can

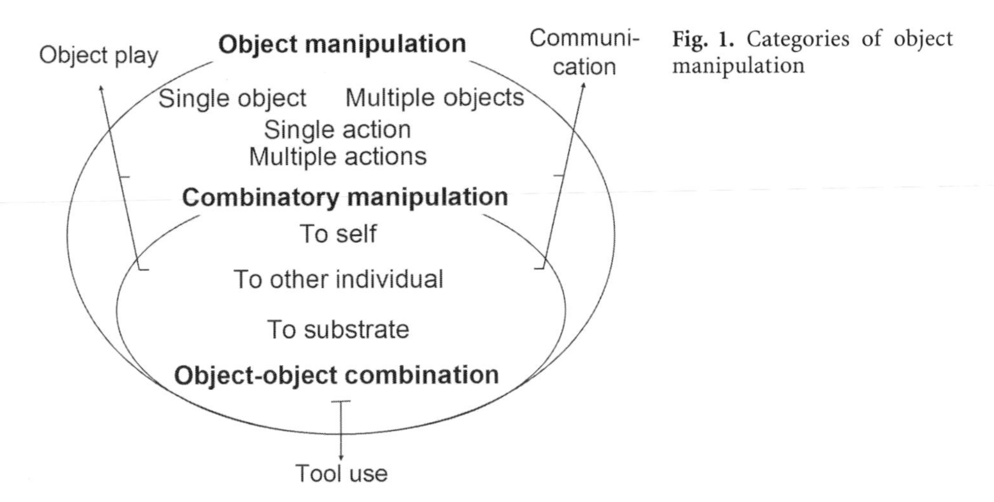

Fig. 1. Categories of object manipulation

also be used as an intermediary for communicating with others or attracting their attention. Object manipulation might be related with both material and social areas of intelligence in humans and nonhuman primates.

Object manipulation develops in the following stages in humans (Ikuzawa 2000). Human infants begin to grab and hold on to objects with one hand at around 3 to 4 months of age and to two objects with both hands at around 6 months. At around 8 months, they start to press or bang two objects together in front of the body using both hands. These manipulations are characterized as moving objects toward or around themselves. The next change occurs at around 10 months of age when human infants begin to move objects toward the surrounding environment; for example, they will insert a block into a container as opposed to the preceding period of simply taking the block out of the container. From around 11 months, infants also start to use objects in social contexts, such as handing an object to another individual. Finally, they start to use objects as tools during their second year of life, followed by development in both the accuracy (Connolly and Dalgleish 1989) and complexity (Tanaka and Tanaka 1984) of tool use.

In this chapter, we use object manipulation as a comparative scale of cognitive development as it can be applied to different species of primates. Based on the accumulated data on human development, the manipulation patterns of ape species are compared with humans to illuminate characteristics of humans and apes from a cognitive–developmental perspective.

2 Comparing the Repertoire of Object Manipulation in Bonobos and Gorillas

Among nonhuman primates, apes show the greatest similarity with humans in terms of their phylogenetic position and intelligence. However, habitual tool use has only been reported in one of the three African great ape species, the chimpanzee. In contrast, wild bonobos and gorillas are not known to habitually use tools. Nevertheless, both bonobos and gorillas show tool-using behavior in captive situations (see review by Tomasello and Call 1997), and gorillas are known to perform highly hierarchical sequences of actions in processing food in the wild (Byrne and Byrne 1993). Although a number of studies have dealt with object manipulation development in chimpanzees as a precursor to tool use, few reports have considered the same issues in infant bonobos and gorillas.

Hayashi and Takeshita (2002) examined the repertoire of object manipulation in bonobos and gorillas from 2 to 4 years of age. The subjects were hand-reared at the Wilhelma Zoo in Stuttgart, Germany. Two male bonobos (Limbuko: 4 years, 9 months; Kuno: 3 years, 8 months) and three gorillas (two females, Luena: 3 years, 7 months; Iringa: 2 years, 6 months; one male, Kumbuka: 2 years, 8 months) were kept in conspecific groups in rooms that were separated from the visitors' area by transparent glass (Fig. 2a,b). We observed free play in the ape infants manipulating three kinds of objects: wooden blocks, nesting cups, and rings and a stalk; these had the advantage of facilitating combinations or relations among the objects. The size of the subjects' behavioral repertoire was measured by focusing on four variables: body parts used, number of objects manipulated, motor patterns, and the targets of combinatory manipulations. Each behavioral pattern was defined by a combination of the four variables and was coded with presence or absence in each subject during the 1-month study period. Both species showed combinatory manipulations comparable to chimpanzee infants of similar ages. The gorillas' repertoire included fewer behaviors that involved the use of the foot than that of the bonobos; they showed a tendency for manipulating single objects using multiple, rather than single, motor patterns.

Hayashi et al. (2003) focused on the manipulation of blocks by the same bonobo and gorilla subjects. In the CD-ROM accompanying their chapter, tables, images, and video clips were used to comprehensively describe the infants' repertoire of block manipulation, with examples illustrating each behavioral category. The three main points that emerged from the results could be summarized as follows. (1) Bonobos used their feet in various ways, such as for grasping blocks with the foot or holding blocks in their groin pocket. In contrast, gorillas seldom used their feet for manipulating blocks other than simply to touch them. (2) Gorillas had a tendency to use block(s) to hit or rub other objects with. They also held two blocks—one in each hand—and hit the floor with both or hit the two together in front of the body. (3) Three of five individuals stacked up blocks in the study period. One gorilla (Luena) was already capable of stacking blocks when testing began. One bonobo (Limbuko) and one gorilla (Iringa) began to stack up blocks during the study period after observing stacking demonstrated by a human caretaker or a conspecific.

This chapter reports block manipulation patterns observed in bonobos and gorillas, focusing on the combinatory manipulations that the subjects exhibited. Table 1 lists the combinatory manipulation repertoire of bonobos and gorillas at 2 to 4 years of age. We excluded combinations to self and to other

Fig. 2. **a** A bonobo (Limbuko) trying to insert a block into a container attached to the climbing frame. **b** A gorilla (Kumbuka) holding two blocks with both hands

Table 1. Repertoire of combinatory manipulation (to substrate and to object) in bonobos and gorillas during block manipulation

	Bonobo		Gorilla		
Category of combinatory manipulation	Limbuko	Kuno	Luena	Kumbuka	Iringa
Put block on object attached to substrate	+	+			+
Put block against mesh and push through an opening	+		+		
Put block(s) into tub of water or container	+		+		
Use block to drink water from tub			+		
Rub substrate with block held in hand(s)	+		+	+	+
Hit substrate with block held in hand			+	+	+
Hit substrate with blocks held in both hands simultaneously or alternately				+	+
Touch two blocks together using both hands	+		+		
Bang two blocks together using both hands			+	+	
Wipe block held in hand with piece of paper held in other hand				+	
Rub object with block held in hand				+	
Hit object or block with block held in hand(s)			+	+	+
Stack up blocks	+		+		+
Put cylindrical block sideways on top of another block	+				+
Put block on top of cylindrical block in sideways position	+				+
Try to put two blocks on top of another block simultaneously	+				+

Some categories, which previous reports differentiated between, are combined here with similar ones

individuals because it was sometimes difficult to evaluate these patterns from our video records. The list includes patterns of combinatory manipulations observed during the manipulation of blocks and shows which of the patterns were recorded from each infant during the 1-month study period.

Both species showed combinatory manipulation using blocks toward both substrates and detached objects. One bonobo (Kuno) showed fewer patterns of combinatory manipulation among objects, although he sometimes combined a block with his own body, such as putting a block to his head or bottom. This tendency might be explained by his subordinate position in the pair and thus limited access to the blocks. Two infants put blocks into water as a part of play and for drinking by licking water off the block. One bonobo (Limbuko) succeeded to stack up seven cubic blocks. However, he failed to stack cylindrical blocks in the correct orientation (with the flat surface facing up), instead attempting to stack them with the round surface up. One gorilla (Luena), who already had the stacking skill when the study began, stacked up five cubic blocks, although she tended to knock her tower down herself when one of the other infants approached. She always used her left hand for positioning the objects, and in addition succeeded to stack up cylindrical blocks by correctly orienting them before stacking. One gorilla (Iringa) stacked up four cubic blocks but failed to stack cylinders.

Gorillas showed particular combinations more often than bonobos: for example, they used block(s) to hit a substrate or another object. Another interesting combination made by gorillas was holding a block in each hand and banging the two blocks together in front of the body. This behavior is often observed in human infants but not in chimpanzees or in bonobos. These patterns may reflect species-specific patterns of object manipulation that differ even among the great apes. More precise developmental data for object manipulation patterns in great apes may be necessary to illuminate the differences and similarities with human development.

3 Developmental Process of Object–Object Combination in Chimpanzees

A number of previous studies have focused on the development of object manipulation in nonhuman primates. The majority of subjects in these studies were human-reared individuals; others were briefly taken from the mother for testing. In the year 2000, three infant chimpanzees were born at the Primate Research Institute, Kyoto University, all of whom were being reared by their biological mothers. On the basis of long-term relationships between the mother chimpanzees and human experimenters, we employed a new method for testing the chimpanzee infants. A human entering a room with a mother–infant chimpanzee pair was able to test the infant's cognitive development in this setting (see details in Matsuzawa 2003). This way of testing is thus comparable to the way in which

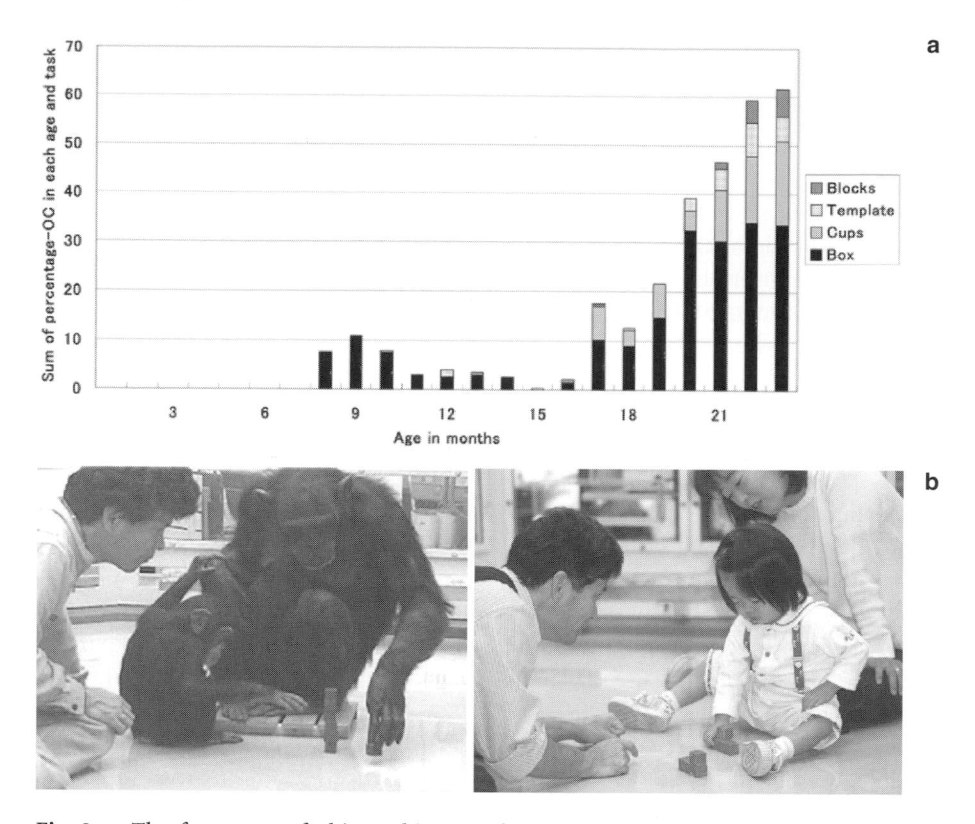

Fig. 3. **a** The frequency of object–object combination in three chimpanzee infants during their first 2 years of life. Hayashi and Matsuzawa (2003) used raw frequencies of object–object combination per session at each age, whereas the present report uses the percentage of object–object combinations at each month of age relative to the total frequency of object–object combinations during the first 2 years in each infant. Data from the three infants were pooled at each age. **b** Chimpanzee and human infants tested in a comparable face-to-face situation. (Figures from Hayashi and Matsuzawa 2003)

human developmental studies are normally carried out (Fig. 3b). Moreover, it allows the observation and testing, under controlled task settings, of infant chimpanzees raised by their own mothers within a social group of chimpanzees. Hayashi and Matsuzawa (2003) presented the same objects that had been used to assess cognitive development in human infants to chimpanzee infants in a face-to-face situation. The chimpanzee mothers were required to participate in the task and to perform object–object combination in demonstrations for the infants. The infants were free to observe the mothers' performance and also to manipulate the objects by themselves. Although previous comparative studies on developmental processes of object manipulation suggested that the onset of object–object combination was delayed in chimpanzees compared to humans, the present study showed that mother-reared chimpanzees began to perform object–object combination at an age comparable to human infants (around 8–11 months of age versus 10 months, respectively). This result seems to suggest that the mother exerts an influence on the development of object manipulation in chimpanzee infants.

The tasks applied to chimpanzees were chosen from the Kyoto Scale of Psychological Development (KSPD), originally designed to assess cognitive development in humans. The scale has been tested on normal human subjects to obtain an average scale for development. The four tasks presented to the chimpanzees were the following: (1) inserting objects into corresponding holes in a box, (2) seriating nesting cups, (3) inserting variously shaped blocks into corresponding holes in a template, and (4) stacking up wooden blocks. Figure 3a shows the frequency of object–object combinations in the three chimpanzee infants during their first 2 years of life. The infants began to exhibit object–object combination before they reached 1 year of age. After this first, early appearance, the frequency of object–object combinations decreased in the first half of their second year of life, then increased again when the infants reached around 1.5 years of age. There were clear task differences in the developmental process of object–object combination. Object–object combination appeared mostly in the "box" task in which the subjects were required to insert a rod or a square block into a hole in the box. In the second phase of object–object combination, which began around 1.5 years of age, there was a rise in combinations observed during the seriating of nesting cups, trying to insert a block into corresponding holes in the plate, and/or the inserting of blocks into a cup. However, the dominant and earliest object–object combination was exhibited in the "box" task. One possible explanation is that chimpanzees may have a strong innate tendency to insert objects into holes or containers. This consideration is concordant with Yamakoshi (2004), who suggests that "insertion feeding" is a feature unique to the great apes among wild nonhuman primates.

Figure 4 pits normal human development against results we obtained from chimpanzee infants using an identical set of object-manipulation tasks. The human data are based on Ikuzawa (2000), showing the ages at which normal human children pass each stage on the developmental scale. Chimpanzee infants are plotted on the same two-dimensional plane, at the age when each infant first succeeded in performing the behavior used in the scale. Some of the behaviors were first exhibited by chimpanzees at around the same age as humans. Others were delayed in chimpanzees compared to humans, such as seriating five nesting cups, inserting three blocks of various shapes in corresponding holes in a plate, and stacking blocks. Humans and chimpanzees may have different developmental pathways (see also Gómez 2004 for a similar discussion). Humans begin to perform a variety of object–object combinations such as inserting or stacking from an early age while chimpanzees concentrate on insertion only. Humans also succeed to produce hierarchical combinations of multiple objects earlier than chimpanzees, as described in the following section.

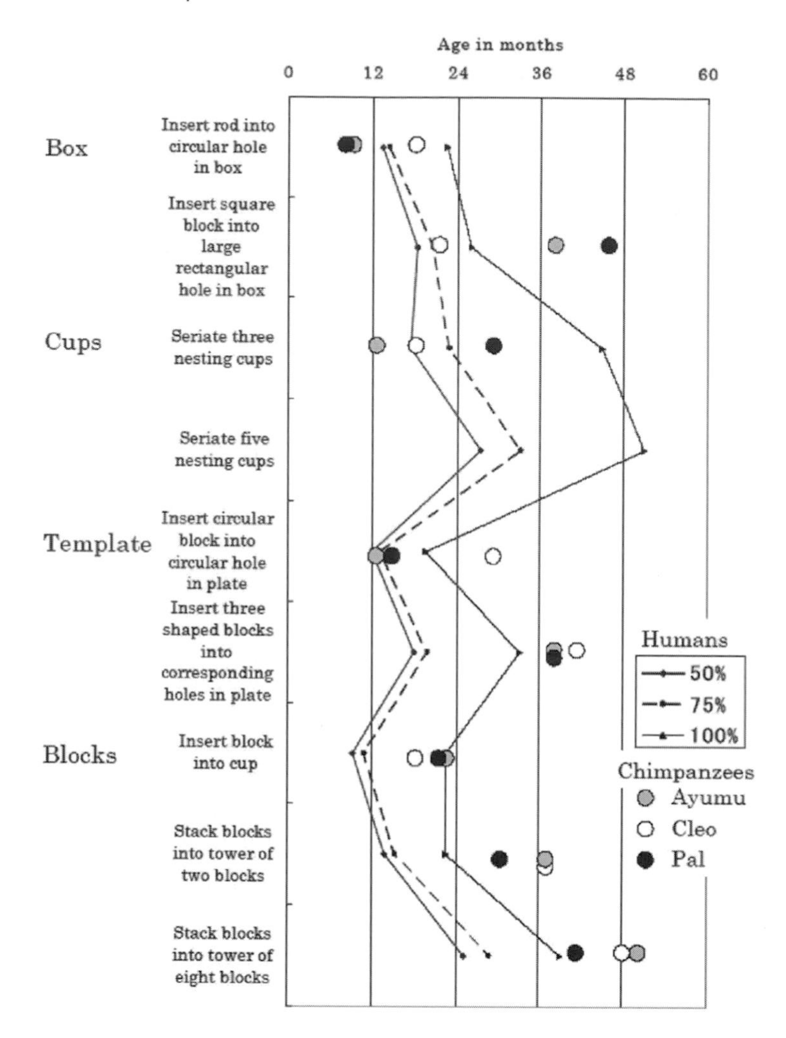

Fig. 4. Comparison between humans and the three chimpanzee infants tested on the four tasks from the Kyoto Scale of Psychological Development (KSPD). Human data are based on Ikuzawa (2000), showing the age at which 50%, 75%, and 100% of normal human children, respectively, pass each stage of the developmental scale. The plots for the chimpanzee infants are based on data from Hayashi and Matsuzawa (2003) and on subsequent developmental records of the three infants up to the age of 5 years. Infant chimpanzees have so far (up to 5 years of age) not succeeded in seriating five nesting cups

4 Describing the Process of Producing Hierarchical Combinations Among Objects

Previous sections focused on the early stages of combinatory manipulation in different ape species and the developmental course of object–object combination in chimpanzees. Ape infants start to exhibit simple combinatory manipulations just like human infants, but humans then proceed to elaborate these by increasing the hierarchical complexity of the combinations. In other words, as they get older, they not only combine two objects together but they also begin to combine three or more. The complexity of object–object combination can be increased by adding more objects and hierarchically combining them into a structure.

Matsuzawa (1996) devised a "tree-structure analysis" for describing the complexity of hierarchical combinations among objects in the context of tool use. He focused on detached objects (targets) in tool use, connecting these one-by-one according to the temporal order in which they are related to each other to produce a cluster. "Level 1" tool use was defined as relating a tool to a target, involving only a single relationship between detached objects. Most of the tool use examples in wild chimpanzees fall into this category. "Level 2" tool use was defined as the use of two tools that were related hierarchically to obtain a goal and involved two kinds of relationships among detached objects. This type of tool use is much less common in wild chimpanzees.

One example of such rare level 2 tool use is nut-cracking behavior, which requires a pair of stones as a hammer and an anvil to crack open hard-shelled nuts. Inoue-Nakamura and Matsuzawa (1997) analyzed developmental processes of stone tool use acquisition. In a wild community of chimpanzees at Bossou, Guinea, infants start to crack nuts (thus successfully combining three objects—anvil, hammer, and nut—in the appropriate manner) at around 3.5 years of age or older. Although they are already able to perform the basic actions involved in nut cracking at 2 years of age, the infants do not succeed to successfully combine these actions in the proper hierarchical sequence until later. Hayashi et al. (2005) reported the first introduction of stones and nuts to naïve adult chimpanzees in a captive situation. One of three chimpanzees tested successfully cracked open nuts after the first demonstration of nut-cracking behavior by a human experimenter. The authors made a precise list of the subjects' manipulation patterns during the first test session. They pointed out that the chimpanzees frequently made combinations of two detached objects though the goal of nut cracking was to combine three detached objects in the appropriate manner. A fine-scale analysis of hierarchical structure in object manipulation by nonhuman primates may shed further light on cognitive development.

Here we report a notation system for describing object manipulation sequences performed by individuals in producing hierarchical combinations among objects. Hayashi (2006) developed this system to describe the flow of manipulative behavior during the nesting-cup task as a list of sequential codes. Nesting cups were first applied in a task for human children by Greenfield et al. (1972) to assess the development of rule-bound strategies that occur in parallel between object manipulation and language acquisition. Several other researchers have used the nesting cup task to assess and compare levels of cognitive development in humans as well as nonhuman primates (DeLoache et al. 1985; Johnson-Pynn and Fragaszy 2001; Johnson-Pynn et al. 1999; Matsuzawa 1991; Takeshita 2001). These previous studies mainly used a set of categories (pairing, pot, and subassembly strategies) as originally defined by Greenfield et al. (1972) to describe cup manipulation. In contrast, the notation system presented here can transcribe the whole sequence of object manipulation and it does not rely on predetermined categories. The notation was devised to illuminate the grammatical rules that exist in the actions performed during object manipulation.

The notation system can be summarized as follows. Each segment of object manipulation was coded in terms of its three main parameters: object, action, and location. These three components of manipulation describe what was being manipulated by the subject, using what kind of action, and where the manipulated object was related toward (if anywhere). A number was assigned to each of the objects used in the task (representing the object and location parts of the code), and letters of the alphabet to behavioral patterns (representing the action). Thus, a segment of object manipulation was coded in the form "n_1 X n_2", indicating object, action, and location, respectively. Each such code representing a single segment of behavior was separated by slashes from those before and after it, such that the entire flow of manipulation could be described in the form of sequential codes.

Using this notation system, we were able to describe cup manipulation comprehensively, including the classification of actions according to the existing categories utilized in previous studies. Besides this advantage, our codes also enabled a more precise analysis of state transitions during the manipulation sequence. In other words, we used the notation system to analyze the dynamic processes involved in the production of hierarchical combinations of objects. This "state transition analysis" looked at the efficiency of manipulation aimed towards producing a fully seriated cup structure, through noting stages of progression and regression in the process. The state of the cups was defined by two parameters: "number of units" and "contiguity." The former corresponded to the number of structures or individual cups that were present in a static condition on the floor whereas the latter indicated how many pairs of cups were seriated in the correct, successive order (i.e., combined with adjacent cups). Taking as an example the seriation of nine cups into one structure, the starting point can be described as 9 for the number of units and 0 for contiguity. The goal, that of a fully seriated structure, can be described as 1 for the number of units and 8 for contiguity. These two measures can then be used as the two axes of a two-dimensional plane for visualizing cup-state transition.

Figure 5 shows the performance of an adult female chimpanzee (named Ai) on her first attempt to seriate nine cups. The most efficient way of combining the nine cups is shown by a line connecting the bottom right and top left corners of the two-dimensional plane. However, Ai performed many progressions and regressions before she eventually achieved the goal of a fully seriated nine-cup structure. Progressive patterns were defined as those in which at least one parameter improved: actions by the subject that meant that the cup state moved closer to the goal in terms of the number of units and/or contiguity. Regressive patterns were those in which at least one parameter deteriorated through manipulation, although this also included the reconstruction process whereby cup

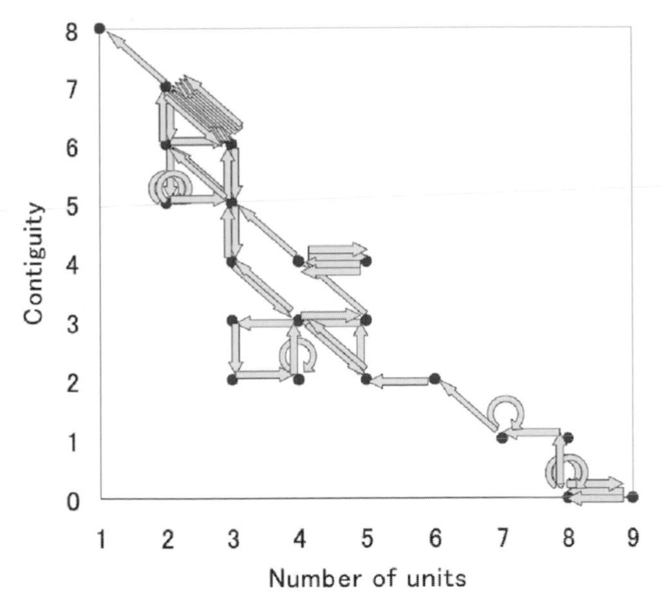

Fig. 5. Efficiency in seriating nine nesting cups by a chimpanzee (Ai) on her first attempt. The *x*-axis shows the number of cups or structures consisting of several cups that were present following each manipulation; the *y*-axis provides a measure of how many correct (adjacent) combinations of cups had been produced

structures with noncontiguous pairs were disassembled and better combinations built. Progressive patterns accounted for 53% of the total number of cup-state transitions, regressive patterns were recorded in 36%, and 11% of moves were neither progressions nor regressions.

As Fig. 5 shows, the process through which the goal of hierarchically combining nine cups was achieved was not straightforward. However, Ai eventually succeeded to combine all nine cups appropriately by modifying cup combinations step by step. She may have been attempting to reduce the number of units and also to increase contiguity, the lack of which she would have been able to detect by the small gaps that remained when a smaller cup was placed inside a nonadjacent larger one. Of the total number of cup-state transitions, 27% were of the most progressive type, reducing the number of units and increasing contiguity at the same time. These results might suggest that the chimpanzee was not merely trying out random combinations but was using a strategy of adjustments to produce a highly hierarchical combination among multiple objects.

5 Discussion and Conclusions

This chapter addressed a variety of issues related to object manipulation in the great apes. We have discussed the repertoire of combinatory manipulations in bonobos and gorillas, the development of object–object combination in chimpanzees, and a novel notation system to describe processes through which hierarchical object combinations are achieved. Object manipulation has emerged as a highly useful objective scale for comparing various different primate species, including humans. Bonobos and gorillas were comparable to chimpanzees in terms of the development of combining objects in a free-play situation. However, the pattern of object manipulation was different between the species: bonobos used their feet for manipulating objects more often than gorillas while gorillas often hit substrates or objects with the objects held in their hand(s) and also banged together two objects, one held in each hand. As the present study does not include data from orangutans, we acknowledge that direct comparisons of all four great ape species, focusing on the development of object manipulation (or more specifically on the development of combinatory manipulation), will be necessary for highlighting species-specific patterns of manipulation. This approach may be informative regarding the ontogeny and phylogeny of human traits and may also provide us with clues in explaining the distribution of habitual tool use among the wild great apes.

The precocious appearance of object–object combination in infant chimpanzees raised by their own mothers seems to confirm the importance of testing ape infants within an appropriate setting. To investigate the natural development of chimpanzees and to draw meaningful comparisons with humans, the present system of testing mother-reared chimpanzees in a face-to-face situation may be the most useful and reliable method. Although comparing the repertoire of object manipulation in free-play situations can also reveal species differences, most human developmental studies use scales of development assessed under controlled task settings. We have just begun to test the mother-reared chimpanzee infants within a face-to-face setting comparable to that used for humans. We need to elaborate the cognitive tasks used to illuminate developmental characteristics of chimpanzees and humans while continuing to use identical objects and test settings for the two species.

One important difference between chimpanzees and humans may pertain to the levels of hierarchical combination achieved. Both species are able to construct highly hierarchical combinations among objects. However, the process through which the goals are achieved needs to be investigated further, by focusing on temporal aspects and grammatical rules that are described by our notation system of object manipulation. Direct and precise comparisons of humans and apes may shed light on the levels of cognitive development in different species.

Acknowledgments

The present study was supported by grants from the Ministry of Education, Science, and Culture in Japan (#12002009 and #16002001), from the biodiversity research of the 21COE (A14), and from the JSPS core-to-core Program HOPE. Preparation of the manuscript was supported by Research Fellowship 16-1059 from the Japan Society for the Promotion of Science for Young Scientists. We thank Masaki Tomonaga, Masayuki Tanaka, Masako Myowa-Yamakoshi, and Yuu Mizuno for their support and advice given in the course of the study. We are also grateful to the staff at Wilhelma Zoo, especially Marianne Holtkötter and Gundi Scharpf, for their helpful support. Thanks are also due to the staff at the Section of Language and Intelligence, Primate Research Institute, Kyoto University, and to the keepers and veterinarians who took care of the chimpanzees at the institute. We thank Dora Biro for help given in the course of revising the manuscript.

References

Byrne RW, Byrne JME (1993) Complex leaf-gathering skills of mountain gorillas (*Gorilla g. beringei*): variability and standardization. Am J Primatol 31:241–261

Connolly K, Dalgleish M (1989) The emergence of a tool-using skill in infancy. Dev Psychol 25:894–912

DeLoache JS, Sugarman S, Brown AL (1985) The development of error correction strategies in young children's manipulative play. Child Dev 56:928–939

Fragaszy DM, Adams-Curtis LE (1991) Generative aspects of manipulation in tufted capuchin monkeys (*Cebus apella*). J Comp Psychol 105:387–397

Gómez JC (2004) Apes, monkeys, children, and the growth of mind. Harvard University Press, Cambridge, pp 83–93

Greenfield PM, Nelson K, Saltzman E (1972) The development of rulebound strategies for manipulating seriated cups: a parallel between action and grammar. Cogn Psychol 3:291–310

Hayashi M (2006) A new notation system of object manipulation in the nesting-cup task for chimpanzees and humans. Cortex (in press)

Hayashi M, Matsuzawa T (2003) Cognitive development in object manipulation by infant chimpanzees. Anim Cogn 6:225–233

Hayashi M, Takeshita H (2002) Object manipulation in infant bonobos and gorillas. Jpn J Anim Psychol 52:73–80 (in Japanese with English summary and tables)

Hayashi M, Takeshita H, Matsuzawa T (2003) Development of object manipulation in bonobo and gorilla infants. In: Tomonaga M, Tanaka M, Matsuzawa T (eds) Development of cognition and behaviors in chimpanzees. Kyoto University Press, Kyoto, pp 374–378 and attached CD-ROM (in Japanese)

Hayashi M, Mizuno Y, Matsuzawa T (2005) How does stone-tool use emerge? Introduction of stones and nuts to naïve chimpanzees in captivity. Primates 46:91–102

Hunt GR (1996) Manufacture and use of hook-tools by New Caledonian crows. Nature (Lond) 379:249–251

Ikuzawa M (2000) Developmental diagnostic tests for children, 2nd edn. Nakanishiya, Kyoto (in Japanese)

Inoue-Nakamura N, Matsuzawa T (1997) Development of stone tool use by wild chimpanzees (*Pan troglodytes*). J Comp Psychol 111:159–173

Johnson-Pynn J, Fragaszy DM (2001) Do apes and monkeys rely upon conceptual reversibility? A review of studies using seriated nesting cups in children and nonhuman primates. Anim Cogn 4:315–324

Johnson-Pynn J, Fragaszy DM, Hirsh EM, Brakke KE, Greenfield PM (1999) Strategies used to combine seriated cups by chimpanzees (*Pan troglodytes*), bonobos (*Pan paniscus*), and capuchins (*Cebus apella*). J Comp Psychol 113:137–148

Matsuzawa T (1991) Nesting cups and metatools in chimpanzees. Behav Brain Sci 14:570–571

Matsuzawa T (1994) Field experiments on use of stone tools by chimpanzees in the wild. In: Wrangham RW, McGrew WC, de Waal FBM, Heltne PG (eds) Chimpanzee cultures. Harvard University Press, Cambridge, pp 351–370

Matsuzawa T (1996) Chimpanzee intelligence in nature and in captivity: isomorphism of symbol use and tool use. In: McGrew WC, Marchant LF, Nishida T (eds) Great ape societies. Cambridge University Press, Cambridge, pp 196–209

Matsuzawa T (2001) Primate foundations of human intelligence: a view of tool use in nonhuman primates and fossil hominids. In: Matsuzawa T (ed) Primate origins of human cognition and behavior. Springer, Tokyo, pp 3–25

Matsuzawa T (2003) The Ai project: historical and ecological contexts. Anim Cogn 6:199–211
McGrew WC (1994) Tools compared. In: Wrangham RW, McGrew WC, de Waal FBM, Heltne PG (eds) Chimpanzee cultures. Harvard University Press, Cambridge, pp 25–39
Parker ST, Gibson KR (1977) Object manipulation, tool use and sensorimotor intelligence as feeding adaptation in cebus monkeys and great apes. J Hum Evol 6:623–641
Povinelli DJ (2000) Folk physics for apes: the chimpanzee's theory of how the world works. Oxford University Press, New York
Takeshita H (1994) Diagnostic tests for orienting manipulation in young chimpanzees (*Pan troglodytes*). Primate Res 10:333–346 (in Japanese with English summary)
Takeshita H (2001) Development of combinatory manipulation in chimpanzee infants (*Pan troglodytes*). Anim Cogn 4:335–345
Takeshita H, van Hooff JARAM (1996) Tool use by chimpanzees (*Pan troglodytes*) of the Arnhem Zoo community. J Psychol Res 38:163–173
Takeshita H, Walraven V (1996) A comparative study of the variety and complexity of object manipulation in captive chimpanzees (*Pan troglodytes*) and bonobos (*Pan paniscus*). Primates 37:423–441
Tanaka M, Tanaka S (1984) Developmental diagnosis of human infants: from 1.5 till 3 years (in Japanese). Ootsuki, Tokyo
Tomasello M, Call J (1997) Primate cognition. Oxford University Press, New York, pp 74–82
Torigoe T (1985) Comparison of object manipulation among 74 species of non-human primates. Primates 26:182–194
van Schaik CP, Deaner RO, Merrill MY (1999) The conditions for tool use in primates: implications for the evolution of material culture. J Hum Evol 36:719–741
Visalberghi E, Limongelli (1994) Lack of comprehension of cause-effect relations in tool-using capuchin monkeys (*Cebus apella*). J Comp Psychol 108:15–22
Weir AAS, Chappell J, Kacelnik A (2002) Shaping of hooks in New Caledonian crows. Science 297:981
Westergaard GC (1992) Object manipulation and the use of tools by infant baboons (*Papio cynocephalus anubis*). J Comp Psychol 106:398–403
Westergaard GC (1993) Development of combinatorial manipulation in infant baboons (*Papio cynocephalus anubis*). J Comp Psychol 107:34–38
Westergaard GC, Suomi SJ (1994) Hierarchical complexity of combinatorial manipulation in capuchin monkeys (*Cebus apella*). Am J Primatol 32:171–176
Yamakoshi G (2004) Evolution of complex feeding techniques in primates: is this the origin of great ape intelligence? In: Russon AE, Begun DR (eds) The evolution of thought: evolutional origins of great ape intelligence. Cambridge University Press, Cambridge, pp 140–171

25
Token Use by Chimpanzees (*Pan troglodytes*): Choice, Metatool, and Cost

CLÁUDIA SOUSA[1] and TETSURO MATSUZAWA[2]

1 Introduction

The use of tokens is a complex behavior involving a set of actions and events, which resemble the use of tools. In other words, a token can be used as a tool, a detached object that is used in some way to arrive at an apparent goal (Matsuzawa 1999a). Thus, both the terms "tool use" and "token use" are applied to the use of a detached and transportable object, used to achieve a clear and defined goal.

The use of tools, as well as their manufacture, was long thought to be one of the hallmarks of our own species, and was thought to have evolved with *Homo habilis* around 3 million years ago (Leakey 1980). The findings of great ape research, however, have shown us differently. The first evidence of the manufacture and use of tools in a wild population of chimpanzees was presented by Jane Goodall in the early 1960s (Goodall 1963). Yamakoshi (2001) listed 54 different tool use patterns gathered from 14 different sites, and more are likely to exist. Without forgetting the importance of tool manufacture, tool use itself is of special importance as it is the most standing example in the discussion of chimpanzee cultural traditions. Chimpanzees exhibit tool use behaviors in many different contexts and often in quite diverse ways (see Beck 1980; Goodall 1970; McGrew 1992).

Several nonhuman primate species use tools, mainly for subsistence purposes. However, chimpanzees regularly and frequently use a greater variety of tools in the wild compared to other primates for both subsistence and nonsubsistence activities. They do so in different contexts and in diverse ways, showing a high degree of flexibility (McGrew 1992; Whiten et al. 1999). Wild chimpanzees select suitable objects well in advance of the anticipated use as a tool (Boesch and Boesch 1984) and sometimes deliberately prepare a tool before carrying it to the site of use (Goodall 1986), suggesting that they possess a mental specification of an appropriate tool (Byrne 2000). Although it may appear that the ape–monkey difference concerning tool use lay in the ability to represent the cause-and-effect

[1]Department of Anthropology, Faculty of Social and Human Sciences, New University of Lisbon, Avenida de Berna, 26-c, 1069-061 Lisbon, Portugal
[2]Primate Research Institute, Kyoto University, 41 Kanrin, Inuyama, Aichi 484-8506, Japan

relations between the tool and the task, the difference may derive from social learning mechanisms (Byrne 2000).

In chimpanzees, tool-using behaviors are acquired through the process of social learning (Matsuzawa et al. 2001). If tool use is to be considered a complex behavioral pattern, then, to acquire it, the learning process would have to work at a higher level in the hierarchical structure of behavior (Byrne 1994 and Byrne and Russon 1998 in Byrne 2000). Recent studies have shown that chimpanzees do tend to copy a demonstrated sequence of actions, but only after repeated observations (Whiten 1998), which suggests that the learning process is dependent on the opportunity to observe the behavior model. Recent laboratory studies have emphasized and reinforced the importance of the opportunity to observe a model in the socially mediated acquisition of knowledge (Hirata and Morimura 2000; Hirata and Celli 2003; Sousa et al. 2003; see Chapter 16 by Hirata, this volume).

Among the various tool use patterns exhibited by wild chimpanzees, some may be said to be more complex in nature than others. For example, the use of a twig to fish for termites (Goodall 1964) requires less complex cognitive abilities than nut cracking, where two stones (anvil and hammer) and a nut have to be combined in the proper order. According to the tree structure analysis methodology, proposed by Matsuzawa (1996, 2001b), of these two examples, termite fishing is considered level 1 and nut cracking is considered level 2. The most complex form of tool use performed by wild chimpanzees is the use of metatools, that is, the use of a tool to produce or enhance the effectiveness of another tool. In nut cracking, Bossou chimpanzees sometimes use a third stone underneath the anvil, which serves as a wedge to stabilize the anvil (Matsuzawa 1991a, 1994). This complex type of tool use is considered level 3 and has been suggested to be near the limit of the cognitive abilities of chimpanzees (Parker and McKinney 1999). The relationship between the complexity of tool use and underlying cognitive abilities is further emphasized by patterns in the ontogeny of tool use. At the age of 2 years, chimpanzees start to perform level 1 tool use, such as the use of leaves to drink water (see Chapter 26 by Biro et al., this volume), whereas only at the age of 3.5 years do they start to perform level 2, such as nut cracking (Matsuzawa 1996; see also Chapter 26 by Biro et al., this volume). Before 3 years of age, chimpanzees are not able to relate more than two objects, which is the reason why they perform inappropriate actions for nut cracking, which require the combination of three objects (Matsuzawa 1994). To date, the youngest chimpanzee that exhibited the only example of level 3 tool use, the metatool, was 6.5 years old (Matsuzawa 1994).

In recent times there has been a growing interest in cognitive abilities underlying complex behavioral patterns such as tool use. But questions such as "What is the limit of chimpanzee cognitive abilities?," "What are the cognitive mechanisms behind complex behaviors like tool use?," or "How do chimpanzees acquire complex behavioral patterns?" cannot be answered through simple naturalistic observation. Experimental studies in captivity must complement observations in the wild to determine not simply what they do but, just as importantly, what they *can* do.

Having these concerns and questions in mind, a new paradigm for the study of chimpanzee cognition was created. This new paradigm draws together, by analogy, the complexity of tool-using behavior and the human monetary system. The human monetary system can be considered symbolic. A symbolic tool is different from the tools used by wild chimpanzees in the sense that the tools that chimpanzees use are in a certain sense all identical in nature, that is, chimpanzee tool use is characterized by a one-to-one correspondence between any particular tool and its target item. In other words, chimpanzees have specific tools for specific goals. A symbolic tool can be used for various goals, has a symbolic relationship to the target, and can be referred to as a "token." Tokens, such as coins, bills, tickets, and cards, can be exchanged for different classes of desired items, such as food, objects, comfort, and services. This "symbolic tool," the token, is unique in several aspects: it can be exchanged for different kinds of items (exchangeability), it is easy to handle and transport (portability), its value remains unchanged for extended periods so that it can be accumulated (saving), and it can be used within a hierarchical system (hierarchy). The new token system created for chimpanzees consisted of a computer-controlled task with token rewards that could be exchanged for items (e.g., food item) by inserting them into a computer-controlled vending machine. The subjects have to understand first the symbolic relationship between the stimuli that compose the task to receive a token, that is, they have to match, for example, a red square with a symbol meaning red. Second, they have to understand that the token that they received after performing the task can be used in a different computer setting than the first one. Finally, they can exchange the token for their preferred food item, performing a binary choice with pictures of the foods. From this perspective, token use consists of a series of complex behavioral events that have to be related in succession, thus simulating the variability of complex tool use in the wild. During the performance of all this sequence of events, the subjects have also to balance the costs and benefits involved choosing a particular type of food or behavioral strategy (see Sousa and Matsuzawa 2001).

Previous studies had already demonstrated that chimpanzees can use tokens in exchange for food rewards (Cowles 1937; Wolfe 1936; Kelleher 1956, 1957a–c, 1958). The main focus of these first studies was the reward aspect of tokens, analyzing essentially their effectiveness. Although these studies showed that chimpanzees can use tokens in exchange for a food reward (Cowles 1937; Wolfe 1936; Kelleher 1956, 1957a–c, 1958), in general, the task that the subjects had to do to receive tokens consisted of simple physical work, such as lifting a weight (Wolfe 1936) or pressing a telephone key (Kelleher 1956, 1957a–c, 1958). This physical work did not require any mental effort or ability besides that of understanding the relationship between the response and the ensuing event. The only exception was Cowles' (1937) study in which the tasks used required cognitive skills to some extent, such as position and visual discrimination.

More recently, the new token system has shown tokens to be equivalent to food rewards in maintaining the subjects working in a matching-to-sample task, eliciting the appearance of a unique behavior among the subjects, the saving of tokens, one of the aspects that characterize the token as a symbolic tool. This study also evidenced another property of the token, its portability. The subjects spontaneous and easily accumulated and transported several tokens before exchanging them for food (Sousa and Matsuzawa 2001).

The new methodology has also allowed the gathering of data about socially mediated acquisition of knowledge. One of the subjects of the study by Sousa and Matsuzawa (2001), Ai, had a son, Ayumu, who she carried to the subsequent experiments. Sousa and colleagues (Sousa et al. 2003) studied and described the behavioral and cognitive development of Ayumu from birth until the age of 2 years and 3 months. The infant spontaneously learned the entire sequence of events in the token experiment performed by his mother. At the age of 9 months and 3 weeks he performed a matching-to-sample trial for the first time, which suggested a spontaneous interest in the mother's activities. For almost 1.5 years he continued to perform matching-to-sample trials, receiving tokens once in a while, but without being able to exchange them for food. This behavior showed a high degree of motivation. But it was only at the age of 2 years and 3 months that he succeeded for the first time in inserting tokens and exchanging them for food. The learning process of Ayumu during this study was characterized by close observation by the infant for extended periods of time, spontaneity and strong motivation for the behavior, and high levels of tolerance from the mother. This successive learning of all the events involved in this complex behavioral pattern, simulating the findings from the wild, emphasizes the importance and validity of 'education by master-apprenticeship' (Matsuzawa 2003; Matsuzawa et al. 2001).

This chapter focuses on three experimental studies that explore the use of tokens by chimpanzees, analyzing the complexity of their cognition and behavior. The research presented in this chapter aims to further research and test the applicability of this new methodology, the token system, to assess chimpanzee cognitive abilities, especially those underlying complex behaviors. If these aims are accomplished, then the work can be stretched further to study the depths of chimpanzee cognitive complexity.

The first two experiments investigate another unique aspect that characterizes a token as symbolic tool, e.g., exchangeability. The exchangeability property of a token is explored by analyzing the choice behavior of two chimpanzees. In experiment 3, the costs and benefits involved in performing the token experiment were analyzed by increasing to two the number of tokens required to obtain one piece of food.

2 General Methods

2.1 Subjects

The subjects were two adult female captive chimpanzees (*Pan troglodytes*), Ai and Pendesa. They were living in a social group, in an outdoor enclosure, in semi-natural conditions. They had extensive experience of matching-to-sample tasks

(Biro and Matsuzawa 2001; Matsuzawa 2001a) and had already received coins as a reward in a previous study (S. Suzuki and T. Matsuzawa, unpublished work; Sousa and Matsuzawa 2001). They were at no time food deprived and were cared for according to guidelines produced by the Primate Research Institute.

2.2 Apparatus

The subjects were tested inside an experimental booth (approximately 180 × 180 × 180 cm) with acrylic panels as walls, and the same apparatus used in a previous experiment by Sousa and Matsuzawa (2001): two touch-sensitive screens (Micro Touch SMT2 and Micro Touch CT-1000) connected to two personal computers, respectively; a vending machine (CZX CONLUX ZD-160-A); three universal feeders (Biomedica universal feeder BFU-310); and three food trays where the rewards were delivered. Japanese 100-yen coins were used as tokens.

2.3 General Procedure

The procedure was similar to the one used in a previous experiment by Sousa and Matsuzawa (2001). The experimental procedure involved two phases: a matching phase and an exchange phase (Fig. 1). The procedure ensured that the number of token rewards received in each session was constant. Each session ended after the subject inserted the final token to get a food reward.

2.3.1 Matching Phase

This phase consisted of a matching-to-sample (MTS) task with two alternatives and token reinforcement. Each trial began with the presentation of a white circle against a black background near the bottom of the touch screen. After the chimpanzee touched this starting stimulus, the circle disappeared and a sample stimulus appeared on the screen. To proceed to the next step, the subject was required to touch this sample stimulus. The touch resulted in the appearance of two choice stimuli while the sample stimulus remained on the screen. The subject was then required to choose and touch on of the two alternatives, which physically matched the sample stimulus. A correct response was followed by a chime sound and the delivery of a token. An incorrect response was not rewarded and was followed by a beep sound and a "time out" (3 s). The starting stimulus then appeared again to mark the start of a new trial. The task used for Pendesa consisted of an identity matching-to-sample (IMTS) task using colors, lexigrams, or kanji. Ai performed both identity (IMTS) and symbolic matching-to-sample (SMTS) tasks, in all nine possible combinations of color, lexigram, and kanji. The subject was allowed to complete as many trials as she wished before proceeding to the exchange phase. The number of tokens accumulated before proceeding to the exchange phase constitutes a bout. The procedure of the matching phase was the same in all three experiments.

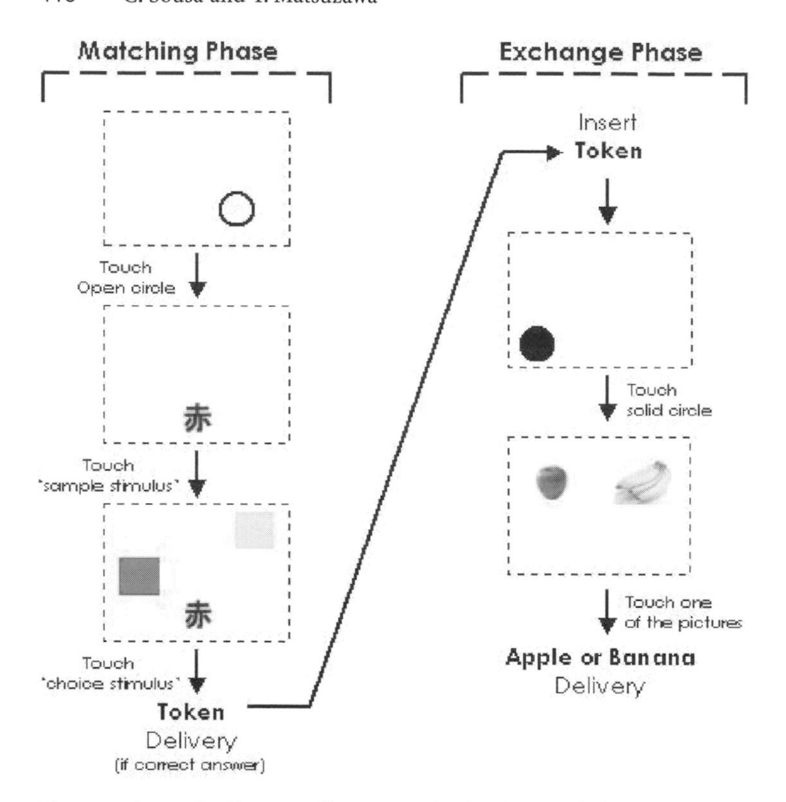

Fig. 1. Schematic diagram illustrating both phases of the token experiment. The matching phase consists of a matching-to-sample task with token rewards. The exchange phase consists of a food choice with pictures on a screen (activated after inserting a token into the vending machine) with one of ten different foods as rewards

2.3.2 Exchange Phase

To exchange each token for a food reward, the subject had to insert the token into the vending machine through a slot in the acrylic panel located to the right of the screen. After each insertion, a solid white circle appeared on the touch monitor. After touching the solid white circle, two pictures appeared on the monitor (Fig. 2). These were randomly chosen from among the food or/and object items (Fig. 3) in use in a given experiment. The subject was required to choose one of the two pictures by touching it and then received the corresponding item. The procedure of the exchange phase had some differences in each of the three experiments according to the aims. The details of these differences are described for each experiment.

2.4 Rewards

According to the aims of each experiment, the rewards for the binary choice in the exchange phase of the experiment consisted of pieces of food or of pieces of food and object items (Fig. 3).

Fig. 2. Ai choosing one of the two food items (blueberry and peanut) presented in experiment 1

Fig. 3. Colored pictures of the 10 food items, 8 of the 34 toy objects, and the 2 keys (with extra value: to open a box with food) used as stimuli for the binary choice procedure

2.5 Stimuli

The stimuli used for the MTS task of the matching phase were 10 colored squares (red, orange, yellow, green, blue, purple, pink, brown, white, gray), 10 visual symbols called lexigrams, and 10 Chinese kanji characters (Fig. 4). The size of the stimuli was 6 cm × 6 cm.

The stimuli for the binary choice of the exchange phase were pictures of the items corresponding to the rewards in use in that experiment (see Fig. 3). The size of the stimuli was 10.3 cm × 8 cm.

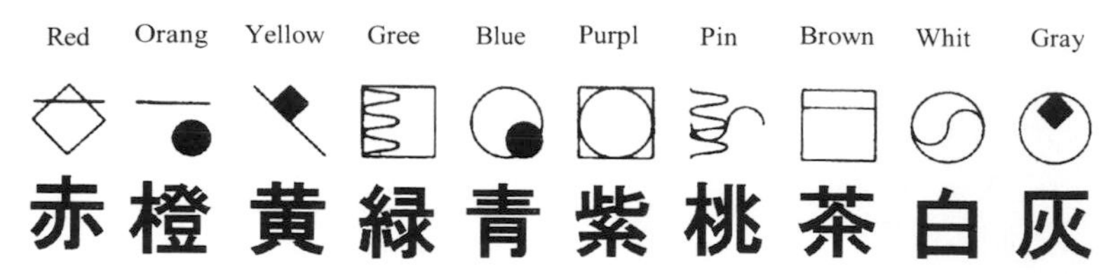

Fig. 4. The 10 colors, 10 lexigrams, and 10 kanji used as stimuli for the matching-to-sample (MTS) tasks

3 Experiment 1: Food Choice

3.1 Purpose and Background

Food is the center of any animal life, and chimpanzees are no exception. In the wild, chimpanzees spend 46% to 60% of their active day feeding (Wrangham 1977). Most of this feeding time occurs in a few uninterrupted bouts, which means that they travel from food patch to food patch. Chimpanzees are omnivorous, that is, they feed on a large variety of foods, but they do not do so randomly. The largest proportion of their diet consists of fruit, followed by leaves and herb piths (Yamakoshi 2001). For example, among the 664 plant species available within their habitat, chimpanzees at Bossou, West Africa, only consume 200 of them (Sugiyama and Koman 1992), suggesting that they carefully choose what they are going to eat. Furthermore, the majority of the examples of chimpanzee tool use are for subsistence purposes, that is, to acquire food and water.

The question of food preference is a particularly interesting one. In general, when we think about the foraging behavior of nonhuman animals, we rationalize it as if it were a simple search for "fuel" necessary for survival. However, studies in captivity with nonhuman primates have shown that subjects exhibit a preferential choice for more palatable foods, when available, suggesting that an immediate response to pleasurable taste stimulation is present in determining food choices (Hladik 1979). If this is true, then foraging activity is likely to be a more complex behavior than previously thought. Individuals must spend their days trying to equilibrate their personal preferences with their nutrient requirements and food availability. Wild chimpanzees are assumed to have specific food preferences based on the time they spend eating a specific food (Wrangham et al. 1996). Without wishing to undermine the veracity of this assumption, we must not forget that the time spent by chimpanzees at a food patch is also determined by the quantity of food items available, the number of competing conspecifics, the size of each food item, and the difficulty in reaching the edible part of the food item, which is often accomplished with the help of a tool (Yamakoshi 2001).

We usually correlate the choice of a specific food with certain nutrient needs; however, previous studies with other nonhuman primates showed that the preferences observed in captivity did not always correspond to the species' diet in the wild, suggesting that some other factor must be involved (Charles-Dominique and Bearder 1979). One of the factors affecting food choice is palatability. The preferential choice of more palatable foods, when available, suggests that the immediate response to pleasurable taste stimulation is not absent in determining food choices (Hladik 1979). There may also be social processes at work (see Chapter 16 by Addessi and Visalberghi, this volume; Visalberghi and Addessi 2001; Wilson 2001). Infant nonhuman primates acquire the knowledge about what foods they should and should not eat by observing the foraging behavior of their conspecifics, and especially their mother (for chimpanzee mother–infant pairs, see Ueno and Matsuzawa 2004; Chapter 10 by Ueno, this volume). In the wild all these factors act in combination, making it impossible to study them independently. For these reasons, food tests in captivity may help to elucidate the importance of various influences upon food selection.

To study chimpanzees' choice behavior, and in specially food choice, it is necessary to find a suitable method to measure those choices. The simplest method of investigating dietary preferences consists in presenting various types of food to a caged animal and recording what it selects and what it rejects (Hladik 1979). However, this method may be less than ideal, given the amount of direct interaction between the experimenter and the subject. As social processes are one of the factors affecting food choice (see Chapter 16 by Addessi and Visalberghi, this volume; Visalberghi and Addessi 2001; Wilson 2001), they could clearly bias the results.

It is common to infer food preference from the time spent by an individual eating each target food item. This approach is particularly frequent in studies in the wild. However, the time allocated for feeding may not simply reflect the preference as consummatory behavior itself also plays a part: the time an individual spends eating a particular food also depends on its size and how difficult it is to access the edible portion. With humans, besides asking verbally which food individuals prefer most, we can also infer preference from the amount of money, from a limited budget, each individual is willing to spend on a particular food item.

In analogy to the human monetary system, a token system can also be used with chimpanzees to measure their food choice. The use of a limited token budget will work in the same way as humans' limited money budget to assess food preferences in chimpanzees. In this sense, a computer-controlled method using tokens that can be exchanged for chosen foods within a binary choice paradigm would constitute a far-superior method to approach questions of preference, giving more accurate and reliable results. In a simple binary choice, human subjects match their response proportions to reinforcement proportions (Fantino 1998). There is no reason to think that chimpanzees will behave in a different way. So, if the reinforcement proportions are the same, the subjects' food preferences alone will determine their choice. Hence, the token system will also constitute a far more efficient method given that no other factor will be at work during the food choice behavior.

In this experiment, we investigated the food preference of the subjects in a free binary food choice situation, using tokens as a measurement scale. At the same time, it has a goal to test the exchangeability characteristic of tokens as symbolic tools. It also serves as a baseline for comparison of behavior with subsequent experiments.

3.2 Method

3.2.1 Subjects

During this experiment, Ai was 23 years old and Pendesa was 22 years old. They were living with nine other adult chimpanzees in a well-established social group.

3.2.2 Apparatus

The apparatus was as described in the general methods section.

3.2.3 Procedure

The procedures of both phases of the experiment were as described in the general methods. A daily experiment usually consisted of 3 sessions. Each session consisted of 40 trials, that is, the subject had to perform the necessary number of trials on the Matching Phase to get 40 tokens. For each token inserted into the vending machine, a binary food choice was presented to the subject. The two pictures that appeared on the monitor were randomly chosen from the 45 possible combinations of pairs taken from among the 10 food items (see Fig. 3) and remained the same throughout a given session. She would then get a piece of food, corresponding to her choice on the monitor of the Exchange Phase. Each session ended after the subject inserted the final, 40th token to get a food reward. Before initiating each session, a piece of each of the two food items being tested was manually delivered to the subject, making sure that the chimpanzee ate the food item. Both subjects participated in a total of 45 sessions.

3.2.4 Rewards

In this experiment the rewards for the food binary choice consisted of pieces of apple, banana, blueberry, carrot, chow, grape, peanut, pistachio nut, potato, and raisin (all approximately the same size: 1 cm × 1 cm × 1 cm, 1.2–1.5 g a piece on average).

3.2.5 Stimuli

The stimuli for the binary choice of the exchange phase were ten pictures of ten different food items (apple, banana, blueberry, carrot, chow, grape, peanut, pistachio nut, potato, and raisin).

3.3 Results

3.3.1 Matching Performance

The results obtained during the MTS task for both subjects, Ai and Pendesa, were calculated in terms of accuracy, for the total of 45 sessions. Accuracy, measured as the percentage of correct responses, was significantly above chance level (50%) both for Ai (average accuracy = 92.1%; binomial test: $z = 36.70$, $P < 0.001$) and for Pendesa (average accuracy = 98.2%; binomial test: $z = 41.24$, $P < 0.001$), and showed little variation over the course of sessions. A more detailed analysis of the subjects' performance in the MTS tasks was not undertaken as that was not the purpose of the present study.

3.3.2 Choice Behavior

The food preference of both subjects was calculated by combining results from all sessions (Fig. 5). Using the total frequencies of choosing a particular food item, an order of preference for each subject was calculated. For Ai, the order of preference was pistachio, blueberry, peanut, banana, chow, raisin, grape, potato, carrot, and apple. For Pendesa, it was blueberry, pistachio, peanut, raisin, chow, grape, potato, banana, apple, and carrot. We calculated the correlation between the orders of preference in the two subjects (Kendall tau = 0.691, $P < 0.01$) and found a strong correlation. The determining coefficient (r^2) was 0.48. The two most preferred food items were blueberry and pistachio nut for both Ai and Pendesa; the two least preferred food items were carrot and apple. Among the ten food items, four (blueberry, peanut, chow, and carrot) were shown to carry no difference in food preference between the two subjects, Ai and Pendesa. The remaining six items (pistachio, raisin, banana, grape, potato, and apple) showed significant differences in individual food preferences. For example, Ai preferred

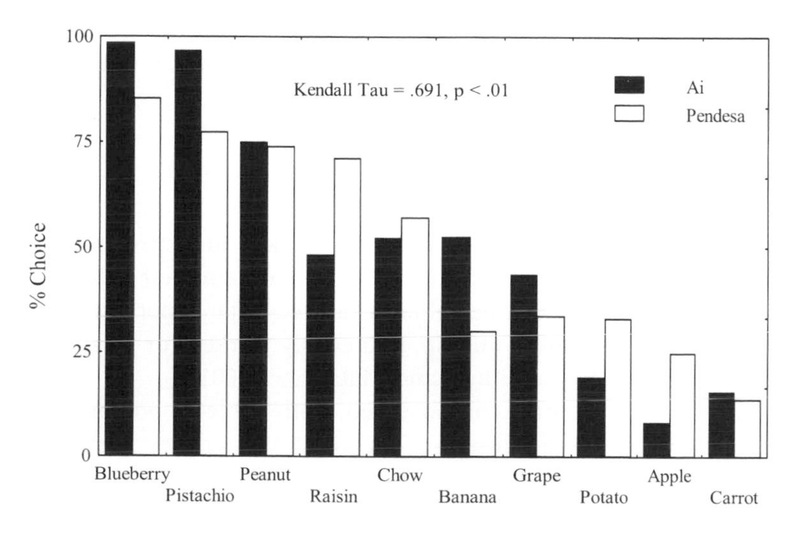

Fig. 5. The percentage of choice of each food item for both subjects in experiment 1

pistachio nuts over all other foods,whereas in Pendesa's rating this was "only" second best.

The Spearman rank-order test was then used to analyze the relationship between food preferences and the energy content of each food item. The result showed no significant correlation between food preference ranking and energy content (Ai: $r_s = 0.539$, $z = 1.81$, $P < 0.11$; Pendesa: $r_s = 0.515$, $z = 1.70$, $P < 0.13$). However, when banana and blueberry were omitted from the food preference ranking, we found significant positive correlation with total energy content (both subjects: $r_s = 0.952$, $z = 7.65$, $P < 0.01$), that is, the chimpanzees clearly preferred foods that are high in energy over foods that are low in energy.

3.3.3 Saving Pattern

The results concerning the saving behavior of both subjects followed the same pattern as in our previous study (Sousa and Matsuzawa 2001). The saving pattern was analyzed with respect to the size of bouts (number of tokens accumulated before proceeding to the exchange phase) following the procedure described in Sousa and Matsuzawa (2001). We had computed a "saving index" based on the weighted probability per opportunity: $I_x = (P_{x-1} \times P_x) \times 100$ (Sousa and Matsuzawa 2001). To compute the saving index (I_x), the probability of collecting each token (P_x), in a rank position (x) in a bout was weighted by the probability of collecting the previous token (P_{x-1}). The probability of collecting each token (P_x) is given by dividing the number of tokens collected by the number of opportunities of collecting them.

Ai's probability of continuing to collect tokens dropped very quickly after receiving the first token. In Pendesa's case, the probability of continuing to collect coins remained at 100% up to the 7th coin and was maintained above 90% up to the 14th coin. It then dropped quickly, but the curve was not as steep as in the case of Ai (Fig. 6). Pendesa's probability of continuing to save never reached 0%, because in four sessions she collected all the coins available (40) in a single bout before exchanging them. The 50% probability of saving was 2.6 tokens on average for Ai, and 24.4 tokens on average for Pendesa.

3.4 Discussion

Our results demonstrated that chimpanzees display marked food preferences. Both subjects worked for tokens and used them as tools to obtain their preferred food item in various choice contexts. The subjects showed a stable performance during the course of this experiment and spontaneously worked for tokens as shown in our previous experiments (Sousa and Matsuzawa 2001).

According to the food choice patterns observed, both subjects possessed a very similar rank order of food preference. The two chimpanzees showed a general tendency to prefer nuts and to rate vegetables and apples less highly. However, as apples had been given as food rewards in most of the past studies that the subjects had participated in, they may simply have become bored with

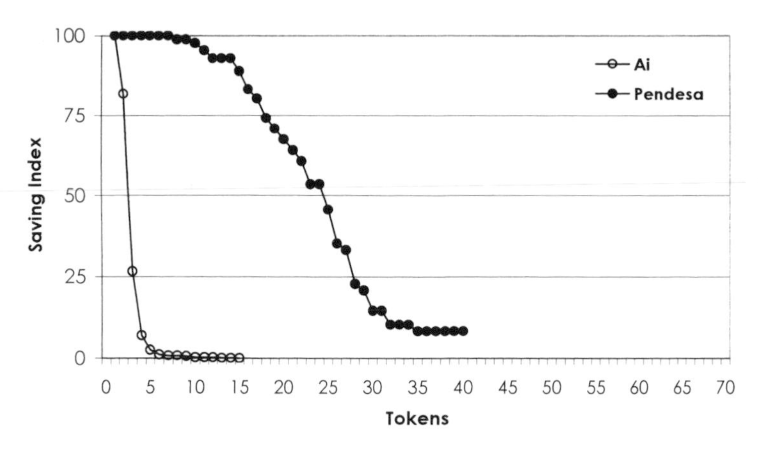

Fig. 6. Saving index of both subjects during experiment 1

apples. The chimpanzees showed interesting differences in food preference. Among the ten food items, six showed slight differences in the preference rank between the two subjects.

Although our findings demonstrate that nutritional value plays an important role in the food preferences of chimpanzees, other factors may also be involved. The choice of banana and blueberry was not correlated with the total amount of energy, suggesting that the palatability of the foods may also affect food choice, as already suggested by Fragaszy and colleagues (Fragaszy et al. 1997).

In summary, the chimpanzees displayed marked food preferences in a computer-controlled two-alternative choice situation, using tokens. The use of tokens provided a sensitive scale of preference for food items in chimpanzees. Further studies are necessary to elucidate additional aspects of chimpanzees' food choice, using tokens as scale indicators.

4 Experiment 2: Object Choice

4.1 Purpose and Background

In addition to the preference for some food items, choice behavior is also present in many other activities of animals, from choosing roots for foraging activities to choosing mating and friendship partners, to other social behaviors. Associated with the choice behavior is the formation of preferences for individuals or for items. This experiment analyzed the preference for nonfood items. Different objects were introduced within the free choice situation, that is, the subjects were required to choose between a food item and an object. Besides this goal, it also aimed to strengthen the fact that a tool is a symbolic tool by showing its exchangeability characteristic. A token can only be considered a symbolic tool if the subjects do not only associate it to food but also can understand its exchangeability for other items.

Experiment 2 tested the choice behavior of the subjects between objects and favorite and nonfavorite food items. It also aimed to further test the exchangeability aspect of tokens. This experiment had two stages. In stage I, the objects used in the binary choice situation did not have any extra value except for the object itself that was delivered to the subject when chosen. In stage II, the objects were keys that could be used to open a box to retrieve a different and bigger food item, that is, they had extra value.

4.2 Method

4.2.1 Subjects

During stage I of this experiment, Ai was 24 years old and Pendesa was 23 years old; during stage II, Ai was 25 years old and Pendesa was 24 years old. They were living in a social group composed of 14 chimpanzees, 3 of which were babies. One of these babies was Ai's son, Ayumu, who was 9 months old (born on April 24, 2000) when this experiment started. Ayumu was raised by his mother, accompanying her at all times, even during experiments. The attention that Ayumu as a baby required from his mother caused some changes on the dynamic of the experiment (e.g., decrease of the number of sessions per daily experiment, increase of the total duration of the experiment). When Ayumu was 9 months and 3 weeks old, during one of the sessions of stage I of this experiment, he touched for the first time the monitor for the MTS task, performing a complete matching-to-sample trial correctly (Sousa et al. 2003). Also during this experiment, stage II, Ayumu started to perform the exchange phase of the token experiment, by inserting the tokens and exchanging them for food rewards. He was 2 years and 3 months when he used a token for the first time (Sousa et al. 2003). This happening caused an increase in his motivation to perform the matching phase of the experiment to get tokens that he could then exchange for food rewards. These facts changed the number of trials performed by Ai on both matching and exchange phases of each experimental session.

4.2.2 Apparatus

The apparatus was as described in the general methods section. During stage I, an extra box was attached to the experimental room. This box had a sliding door that allowed the experimenter to deliver objects to the subject when it constituted her choice. During stage II, two extra boxes were attached to the experimental room. Each of these boxes had a door with a lock that could be opened by the subject with a key to retrieve a big piece of food.

4.2.3 Procedure

A daily experiment usually consisted of one session, but depending on the willingness of the subjects to continue, they could be presented with two or three sessions. Although both subjects were presented with the same number of trials in each session (40), the total number of trials performed by Ai was different from session to session given that she was accompanied by her son Ayumu, who almost always did a couple of trials. The procedure of the exchange phase was the same as in experiment 1, except for the fact that on the last trial of the exchange phase, that is, after the insertion of the 40th token, a picture of an object was displayed together with the picture of one of the two food items used during that session (Fig. 7A). If the subject touched the picture of the object, the corresponding object would be delivered through the delivery box. In stage II, the object delivered (a red or a blue key) could then be used to open a box (Fig. 7) that contained a different food item. Both subjects participated in 38 sessions in stage I and 60 sessions in stage II.

4.2.4 Rewards

The food rewards used for the binary choice were the same as used during experiment 1. Besides those food rewards, object rewards were also used. In stage I, the objects used as rewards in the binary choice situation were the following: bicycle bell, creaking ball, doll, small container, handgrip, green box, towel, ball float, magnet, doll pot, brush, yellow ball, red pen, mini disc, notebook, board magnet, big clip, blue ball, envelope, yellow container, blue glove, blue box, toothbrush, pink towel, blue bear, chicken, gray seal, slide box, pencil, wood spoon, green key, grey key, pink key. In stage II, two keys (blue and red) were also used as rewards in the binary choice situation. These keys could be used to open a box where one of the following fruit items would be placed: apple, banana, fig, grapefruit, kiwi, lemon, mandarin orange, orange, peach, pear, persimmon fruit, pineapple, plum, strawberries, and watermelon. These food rewards were given as either the entire fruit (e.g., one apple or two strawberries) or a portion of the fruit (e.g., two slices of pineapple).

4.2.5 Stimuli

The stimuli for the binary choice of the exchange phase of the experiment were the same 10 pictures of food items used in experiment 1 and pictures of the objects used as rewards (see Fig. 3). In stage I, 34 pictures of different objects were used. In stage II, a picture of a blue key and a picture of a red key were used as stimuli.

4.3 Results

4.3.1 Matching Performance

The accuracy in the MTS task was calculated as in experiment 1. In the case of Ai, only the trials performed by her where considered. As Ayumu performed a variable numbers of MTS trials, the number of trials performed by Ai was not constant through the sessions. The results obtained during the MTS task

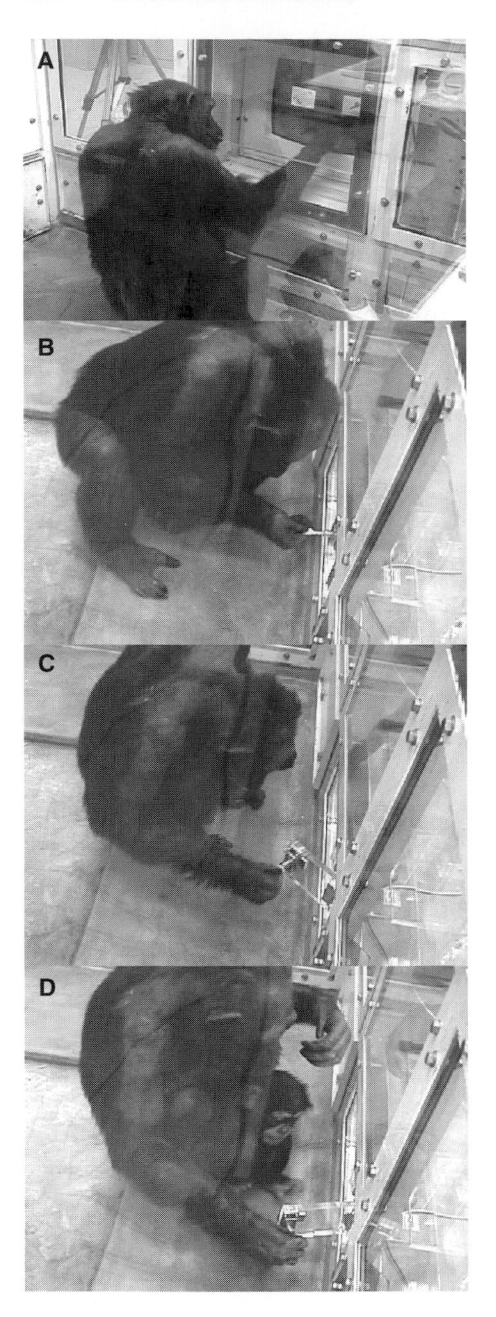

Fig. 7. Ai choosing a key (over *banana*) and using it to open the box, in stage II of experiment 2

did not differ from the ones in experiment 1, were significantly above chance level (50%) both for Ai and for Pendesa ($P < 0.001$ in binomial test), and showed little variation over the course of sessions. As in the previous experiment, a more detailed analysis of the subjects' performance in the MTS tasks was not undertaken as that was not the purpose of the present study.

Ai

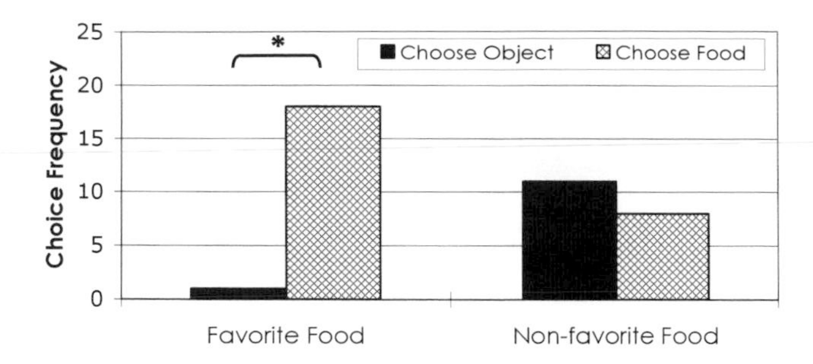

Pendesa

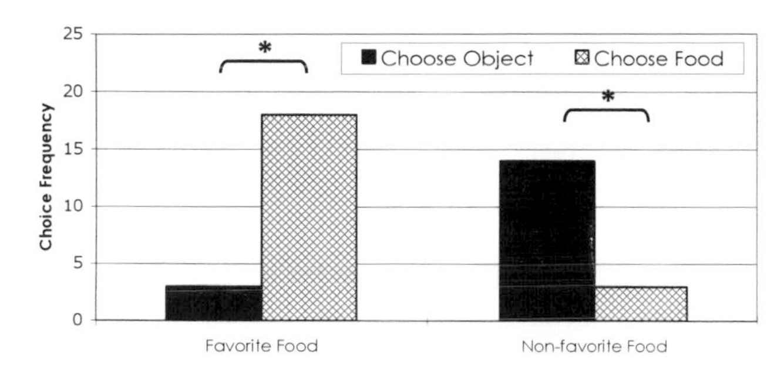

Fig. 8. The frequency of choosing objects or food items for both subjects in experiment 2, stage I. The data are presented according to the context of the choice, that is, if the subject was presented with a favorite food item or with a nonfavorite food item (*$P < 0.01$)

4.3.2 Choice Behavior

4.3.2.1 Stage I: Object Choice

The results of this experiment were grouped taking into consideration the subject's preference for the food item presented together with the object. The food preference was considered session-by-session, through the total number of choices of the subject in that specific session. Given this, the results of choosing an object or a piece of food were analyzed taking into consideration whether that object was presented together with a favorite or a nonfavorite food item (Fig. 8). When the choice alternative to the object was a favorite food, both subjects preferred to choose the food (proportion test, $P < 0.001$). When the choice alternative to the object was a nonfavorite food, Ai did not have a clear preference (proportion test, $P = 0.648$), whereas Pendesa preferred to choose the object (proportion test, $P < 0.013$).

4.3.2.2 Stage II: Object Choice with Extra Value

The results of this experiment were analyzed by comparing with the total number of choices of object and food items in Stage I (Fig. 9). Both subjects

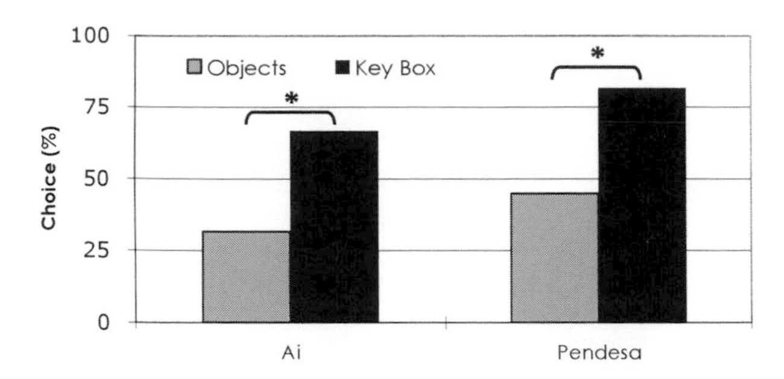

Fig. 9. The percentage of choice of objects or keys (which could be used to open a box with food) for both subjects in experiment 2, stage II

chose more frequently the objects (keys to open the box) in stage II than in stage I (two-proportion test, $P < 0.001$).

4.3.3 Saving Pattern

Figure 10 compares the saving index of both Ai and Pendesa during experiment 1 and both stages of experiment 2.

4.3.3.1 Stage I: Object Choice

Ai's probability of continuing to collect tokens dropped very quickly after receiving the second token, but in general her saving patterns did not differ much from that in experiment 1 (Wilcoxon: $z = -0.385, P < 0.70$). In Pendesa's case, the probability of continuing to collect coins remained at 100% up to the 7th coin and was maintained above 90% up to the 19th coin. It then dropped quickly, following a pattern similar to that of experiment 1, although significantly different (Wilcoxon: $z = -2.619, P < 0.01$). The 50% probability of saving was 3 tokens on average for Ai and 25.5 tokens on average for Pendesa.

4.3.3.2 Stage II: Object Choice with Extra Value

Ai's probability of continuing to collect tokens dropped very quickly after receiving the 2nd token, showing a saving pattern significantly different from that in experiment 1 (Wilcoxon: $z = -4.618, P < 0.001$). While in experiment 1 Ai's probability to continue to collect tokens reached zero after the 14th token, here it reached zero only after collecting 28 tokens. In Pendesa's case, the probability of continuing to collect tokens started to drop after the 1st token, although slowly, remaining above 90% up to the 4th token. It then dropped more quickly than in the previous experiments, following a pattern significantly different (Wilcoxon: $z = -5.318, P < 0.001$). The 50% probability of saving was also 3 tokens on average for Ai, but 16.1 tokens on average for Pendesa.

4.4 Discussion

The results of stage I demonstrated that the subjects' preference continued to be for food when objects were introduced within the binary choice situation. In

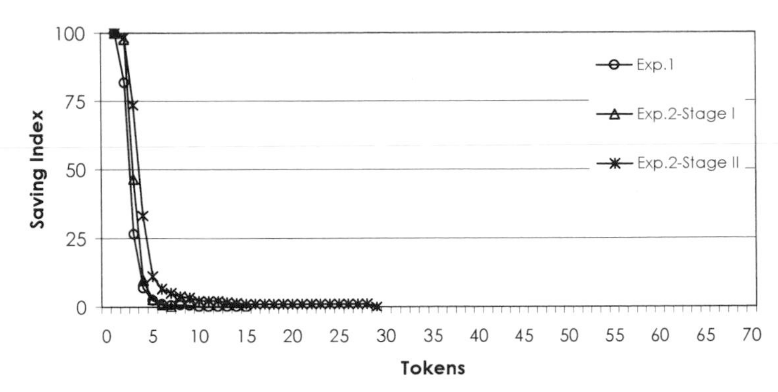

Pendesa

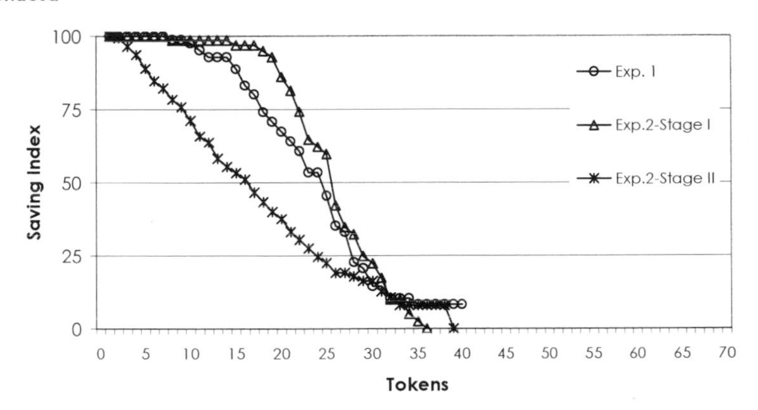

Fig. 10. Comparison of the saving index of both subjects during experiment 1 and stages I and II of experiment 2

general, the subjects only chose the objects when the other choice alternative was a nonfavorite food item. However, when the objects (keys) within the binary choice situation acquired an extra value, that is, they could be used to open a box and get access to a bigger and different food item, the choice pattern changed completely. In this situation, both subjects started to choose the keys over the food item presented. This result shows that the subjects understood the extra value acquired by the keys and that they could use them to get a bigger and different piece of food. In other words, the results suggest that the subjects were evaluating the costs and benefits of their actions, maximizing their choices.

A key that can be used to open a box and get access to a food item is a tool. The subjects were using a token to get a key to get food. This sequence of actions can be compared to the use of a metatool, a tool for another tool, by chimpanzees. Bossou chimpanzees sometimes placed a third stone beneath their anvil stone as a wedge, stabilizing the anvil and maintaining a flat, horizontal upper surface, suitable for nut-cracking (Matsuzawa 1994; see also Chapter 26 by Biro et al., this volume). In this group, a 4-year-old female chimpanzee was observed using a dead twig to push a leaf-tool deep into a tree hole, and then use the twig to retrieve it, before putting it into her mouth and drinking water (Matsuzawa 1991b; Sugiyama 1995). These examples are described as a use of metatool or the use of a tool composite. The subjects of this experiment, Ai and Pendesa, also used a token as a metatool to get a key that they could use to get a bigger food item. This situation happened in several instances, repeatedly over the sessions, showing their understanding of this complex sequence of actions to attain a clear and defined goal, the food. The sequence of actions performed by the subjects in this experiment can then be compared to the used of a metatool or a tool composite.

The introduction of objects also caused changes in the saving pattern of the subjects. Although both subjects maintained the general pattern of saving tokens, that is, one of them (Ai) saving a small amount of tokens and the other one (Pendesa) saving a higher amount of tokens, significant changes were observed when comparing the results of experiment 1 to those of experiment 2. Ai increased the number of tokens collected per bout before proceeding to the exchange phase, in both stages of the experiment, although the differences were higher in stage II, when the choice was between a food item and a key that could be used to get access to a bigger piece of food. Pendesa, in stage I, slightly increased the number of tokens collected per bout before proceeding to the exchange phase. However, her behavior changed drastically in stage II, when the choice was between a food item and a key that could be used to get access to a bigger piece of food. In this case, Pendesa showed an enormous decrease in the number of tokens collected per bout.

The results of this experiment further reinforces the exchangeability characteristic of the tokens, because the subjects were able to understand that they could use the same detachable object, the token, to choose between different kinds of items, that is, between food and an object.

5 Experiment 3: Double Price

5.1 Purpose and Background

Behind the establishment of a preference is the formation of the concept of value of that item (Brosnan and de Waal 2004). This idea implies some simple economy at work, by comparing items based on their value. This economy, or the evaluation of the item value in a decision-making situation, can also be influenced by a balance of costs and benefits. The costs and benefits can be related with time spent, energy spent and acquired, and also with other factors of social life, especially if we are talking about social animals such as primates. The saving behavior described by Sousa and Matsuzawa (2001) might be related to the discussion of the evaluation of costs and benefits, based on the time spent to perform the behavior, and consequently the latency to obtain the food reward. In experiment

3, the costs and benefits involved in performing the token experiment were analyzed by increasing to two the number of tokens required to obtain one piece of food.

5.2 Method

5.2.1 Subjects

During this experiment, both Ai and Pendesa were 26 years old. As in experiment 2, they were living in a social group composed of 14 chimpanzees, 3 of which were babies. One of these babies was Ai's son, Ayumu, who was 3 years and 2 months when this experiment started.

5.2.2 Apparatus

The apparatus was as described in the general methods section.

5.2.3 Rewards

The rewards used were the same ten food items used in experiment 1.

5.2.4 Stimuli

The stimuli were the same as in experiment 1.

5.2.5 Procedure

The procedure of the exchange phase was similar to that in experiment 1, but the number of tokens required to receive one piece of food was doubled, that is, the subject had to insert two tokens into the vending machine to have the two pictures of food displayed on the monitor (Fig. 11), and then choose and receive one of them. Stage I consisted of 12 sessions of 40 trials each, meaning that in each session the subject would get 40 tokens, but only 20 pieces of food. In stage II the number of trials was doubled, so that at the end of each session the subject would get 40 pieces of food as in previous experiments. As in experiment 2, stage II, the total number of trials performed by Ai was different across the session given that her son Ayumu almost always did a couple of trials.

5.3 Results

5.3.1 Matching Performance

The accuracy in the MTS task was calculated as in experiment 1. In the case of Ai, as in experiment 2, only the trials performed by her where considered, resulting in an irregular number of trials per session. The results obtained during the MTS task did not differ from those in experiment 1 or 2 and were significantly above chance level (50%) both for Ai and for Pendesa ($P < 0.001$ in binomial test) and showed little variation over the course of sessions.

Fig. 11. Schematic diagram illustrating the exchange phase of experiment 3. The experiment consists of a food choice with pictures on a screen (activated after the subjects insert two tokens into the vending machine) with one of ten different foods as rewards

5.3.2 Saving Pattern

Figure 12 compares the saving index of both Ai and Pendesa during experiment 1 and both stages of experiment 3.

5.3.2.1 Stage I: Double Price with 40 Trials and 20 Food Rewards

Ai's probability of continuing to collect tokens dropped very quickly after receiving the third token, showing a saving patterns significantly different from that in experiment 1 (Wilcoxon: $z = -3.110$, $P < 0.01$). In Pendesa's case, the probability of continuing to collect tokens started to drop after the 1st token, although slowly, remaining above 90% up to the 6th token. It then dropped quickly, following a pattern significantly different (Wilcoxon: $z = -5.445$, $P < 0.001$). The 50% probability of saving was 4.8 tokens on average for Ai, and 12.1 tokens on average for Pendesa.

5.3.2.2 Stage II: Double Price with 80 Trials and 40 Food Rewards

Ai's probability of continuing to collect tokens dropped very quickly after receiving the third token, showing a saving patterns also significantly different from that in experiment 1 (Wilcoxon: $z = -7.390$, $P < 0.001$). In Pendesa's case, the probability of continuing to collect tokens remained at 100% up to the 7th token

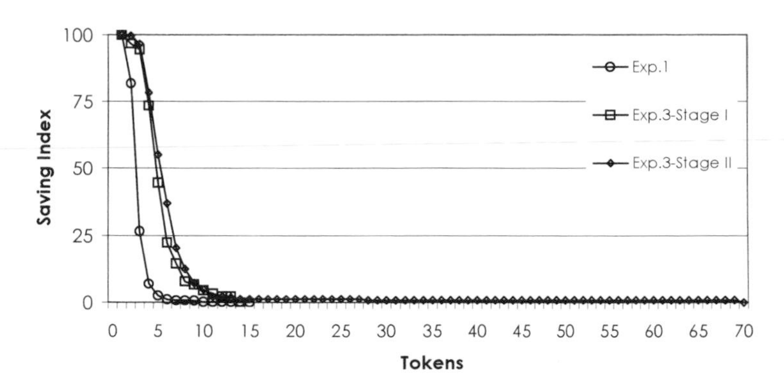

Fig. 12. Comparison of the saving index of both subjects during eperiment 1 and stages I and II of experiment 3

and was maintained above 90% up to the 14th token, as in experiment 1. It then dropped quickly, following a pattern similar to that of experiment 1 (Wilcoxon: $z = -1.600$, $P < 0.11$). The 50% probability of saving was 5.3 tokens on average for Ai, and 23.1 tokens on average for Pendesa.

5.3.3 Emergence of New Behaviors

During experiment 3, two new behaviors emerged in both subjects. As this experiment required two tokens for each piece of food, two tokens would have to be inserted consecutively into the vending machine before the software allowed any action from the subject, that is, to touch the white solid circles to proceed to the binary food choice. As the subjects were used to touch each circle after each token insertion and no training was provided, in the beginning both of them continued to show this behavior. However, both subjects quickly started not to touch the circle after the insertion of the first token; Ai started on the 3rd session and Pendesa on the 2nd session. Both subjects increased the frequency with which this behavior happened, as shown in Fig. 13.

Also, because this experiment required two tokens for each piece of food, to utilize all the tokens in a bout it would have to be even, otherwise the last token in the bout could not be used until the subject got more tokens. Also in this case, both subjects started not to insert the last token when the bout had an odd number of tokens, carrying it with them to collect more tokens. Ai started this behavior on the 2nd session of stage I, while Pendesa started it on the 4th session of stage II (Fig. 13).

5.4 Discussion

The increase of the number of tokens required to get one piece of food caused changes in the saving pattern of the subjects. Although both subjects maintained

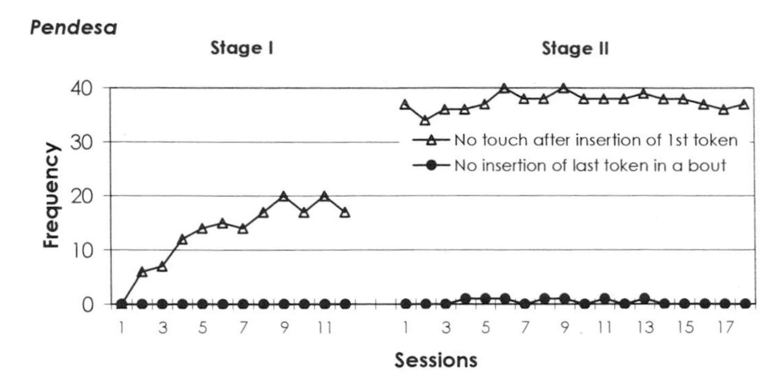

Fig. 13. Frequency of two kinds of events per session during both stages of experiment 3 for both subjects: not touching the circle on the screen after the insertion of the first token, and continuing to insert the last token in a bout when there is an odd number of tokens

the general pattern of saving tokens, significant changes were observed when the results of experiment 3 were compared to those of experiment 1. In both stages, Ai increased the number of tokens collected per bout before proceed to the exchange phase. In stage I, Pendesa showed a decrease of the number of tokens collected per bout before proceeding to the exchange phase. In stage II, Pendesa showed an increase of the number of tokens collected per bout to values close to those showed during experiment 1. To increase the number of tokens per bout, as Ai did, means to save time, because the vending machine was 1.9 m away from the place where the tokens were delivered and so it takes time to cover that distance. If the subject collects a higher number of tokens per bout she has to cover that distance fewer times and then saves time to get to the total amount of rewards in an experiment. This result suggests that both subjects were evaluating costs and benefits in their behavioral strategy and adjusting their behavior accordingly.

The present experiment provided evidence of a unique behavior, "saving marks," which corresponds to the accumulation of tokens inside the vending machine without immediately exchanging them for food. Both subjects spontaneously started to insert two or more tokens consecutively into the vending machine without trying to exchange them for food. This interesting and unique behavior was named saving marks in a previous experiment, where the behavior spontaneously emerged in one of the subjects (Sousa and Matsuzawa 2001). Although both Ai and Pendesa had participated in that previous experiment, neither of them showed the behavior at that time (Sousa and Matsuzawa 2001), not even during the experiments that followed (see experiments 1 and 2 in this chapter). This behavior shows that the subjects understood that they could not get a food reward after the insertion of only one token into the vending machine. Besides that, it also demonstrates their capacity to adjust and change their previous behavior when facing a new situation. This change of the saving behavior can also be related with the discussion of costs and benefits involved in the token experiment. Touching the mark on the monitor after inserting only one token would not have any result because two tokens were required to perform the food choice. However, it would mean spending more time to get the food reward.

When the number of tokens in a collected bout was odd, the subjects started not to insert the last token, taking it with them to get more tokens in the matching phase. This behavior in both subjects suggests that they understood that when only one token was left they could not use it to acquire a piece of food, and so they kept it with them while working for more tokens. In the case of Ai this behavior appeared before it did in Pendesa. Ai was sharing the experiments with her son Ayumu; if she left a token inside the vending machine, her son could easily use it by inserting another token he had collected, reducing the number of food rewards she could get. This behavior, besides showing that the subjects understood that they could not get a food reward after the insertion of only one token, also showed that Ai understood that without this behavior she could lose some food rewards.

6 General Discussion

The token system created has proven to be an efficient method to assess chimpanzee cognitive abilities underlying complex behavioral patterns. Previous studies had shown that tokens were almost equivalent to food rewards in their capacity to maintain adult chimpanzees working in intellectually costly tasks. Tokens were as effective as direct food rewards and were seen as an item exchangeable for food by the chimpanzees (Sousa and Matsuzawa 2001). The token system also created an opportunity to study and analyze the social transmission of skills from mother to infant (Sousa et al. 2003), allowing the observation of all steps involved in the learning process.

The use of tokens created a context for the emergence of unique behaviors. Chimpanzees spontaneously started to accumulated tokens during the matching phase of the experiment—saving tokens (Sousa and Matsuzawa 2001), and also began to accumulate tokens inside the vending machine before exchanging them for food—savings marks (experiment 3). Also during experiment 3, when the price was doubled and the subjects only had one token left in their hand or mouth, they started to transport it back to get more tokens. This fact further reinforces the portability characteristics of tokens already shown by Sousa and Matsuzawa (2001). These facts demonstrate that a token is a symbolic tool because it can be exchanged for different kinds of items—exchangeability (experiments 1 and 2), it is easy to handle and transport—portability (Sousa and Matsuzawa 2001, and experiment 3), and its value remains unchanged for extended periods so that it can be accumulated—saving of tokens and marks (Sousa and Matsuzawa 2001, and experiment 3). The fourth property of tokens as a symbolic tool, hierarchy, remains to be analyzed.

The studies presented in this chapter further show that the token system is also an efficient and reliable method to study food choice and preference in chimpanzees. Both subjects showed a clear order of food preference. The computer-controlled binary choice situation mediated by tokens provided a sensitive scale of preference for food items in chimpanzees. The token system was also a reliable method to analyze choice behavior in chimpanzees. The subjects understood the extra value acquired by the keys, and that they could use them to get a bigger and different piece of food.

Besides analyzing the choice behavior of chimpanzees, the present set of experiments also show a higher level of cognitive ability in chimpanzees, in the sense that the subjects had to relate several events in the flow of the experiment. When they use a token to choose a key to open a box with food they are actually using a tool for another tool, much as wild chimpanzees do when using a metatool (see Matsuzawa 1991a, 1994).

The variations in the saving behavior of the subjects suggest that they were balancing the costs and benefits of their actions, by changing their pattern of saving. The saving behavior means to spend less time in some steps of the experiment to acquire the food item more quickly. The discussion about the evaluation of costs and benefits by chimpanzees can also be extended to the preference of a key over a piece of food. The preference for a key that opens a box with food suggests that the subjects were again evaluating the costs and benefits of their actions, maximizing their choices. Some complex cognitive abilities are at work behind the decision making involved in choice behavior, and the chimpanzees were constantly evaluating the contexts of that choice and adjusting their behavior.

As these studies show, the token system comprises a very effective method to further explore the cognitive abilities underlying other complex behavioral patterns in chimpanzees. Overall, laboratory studies of behavior and cognition are important in complementing studies in the species' natural habitat and increase our knowledge about the depths of chimpanzee cognitive abilities. As chimpanzees are our closest extant relatives, an understanding of their behavior and cognition is indispensable for comprehending the evolution of human behavior and cognition.

Acknowledgments

This study was conducted at the Primate Research Institute of Kyoto University and was financially supported by the Ministry of Education, Science and Culture of Japan (Grant-in-Aid for Scientific Research nos. 07102010 and 12002009 to Tetsuro Matsuzawa). We would like to express our thanks to Sumiharu Nagumo of Kyoto University for his technical assistance. Special thanks are also due to Sanae Okamoto of Nagoya University, Dora Biro of Oxford University and Cristina Santos of University of the Azores for their suggestions and help.

References

Beck B (1980) Animal tool behavior. Garland Press, New York

Biro D, Matsuzawa T (2001) Chimpanzee numerical competence: cardinal and ordinal skills. In: Matsuzawa T (ed) Primate origins of human cognition and behavior. Springer, Tokyo, pp 199–225
Boesch C, Boesch H (1984) Mental map in wild chimpanzees: an analysis of hammer transports for nut cracking. Primates 25(2):160–170
Brosnan SF, de Waal FBM (2004) Socially learned preferences for differentially rewarded tokens in the brown capuchin monkey (*Cebus apella*). J Comp Psychol 118(2): 133–139
Byrne RW (2000) Evolution of primate cognition. Cogn Sci 24(3):543–570
Charles-Dominique P, Bearder SK (1979) Field studies of lorisid behavior: methodological aspects. In: Doyle GA, Martin RD (eds) The study of prosimian behavior. Academic Press, New York, pp 567–629
Cowles JT (1937) Food-tokens as incentives for learning by chimpanzees. Comp Psychol Monogr 23:1–96
Fantino E (1998) Behavior analysis and decision making. J Exp Anal Behav 69(3):355–364
Fragaszy D, Visalberghi E, Galloway A (1997) Infant tufted capuchin monkeys' behaviour with novel foods: opportunism, not selectivity. Anim Behav 53:1337–1343
Goodall J (1963) Feeding behaviour of wild chimpanzees: a preliminary report. Symp Zool Soc Lond 10:39–48
Goodall J (1964) Tool-using and aimed throwing in a community of free-living chimpanzees. Nature (Lond) 201:1264–1266
Goodall J (1970) Tool-using in primates and other vertebrates. Adv Study Behav 3:195–250
Goodall J (1986) The chimpanzees of Gombe: patterns of behaviour. Harvard University Press, Cambridge
Hirata S, Celli ML (2003) Role of mothers in the acquisition of tool-use behaviours by captive infant chimpanzees. Anim Cogn 6:235–244
Hirata S, Morimura N (2000) Naïve chimpanzees' (*Pan troglodytes*) observation of experienced conspecifics in a tool-using task. J Comp Psychol 114:291–296
Hladik CM (1979) Diet and ecology of prosimians. In: Doyle GA, Martin RD (eds) The study of prosimian behavior. Academic Press, New York, pp 307–357
Kelleher RT (1956) Intermittent conditioned reinforcement in chimpanzees. Science 124:279–280
Kelleher RT (1957a) A comparison of conditioned and food reinforcement with chimpanzees. Psychol Newsl 8:88–93
Kelleher RT (1957b) A multiple schedule of conditioned reinforcement with chimpanzees. Psychol Rep 3:485–491
Kelleher RT (1957c) Conditioned reinforcement in chimpanzees. J Comp Physiol Psychol 50:571–575
Kelleher RT (1958) Fixed-ratio schedules of conditioned reinforcement with chimpanzees. J Exp Anal Behav 1:281–289
Leakey R (1980) The making of mankind. Book Club Associates, London
Matsuzawa T (1991a). Nesting cups and metatools in chimpanzees. Behav Brain Sci 14(4):570–571
Matsuzawa T (1991b) Chimpanzee mind (in Japanese). Iwanami, Tokyo
Matsuzawa T (1994) Field experiments on use of stone tools by chimpanzees in the wild. In: Wrangham RW, McGrew WC, de Waal FBM, Heltne PG (eds) Chimpanzee cultures. Harvard University Press, Cambridge, pp 351–370
Matsuzawa T (1996) Chimpanzee intelligence in nature and in captivity: isomorphism of symbol use and tool use. In: McGrew WC, Marchant LF, Nishida T (eds) Great ape societies. Cambridge University Press, Cambridge, pp 196–209
Matsuzawa T (1999) Communication and tool use in chimpanzee: cultural and social contexts. In: Hauser M, Konishi M (eds) The design of animal communication. MIT Press, Cambridge, pp 645–671
Matsuzawa T (2001a) Primate origins of human cognition and behavior. Springer, Tokyo
Matsuzawa T (2001b) Primate foundations of human intelligence: a view of tool use in nonhuman primates and fossil hominids. In: Matsuzawa T (ed) Primate origins of human cognition and behavior. Springer, Tokyo, pp 3–25

Matsuzawa T (2003) The Ai project: historical and ecological contexts. Anim Cogn 6(4): 199–211

Matsuzawa T, Biro D, Humle T, Inoue-Nakamura N, Tonooka R, Yamakoshi G (2001) Emergence of culture in wild chimpanzees: education by master-apprenticeship. In: Matsuzawa T (ed) Primate origins of human cognition and behavior. Springer, Tokyo, pp 557–574

McGrew WC (1992) Chimpanzee material culture: implications for human evolution, Cambridge University Press, Cambridge

Parker ST, McKinney ML (1999) Origins of Intelligence: the evolution of cognitive development in monkeys, apes, and humans. Johns Hopkins University Press, Baltimore

Sousa C, Matsuzawa T (2001) The use of tokens as rewards and tools by chimpanzees. Anim Cogn 4:213–221

Sousa C, Okamoto S, Matsuzawa T (2003) Behavioural development in a matching-to-sample task and token use by an infant chimpanzee reared by his mother. Anim Cogn 6(4):259–267

Sugiyama Y (1995) Drinking tools of wild chimpanzees at Bossou. American Journal of Primatology 37(3):263–269

Sugiyama Y, Koman J (1992) The flora of Bossou: its utilization by chimpanzees and humans. Afr Stud Monogr 13(3):127–169

Ueno A, Matsuzawa T (2004) Food transfer between chimpanzee mothers and their infants. Primates 45:231–239

Visalberghi E, Addessi E (2001) Acceptance of novel foods in capuchin monkeys: do specific social facilitation and visual stimulus enhancement play a role? Anim Behav 62:567–576

Whiten A (1998) Imitation of the sequential structure of actions by chimpanzees (*Pan troglodytes*). J Comp Psychol 112:270–281

Whiten A, Goodall J, McGrew WC, Nishida T, Reynolds V, Sugiyama Y, Tutin CEG, Wrangham RW, Boesch C (1999) Cultures in chimpanzees. Nature (Lond) 399:682–685

Wilson CS (2001) Reasons for eating: personal experiences in nutrition and anthropology. Appetite 38(1):63–67

Wolfe JB (1936) Effectiveness of token-rewards for chimpanzees. Comp Psychol Monogr 12:1–72

Wrangham R (1977) Feeding behaviors of chimpanzees in Gombe National Park, Tanzania. In: Clutton-Brock TH (ed) Primate ecology: studies of feeding and ranging behavior in lemurs, monkeys and apes. Academic Press, New York, pp 503–538

Wrangham R, Chapman CA, Clark-Arcadi A, Isabirye-Basuta G (1996) Social ecology of Kanyawara chimpanzees: implications for understanding the costs of great ape groups. In: McGrew WC, Marchant LF, Nishida T (eds) Great ape societies. Cambridge University Press, Cambridge, pp 45–57

Yamakoshi G (2001) Ecology of tool use in wild chimpanzees: towards reconstruction of early hominid evolution. In: Matsuzawa T (ed) Primate origins of human cognition and behavior. Springer, Tokyo, pp 537–556

26
Behavioral Repertoire of Tool Use in the Wild Chimpanzees at Bossou

Gaku Ohashi

1 Introduction

Since Goodall (1963a,b) reported with photos and her detailed descriptions that wild chimpanzees fished for termites with a grass stalk, tool use has been the focus of intensive investigations at wild chimpanzee study sites across Africa. As long-term studies on wild chimpanzees have been carried out, many differences in behavioral repertoires have been found between subspecies (only the West African subspecies, *Pan troglodytes verus*, crack open a nut with a pair of stones as a hummer and anvil; Biro et al. 2003; Boesch et al. 1994; Sugiyama 1993; and only the Central African subspecies, *Pan troglodytes troglodytes*, use a tool set composed of stout sticks and slender fishing probes to extract termites from their nests; Sabater Pi 1974; Sanz et al. 2004) and even between adjacent communities (Mahale chimpanzees have never been observed to dip for commonly found driver ants, although Gombe chimpanzees, 170 km to the north, regularly do so; McGrew 1974; Nishida 1987).

It has been claimed that environmental differences can explain some of the variation. However, there are also many behavioral differences that ecological features fail to explain. An exhaustive comparison of seven long-term wild chimpanzee studies identified regional variations in 39 behavioral patterns (of which 23 involved tool use) that cannot be explained by differences in local environmental conditions (Whiten et al. 1999). The regional variations have been discussed as cultural behaviors (McGrew 1992, 2004; Yamakoshi 2001; Whiten 2005; Whiten et al. 1999).

Matsuzawa (1999) defined "culture" as a set of behaviors that are shared by members of a community and are transmitted from one generation to the next through nongenetic channels. Based on this point of view, longitudinal studies about acquisition process of tool use behaviors have been conducted. Especially, nut-cracking behavior has been well studied under experimental condition since 1987 (Sakura and Matsuzawa 1991; Chapter 28). On the other hand, records of anecdotal tool use behaviors also have been accumulated. Patterns in the presence or absence of tool use behaviors at particular sites have been discussed as

Primate Research Institute, Kyoto University, 41 Kanrin, Inuyama, Aichi 484-8506, Japan

features of chimpanzee cultures. However, the explanations of some tool use behavior patterns are ambiguous. This chapter introduces behavioral repertoires of the tool use of Bossou chimpanzees, with photographs, and clarifies the characteristics of the behaviors.

2 Behavioral Repertoire of Tool Use in the Wild Chimpanzees at Bossou

Bossou is located in the southeastern corner of the Republic of Guinea, West Africa (7°39′ N and 8°30′ W). A group of wild chimpanzees at Bossou has been studied since 1976 (Sugiyama and Koman 1979a). Their home range covers about 15 km^2 of primary and open secondary forests that are surrounded by cultivated and abandoned fields. All the individuals have been identified since the start of study. The group size has fluctuated between 12 and 22 (Sugiyama 2004; also see Chapter 1, this volume).

There have been reviews on the tool use of Bossou chimpanzees (Matsuzawa 1999; Sugiyama 1997). This chapter aims to describe the behavioral repertoire of tool use of Bossou chimpanzees, based on direct observations by the author from July 2001 to November 2001 (73 days of observations), from July 2002 to March 2003 (135 days of observations), and from April 2004 to September 2004 (87 days of observations). In 2001, I recorded activities of individuals with a 10-min scan sampling method by subgroup following because the observation conditions were too difficult to use focal animal sampling. I recorded tool use behaviors also by video camera. From 2002, I followed the males all day via focal animal sampling. I recorded the activities of the individuals with all-occurrence sampling. The positions of the focal individuals were recorded by GPS. Although a local assistant followed other individuals simultaneously with GPS and recorded their activities in this period, only my observation data are used here. Tool use behaviors by individuals near the focal animals were also recorded.

2.1 Cracking Hard Nuts with a Pair of Stones

Chimpanzees at Bossou are known to use a pair of stones as a hummer and anvil to crack open oil-palm nuts, *Elaeis guineensis* (Biro et al. 2003; Inoue-Nakamura and Matsuzawa 1997; Matsuzawa 1994, 1999; Matsuzawa et al. 2001; Sakura and Matsuzawa 1991; Sugiyama and Koman 1979b). As an anvil, they sometimes use a rock that is an outcropping on the ground (Matsuzawa 1994). When the surface of an anvil was slanted, a chimpanzee was observed to put a small stone under its lower part to make the surface level (Matsuzawa 1994). Field experiments have been conducted to reveal the characteristics of the technique in detail and its acquisition process since 1987 (Biro et al. 2003; Inoue-Nakamura and Matsuzawa 1997; Matsuzawa 1994, 1999; Matsuzawa et al. 2001; Sakura and Matsuzawa 1991). Detailed analysis showed that each individual chimpanzee

Fig. 1. Cracking hard nuts with a pair of stones

always used the same hand as the hammer-holding hand (Matsuzawa 1994; Sugiyama 1993). Moreover, hand preference of nut cracking was congruent among siblings (Matsuzawa 1999). In natural conditions, the number of stones is retained. The chimpanzees normally secure a pair of stones first. They often transport the stones about 1 to 5 m (Sakura and Matsuzawa 1991). The chimpanzees collect the oil-palm nuts under a tree. They put the nut on the stone by hand, pick up another stone with the other hand, raise it about 30 cm, and strike the nut with the stone hammer repeatedly until the nut is cracked (Fig. 1).

2.2 Dipping for Ants with a Wand

The chimpanzees at Bossou dip for arboreal ants (*Dorylus* spp.) with a wand (Humle and Matsuzawa 2002; Sugiyama 1995a; Yamakoshi and Myowa-Yamakoshi 2004; Chapter 27). An old female chimpanzee was once observed to pound or dig the ground with a long thick stick (Sugiyama 1995a), but the chimpanzees at Bossou usually excavate ant nests with their hands directly. They uproot a stem from some nearby vegetation with their hand and make a wand by tearing the leaves off with the hand or mouth. Then they dip for migrating ants on the ground or insert the wand into a hole in the ant nest. Bossou chimpanzees exhibit two ant-dipping techniques: one is direct mouthing and the other is pull-through (Yamakoshi and Myowa-Yamakoshi 2004). The direct mouthing technique was more frequently observed at Bossou than the pull-through technique. When the chimpanzees dip for ants with a wand, they leave the wand or often thrust it back and forth and then slowly withdraw it with a large mass of ants. They bring the proximal end of the wand to their mouth, and quickly pull the wand so that it passes through their lips or teeth from proximal to distal end (Fig. 2). The pull-through technique was observed when the ant nest was relative large. The chimpanzees insert a wand into a hole in the ant nest

Fig. 2. Direct mouthing technique of ant dipping

Fig. 3. Pull-through technique of ant dipping

with their hand and swipe the length of the wand to gather the ants in the other hand before rapid transfer to the mouth. The wands were longer than those used in the direct mouthing technique (Humle and Matsuzawa 2002). They sometimes sifted the wand from the hand to the hindleg before swiping (Fig. 3).

2.3 Extracting Water from a Tree Hole with a Leaf

The chimpanzees at Bossou extract water from a tree hole with a leaf (Sugiyama 1995b; Tonooka 2001; Tonooka et al. 1995). Field experiments revealed that Bossou chimpanzees exhibit three types of techniques using a leaf, called "leaf sponge," "leaf spoon," and "leaf folding" (Tonooka et al. 1995; Chapter 28). In experimental conditions, the leaf folding technique was more frequently observed than the other techniques (Tonooka 2001). They break off a leaf, put it into their mouth with a hand, fold it with the roof of their mouth, take it out, insert it into the tree hole, then put the folded leaf with water back into the mouth. When the entrance of the tree hole was small and water was deep in the bottom of the hole, one chimpanzee pushed a leaf into the water with a stick and pulled out the leaf to drink the water (Matsuzawa 1991). Some locations are usually used by the chimpanzees for drinking water (Sugiyama 1995b). However, I did not observe chimpanzees drinking water with a tool during my observation periods. They regularly pass across streams, and in the dry season they were observed to drink water from the stream directly with their mouth almost every day. They also seem to receive hydration by feeding on juicy herbaceous plants, such as *Costus afer*.

Drinking tools are often observed being used for palm wine that is made from *Raphia gracilis* sap. *Raphia* trees are planted in swamps and gallery forest. Local people set a bottle on the crown of a *Raphia* tree in the morning and recover the bottle filled with the wine in the evening. There are not so many *Raphia* trees in the core area of the chimpanzees. On the other hand, many *Raphia* trees are found in the peripheral area, where chimpanzees visit in consortship periods, in which a male lead estrous female away from the other males. On August 8, 2004, for example, FF (an adult male), Pm (an adult female), and PE (a 6-year-old male) visited gallery forest (7°40′ 17″ N, 8°28′ 55″ W) about 3 km away from their ordinal home range. Once the chimpanzees found a bottle on the *Raphia* tree, they climbed up the tree. They broke off a nearby leaf, put it into their mouth with one hand, chewed the leaf repeatedly in the mouth, took it out of the mouth, and soaked it in the wine in the bottle. They then picked up the leaf and sucked the wine from the leaf. They spent about 12 min drinking the wine with these tools (Fig. 4).

Fig. 4. Extracting *Raphia* wine from a bottle with a leaf

2.4 Pounding the Apical Meristem of an Oil Palm Using a Palm Frond as a Pestle

The chimpanzees at Bossou use a palm frond to pound the apical meristem of the palm tree (Sugiyama 1994; Yamakoshi and Sugiyama 1995). The behavioral components were well described by Yamakoshi and Sugiyama (1995). The chimpanzees climb up the crown of a palm tree and stand up on the center of the crown. They spread out the radiating mature leaves using their hands and feet to expose the base of central young shoots or spear leaves. Then, they repeatedly try to pull the shoots out. This behavior requires much force, so they often fail to pull them out. When they succeed, they chew and eat the soft white base of the shoots. At the site where the young shoots have been pulled out, a vertical hole is left at the heart of the crown. Picking up a discarded leaf, the chimpanzees use its petiole as a pestle to repeatedly pound and deepen the hole. Then they lick the edge of the petiole and put it aside, and insert their arm into the hole to extract and eat the juicy fibrous products (Fig. 5). Young chimpanzees cannot pull the shoots out. They often climb up the same tree where adult chimpanzees feed and wait on the mature leaves. When the adult chimpanzees climb down, the young individuals start to pound with the petiole that the adult chimpanzee used (Hirata and Morimura 2000; Hirata and Celli 2003).

At present, this behavior has been habitual among the Bossou chimpanzees, but some chimpanzees have not been observed to use the tool. For example, YL (an adult male) could not use a petiole as a pestle to pound the hole. He could pull the shoots out, and often ate the base of the shoots. On November 13, 2001, YL (when he was 10 years old) was observed to pound the hole on a palm tree. YL was feeding on the red pericarp of palm fruits with two adult males. On the same palm tree, Nn (an adult female) was pounding the hole of the crown with a petiole. After Nn climbed down the tree, YL went up to the crown. He picked

Fig. 5. Pestle-pounding behavior

Fig. 6. This adolescent male snapped the petiole into two and inserted the stick into the hole, as if dipping for ants

up the petiole that Nn had used as a pestle, snapped it in two, and inserted the petiole half into the hole, as if he dipped for ants. This behavior was continued for about 10 min. He licked the stick tool and ate the juicy products. The future acquisition of the pestle-pounding technique was expected, but YL has never observed to pound the apical meristem with a palm petiole after this observation (Fig. 6).

2.5 Scooping Algae from Water with a Wand

There are many swamps around the Bossou hills. Bossou chimpanzees use a grass stem to scoop algae (*Spirogyra* sp.) floating on pond surfaces (Matsuzawa 1999; Matsuzawa et al. 1996; Yamakoshi 1998). They uproot a grass stem from some nearby vegetation with their hand and make a wand by tearing off the leaves with their hand or mouth. They insert the wand into the pond and scoop the algae, then put the proximal end of the wand to their mouth and pull the wand so that it passes through their lips from proximal to distal end (Fig. 7).

2.6 Fishing for Termites with a Stick

Sugiyama and Koman (1979b) reported that two adult chimpanzees pounded a tree hole with a stick and ate a few termites. They took a small twig 5 to 15 cm in length, removed the side branches and leaves, and pounded the bottom of the hole several times. On pulling the stick out, a few termites were attached to it, mostly broken and adherent. The chimpanzees licked them off and again tried to pound the bottom of the hollow. They attempted this behavior for 30 min each, but succeeded in retrieving only a few termites.

On the other hand, the chimpanzees at Bossou had never been observed to fish termites with a stick, although the terrestrial mounds are abundant in their home range. However, termite fishing is not absent at Bossou. Two chimpanzees,

Fig. 7. Algae-scooping behavior

Yo (an adult female) and YL (a 6-year-old male), were observed to fish for termites with a grass stalk (Humle 1999). They dug a hole with their thumb and index fingers into the earthen mound, then bit off the distal end of the stalk and put it into the mound. The termites attacked the stalk by clamping onto it with their jaws. The chimpanzees then withdrew the tool and used their lips to nip the termites from it. However, termite fishing is not habitual behavior. In 2004, for example, I followed YL as a focal animal for 27 days from May to September and observed YL eating termites 22 times. He picked up the termites directly from the mounds and never used a tool.

2.7 Extracting Insect Larvae from a Nest Tunnel with a Stick

The chimpanzees at Bossou extract insect larvae from a nest tunnel with a stick. They often break off a branch or woody vine to look for insect larvae. When they find a nest tunnel of the insect, they sometimes use a tool. They break off a small branch from nearby, insert the stick into the tunnel, and extract the insect larvae. This stick is relatively short. Four sticks were collected and measured from my observations in 2001, measuring 24.0, 24.2, 26.0, and 29.1 cm. (Fig. 8)

2.8 Dipping for Honey or Gum with a Stick

When the chimpanzees find a beehive on a tree, they often knock off the hive with their hands. They break the hive on the ground and chew a piece of the hive. On August 8, 2002, PO and PE were observed to dip for honey with a stick. First, Pm (an adult female) knocked the hive off the tree. Pm, FF (an adult male), and PE (a 4-year-old male) approached the hive on the ground and chewed a piece of hive, uttering a food-grunt. PO (a 9-year-old male) also tried to

Fig. 8. Extracting insect larvae

Fig. 9. Honey dipping

approach the hive, but Pm drove him away. After FF and Pm left there with a mass of hive, PO approached again, and uprooted a stem from the nearby vegetation with his hand. PO prepared a stick, and dipped for honey with the stick for a while. When PO abandoned the stick and started to eat the beehive directly with his hands, PE picked up the stick and dipped for honey with it (Fig. 9).

The chimpanzees at Bossou also eat the honey of stingless bees. The stingless bees make nests in a small hollow of a tree. October 28, 2001, three adolescent chimpanzees inserted their fingers into the hollow between a fig tree and a palm tree. They licked their fingers, to which the honey adhered. Once Nt (an 8-year-old female) found a thin vine, broke it off, inserted it into the hollow, and licked the vine (Fig. 10).

Sugiyama and Koman (1979b) reported that young chimpanzees licked resin with a stick. They broke off a branch, removed the side branches and leaves with

Fig. 10. Dipping for honey from a nest of stingless bees with a vine

Fig. 11. Animal probing

their teeth, and inserted the stick into the tree hollow. They pulled the stick up, and licked a brown-colored resin that coagulated on the stick.

2.9 Probing Animals in Tree Holes with a Stick

Bossou chimpanzees do not eat a number of animal species (hyrax; Hirata et al. 2001). Other animal species were also observed to be abandoned by chimpanzees during my observation periods, although the chimpanzees had successfully caught them (civet on October 29, 2001; mongoose on May 8, 2004; squirrel on August 11, 2004; bird on August 4, 2004). On the other hand, they eat pangolin meat (Sugiyama 1981). They sometimes probe animals in tree holes with a stick. They break off a branch from nearby with their hand and make a stick by tearing the leaves off with their hand or mouth, then insert the stick to probe for animals (Fig. 11).

2.10 Probing Animal or Human Traces on the Ground with a Stick

Bossou chimpanzees sometimes use a stick to probe something smelly on the ground. On May 14, 2004, for example, YL stopped moving and looked down to the ground, broke off a shrub, and made a long stick. He poked on the ground with the stick and sniffed at it. No insects were found around the spot.

When the chimpanzees encounter a snare, they try to break it with their hands. The chimpanzees were observed twice to successfully deactivate snares (August 1, 2002, by FF, an adult male; July 29, 2004, by PE, a 6-year-old male).

3 Discussion

At Bossou, a number of tool use behaviors are observed. Matsuzawa (1999) reported that hand preference of nut cracking was consistent among siblings. This agreement may suggest that some social learning affects the acquisition of tool use techniques. On the other hand, the chimpanzees at Bossou seem to use tools flexibly depending on the situation. Especially, they can easily apply stick use technique toward various objects. They can also change the stick length easily. However, most of the anecdotal behaviors are seen only in young chimpanzees; adult chimpanzees seldom do these. Why do adult chimpanzees not engage in these behaviors? One possibility is that ineffective behaviors become reduced by individual learning as they get older. They may improve their feeding technique day by day. Another scenario is also possible. Especially, young male chimpanzees spend longer times with adult male chimpanzees as they get older. When the adult male chimpanzees start to move, young male chimpanzees often follow them. If the adult chimpanzees neglect some food items, young chimpanzees may abandon those food items. For a tool use behavior to become fixed in the community, it may be necessary that other adult chimpanzees are also interested in the behavior. It is possible that not only acquisition but also restriction of tool use behaviors is affected socially.

Video clips of tool use behaviors are available at the following Web site: http://www.pri.kyoto-u.ac.jp/chimp/index.html

Acknowledgments

I am grateful to the Direction Nationale de la Recherche Scientifique et Technique, and Institut Recherche Environnementale de Bossou in Guinea for granting me permission to carry out this research. I thank Professor T. Matsuzawa for his supervision and advice; Professor Y. Sugiyama, Dr. G. Yamakoshi, and anonymous colleagues for their support at the study site; and P. Goumy, P. Cherif, J. Dore, B. Zogbila, M. Dore, and H. Gbelegbe for their field assistance at Bossou. This study was financed by grants of the Ministry of Education, Culture, Sports, Science, and Technology, Japan (nos. 12002009 and 16002001 to T. Matsuzawa), a grant under Research Fellowships of the Japan Society for Promotion of Science for Young Scientists (no. 160896 to G. Ohashi), and a grant of The Kyoto University Foundation to G. Ohashi.

References

Biro D, Inoue-Nakamura N, Tonooka R, Yamakoshi G, Sousa C, Matsuzawa T (2003) Cultural innovation and transmission of tool use in wild chimpanzees: evidence from field experiments. Anim Cogn 6:213–223

Boesch C, Marchesi P, Marchesi N, Fruth B, Joulian F (1994) Is nut cracking in wild chimpanzees a cultural behavior? J Hum Evol 26:325–338

Goodall J (1963a) Feeding behaviour of wild chimpanzees: a preliminary report. Symp Zool Soc Lond 10:39–48

Goodall J (1963b) My life among wild chimpanzees. Natl Geogr 124(2):272–308

Hirata S, Celli ML (2003) Role of mothers in the acquisition of tool-use behaviours by captive infant chimpanzees. Anim Cogn 6:235–244

Hirata S, Morimura N (2000) Naïve chimpanzees' (*Pan troglodytes*) observation of experienced conspecifics in a tool-using task. J Comp Psychol 114:291–296

Hirata S, Yamakoshi G, Fujita S, Ohashi G, Matsuzawa T (2001) Capturing and toying with hyraxes (*Dendrohyrax dorsalis*) by wild chimpanzees (*Pan troglodytes*) at Bossou, Guinea. Am J Primatol 53:93–97

Humle T (1999) New record of fishing for termites (*Macroyermes*) by the chimpanzees of Bossou (*Pan troglodytes verus*), Guinea. Pan Africa News 6:3–4

Humle T, Matsuzawa T (2002) Ant-dipping among the chimpanzees of Bossou, Guinea, and some comparisons with other sites. Am J Primatol 58:133–148

Inoue-Nakamura N, Matsuzawa T (1997) Development of stone tool use by wild chimpanzees (*Pan troglodytes*). J Comp Psychol 111:159–173

Matsuzawa T (1991) Chimpanzee mind (in Japanese). Iwanami, Tokyo

Matsuzawa T (1994) Field experiments on use of stone tools in the wild. In: Wrangham RW, McGrew WC, de Waal FBM, Heltone PG (eds) Chimpanzee cultures. Harvard University Press, Cambridge, pp 351–370

Matsuzawa T (1999) Communication and tool use in chimpanzees: cultural and social contexts. In: Hauser M, Konishi M (eds) The design of animal communication. Cambridge University Press, New York, pp 211–232

Matsuzawa T, Yamakoshi G, Humle T (1996) A newly found tool-use by wild chimpanzees: algae scooping (abstract). Primate Res 12:283

Matsuzawa T, Biro D, Humle T, Inoue-Nakamura N, Tonooka R, Yamakoshi G (2001) Emergence of culture in wild chimpanzees: education by master-apprenticeship. In: Primate origins of human cognition and behavior. Springer, Tokyo, pp 557–574

McGrew WC (1974) Tool use by wild chimpanzees in feeding upon driver ants. J Hum Evol 3:501–508

McGrew WC (1992) Chimpanzee material culture: implications for human evolution. Cambridge University Press, Cambridge

McGrew WC (2004) The cultured chimpanzee: reflections on cultural primatology. Cambridge University Press, Cambridge

Nishida T (1987) Local traditions and cultural transmission. In: Smuts BB, Cheney DL, Seyfarth RM, Wrangham RW, Struhsaker TT (eds) Primate societies. University of Chicago Press, Chicago, pp 462–474

Sabater Pi J (1974) An elementary industry of the chimpanzees in the Okorobiko Mountains, Rio Muni (Republic of Equatorial Guinea), West Africa. Primates 15:351–364

Sakura O, Matsuzawa T (1991) Flexibility of wild chimpanzee nut-cracking behavior using stone hammers and anvils: an experimental analysis. Ethology 87:237–248

Sanz C, Morgan D, Gulick S (2004) New insights into chimpanzees, tools, and termites from the Congo Basin. Am Nat 164:567–581

Sugiyama Y (1981) Observation on the population dynamics and behavior of wild chimpanzees at Bossou, Guinea, 1979–1980. Primates 22:435–444
Sugiyama Y (1993) Local variation of tools and tool use among wild chimpanzee populations. In: Berthelet A, Chavaillion J (eds) The use of tools by human and non-human primates. Clarendon Press, Oxford, pp 175–187
Sugiyama Y (1994) Tool use by wild chimpanzees. Nature (Lond) 367:327
Sugiyama Y (1995a) Tool-use for catching ants by chimpanzees at Bossou and Monts Nimba, West Africa. Primates 36:193–205
Sugiyama Y (1995b) Drinking tools of wild chimpanzees at Bossou. Am J Primatol 37:263–269
Sugiyama Y (1997) Social tradition and use of tool-composites by wild chimpanzees. Evol Anthropol 6:23–27
Sugiyama Y (2004) Demographic parameters and life history of chimpanzees at Bossou, Guinea. Am J Phys Anthropol 124:154–165
Sugiyama Y, Koman J (1979a) Social structure and dynamics of wild chimpanzees at Bossou. Primates 20:323–339
Sugiyama Y, Koman J (1979b) Tool-using and making behavior in wild chimpanzees at Bossou, Guinea. Primates 20:513–524
Tonooka R (2001) Leaf-folding behavior for drinking water by wild chimpanzees (*Pan troglodytes verus*) at Bossou, Guinea. Anim Cogn 4:325–334
Tonooka R, Inoue N, Matsuzawa T (1995) Leaf-folding behavior for drinking water by wild chimpanzees at Bossou, Guinea: a field experiment and leaf selectivity (in Japanese with English summary). Primate Res 10:307–313
Whiten A (2005) The second inheritance system of chimpanzees and humans. Nature (Lond) 437:52–55
Whiten A, Goodall J, McGrew WC, Nishida T, Reynolds V, Sugiyama Y, Tutin CEG, Wrangham RW, Boesch C (1999) Cultures in chimpanzees. Nature (Lond) 399:682–685
Yamakoshi G (1998) Dietary responses to fruit scarcity of wild chimpanzees at Bossou, Guinea: possible implications for ecological importance of tool use. Am J Phys Anthropol 106: 283–295
Yamakoshi G (2001) Ecology of tool use in wild chimpanzees: toward reconstruction of early hominid evolution. In: Matsuzawa T (ed) Primate origins of human cognition and behavior. Springer, Tokyo, pp 537–556
Yamakoshi G, Myowa-Yamakoshi M (2004) New observations of ant-dipping techniques in wild chimpanzees at Bossou, Guinea. Primates 45:25–32
Yamakoshi G, Sugiyama Y (1995) Pestle-pounding behavior of wild chimpanzees at Bossou, Guinea: a newly observed tool using behavior. Primates 36:489–500

27
Ant Dipping in Chimpanzees: An Example of How Microecological Variables, Tool Use, and Culture Reflect the Cognitive Abilities of Chimpanzees

Tatyana Humle

1 Culture and Ant Dipping

In the 1950s, Japanese primatologists described the social transmission of sweet-potato washing in a population of Japanese macaques (*Macaca fuscata*) on Koshima island (Hirata et al. 2001; Kawai 1965). Their observations played a key role in drawing scientists' attention to the issue of culture in nonhuman animals. The concept of culture in chimpanzees (*Pan troglodytes*) and other animals has since been a source of much debate and controversy. Culture has, nevertheless, recently been operationally defined to encompass behaviors that are socially transmitted within and between generations in groups and populations of the same species (Laland and Hoppitt 2003; Parker and Russon 1996). As defined, the concept of culture has stimulated in recent years a multitude of studies on a wide range of animal taxa, ranging from insects, fish, birds, and cetaceans to primates, both in the laboratory or in the field (Fragaszy and Perry 2003).

In the field, the identification of a cultural variant in nonhuman animals is established on the basis of a set of indicators. These indicators include (1) a patchy geographical distribution of the behavior, (2) its habitual and customary occurrence at sites where it has been confirmed, (3) its persistence across generations, and (4) the unlikely attribution of its occurrence to ecological or genetic differences between sites and its absence across dispersal barriers. Because of the difficulty in identifying social learning processes involved in the transmission of behavior in natural settings and in firmly excluding environmental differences as an explanation for observed variations in behavior, this approach has been criticized by skeptics who challenge the ascription of culture to nonhuman animals (Galef 1992; Tomasello 1994, 1999; Tomasello et al. 1993). Nevertheless, four decades of field studies of wild chimpanzees in Africa have revealed substantial differences in behavioral repertoires at the subspecies, population, and community level (for reviews, see Whiten et al. 1999, 2001; Yamakoshi 2001). The list of differences is extensive and comprises a multitude

Department of Psychology, University of Wisconsin, Madison, 250 N. Mills St., Madison, WI 53706, USA

of behaviors encompassing tool use, feeding, and social and communication domains (Humle and Matsuzawa 2004; McGrew 1985, 1992, 1998; McGrew et al. 1979; Nishida 1987; Nishida et al. 1983; Sugiyama 1993, 1997). Whiten et al. (2001) identified 39 candidate behavioral patterns as potential cultural variants on the grounds that they occurred sufficiently frequently at one or more site(s) to be consistent with social transmission, yet were absent at one or more other(s) and where environmental explanations could be rejected. As more detailed data are being gathered and compiled across the chimpanzee field sites, it has become apparent that chimpanzees potentially exhibit a much greater number of candidate cultural variants than previously reported (Whiten et al., in preparation).

Probe-using behavior is one of the most prominent and diversified forms of tool use among chimpanzees in their natural habitat. Based on data from long-term field sites, stick- or stalk-using for catching social insects on the ground and/or in trees is common to chimpanzees throughout their range, with the exception of Budongo, Uganda (Whiten et al. 1999). However, the prevalence of each type of behavior differs by locality, implying cultural differences across chimpanzee communities (McGrew 1992; Whiten et al. 1999; Yamakoshi 2001). The ubiquity of stick- or stalk-using behaviors has been linked to the ready availability of diverse materials for tool making and the presence of potential target prey in all habitats in which chimpanzees live (McGrew and Collins 1985; Collins and McGrew 1987). In most cases, a tool is used to gain access to the social insect prey within a protected structure, that is, the nest, as in termite fishing or ant fishing. Ant dipping, the tool-use behavior I focus on in this chapter, is aimed at driver ants (*Dorylus* spp.). These ants often migrate on the ground or move among low terrestrial herbaceous vegetation in great numbers, up to several million individuals, hunting for prey. They construct tunnel nests underground that they use as a temporary bivouac. The entrance of the nest is often covered by a layer of fallen leaves and/or soil and can be readily penetrated manually. The reliance on a tool for ant dipping by chimpanzees has been proposed to allow for more efficient and less painful harvesting of these biting ants, rather than taking them directly by hand or mouth (McGrew 1974).

Ant dipping has only been described in detail at three long-term study sites: Gombe in Tanzania; Taï in Côte d'Ivoire, and Bossou in Guinea. The early descriptions of this tool-use behavior emerging from Gombe and Taï soon revealed that the chimpanzees at these two sites employ tools of significantly different lengths when dipping (Boesch and Boesch 1990; Goodall 1986). Tools used by chimpanzees at Gombe ($n = 13$; mean = 66 cm; range, 15–113 cm (cf. McGrew 1974)) are indeed significantly longer than those used at Taï ($n = 35$; mean = 23.9 cm; range = 11–50 cm (cf. Boesch and Boesch 1990)). Differences in ant dipping between Gombe and Taï are not restricted to tool length but also concern the technique employed in consuming the ants from the tool. At Gombe, chimpanzees use one hand to hold the stick among the attacking ants and, once these have swarmed about halfway up the tool, the chimpanzee usually withdraws the stick and sweeps it through the closed fingers of its free hand, a technique known as pull-through. The mass of ants is then rapidly transferred to the mouth and chewed (McGrew 1974). Chimpanzees at Gombe on rarer occasions take ants directly from the tool by direct mouthing, that is, by directly pulling the tool sideways through the lips (McGrew 1974). At Taï, on the other hand, the chimpanzee holds the stick among the soldier ants with one hand until they have swarmed about 10 cm up the tool (Boesch 1996). On withdrawal of the tool, the chimpanzee then typically twists the hand holding the tool and directly nibbles off the ants with the lips, thus always performing a frontal version of direct mouthing (Yamakoshi and Myowa-Yamakoshi 2003). It has been suggested that the differences in ant-dipping technique and tool length between Gombe and Taï are based on social learning and reflect cultural variation among chimpanzees (Boesch and Boesch 1990; McGrew 1992).

At Bossou, Sugiyama (1995) reported that chimpanzees employ a direct mouthing technique when dipping for ants, similar to that observed on rare occasions at Gombe. In contrast to the frontal version of direct mouthing observed at Taï, when employing this technique Bossou chimpanzees nearly exclusively pull the tool sideways through the lips to remove the ants. Bossou chimpanzees only more rarely perform a frontal version of this technique. Recent observations of ant dipping from Bossou indicate that some members of the community also occasionally employ another technique, that is, the pull-through technique observed at Gombe (Humle and Matsuzawa 2002; Yamakoshi and Myowa-Yamakoshi 2003). The pull-through technique at Bossou was first noticed in 1997 by the author in a juvenile individual named Fotayu, aged 6 years at the time.

Several hypotheses have been put forth in explaining the differences in tool length and technique between Bossou, Gombe, and Taï. Because Bossou chimpanzees exhibit both the direct mouthing and the pull-through technique, this community offers the potential to explore variables that might influence tool length and technique employed by chimpanzees during ant dipping. Sugiyama (1995, p. 203) proposed that differences in ant-dipping techniques, tool length, dipping posture, and material selection may depend on variations in prey characteristics, most particularly the aggressiveness of the prey, *Dorylus* spp., across these different study sites, and "may (also) to some extent reflect a tradition in the chimpanzee community." Hashimoto et al. (2000) further suggested that differences in the length of tools might reflect the difference in techniques used for catching ants. In this chapter, I present data testing this latter hypothesis, as well as the influence of the behavior of the driver ant species targeted and the dipping context (location and status of the ants) on the ant-dipping behavior of the Bossou chimpanzees. I address the extent to which microecological variables such as driver ant aggressiveness and/or gregariousness might explain variations in tool length or dipping position in individual chimpanzees at Bossou. Finally, I discuss the results in light of the cultural hypothesis currently proposed in explaining the observed differences in ant-dipping behavior between Taï and Gombe chimpanzees.

2 Cognitive and Social Influences on the Development of Behavior

When exploring the characteristic patterns of a behavior, particularly one as complex as tool use, it is essential to adopt a developmental approach when analyzing its acquisition in young. Ontogeny is a very different process for different animal species. For some species, it is essential that the young be almost fully functional from birth to maximize their chances of surviving to the age of reproduction, whereas for other species a long ontogeny, with a combination of individual and social learning, is the optimal life history strategy (Tomasello 1999). In species with a long period of development, such as the chimpanzee, age and critical learning periods are both likely to affect transmission of different behavioral traits. These windows of cognitive receptivity to learning experiences appear to be directly related to the cognitive development of the young within its social and physical environment. Matsuzawa (1994, 1999) provided some evidence that the age of acquisition of a tool use behavior in wild chimpanzees depends on the tool task and the level of complexity involved. Inoue-Nakamura and Matsuzawa (1997) also demonstrated that critical learning periods might be essential in determining the ultimate ability of young chimpanzees in performing a complex manipulative task. Indeed, through their longitudinal study of the development of nut-cracking behavior among Bossou chimpanzees, Inoue-Nakamura and Matsuzawa (1997) found that the critical learning period for the acquisition of this complex tool-use behavior lies between the ages of 3 and 5 years. Beyond this age, acquisition appears less probable, as demonstrated by the reported inability of some adolescents (8–11 years old) or adults (>11 years old) of this same community to demonstrate this behavior.

For the young chimpanzee, as in humans and many other primate species, the primary socializing agent is the mother. For at least the first 5 years of life, the vast majority of the chimpanzee infant's social interactions are with the mother (McGrew 1977). This period of prolonged dependency ensures that the infant (0–3 years of age) is exposed to all the mother's feeding and social activities at close range. Therefore, one would expect the mother to act as the prime model for the infant, providing the latter with exposure and opportunities for practicing a given behavior (van Schaik et al. 2003). However, it still remains unclear, in the context of many tool-use behaviors observed in wild chimpanzees, how much influence the mother's behavior and the social and physical environment might have on her offspring's behavioral acquisition and performance.

Matsuzawa et al. (2001) characterized the process of how young chimpanzees in the wild learn nut-cracking behavior as "education by master-apprenticeship." During this process, no active teaching occurs between master and apprentice, that is, there is no shaping or molding of the apprentice's behavior by the master. Instead, the apprentice acquires the skill through repeated observations of the master, and the master, in turn, exhibits high levels of tolerance for the close proximity of the apprentice. Active demonstration and assistance in canalizing an immature's acquisition of a complex tool-use behavior, such as nut cracking, have rarely been observed (see Chapter 28; Biro et al. 2003; Boesch 1991; Inoue-Nakamura and Matsuzawa 1997; Matsuzawa et al. 2001).

Coussi-Korbel and Fragaszy (1995) proposed that social dynamics among group members are more important than cognitive ability or phylogenetic lineage in predicting social learning. They argue that individuals in more egalitarian and tolerant social groups are more likely to learn socially and exhibit homogeneity in behavior because they experience more opportunities for close behavioral coordination in space and time with other group members. Chimpanzees live in a fission–fusion social structure (Nishida 1968), which implies that at any time temporary and unstable parties are formed representing only a subset of the whole community. Chimpanzees exhibit a strong dominance hierarchy among males and variable levels of affiliation among females (Goodall 1968; Sugiyama 1988). In addition, tolerance to immatures by adult members of the community other than the mother differs with the age of the offspring. In a study of the development of nut-cracking behavior, Inoue-Nakamura and Matsuzawa (1997) showed that only infants really have the opportunity to freely access stones and nuts from other individuals, thus gaining ample opportunity for individual practice with manipulating stones and nuts and close observation of competent models. Adults are usually less tolerant of juveniles (4–7 years of age) behaving in the same fashion. Thus young chimpanzees are faced with varying levels of tolerance by members of their social group during the course of their development.

Finally, these changes in social interactions and tolerance patterns, as well as the cognitive development of the young, constitute a complex web of interconnected variables influencing the acquisition of behavior in chimpanzees. This acquisition process frequently occurs under heterogeneous environmental conditions, influenced by seasonal and spatial variation in target location or even in prey behavior, as we shall see for ant dipping, that irrevocably add an extra layer of complexity to this whole developmental process. In this chapter, I therefore also present some preliminary data on the development of ant-dipping behavior among subadults (individuals between 4 and 11 years of age) from the Bossou community, thus examining at least one layer of this complex web.

3 Site and Methods

3.1 Bossou Site

The village of Bossou (7°39′ N and 8°31′ W) is situated in the southeastern region of Guinea, West Africa, about 6 km from the foot of the Nimba Mountains on the border with Côte d'Ivoire and Liberia. Bossou was established as a chimpanzee field site in 1976 (Sugiyama and Koman 1979; Sugiyama 1981). Since then, this community of wild chimpanzees has been habituated to observers, without provisioning, and individuals can be monitored on a daily basis at distances ranging from 5 to 15 m. Presently, a population of 12 chimpanzees (*Pan troglodytes verus*) inhabits the forest surrounding the village of Bossou, and group size has fluctuated between 12 and 23 individuals since 1976 (Sugiyama 1981, 1984, 1999). For further details about this field site, refer to Sugiyama (1999), Humle and Matsuzawa (2001), and Matsuzawa et al. (2001).

3.2 Methods and Approaches to the Study of Ant Dipping

3.2.1 Video Recording: A Tool for Detailed Behavioral Recording

Video recording of chimpanzees in the field is a useful tool that allows for detailed analysis of behavior and for the archiving of the performance of an individual over the course of its lifetime (Matsuzawa et al. 2001). I filmed ant-dipping behavior among the Bossou chimpanzees using a Sony DCRTRV20 digital camera in June–September 2000 and June–September 2001, and using a Sony Hi8 video camera in October 1997. G. Yamakoshi gathered video data between August and October 1999 using a Sony DCR-TRV9 digital camera, and one session was recorded in August 2001 by G. Ohashi. More than 10h of video data were amassed, encompassing 24 ant-dipping sessions. A session was defined as a period during which at least one chimpanzee was engaged in tool behavior; the session was terminated when the last remaining chimpanzee of the subgroup ended its tool-use activity. After each filming session, the ant species dipped for was collected for subsequent identification, and the condition of the ants (nest or migrating/foraging) was noted.

I analyzed all the video data twice, and 58% of the sessions were analyzed once by a second observer blind to the hypotheses being tested. Any divergences in scoring were reviewed by both observers until a consensus was reached. During the video analysis, tool length was recorded for each individual as either <50 cm or ≥50 cm. This 50-cm demarcation was based on the average between mean tool length reported by Sugiyama (1995) (46.7 cm) and that found in the tool sample set gathered during the course of this study (n = 189; mean = 53.7 cm; range = 23–154 cm; SD = 21.01; median = 48.2 cm), as well as the reported maximum tool length reported from Taï, that is, 50 cm. Ascription to these two categories was based on either precise tool length data when available from direct observations of the behavior (48.5% of tools) or simply comparing the length of the tool with objects of known length in the environment. The position of the tool user was also noted as above ground (i.e., sitting on a liana or a bentover sapling, or hanging from a liana or a branch) or on the ground (i.e., sitting or standing at ground level).

Tools were collected over four study periods: July–October 1997; July–September 1999, June–September 2000, and June–September 2001. Systematic length measurements were obtained for all tools collected, along with information about tool-user identity, whenever possible.

3.2.2 Human Ant-Dipping Experiment

I conducted the following ant-dipping experiment in September 2001 using measures based on 89 tools I collected at Bossou during the first three study periods. A human dipped for ants using tools of three different lengths: (1) mean lower quartile length (28.1 cm), (2) mean length (55.3 cm), and (3) mean upper quartile length (101.7 cm). Each tool was made from *Maranthacloa* spp., a commonly selected plant species used for tool making at Bossou (Humle 2003).

The most recent taxonomic revision of *Dorylus* ants at Bossou was conducted by C. Schöning (Schöning et al., in preparation). Schöning revealed that five species are consumed by Bossou chimpanzees. At the onset of this study of ant-dipping behavior among the Bossou chimpanzees, these species were classed into two distinct classes, which I refer to subsequently as (1) the red type, which actually comprises three species: *D. (Anomma) emeryi* (Mayr), *D. (Anomma) gerstäckeri* (Emery), and *D. (Anomma) mayri* (Santschi); and (2) the black type, which consists of two morphologically and behaviorally very similar species: *D. (Anomma) nigricans* (Illiger) and *D. (Anomma) arcens* (Westwood). Because chimpanzees at Bossou consume several species of driver ants, the human dipping sessions were performed on representative species of the red type and one species representing the black type, in both nesting and migrating conditions, thus creating four conditions.

For each tool used and on a random basis over a total of eight sessions, one person dipped for ants using different set bout durations (range, 2–120 s), averaging 37 dips per tool for each session. The ants harvested from each dip were placed into a sealable polythene bag and counted. Bout duration corresponded to the time from when the tool made contact with the ants to when the tool was just being inserted into the sealable bag. One person timed the experiment while another (the same throughout the experiment) dipped for the ants in a fashion similar to that observed among Bossou chimpanzees, making slight regular back-and-forth movements of the tool to stimulate swarming of the ants. A new tool was made for each session. The purpose of the experiment was to assess differences in prey density and prey belligerence across ant condition (nest versus migrating or foraging) and the two types of *Dorylus* ants. In addition, I was able to acquire a measure of the number of ants harvested across tools of different length under these four conditions.

3.3 Avoiding the "Pooling Fallacy"

When analyzing tool length data in the context of chimpanzee ant dipping, I carefully considered whether the data points were independent of one another. I therefore avoided the "pooling fallacy" (Martin and Bateson 1993) by assigning a single data point each time a new tool was used and each time the tool was modified in length during its use. For technique used, that is, direct mouthing or pull-through, a single data point was recorded each time there was a switch in technique employed or in tool being used. Therefore, if a chimpanzee dipped with the same tool 20 consecutive times and each time was observed directly mouthing, this was scored as a single mouthing data point. Postural data during ant dipping were scored each time the chimpanzee changed position from above ground to ground level or vice versa.

4 Results: Response to Risk

4.1 Tool Length Relative to Risk Incurred

The human ant-dipping experiment revealed that significantly more ants were harvested when dipping at the nest than on migrating or foraging ants (Mann–Whitney U test: $z = -9.137$; $P < 0.001$) (Fig. 1). This result is independent of the type of *Dorylus* (Humle and Matsuzawa 2002). This finding, therefore, clearly suggests that driver ants are found at a greater density and/or are more belligerent at the nest than while progressing on the ground. Based on this result, I predicted that the chimpanzees would use longer tools when dipping at a nest site than on migrating or foraging ants to minimize the risk of getting bitten. Bossou chimpanzees do indeed use significantly longer tools when dipping at the nest than when dipping on migrating or foraging ants (Mann–Whitney U test: $z = -5.383$; $P < 0.001$) (Fig. 2). This result supports the hypothesis that the chimpanzees are responding to ant aggressiveness and/or gregariousness by using longer tools when targeting ants at the nest site, thus reducing the risk posed by these biting ants.

Similarly, during the human ant-dipping experiment, significantly more black than red ants were harvested (Mann–Whitney U test: $z = -4.783$; $P < 0.001$) (Fig. 3). This pattern was consistent whether analyzing the nest or the migrating and foraging conditions separately (Humle and Matsuzawa 2002). This result suggests that the black type is either more aggressive or present at higher densities than the species of the red type, thus presenting the chimpanzees with a higher-risk situation when ant dipping. I therefore predicted that the chimpanzees would rely on longer tools when targeting the black type than when targeting

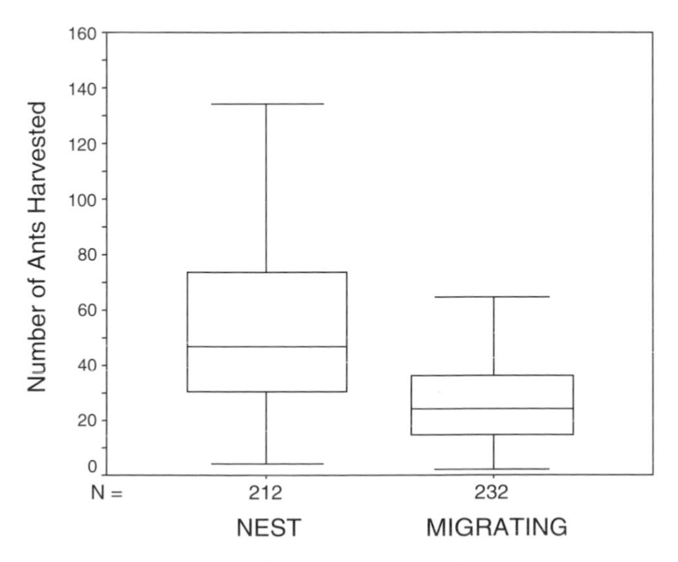

Fig. 1. Number of *Dorylus* ants harvested during the human ant-dipping experiment either at the nest or on migrating or foraging ants (*N*, number of dips upon which the data are based)

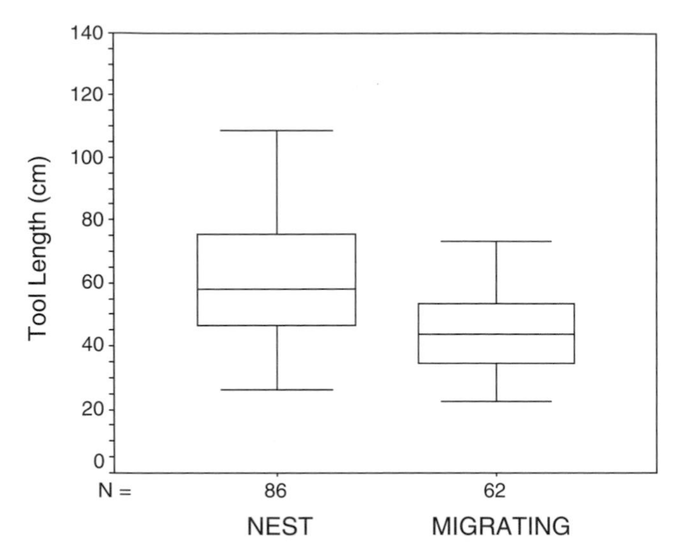

Fig. 2. Tool length employed by Bossou chimpanzees when ant dipping either at the nest or on migrating or foraging ants (*N*, number of tools upon which the data are based)

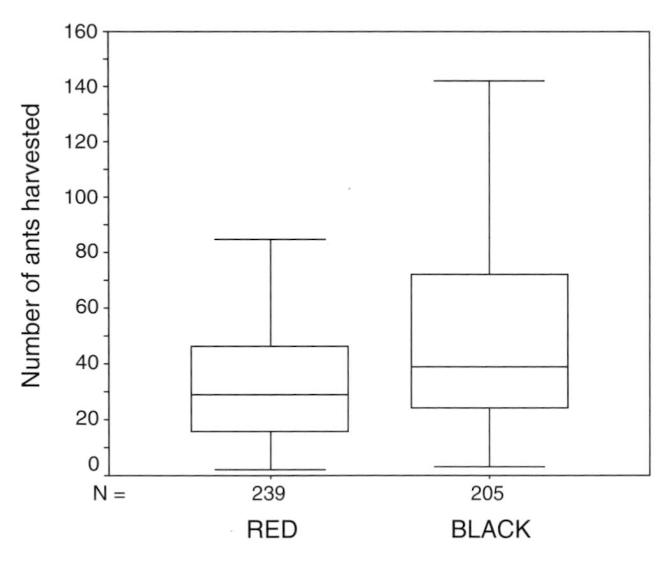

Fig. 3. Number of *Dorylus* ants of either the black or red type harvested during the human ant-dipping experiment (*N*, number of dips upon which the data are based)

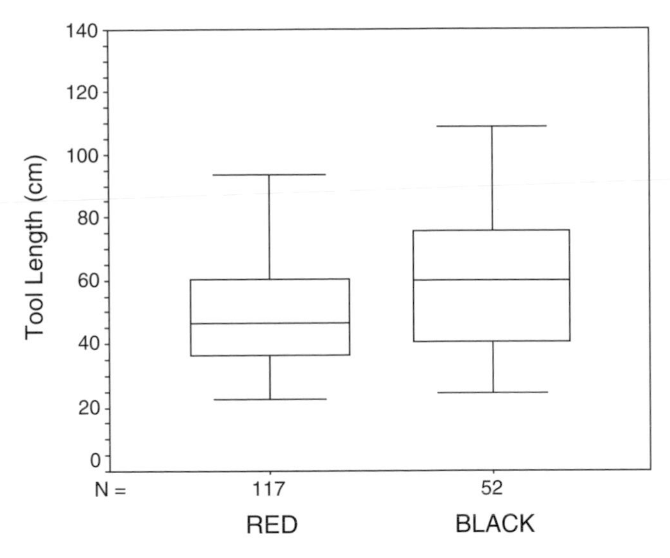

Fig. 4. Tool length employed by Bossou chimpanzees when ant dipping on either the black or the red type of *Dorylus* ants (*N*, number of tools upon which the data are based)

species of the red type. Chimpanzees were indeed found to employ longer tools when dipping for the black type than for the other three species of the red type (Mann–Whitney U test: $z = -2.802$; $P < 0.01$) (Fig. 4).

4.2 Position of Ant-Dipping Chimpanzees

Overall, at the nest or on migrating or foraging ants, the position of the chimpanzee while dipping was independent of tool length, technique used, and ant type dipped for (Humle and Matsuzawa 2002). However, a chimpanzee ant dipping at the nest site was significantly more likely to be positioned above ground than when dipping on migrating or foraging ants (two-tailed sign test: $n = 9$ individuals; $P < 0.05$) (Fig. 5). This result again suggests that the chimpanzees are responding to the risk of getting bitten because ants at the nest are more aggressive or are found at greater densities than while migrating or foraging, possibly because they are defending the colony as a whole.

4.3 Technique Used in Relation to Tool Length

The direct mouthing technique was more frequently observed at Bossou than the pull-through technique (Fig. 6). However, individual chimpanzees were significantly more likely to employ the pull-through technique at the nest site than on migrating or foraging ants (two-tailed sign test: $n = 9$ individuals; $P < 0.05$) (see Fig. 6). A single juvenile individual, Juru, was responsible for the rare instances of pull-through observed on migrating ants (see Fig. 6).

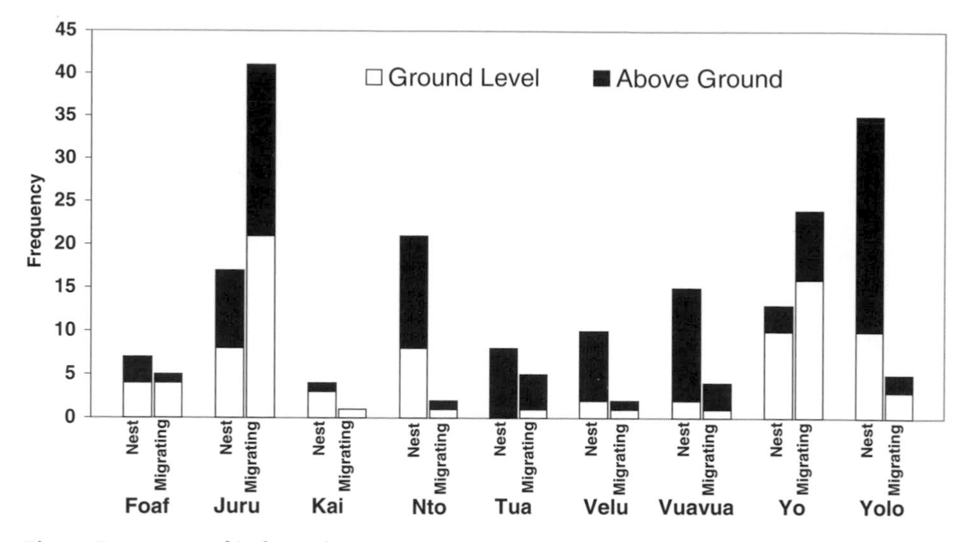

Fig. 5. Frequency of independent position scores (above ground or ground level) as a function of ant condition across individual members of the Bossou community for which data were obtained under both ant conditions

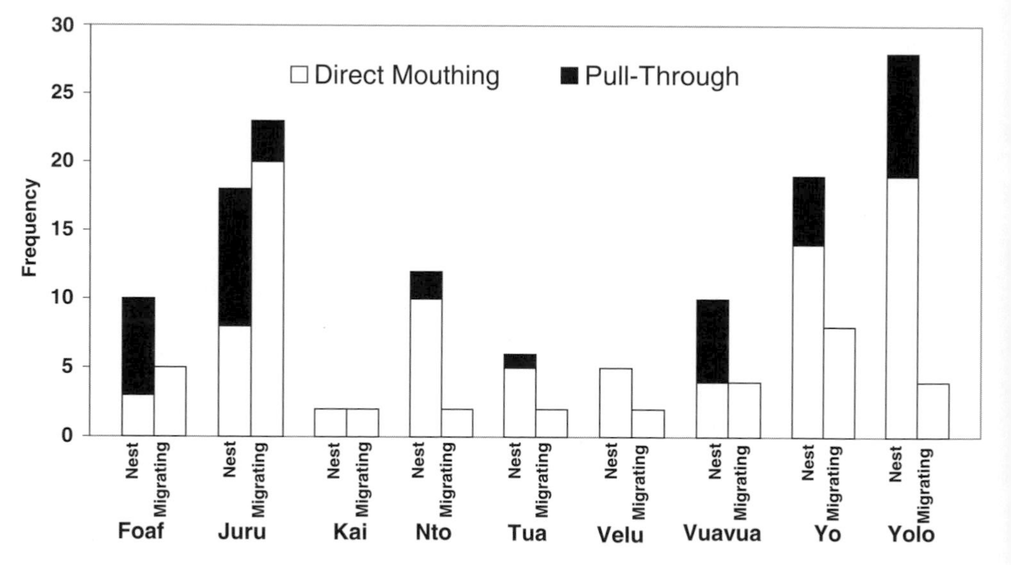

Fig. 6. Ant dipping at the nest site and on migrating or foraging ants and associated technique (based on independent data points) for each individual chimpanzee for which data were obtained under both ant conditions

Although pull-through was more frequently observed with dipping at the nest site, individual variation among chimpanzees was observed (see Fig. 6). For example, for those individuals for which there are data under both ant conditions, two adult females, Kai and Velu, were never seen employing the pull-through technique while dipping on ants at the nest. However, the latter technique was observed in seven other individuals: these included adults (>11

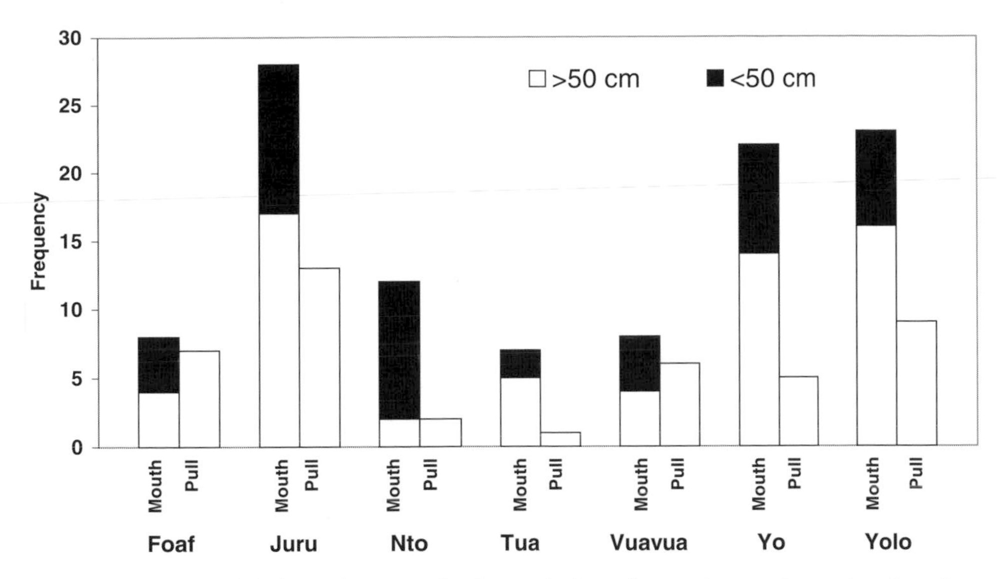

Fig. 7. Frequency of independent ant-dipping technique data points and associated tool length for individual Bossou chimpanzees that have so far been observed employing both techniques

years old) (Foaf, Tua, and Yo), as well as adolescents (8–11 years old) (Yolo, Vuavua, and Nto) and juveniles (Juru, Nto) of both sexes (see Fig. 6). Yamakoshi and Myowa-Yamakoshi (2003) also confirmed observing Fotayu (juvenile) demonstrate the pull-through technique in 1999.

Finally, the pull-through technique emerged to be significantly associated with the use of tools more than 50 cm long (Fig. 7). For tools less than 50 cm long, Bossou chimpanzees only employed the direct mouthing technique when consuming harvested ants (Fig. 7).

5 Efficiency in Ant Dipping

5.1 Dipping Time and Efficiency

The human ant-dipping experiment revealed no correlation between dipping time and quantity of ants harvested on migrating or foraging ants ($Rs = -0.102$; $n = 232$; n.s.). However, there was a significant positive correlation between dipping time and the number of ants collected during the experiment when ants were dipped for at the nest ($Rs = 0.316$; $n = 212$; $P < 0.001$) (Fig. 8). Thus, longer dipping times at the nest might yield an enhanced ant harvest for the dipping chimpanzee.

Dipping time in seconds in the chimpanzees was assessed from the video records. Dipping time refers to the time elapsed between when the chimpanzee places its tool into the mass of ants and when it starts to ingest the ants. Dipping

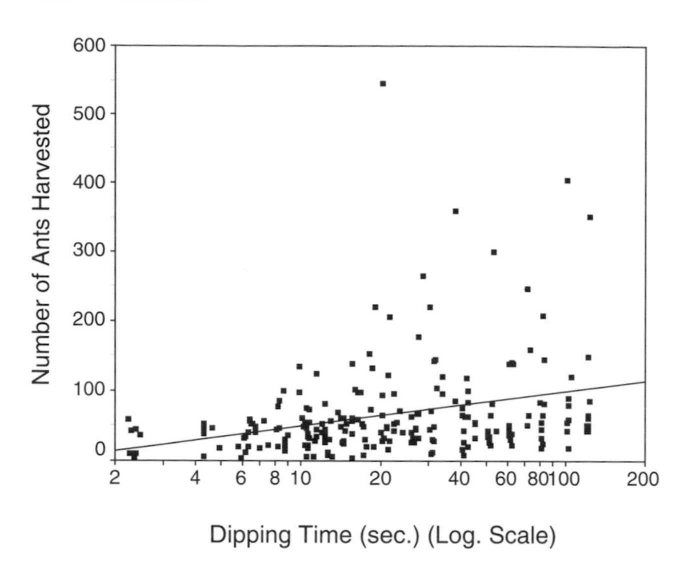

Fig. 8. Dipping time in seconds (log scale) against the numbers of ants harvested at the nest site during the human ant-dipping experiment

times in the chimpanzees were significantly longer when pulling-through than when direct mouthing (Wilcoxon signed-ranks test: $n = 8$ individuals, $z = -2.100$, $P < 0.05$).

5.2 Tool Length and Efficiency

During the human ant-dipping experiment, a significant difference in the amount of ants gathered at the nest site emerged across the three different tool lengths employed (Kruskal–Wallis test: $\chi^2 = 8.521$; $df = 2$; $P < 0.05$) (Fig. 9). Dunn's post hoc test revealed that the long tool yielded more ants than either the short tool or the medium length tool; however, there was no difference between the latter two. No difference across the three tools occurred for migrating or foraging ants (Kruskal–Wallis test: $\chi^2 = 1.747$; $df = 2$; n.s.).

Finally, the pull-through technique is associated with the use of longer tools and dipping at the nest site, and longer dipping times in the chimpanzees at the nest. It is, therefore, possible that the use of longer tools at the nest site is not simply a response to the greater biting risk, but also an adaptation for greater efficiency in prey procurement in this particular context.

5.3 Subadults Versus Adults

Based upon 1104 successful dips, chimpanzees at Bossou performed the dipping movement on average 2.37 times per minute (SD = 2.7). Based upon 610 successful dips, adult chimpanzees (12 years old and older) performed dipping acts

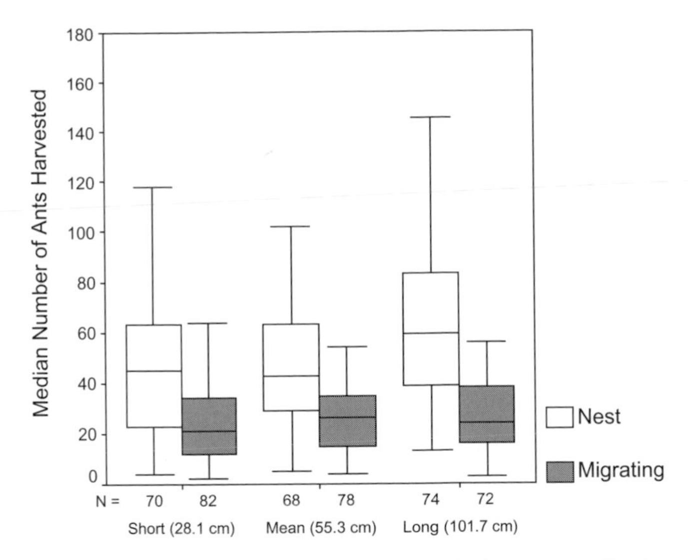

Fig. 9. Number of ants harvested during the human ant-dipping experiment under both ant conditions and for the three different tool lengths used

2.6 times per minute (SD = 2.3). Based on 444 dips across both types of *Dorylus* and under both ant conditions, the overall mean number of ants harvested during the human ant-dipping experiment was 50.24 per dip (SD = 63.1). Extrapolating from these figures, overall, chimpanzees at Bossou gathered on average 119 ants per minute (SD = 105.1) during an average dipping session, and adults gathered an average of 131 ants per minute (SD = 77.6).

An analysis of relative efficiency in behavior comparing adults (>11 years old, $n = 7$) and subadults (including juveniles and adolescents, as in individuals between 4 and 11 years of age, $n = 5$) revealed that adults displayed a significantly greater dipping rate (dips/minute) than did subadults [independent samples t test: $t(10) = 3.964$; $P < 0.01$] (Table 1). In addition, adults gathered significantly more ants per minute [independent samples t test: $t(10) = 4.248$; $P < 0.01$] and were less error prone than subadults [independent samples t test: $t(10) = -3.229$; $P < 0.01$] (see Table 1). None of the subadults exceeded adults in dipping rate or ants gathered per minute, and subadults were overwhelmingly more error prone than adults (see Table 1). Finally, it takes subadults years of practice before they attain an adult's level of efficiency. This observation yields the question as to what are the characteristic stages of development of ant dipping in youngsters and whether driver ant species and dipping context might influence this developmental process, considering the risk posed by these biting ants for young learners.

Table 1. Ant-dipping performance measures for each individual chimpanzee at Bossou

Name	Age class[a]	Sex	Dips/min[b]	Ants/min[b]	Error rate (%)[b]
Tua	Adult	Male	2.7	133.2	1.5
Velu	Adult	Female	2.5	124.2	0.0
Kai	Adult	Female	2.4	118.6	0.0
Yo	Adult	Female	2.1	101.8	1.4
Nina	Adult	Female	1.7	86.5	0.0
Jire	Adult	Female	1.7	85.6	0.0
Foaf	Adult	Male	1.7	83.9	0.0
Yolo	Adolescent	Male	1.6	74.6	3.9
Vuavua	Adolescent	Female	1.1	54.9	3.0
Nto	Adolescent	Female	1.1	50.5	6.2
Fotayu	Adolescent	Female	0.6	31.0	0.0
Juru	Juvenile	Female	1.5	71.6	3.1
Jeje	Infant	Male	0.9	21.9	54.5

To avoid small sample size effects, only data from individuals for which a minimum of 6 bouts and 20 dips were recorded were retained for analysis.
A bout is a period during which an individual is engaged in tool use, separated by intervals when no tool is held (when relevant), the hand performs an intervening activity, such as self grooming or suckling, or when the chimpanzee changes position (McGrew and Marchant 1992, p 115).
[a]Age class: adult, >11 years old; adolescent, 8–11 years old; juvenile, 4–7 years old; infant, <4 years old (based on Sugiyama 1999)
[b]Note on measures: Dips/min, the rate of dipping was calculated as the ratio of the total number of dips over the total time spent ant dipping in minutes; Ants/min, ant-dipping efficiency was calculated for each individual chimpanzee based on the assumption that each dip yielded an average of 50 ants (refer to Section 5.3 for origin of this value). The equation used was as follows: Ants/Min. = (Dips/Min.)*[(Successful dips*50)/(Total no. of Dips)]; Error rate (%) = [(Total no. of withdrawals yielding no ants)/ (Total no. of withdrawals)]*100

6 The Development of Ant Dipping in Subadults

The youngest member of the Bossou community observed ant dipping was a 32-month-old male infant (Jéjé), who dipped for migrating ants with a short tool (32 cm) while hanging from a vine and sitting on the back of his mother. The latter was also engaged in dipping when standing tripedally on the ground. The behavioral sequence performed by Jéjé was similar to that of adults, although it was apparent that the gain was small and the competence level relatively poor (see Table 1).

The pattern of development of ant dipping at Bossou is similar to that observed of Gombe (McGrew 1977). However, youngsters at Bossou appear to begin dipping at an earlier age than at Gombe, where chimpanzees only start at around 5 years of age (McGrew 1977). McGrew (1977) pointed out that infants at Gombe never tried to dip for driver ants, but performed some elements of this tool behavior in isolation, whereas Jéjé, a 32-month-old male infant from Bossou, was observed correctly producing the ant-dipping sequence on migrating ants, although not very effectively. The youngest Gombe chimpanzee observed dipping was a 46-month-old female. Given that infants and juveniles are clearly vulnerable to getting bitten by driver ants, the difference in the onset of ant dipping between the two sites could be because *Dorylus* ants at Bossou are less aggressive than those consumed at Gombe.

By the age of 6, although less efficient than adults, Bossou chimpanzees are able ant dippers. When engaged in tool-use behavior, young chimpanzees often behave in a less stereotyped fashion than adults (McGrew 1977; Inoue-Nakamura and Matsuzawa 1997). Juru's occasional pulling-through on migrating ants as a juvenile could well reflect a lack of stereotypy in behavior, or a lack of experience, or represent a different threshold to exposure to discomfort when compared to adults.

The influence of exposure to discomfort on the ant-dipping habits of young Bossou chimpanzees was reflected by the fact that no adolescents were ever observed dipping on the more gregarious/aggressive *Dorylus nigricans* at the nest site, whereas both adults and juveniles have been observed dipping in this context, although rarely, that is, less than 10% of all sessions that each individual was observed dipping. In addition, a significant association between position, whether above ground or at ground level, and age-class emerged overall (chi-square test: $\chi^2(2) = 6.282$; $P < 0.05$) and when ants were dipped for at their nest (chi-square test: $\chi^2(2) = 6.383$; $P < 0.05$), whereas there was none when the behavior was targeted at migrating or foraging ants (chi-square test: $\chi^2(2) = 3.170$; n.s.). A post hoc analysis revealed that, overall, and at the nest, adolescents positioned themselves significantly more often above ground than adults (overall: adolescent/above ground: 72/113; adult/above ground: 40/87; two-tailed Z test: $z = 2.53$; $P < 0.05$; nest: adolescent/above ground: 58/80; adult/above ground: 25/47; two-tailed Z test: $z = 2.19$; $P < 0.05$).

Thus, adolescent chimpanzees showed a more cautious approach to dipping, particularly on nesting ants, compared with adults and juveniles. However, juveniles focused their dipping more on migrating ants (15/18 individual sessions) than did adolescents (13/21). Adults dipped on nesting and migrating *Dorylus* indiscriminately ($n_{\text{nesting}} = 18$; $n_{\text{migrating}} = 20$; binomial test, n.s.) and expressed no significant preference for either ant type, although they tended to dip more frequently for the red type ($n_{\text{red}} = 24$; $n_{\text{black}} = 14$; binomial test; n.s.). Nevertheless, at the nest, adults dipped significantly more often on the red species (16/18) than on the black (2/18) (two-tailed Z test comparing two proportions: $z = 7.42$; $P < 0.001$).

We can distinguish five characteristic stages in the development of ant-dipping behavior in young chimpanzees: (i) manipulatory play and (ii) tool manufacture in infants, (iii) motor skill of tool use in infants and juveniles, (iv) knowledge of the quality of the tool and efficiency of its use in both juveniles and adolescents, and (v) refinement of motor skill in response to the antipredator behavior of the ants and increased instances of dipping in similar contexts chosen by adults in adolescents. Indeed, both juveniles and adolescents practice and perform ant dipping under both ant conditions. However, juveniles tend to dip in contexts that present less risk. Adolescents dip both at the nest and on migrating or foraging ants, but while dipping they exhibit a more-cautious approach by positioning themselves more often above ground. Adolescents are thus able to increase their understanding of the relationship between tool length, the effectiveness and suitability of a technique, the biting risk posed by the ants, and the overall efficiency of their prey procurement.

There was no obvious link between mother and offspring in the repertoire of techniques each displayed under similar ant-dipping conditions. Three of four mothers have so far never been observed pulling through while their offspring perform both techniques. Considering that the mother plays a vital role in the transmission of behavior to her offspring (McGrew 1977), this observation suggests that technique employed is likely to be acquired via individual learning rather than social learning, unless the postweaning environment offers the youngster exposure to alternative influential models. Finally, at Bossou, the prime period for the development of ant dipping in the young chimpanzee is as a juvenile (4–7 years old), and perfection in efficiency in this skill continues throughout adolescence (until 11 years of age).

7 Comparison with Other Sites

Differences in prey aggressiveness and behavior may lead to differences in tool length within and between communities of chimpanzees. However, as suggested by Hashimoto et al. (2000), differences in tool length may also reflect the different techniques used for catching ants. Indeed, it is also possible that pull-through may simply be the most effective and functional method of gathering ants off a long tool, which would then explain the predominance of this technique at Gombe. During pull-through, the gathered mass of ants is crumpled and jumbled so that few can bite the chimpanzee before they are consumed, whereas they might pose a greater biting risk to the chimpanzee if the long tool is mouthed.

To what extent can the results that have emerged from Bossou apply to what is observed at Taï and Gombe? Gombe chimpanzees rarely extract *Dorylus* ant larvae and eggs directly by hand from the nest, whereas Bossou chimpanzees occasionally do so. In addition, Gombe chimpanzees stay off the ground in 74% of ant-dipping episodes, while Bossou chimpanzees do so only 55.9% of the time (McGrew 1974; Humle and Matsuzawa 2002). These latter observations suggest that Gombe chimpanzees either behave more cautiously toward the driver ants than Bossou chimpanzees or are dealing with more gregarious or aggressive species of *Dorylus* ants. So, extrapolating from the results obtained at Bossou, we would expected Gombe chimpanzees to use longer tools than at Bossou and to exhibit the pull-through technique, and this is indeed what is observed at this site (McGrew 1974).

Taï chimpanzees, on the other hand, perform ant-brood extraction on a frequent basis, suggesting that they have either developed an efficient strategy in coping with biting ants, or are less susceptible to their bite, or that the *Dorylus* ant species targeted might not pose as great a risk to the dipping chimpanzee (Boesch 1996). We might in any case expect them to employ shorter tools. Indeed, Taï chimpanzees only employ tools shorter than 50 cm (Boesch and Boesch 1990). On the basis of our results from Bossou, we would consequently predict that the Taï chimpanzees would only perform the direct mouthing technique, which is indeed consistent with observations of ant dipping at this site (Boesch and Boesch 1990). A puzzling recent finding has been that Taï chimpanzees actually consume the same species of *Dorylus* ants as at Bossou (Schöning et al., in preparation.). At Taï, however, only young chimpanzees dip on migrating or foraging ants, and the eggs and larvae of *D. nigricans* are favored by the chimpanzees. In addition, adult Taï chimpanzees apparently mainly dip on the red type at the nest site (C. Boesch, personal communication). These variations in prey emphasis and technique used suggest that social learning might still explain most of the between-site variation observed in ant-dipping behavior and prey emphasis. The species ingested at Gombe still remain to be identified and described.

Finally, the transmission of cultural traits and preferences within and between adjacent chimpanzee communities is likely to be affected by developmental, experiential, social, and ecological factors (Humle and Matsuzawa 2004). However, these factors cannot readily be isolated, and therefore we have a huge challenge before us if we are to understand the interactions between these processes and variables.

8 Learning Mechanisms at Work

Our study of ant dipping at Bossou so far cannot distinguish between the different learning mechanism(s), whether social or individual, involved in the acquisition of ant dipping. Ant dipping at Bossou is, nevertheless, commonly displayed by all able-bodied members of the community and is perpetuated from one generation to the next. Moreover, some chimpanzee communities, such as Mahale, Lopé, and Budongo, do not exhibit this behavior, although driver ants are available and stick- or stalk-tool uses have been observed in other contexts at two of these sites at least. Ant dipping, therefore, still exhibits strong cultural patterns and appears to remain a good example of culture in chimpanzees.

In previous studies of culture, it has been argued that environmental explanations exclude a given behavior from being considered cultural. However, these studies are based on evidence from between-site differences and often ignore the link between behavioral variation and environmental variables within a given site. Therefore, a key result of my study is that microecological variables do significantly influence the position and tool length employed in ant dipping in individual chimpanzees. These microecological factors may also play a role in the process of acquisition and the performance of this tool-use behavior in subadults. In addition, I have shown here that variations in technique employed at the individual level are to a large extent influenced by tool length and dipping time, but also possibly reflect idiosyncratic preferences (i.e., some individuals were never observed pulling though and always relied upon the direct mouthing technique). These findings suggest that individual chimpanzees adapt their behavior not only to minimize the risk of getting bitten but also potentially to increase their own behavioral efficiency. These results are not surprising considering chimpanzees' cognitive ability to demonstrate plasticity in behavior and to adapt to variable environmental or social conditions.

Despite the importance of environmental variables, is it still possible that the more intricate details of ant-dipping behavior, such as tool length and technique employed, be socially learned? In their study of factors influencing imitation of manipulatory actions in captive chimpanzees, Myowa-Yamakoshi and Matsuzawa (1999) demonstrated that chimpanzees can "imitate" others' actions by reproducing the final state of target objects, or movement of tools and/or target objects, but not the action itself. Moreover, other studies indicate that chimpanzees do not readily copy manipulatory actions modeled by human subjects (Nagell et al. 1993; Custance et al. 1995). The only two studies showing that chimpanzees are able to do so were conducted on "enculturated" chimpanzees, that is, reared by humans in a relatively enriched and stimulating environment (Hayes and Hayes 1952; Tomasello et al. 1993). This enculturation of chimpanzees at an early age may influence social learning abilities to an extent that is not observed among wild chimpanzees. Although the results of most studies in captivity suggest that the precise techniques and tool lengths employed by wild chimpanzees during ant dipping are unlikely to be socially learned, the general sequence of the behavior almost certainly involves the input of some social learning mechanism (Byrne and Russon 1998).

So, although the involvement of social learning in the acquisition of ant dipping in chimpanzees has yet to be ascertained, it appears that social learning and individual learning may act in concert, allowing efficiency in performance and flexibility in behavior in the face of variable spatial and temporal conditions and potential exposure to risk.

9 Future Directions

The social contribution to the transgenerational maintenance of ant dipping and to the observed variations in technique(s) and tool length employed by chimpanzees across study sites clearly warrants further exploration. On the one hand, we can investigate further the extent to which microecological variables account for differences in driver ant foraging between sites. This question has to be addressed fundamentally, especially because the species of *Dorylus* ants present at Bossou and Taï are identical (Schöning, in preparation). Could there still be differences in the aggressiveness, gregariousness, and general antipredatory behavior of these species that could explain the variations in prey emphasis and behavioral pattern in skill performance? Field experiments such as the human ant-dipping experiment carried out during this study might prove a very useful approach for clarifying this latter question.

On the other hand, while acknowledging limitations of working in a natural setting, rather than a controlled laboratory environment, is it possible to gather data on variables that might affect and stimulate learning of ant dipping in young chimpanzees? What is the early influence of the mother in providing opportunities for practice in her offspring and facilitating the acquisition of this behavior in terms of motor-skill development and efficiency in the task? Through "opportunity teaching," the mother may intervene by "creating a discovery environment," by placing its offspring "in a situation conducive to learning a new skill or acquiring new knowledge" (Caro and Hauser 1992, p. 166). However, any potential benefit will depend on the infant's response expressed, for example, through close observation of competent models, including its mother, and practice of the behavior. In addition, during the course of development, what is the contribution of scaffolding behaviors, whereby adults modify their behavior, thus promoting learning (Bruner 1982; Wood 1980), to the acquisition of ant dipping in young? To date, few studies of primates (human or nonhuman) have attempted explicitly to assess the role of "opportunity teaching" or scaffolding in the acquisition of complex tasks, particularly in relation to the social and environmental context in which the behavior takes place and to efficiency in performance. Ant dipping, a hazardous foraging tool-use behavior occurring in varying contexts, constitutes in that sense a good focal behavior for such an investigation.

In her study of the role of mothers in the acquisition of termite-fishing behaviors in wild chimpanzees at Gombe, Lonsdorf (2005) found limited evidence for the importance of learning opportunities provided by the mother on the acquisition speed of termite fishing of their offspring. However, in a study of the acquisition of a honey-fishing task in three captive chimpanzee mother–infant pairs, Hirata and Celli (2003) demonstrated that infants observed their mother's performances in detail, as well as that of other adults. The infants in this captive study acquired the skill between the ages of 20 and 22 months, which is much earlier than reports of stick- or stalk-tool-use skill acquisition in the field. Hirata and Celli (2003) suggest that the captive environment may have accelerated the development of the skill by providing more opportunities for object manipulation and observation than in the field. Lonsdorf et al. (2004) also reported finding distinct sex differences in the development of termite fishing in chimpanzees at Gombe. Although, when at the termite mound, mothers were equally tolerant of their offspring regardless of their sex, young females spent significantly more time watching others than did young males. Female offspring actually developed the skill more quickly than males, were more proficient at the skill once it was acquired, and their choice of tool length resembled their mothers' more so than that of males. Will similar patterns emerge for ant dipping at Bossou or will acquisition patterns differ because of the different nature of the target prey and more variable contexts in which this behavior takes place or differences in social dynamics between chimpanzees at Gombe and Bossou?

Issues that have also rarely been empirically addressed in primates are whether there are costs associated with social learning and how reliance on individual and social learning may vary with environmental circumstances (Laland et al. 1993). There is growing evidence among various animal species that a strong reliance on social learning can result in maintenance of maladaptive or suboptimal behaviors (rats: Galef 1986; guppies: Laland and Williams 1998; cotton-top tamarins: Snowdon and Boe 2003). A longitudinal and developmental approach to the study of ant dipping could provide further insight into this topic by exploring the relationship between efficiency in behavior, environmental context, task complexity, and the interaction between individual and social learning.

Acknowledgments

I thank the Ministère de l'Enseignement Supérieur et de la Recherche Scientifique, in particular the Direction Nationale de la Recherche Scientifique and l'Institut de Recherche Environnementale de Bossou (IREB), for granting me the permission to carry out research at Bossou. I am particularly grateful to Tetsuro Matsuzawa, Charles Snowdon, and William McGrew for their advice and support and to Gen Yamakoshi and Gaku Ohashi for contributing some of their video recordings of ant dipping. I am also very grateful to Caspar Schöning for his revision of the taxonomy of driver ants consumed by Bossou chimpanzees and to Rosamunde Almond and Charles Snowdon for their critical comments on this chapter. The present study was supported in part by grants from the Ministry of Education, Science, and Culture, Japan (nos. 07102010, 12002009 to T. Matsuzawa, and 10CE2005 to O. Takenaka), a Leakey Foundation Grant, and an NIH Kirschstein-NRSA Postdoctoral Fellowship (no. MH068906-01) to T.H.

References

Biro D, Inoue-Nakamura N, Tonooka R, Yamakoshi G, Sousa C, Matsuzowa T (2003) Cultural innovation and transmission of tool use in wild chimpanzees: evidence from field experiments. Anim Cogn 6:213–223

Boesch C (1991) Teaching among wild chimpanzees. Anim Behav 41:530–532

Boesch C (1996) The emergence of cultures among wild chimpanzees. Proc Br Acad 88:251–268

Boesch C, Boesch H (1990) Tool use and tool making in wild chimpanzees. Folia Primatol 54:86–99

Bruner J (1982) The organization of action and the nature of the adult-infant transaction. In: Tronick EZ (ed) Social interchange in infancy. University Park Press, Baltimore, pp 23–35

Byrne RW, Russon AE (1998) Learning by imitation: a hierarchical approach. Behav Brain Sci 21:667–721

Caro T, Hauser MD (1992) Is there teaching in non-human animals? Q Rev Biol 67:151–171

Collins DA, McGrew WC (1987) Termite fauna related to differences in tool-use between groups of chimpanzees (*Pan troglodytes*). Primates 28:457–471

Coussi-Korbel S, Fragaszy DM (1995) On the social relation between social dynamics and social learning. Anim Behav 50:1441–1453

Custance DM, Whiten A, Bard KA (1995) Can young chimpanzees (*Pan troglodytes*) imitate arbitrary actions? Hayes & Hayes (1952) revisited. Behaviour 132:837–859
Fragaszy DM, Perry S (eds) (2003) The biology of traditions: models and evidence. Cambridge University Press, Cambridge
Galef BG Jr (1986) Social interaction modifies learned aversions, sodium appetite, and both palatability and handling time induced dietary preference in rats. J Comp Psychol 100:432–439
Galef BG Jr (1992) The question of animal culture. Hum Nat 3:157–178
Goodall J (1968) The behaviour of free-living chimpanzees in the Gombe Stream Reserve. Anim Behav Monogr 1:161–311
Goodall J (1986) The chimpanzees of Gombe. Belknap Press, Cambridge
Hashimoto C, Furuichi T, Tashiro Y (2000) Ant dipping and meat eating by wild chimpanzees in the Kalinzu forest, Uganda. Primates 41:103–108
Hayes KJ, Hayes C (1952) Imitation in a home-raised chimpanzee. J Comp Physiol Psychol 45:450–459
Hirata S, Celli ML (2003) Role of mothers in the acquisition of tool-use behaviors by captive infant chimpanzees. Anim Cogn 6:235–244
Hirata S, Watanabe K, Kawai M (2001) "Sweet-potatoe washing" revisited. In: Matsuzawa T (ed) Primate origins of human cognition and behavior. Springer, Tokyo, pp 487–508
Humle T (2003) Culture and variation in wild chimpanzee behaviour: a study of three communities in West Africa. PhD dissertation, University of Stirling, Scotland, UK
Humle T, Matsuzawa T (2001) Behavioural diversity among the wild chimpanzee populations of Bossou and neighbouring areas, Guinea and Cote d'Ivoire, West Africa. Folia Primatol 72:57–68
Humle T, Matsuzawa T (2002) Ant dipping among the chimpanzees of Bossou, Guinea, and some comparisons with other sites. Am J Primatol 58:133–148
Humle T, Matsuzawa T (2004) Oil palm use by adjacent communities of chimpanzees at Bossou and Nimba Mountains, West Africa. Int J Primatol 25:551–581
Inoue-Nakamura N, Matsuzawa T (1997) Development of stone tool use by wild chimpanzees (*Pan troglodytes*). J Comp Psychol 111:159–173
Kawai M (1965) Newly acquired pre-cultural behavior of the natural troop of Japanese monkeys on Koshima islet. Primates 6:1–30
Laland KN, Hoppitt W (2003) Do animals have culture? Evol Anthropol 12:150–159
Laland KN, Williams K (1998) Social transmission of maladaptive information in the guppy. Behav Ecol 9:493–499
Laland KN, Richerson PJ, Boyd R (1993) Animal social learning: toward a new theoretical approach. In: Klopfer PH, Bateson PP, Thomson N (eds) Perspectives in ethology, vol 10. Behaviour and evolution. Plenum Press, New York, pp 249–277
Lonsdorf EV (2005) What is the role of mothers in the acquisition of termite-fishing behaviors in wild chimpanzees (*Pan troglodytes schweinfurthii*)? Anim Cogn (Advance online publication): online 1–11
Lonsdorf EV, Pusey AE, Eberly L (2004) Sex differences in learning in chimpanzees. Nature (Lond) 428:715–716
Martin P, Bateson P (1993) Measuring behaviour: an introductory guide, 2nd edn. Cambridge University Press, Cambridge
Matsuzawa T (1994) Field experiments on use of stone tools by chimpanzees in the wild. In: Wrangham RW, McGrew WC, de Waal FBM, Heltne PG (eds) Chimpanzee cultures. Harvard University Press, Cambridge, pp 352–370
Matsuzawa T (1999) Communication and tool-use in chimpanzees: cultural and social contexts. In: Hauser M, Konishi M (eds) Neural mechanisms of communication. MIT Press, Cambridge, pp 645–671
Matsuzawa T, Biro D, Humle T, Inoue-Nakamura N, Tonooka R, Yamakoshi G (2001) Emergence of culture in wild chimpanzees: education by master-apprenticeship. In: Matsuzawa T (ed) Primate origins of human cognition and behavior. Springer, Tokyo, pp 557–574
McGrew WC (1974) Tool use by wild chimpanzees in feeding upon driver ants. J Hum Evol 3:501–508

McGrew WC (1977) Socialization and object manipulation of wild chimpanzees. In: Chevalier-Skolniikoff S, Poirier F (eds) Primate biosocial development. Garland Press, New York, pp 261–288

McGrew WC (1985) The chimpanzee and the oil palm: patterns of culture. Soc Biol Hum Affairs 50:7–23

McGrew WC (1992) Chimpanzee material culture: implications for human evolution. Cambridge University Press, Cambridge

McGrew WC (1998) Behavioural diversity in populations of free-ranging chimpanzees in Africa: is it culture? Hum Evol 13:209–220

McGrew WC, Collins DA (1985) Tool-use by wild chimpanzees (*Pan troglodytes*) to obtain termites (*Macrotermes herus*) in the Mahale Mountains, Tanzania. Am J Primatol 9:47–62

McGrew WC, Marchant LF (1992) Chimpanzees, tools and termites: hand preference or handedness? Curr Anthropol 33:114–119

McGrew WC, Tutin CEG, Baldwin PJ (1979) Chimpanzees, tools and termites: cross-cultural comparisons of Senegal, Tanzania and Rio Muni. Man 14:185–214

Myowa-Yamakoshi M, Matsuzawa T (1999) Factors influencing imitation of manipulatory actions in chimpanzees (*Pan troglodytes*). J Comp Psychol 113:128–136

Nagell K, Olguin RS, Tomasello M (1993) Processes of social learning in the tool-use of chimpanzees (*Pan troglodytes*) and human children (*Homo sapiens*). J Comp Psychol 107:174–186

Nishida T (1968) The social group of wild chimpanzees in the Mahale Mountains. Primates 9:167–224

Nishida T (1987) Local traditions and cultural transmission. In: Smuts BB, Cheney DL, Seyfarth RM, Wrangham RW, Struhsaker TT (eds) Primate societies. University of Chicago Press, Chicago, pp 462–474

Nishida T, Wrangham RW, Goodall J, Uehara S (1983) Local differences in plant-feeding habits of chimpanzees between the Mahale Mountains and Gombe National Park. J Hum Evol 12:467–480

Parker ST, Russon AE (1996) On the wild side of culture and cognition in the great apes. In: Russon AE, Bard KA, Parker ST (eds) Reaching into thought: the minds of the great apes. Cambridge University Press, Cambridge, pp 430–450

Snowdon CT, Boe CY (2003) Social communication about unpalatable foods in tamarins (*Saguinus oedipus*). J Comp Psychol 117:142–148

Sugiyama Y (1981) Observations on the population dynamics and behavior of wild chimpanzees at Bossou, Guinea, 1979–1980. Primates 22:435–444

Sugiyama Y (1984) Population dynamics of wild chimpanzees at Bossou, Guinea, between 1976–1983. Primates 25:391–400

Sugiyama Y (1988) Grooming interactions among adult chimpanzees at Bossou, Guinea, with special reference to social structure. Int J Primatol 9:393–407

Sugiyama Y (1993) Local variation of tools and tool-use among wild chimpanzee populations. In: Berthelet A, Chavaillon J (eds) The use of tool by humans and non-human primates. Clarendon Press, Oxford, pp 175–187

Sugiyama Y (1995) Tool-use for catching ants by chimps at Bossou and Monts Nimba. Primates 36:193–205

Sugiyama Y (1997) Social traditions and the use of tool-composites by wild chimpanzees. Evol Anthropol 6:23–28

Sugiyama Y (1999) Socioecological factors of male chimpanzee migration at Bossou, Guinea. Primates 40:61–68

Sugiyama Y, Koman J (1979) Social structure and dynamics of wild chimpanzees at Bossou, Guinea. Primates 20:323–339

Tomasello M (1994) The question of chimpanzee culture. In: Wrangham RW, McGrew WC, de Waal FBM, Heltne PG (eds) Chimpanzee cultures. Havard University Press, Cambridge, pp 301–317

Tomasello M (1999) The cultural origins of human cognition. Harvard University Press, Cambridge

Tomasello M, Savage-Rumbaugh S, Kruger A (1993) Imitative learning of actions on objects by children, chimpanzees, and enculturated chimpanzees. Child Dev 64:1688–1705
van Schaik CP, Fox EA, Fechtman LT (2003) Individual variation in the rate of use of tree-hole tools among wild orangutans: implications for hominin evolution. J Hum Evol 44:11–23
Whiten A, Goodall J, McGrew WC, Nishida T, Reynolds V, Sugiyama Y, Tutin CEG, Wrangham RW, Boesch C (1999) Cultures in chimpanzees. Nature (Lond) 399:682–685
Whiten A, Goodall J, McGrew WC, Nishida T, Reynolds V, Sugiyama Y, Tutin CEG, Wrangham RW, Boesch C (2001) Charting cultural variation in chimpanzees. Behaviour 138:1481–1516
Wood DJ (1980) Teaching the young child: some relationships between social interaction, language and thought. In: Olson DR (ed) The social foundations of language and thought. Norton, New York, pp 280–298
Yamakoshi G (2001) Ecology of tool use in wild chimpanzees: towards reconstruction of early hominid evolution. In: Matsuzawa T (ed) Primate origin of human cognition and behavior. Springer, Tokyo, pp 537–556
Yamakoshi G, Myowa-Yamakoshi M (2003) New observations of ant dipping techniques in wild chimpanzees at Bossou, Guinea. Primates 45:25–32

28 Ontogeny and Cultural Propagation of Tool Use by Wild Chimpanzees at Bossou, Guinea: Case Studies in Nut Cracking and Leaf Folding

Dora Biro[1], Cláudia Sousa[2], and Tetsuro Matsuzawa[3]

1 Introduction

The discovery more than four decades ago that wild chimpanzees habitually made and used tools (Goodall 1964) helped to put a fairly abrupt end to the notion that tool use was a defining characteristic unique to humans. Since then, reports of the skilful use of tools from a wide variety of primate and non-primate species have been accumulating steadily. As somewhat of a parallel, initial observations on the establishment and spread of sweet-potato washing behaviour by Japanese monkeys on Koshima island (Kawai 1965) as well as McGrew and Tutin's (1978) original report on regional differences in wild chimpanzee behaviour have been elaborated to such an extent since then (McGrew 1992; Whiten et al. 1999, 2001) that the issue of "culture" in nonhuman primates has become one of the hottest topics in current primatology. The debate centres on behaviours spanning the tool-using, self-maintenance, and social domains, and which are shared by individuals within specific communities but are known to be absent from or assume different forms in other communities. Such regional variation, when it cannot be explained by ecological or genetic factors, gives rise to questions about processes underlying the emergence, maintenance, and propagation of community-specific behaviours as well as the terminology used to describe them.

Do nonhuman animals possess culture? As is often the case with questions of this sort, the answer depends on what we understand to constitute "culture"; different definitions will yield more or less inclusive pictures of how widespread the phenomenon is across the animal kingdom (see McGrew 2004 for a comprehensive review of the controversy regarding membership in the "Culture Club"). For the purposes of this chapter, we rely on a useful working definition provided by Matsuzawa (1999): culture can be thought of as "a set of knowledge, techniques, and values that are shared by members of a community and are transmitted from one generation to the next through non-genetic channels." We

[1]Department of Zoology, University of Oxford, South Parks Road, Oxford OX1 3PS, UK
[2]Department of Anthropology, Faculty of Social and Human Sciences, New University of Lisbon, Avenida de Berna, 26-c, 1069-061 Lisbon, Portugal
[3]Primate Research Institute, Kyoto University, 41 Kanrin, Inuyama, Aichi 484-8506, Japan

focus on two tool-using behaviours which, although in general aspects not unique to our study site at Bossou, Guinea, are not found universally across all chimpanzee populations in Africa. One, the cracking of hard-shelled oil palm nuts (*Elaeis guineensis*) with the aid of a pair of stones as hammer and anvil, is restricted to West African chimpanzees, which is surprising as both nuts and stones are readily available in the habitats of Central and East African populations. The other, the use of leaves for drinking water, has been observed at many sites; however, the precise techniques used vary considerably across populations. At Bossou, leaf folding (the use of leaves that are folded, accordion-like, inside the mouth before being dipped into water and retrieved) dominates over other forms of leaf use in drinking (such as leaf sponging or leaf spooning). For both these behaviours we examine general features common to skilful users of the tools, such as tool selectivity and technique, as well as developmental aspects involved in the acquisition of the skill by young members of the group. Our setup provides us with a unique window of observation, allowing us to study the two behaviours side by side: at the same place, at the same time, and in the same individuals across several years. In accordance with the definition of culture outlined here, we examine what forms of social transmission may be responsible for the maintenance of these behaviours within the Bossou group, leading ultimately to the kind of community specificity that may be regarded as a hallmark of primate cultures.

2 The Study Site

2.1 Bossou

Bossou, located in the southeastern corner of the Republic of Guinea and home to a small group of chimpanzees of the Western subspecies (*Pan troglodytes verus*), is one of eight major long-term chimpanzee research sites around Africa. Study at the site began in 1976 and is about to enter its fourth decade. Research has focused on a variety of disciplines, including chimpanzee ecology, behaviour, genetics, physiology, and conservation. Until 2003, the size of the Bossou community had remained relatively stable around 20 individuals (minimum 16, maximum 22). However, a flu-like epidemic at the end of 2003 took the lives of 5 community members (Matsuzawa et al. 2005), and the disappearance (probable emigration) of 3 more individuals in 2004, followed by the birth of a single infant, means that the group currently numbers only 12 individuals, the lowest in the past 29 years.

The core area of the Bossou community measures about 5 to 6 km^2, consisting mainly of primary and secondary forest. This core area is surrounded by savanna and cultivated fields, which the chimpanzees do not commonly traverse. Beyond a stretch of about 3 to 4 km of this savanna lie the Nimba Mountains, West Africa's largest mountain range and home to a large number of chimpanzees.

Because of the isolated nature of Bossou, both immigration and emigration—common features of wild chimpanzee societies—have been rare. Only three cases of transient immigration have been recorded; none of these individuals remained permanently at Bossou (the three visits lasted 1 day, 20 days, and somewhere between 3 months and a year, respectively; Sugiyama 1999). Emigration has likely occurred more often, as several community members (mainly adolescents or young adults of both sexes) have disappeared, although in none of these cases is it known whether these individuals successfully joined adjacent communities because their presence at neighbouring sites has never been directly confirmed.

Bossou chimpanzees are known to utilize a variety of tools in feeding contexts; these include nut cracking, the use of leaves for drinking water, ant dipping, termite fishing, algae scooping, and pestle pounding (see Sugiyama 1998; Matsuzawa 1999, for extensive reviews). In addition, examples of tool use in non-feeding contexts have also been reported (Hirata et al. 1998, 2001b; Matsuzawa 1997).

2.2 Outdoor Laboratory at Bossou

Witnessing tool-using behaviours in the chimpanzee's natural habitat, particularly obtaining longitudinal records on specific individuals, is often complicated by the unpredictability of encounters with community members as well as the often dense vegetation through which the behaviours must be observed. In 1988, T. Matsuzawa set up a facility for the intensive observation of tool-using behaviours at Bossou (Matsuzawa 1994). In a clearing at the top of one of the hills within the Bossou group's core area, an "outdoor laboratory" was established, with the aim of increasing rates of encounters with all members of the community as well as the opportunity to observe tool-using behaviours in a visually uncluttered environment (Fig. 1). The laboratory is opened once each year for a period of approximately 1 to 2 months during the dry season (in December, January, or February), during which time researchers control the availability of various items inside the clearing. The location of the clearing is such that it is at the crossroads of several paths used frequently by all members of the Bossou group, and as a result chimpanzee parties of various sizes visit the outdoor laboratory on average once a day. Observers, hidden behind a grass screen at one end of the clearing, monitor the site from 0700 until 1800 each day and video record all visits by chimpanzees from at least two different angles simultaneously.

Besides easy and regular visual access to individuals, there is another important advantage associated with the outdoor laboratory. The setup facilitates extremely detailed observation of the same community members across many years, focusing not only on a single behaviour but on various different skills. These skills can be observed at the same place, often within no more than a few minutes of each other, as is the case with nut cracking and leaf folding. For

Fig. 1. "Outdoor laboratory" where intensive observations of tool-using behaviours were conducted. The *arrow on the left* points to a tree with an enlarged natural hollow containing water (only the forward-facing hole is visible; a second hole is located in the side of the tree, facing right; see Fig. 2B), from which chimpanzees drank with the aid of folded leaves. An adult female can be seen putting a leaf tool into her mouth, having just retrieved it after dipping it into the water inside the tree hollow. The *arrow on the right* shows the location of about 50 stones laid out within a small rectangular area (see Fig. 2A); from among these, chimpanzees selected their tools for use in nut cracking. Piles of oil-palm nuts were scattered on the ground within the clearing before the group's arrival, and several individuals can be seen performing the behaviour (most prominently an adult male in the centre). Photograph was taken from behind a grass screen that separated the observers from the chimpanzees inside the outdoor laboratory

example, we might observe a 3-year-old infant chimpanzee rolling stones and nuts on the ground and scrounging freshly extracted kernels from his mother, then shortly thereafter picking up a leaf tool discarded by an older individual and dipping it into a tree hollow to retrieve water. In a couple of years' time, we may see the same individual successfully cracking nuts by himself, then not only using but also making his own drinking tool. Such longitudinal records provide us with data on individual acquisition of skills, as well as the relative course of development of different behaviours within individuals.

2.2.1 Nuts and Stones for Cracking

The initial focus of the outdoor laboratory studies was nut cracking: researchers provided a set of numbered stones of known weights and dimensions (while clearing the area of all other naturally occurring stones) as well as 2 to 5 kg of oil-palm nuts laid out in several piles (Fig. 2A). Chimpanzees that visited the

Fig. 2. Materials at the outdoor laboratory used in the nut-cracking and leaf-folding studies reported here. **A** Stones available within the clearing as arranged before the arrival of each party of chimpanzees. A pile of oil-palm nuts is visible to the right of the top-right corner of the stone matrix; another can be seen above the top-left corner. **B** The tree at the back of the outdoor laboratory from which chimpanzees drank using leaf tools. The two holes are at approximately right angles to each other

clearing utilized the nuts and stones present. After each visit, the supply of oil-palm nuts was replenished and the stones were returned to their original positions. Analyses of these episodes provided data on various aspects of the behaviour. For example, in the following sections we explore topics such as the characteristics of objects that were preferentially selected as hammers or anvils, individual differences in technique, and stages of development that young chimpanzees pass through before they acquire the skill.

2.2.2 Artificial Tree Hollow for Drinking

In addition to nut cracking, the outdoor laboratory also provided an excellent opportunity to observe leaf-folding behaviour. Particularly in the dry season when the nut-cracking experiments were conducted, water is a relatively scarce resource and the provision of water at this site meant that chimpanzees often drank here with the aid of leaves gathered inside or at the edge of the clearing. A natural hollow in a large tree (*Richinodendron heudelotii*) at the back of the clearing was cleaned and filled with clear water, the natural entry hole enlarged, and an extra hole drilled (Fig. 2B). The hollow had a capacity to hold about 12 to 17 l water. The amount of water drunk during each visit by a party of chimpanzees was measured by refilling the water to the brim once all chimpanzees had exited the clearing. At the same time, the locations of all discarded tools (clumps of folded leaves) on the ground were noted for subsequent matching to the user based on video records, then at the end of the day (or, in the earlier years of the study, just after the party's departure) removed from the outdoor laboratory for further analysis.

3 Nut Cracking and Leaf Folding: Techniques and Individual and Age-Related Differences

In this section, we summarize what long-term records have revealed about the characteristics of nut cracking and leaf folding, focusing on both the details of techniques that skilled adult performers apply in the tasks and the developmental stages in young chimpanzees' acquisition of the skill. Figures 3 and 4 illustrate the two tasks and provide outlines of the different behavioural components involved in their performance.

3.1 Skilled Performers

Nut cracking by chimpanzees using a pair of stones as hammer and anvil was first reported by Sugiyama and Koman (1979) and is regarded as one of the most complex forms of tool use found in the wild. According to Matsuzawa's (1996) scheme, nut cracking constitutes "level 2" tool use: three objects must be related to each other in a specific temporal and spatial pattern. Thus, a typical bout of nut cracking would proceed as follows. First, a nut (object 1) must be stably

Fig. 3. Chimpanzees at Bossou performing the tool-using tasks examined in the present chapter. **A** Two adult females, Velu and Jire, crack oil-palm nuts with the aid of a pair of stones. Velu (*left*) has just placed a nut on an anvil stone with her left hand and is about to strike it with a hammer stone held in her right hand. Jire (*right*) has just struck the nut with a hammer held in her left hand, while her infant observes her actions from close range. **B** An adult female, Jire, drinks water from the artificial tree hollow with the aid of leaves. The leaves are visible in Jire's right hand, held between the index and middle fingers, being withdrawn from the hole. A younger female, Vuavua, is visible on the *right*, observing Jire's actions

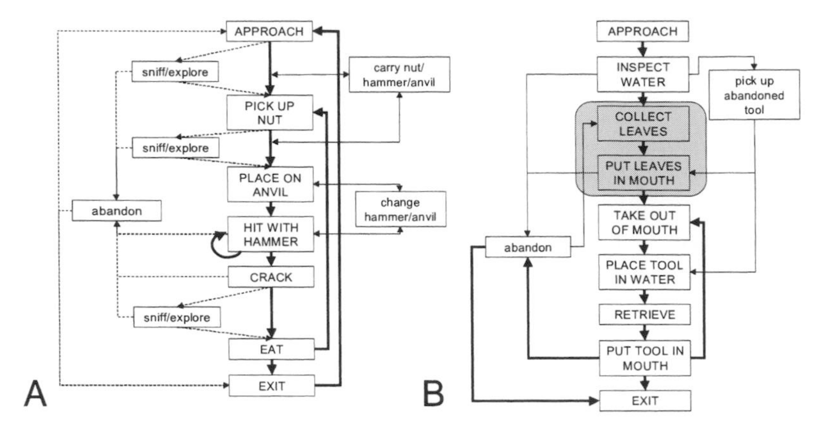

Fig. 4. Flowcharts illustrating main behavioural stages in (**A**) nut cracking and (**B**) leaf folding for drinking water. *Capital letters* and *thick arrows* correspond to a typical bout as performed by adult users of the tools; *lowercase letters* and *thin arrows* correspond to variations less frequently observed in adults and more common in infants and juveniles. *Dashed arrows* in **A** show behaviours observed during the presentation of unfamiliar species of nuts (see Section 4.2), not normally seen in the case of oil-palm nuts. *Grey box* in **B** indicates the tool-making phase, observed from the age of about 3.5 years (younger individuals use tools abandoned by others only; see Section 3.3.1)

placed on an anvil stone (object 2). Next, a hammer stone (object 3) must be used to strike the nut with a force appropriate for cracking the shell, but leaving the contents (the kernel) relatively unharmed. The kernel can then be retrieved and eaten, and the process repeated with the next nut.

Leaf folding, in contrast, is an example of a "level 1" tool use. One object (crumpled leaf or, more often, two or more leaves folded in parallel) is related to another (water). However, unlike nut cracking, leaf folding incorporates a tool-manufacturing as well as a tool-using phase, both of which the individual must perform to succeed. A typical bout of drinking water with the aid of folded leaves may begin by a chimpanzee approaching a drinking source (tree hollow filled with water, in our experiments) and inspecting it. He will then retreat to collect from one to about four leaves of a plant (generally one fairly close by, up to about 2 m away, although individuals sometimes move further away or even arrive with a tool already in the mouth) (Sugiyama 1995; Tonooka 2001). The leaves will then be put into the mouth, and parts of the leaves may be trimmed away with the hand and teeth. The leaves inside the mouth are occasionally chewed, but mainly just laid parallel and folded accordion-like (Fig. 5) with 2 to 3 cm between ridges. The tool thus made is then taken out of the mouth and, held between the index and middle fingers, inserted into the tree hollow, dipped into the water, then retrieved and returned to the mouth, and the water carried between the folds of the leaves is drunk. The tool can be reused several times by reinserting it into the water, and is eventually dropped when the drinking bout is over or if the individual begins to make another tool from fresh leaves.

Fig. 5. Leaf tool (*Hybophrynium braunianum*, the most commonly used species in the outdoor laboratory) for retrieving water from a tree hollow. **A** The tool as it was found immediately after the user had dropped it following a bout of drinking. **B** The same tool unfolded to reveal the characteristic accordion-like shape, achieved by folding several leaves inside the mouth

3.1.1 Tool Selectivity

Are chimpanzees specific in their choice of tools they use for cracking nuts and drinking water? In both cases, it is clear that certain objects will make more efficient tools than others: for example, a stone that is too large will not make a good hammer if the individual has difficulty lifting it; leaves that are too small or too fragile will reduce the volume of water retrieved.

We examined tool choice in nut cracking in the outdoor laboratory, where the weight and dimensions of each available stone were known (52 stones, ranging in weight from 0.2 to 5 kg). During observation at the outdoor laboratory, as well as from video data, we recorded where possible the identity of the stones being used as hammer and anvil, as well as the chimpanzee who used them. From a total of 550 such stone identifications (between 1999 and 2002), we calculated average hammer weights to be 1.0 kg and anvil weights to be 2.5 kg. This tendency was clear at the individual level as well: all individuals observed nut cracking showed a preference for heavier anvils than hammers on average. Of all hammer–anvil sets where both stones could be identified, only in 12% of cases was the hammer heavier than the anvil.

Figure 6 shows patterns in the use of stones of different sizes as hammer and anvil. The data confirm that while small stones tended to be used primarily as hammers and large stones as anvils (indeed, anything over 2.5 kg almost exclusively so), stones of intermediate sizes were used as both anvil and hammer with comparable frequencies. Seven stones were never used; all of these weighed 0.5 kg or less. Weight was not the only factor determining use, however, as several stones around 0.5 kg were in fact quite popular: these tended to be stones with a more compact shape rather than elongated, with one or more wide, flat surfaces.

In a similar vein, chimpanzees showed selectivity in the species of plants whose leaves they use in manufacturing drinking tools. Leaf tools generally consist of several leaves (usually between one and four) of the same species, torn

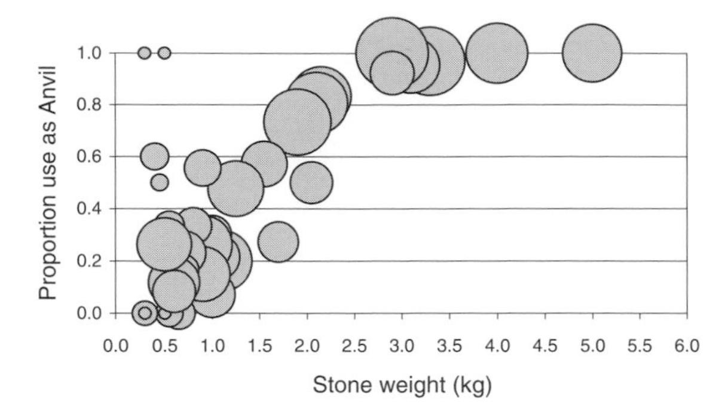

Fig. 6. Use of stones provided at the outdoor laboratory as hammer and anvil as a function of weight. Each *circle* represents a single stone, with the diameter of the circles corresponding to the relative frequencies with which the stones were used (the smallest circles indicate a single use; the largest represents 35 uses). Height on the y-axis corresponds to the ratio of use of the stone as anvil; i.e., 1 = exclusive use as anvil, 0 = exclusive use as hammer

off at the same time by grasping them in one hand and pulling them off a live branch. We found that in the outdoor laboratory the majority (about 75% in 2002 and 50% in 2003) of the tools were made from *Hybophrynium braunianum* leaves, with *Napoleona vogelii* and *Carapa procera* becoming more common over the course of the 2003 field season (M. Hayashi, personal communication). In Tonooka's (2001) study, carried out at a different tree outside the outdoor laboratory, the majority (76%) of leaf tools belonged to *H. braunianum*, with three more species (*Aningueria robusta*, *Blighia sapida*, and *Thaumatococcus daniellii*) making up most of the rest. Sugiyama (1995) also reports that most drinking tools were made from *H. braunianum* and *C. procera*. Thus, in all these studies, *H. braunianum* seemed to be a strongly preferred candidate for toolmaking over other species that were equally accessible around the vicinity of the drinking sites. The leaves of this species are fairly soft and are characterised by a large, smooth, hairless surface. Leaves of the other species listed are smaller, but are also hairless, smooth, and soft. It is likely that such characteristics contribute to tool effectiveness, and empirical evidence regarding, for example, the amount of water that comparably sized tools made from different species can hold is currently being evaluated.

3.1.2 Tool Fidelity

Within a nut-cracking bout, adult chimpanzees rarely exchanged their hammers or anvils for other stones. When individuals entered the outdoor laboratory, they often approached the stone matrix first, selected two stones, and transported them to one of the piles of nuts scattered around the clearing. Once that pile was exhausted, chimpanzees often transported both hammer and anvil across the clearing to the next pile. Such transports have been studied extensively at the Taï Forest, Ivory Coast (Boesch and Boesch 1984), where chimpanzees have been observed to carry stones over distances of several tens or even hundreds of metres when nut trees of a rarer species did not have any suitable stones around them. In general, in Bossou such transports tend to be much shorter, as stones are readily available throughout the forest (Sakura and Matsuzawa 1991).

Similarly, adult chimpanzees began a drinking bout by manufacturing a tool, which they then continued to reuse until the end of the bout (on average, about 25–30 times in roughly 10-min bouts). Additional leaves were sometimes added to the existing tool (26% of bouts), but rarely did they drop the current tool to make a new one to continue drinking (Tonooka 2001). In both cases, the fate of the tools left behind by adults (whether a pair of stones balanced on top of or next to each other, or a clump of folded leaves) is extremely interesting in itself. We return to this point in more detail in Section 3.3.1.

3.1.3 Metatools

In addition to serving as hammers and anvils, stones in the outdoor laboratory very occasionally took on another role as well: that of a meta-tool. Matsuzawa (1994) reports three instances (an adult female in 1991, and an adult and a juvenile male in 1992) when chimpanzees at Bossou placed a third stone beneath their anvil as a wedge, stabilizing the anvil and maintaining a flat, horizontal upper surface. We have observed such cases in subsequent years as well (Fig. 7); however, it is sometimes difficult to ascertain to what extent individuals understand the function of the metatool. We suspect that the construction in Fig. 7, for example, was accidental rather than deliberate, as it came about after the individual turned the anvil stone over several times, probably in an attempt to stabilize the upper surface, until it rolled onto the stone that then served as the wedge. Nevertheless, genuine metatools in the nut-cracking context raise the stakes considerably in the search for the most complex tool use found in the wild: at present the wedge is the only example of a "level 3" tool (Matsuzawa 1996).

An analogous example in the case of the use of leaves for drinking water has been reported (Matsuzawa 1991; Sugiyama 1995), this time performed by a 4-

Fig. 7. Use of a "metatool" in nut cracking. A wedge stone (number *8*) is inserted under the anvil stone (number *67*), which in turn now has a flat upper surface. Note the nutshell leftovers on top of the anvil; photograph was taken immediately after the chimpanzee user exited the scene

year old female. This individual was seen to use a dead twig to push a leaf tool deep into a tree hole, then use the twig to retrieve it, before putting it into her mouth and drinking.

3.1.4 Laterality

One of the benefits of the outdoor laboratory setup is that individuals' performance in the tool-using tasks can be followed within a bout of tool use, across a field season, and over several years. This advantage has allowed us to collect data on longitudinal aspects of the behaviours. Table 1 shows, for each of the individuals at Bossou since the start of the outdoor laboratory experiments in 1988, records of nut-cracking ability and the hand used in hammering. Of these 34 chimpanzees, 11 were never seen to succeed at cracking nuts: 2 were adult females, 2 were infants who disappeared together with their mothers, 4 were infants who died, 1 is a current infant, and 2 were juveniles who both disappeared before they reached adulthood although they had already shown some form of hammering action, albeit without success. The remaining 23 chimpanzees show a striking pattern: they exhibit perfect laterality on the individual level. The hand used to hold the hammer stone is consistent from nut-to-nut, day-to-day, and year-to-year. The only two exceptions are Fana, who in 1996 switched from the left to the right hand for hammering following paralysis of the left arm, and Jeje, who has just begun to crack nuts successfully and currently uses both hands alternately (a phase observed in other new crackers as well; see Section 3.2.3).

Taking all crackers (excluding Fana and Jeje) into consideration, right-handers outnumber left-handers at 62% versus 38%, respectively, but this does not correspond to a significant community-level bias (binomial test; $n = 22$, $P = 0.097$). It has recently been argued (Lonsdorf and Hopkins 2005) that given larger sample sizes, such differences should emerge as significant, demonstrating population-level handedness in nut cracking as well as other tool-using tasks, such as termite fishing and leaf sponging (a variation on leaf folding where leaves are chewed to produce a sponge-like wadge which is then dipped into water). This intriguing possibility remains to be seen at Bossou.

Other interesting patterns in Table 1 concern the distribution of right- and left-handed crackers within matrilines. Arguing against a genetic explanation for handedness, congruence in laterality is found in only 4 of the 16 mother–infant pairs where the infant's handedness was known. In 6 cases, laterality was incongruent, while in the remaining 6 pairs nut-cracking offspring belonged to non-nut-cracking mothers. On the other hand, between siblings we found near-perfect congruence in handedness: only a single infant (Peley) developed a cracking hand different from his siblings. Note that Peley and his siblings were descended from a non-nut-cracking mother, Pama. The implications of this interesting pattern are discussed in Section 3.3.3.

In terms of general patterns in laterality, leaf-folding behaviour presents a contrast. Figure 8 shows data collected in 2000 on individuals' use of the left and

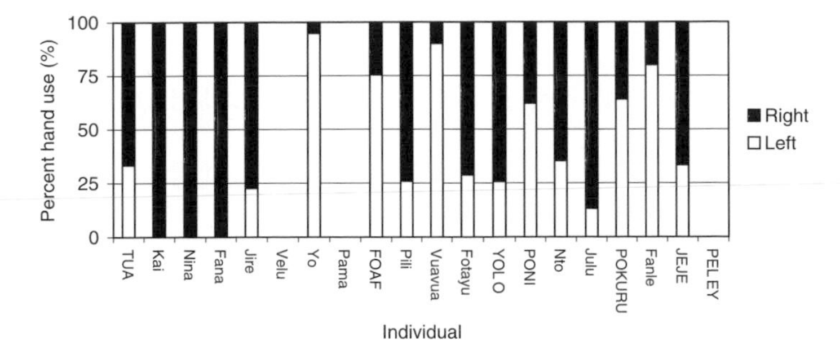

Fig. 8. Laterality during use of folded leaf tools. The hand used to dip the tool into the water, retrieve it, and place it into the mouth was recorded for each dipping action. Males are in *capitals*, females in *lowercase*. Data were collected in 2000; individuals for whom no data were obtained during the period of observation that year (Velu, Pama, and Peley) were seen performing the behaviour in other years

right hand in dipping leaf tools into the tree hollow. Although a few individuals do show consistency in the use of hand, the majority have been observed to use both the left and the right hand to perform the action. Even within the same drinking bout, chimpanzees will change the hand used to dip the tool into the water (adults at a rate of about 0.27 times per bout). It is more difficult to assign clear left- or right-handedness to individuals than in the case of nut cracking, although some degree of preference for a particular hand is shown by most individuals.

There are several factors that may contribute to the lack of perfect laterality in individuals performing the drinking task. Tree holes are often located in hard-to-reach places where the orientation of the hole and nearby branches which chimpanzees grasp for stability may determine which hand is free for performing the task. At the tree in the outdoor laboratory, two individuals could drink from the hollow simultaneously, which meant that the hand used in the dipping action was strongly influenced by the hole next to which the individual happened to be or was forced to assume position.

Table 2 examines concordance in hand preference between nut cracking and leaf use within individuals and within mother–infant pairs. Neither shows a clear pattern, with about half the individuals preferring the same hand in both tasks, while the other half use different hands, and about half of the mother–infant pairs showing concordance in the preferred hand while the other half do not. This observation argues both in favour of "handedness" being task specific rather than task independent and against a genetic explanation for individual patterns.

3.2 Ontogeny

The longitudinal data on nut cracking presented in Table 1 are clearly also informative regarding the processes of acquisition. None of the infants born at

Table 1. Longitudinal record of stone-tool use by chimpanzees at Bossou, Guinea

Name	Sex	Age	Mother	Year observed															
				88	90	91	92	93	94	95	96	97	98	99	00	02	03	04	05
Tua	m	Adult	N.A.	?	L	L	L	L	L	L	L	L	L	L	L	L	L	L	L
Kai†	f	(Adult)	N.A.	?	R	R	R	R	R	R	R	R	R	R	R	R	R	—	—
Kie*	f	29	Kai	?	R	R	—	—	—	—	—	—	—	—	—	—	—	—	—
Kakuru*	f	18	Kie	?	A	R	—	—	—	—	—	—	—	—	—	—	—	—	—
Nina*	f	(Adult)	N.A.	?	X	X	X	X	X	X	X	X	X	X	X	X	X	—	—
Na*	m	19	Nina	?	R	R	R	R	R	R	—	—	—	—	—	—	—	—	—
Nto*	f	11	Nina	—	—	—	—	—	X	X	X	R	R	R	R	—	—	—	—
Fana	f	Adult	N.A.	?	L	L	L	L	L	L	R	R	R	R	R	R	R	R	R
Foaf	m	24	Fana	?	R	R	R	R	R	R	R	R	R	R	R	R	R	R	R
Fotayu*	f	13	Fana	—	—	—	X	X	X	AR	R	R	R	R	R	R	R	R	—
Fokaiye*	m	3.5	Fotayu	—	—	—	—	—	—	—	—	—	—	—	—	X	X	X	—
Fanle	f	7	Fana	—	—	—	—	—	—	—	—	—	X	X	X	R	R	R	R
Jire	f	Adult	N.A.	?	L	L	L	L	L	L	L	L	L	L	L	L	L	L	L
Ja*	f	21	Jire	?	R	R	R	R	—	—	—	—	—	—	—	—	—	—	—
Jokro†	f	(3)	Jire	—	X	X	X	—	—	—	—	—	—	—	—	—	—	—	—
Juru*	f	11	Jire	—	—	—	—	—	X	X	X	X	r	r	r	—	—	—	—
Jeje	m	7	Jire	—	—	—	—	—	—	—	—	—	X	X	X	X	X	X	A
Jimato†	m	(1.5)	Jire	—	—	—	—	—	—	—	—	—	—	—	—	X	X	—	—
Joya	f	0.5	Jire	—	—	—	—	—	—	—	—	—	—	—	—	—	—	—	X

Velu	f	Adult	N.A.	?	R	R	R	R	R	R	R	R	R	R	R	R	R	R	R
Vube*	f	22	Velu	?	L	—	—	—	—	—	—	—	—	—	—	—	—	—	—
Vui*	m	18	Velu	?	X	X	L	L	L	L	L	L	L	L	—	—	—	—	—
Vuavua*	f	13	Velu	—	—	—	X	X	X	AL	L	L	L	L	L	L	L	L	—
Veve†	f	(2.5)	Vuavua	—	—	—	—	—	—	—	—	—	—	—	—	X	X	—	—
Yo	f	Adult	N.A.	?	L	L	L	L	L	L	L	L	L	L	L	L	L	L	L
Yunro*	f	20	Yo	?	X	X	X	l	—	—	—	—	—	—	—	—	—	—	—
Yela†	m	(0.5)	Yo	—	X	—	—	—	—	—	—	—	—	—	—	—	—	—	—
Yolo	m	13	Yo	—	—	—	X	X	X	X	L	L	L	L	L	L	L	L	L
Pama	f	Adult	N.A.	?	X	X	X	X	X	X	X	X	X	X	X	X	X	X	X
Puru*	m	24	Pama	R	R	R	—	—	—	—	—	—	—	—	—	—	—	—	—
Pili*	f	17	Pama	?	X	R	R	R	R	R	R	R	R	R	R	—	—	—	—
Pokru*	m	8	Pili	—	—	—	—	—	—	—	—	X	X	X	X	—	—	—	—
Poni†	m	(9)	Pama	—	—	—	—	X	X	X	R	R	R	R	R	R	R	—	—
Peley	m	6	Pama	—	—	—	—	—	—	—	—	—	—	X	X	AL	L	L	L

Individuals are sorted according to matrilinial kinship

Oil-palm-nut-cracking ability and hand used to hold hammer stone recorded for all individuals since 1988

Age represents estimate in years as of January 2005; numbers in parentheses represent age in years at which individual died

*Individual disappeared before January 2005

†Individual confirmed dead

L, always uses left hand for hammer; R, always uses right hand for hammer; A, ambidextrous use of hammer; l, uses left hand to pound nut without a hammer; r, uses right hand to pound nut without a hammer; X, no successful hammer use but eating nuts cracked by others; ?, data unavailable because of lack of observation; —, data unavailable as subject had not yet been born, had disappeared, or died; N.A., not available

Table 2. Comparison of hand preference in nut cracking and the use of leaves for drinking water

			Hand preference				
						Mother–infant concordance	
Name	Sex	Mother	(a) Nut cracking	(b) Leaf folding	(a)–(b) concordance	(a)	(b)
Tua	m		L	R	✗		
Kai	f		R	R	✓		
Nina	f		—	R			
Fana	f		R	R	✓		
Jire	f		L	R	✗		
Velu	f		R	—			
Yo	f		L	L	✓		
Pama	f		—	—			
Foaf	m	Fana	R	L	✗	✓	✗
Pili	f	Pama	R	R	✓	(✗)	
Vuavua	f	Velu	L	L	✓	✗	
Futayu	f	Fana	R	R	✓	✓	✓
Yolo	m	Yo	L	R	✗	✓	✗
Poni	m	Pama	R	L	✗		
Nto	f	Nina	R	R	✓	(✗)	✓
Juru	f	Jire	—	R			✓
Pokuru	m	Pili	—	L			✗
Fanle	f	Fana	R	L	✗	✓	✗
Jeje	m	Jire	A	R	?	?	✓
Peley	m	Pama	L	—		(✗)	

Column for hand preference in leaf folding is based on hand used more often during the task
Data are for individuals present at Bossou in 2000 (arranged in order of decreasing age)
Hammering hand for the youngest individuals not yet able to nut crack in 2000 was determined in subsequent years
(✗), concordance was not possible to evaluate as the mother never engaged in nut cracking
?, Jeje is at present ambidextrous in nut cracking, hence concordance cannot yet be determined

Bossou during the period covered by the outdoor laboratory studies was able to successfully crack nuts before the age of 3 to 3.5 years. In addition, individuals who did not begin to crack by the age of about 7 years were never seen to acquire the skill later in life, which has led to the proposition that there exists a critical period for learning the skill (Matsuzawa 1994). Similarly, no infant younger than about 2 years has been seen drinking water with the aid of folded leaves, and none younger than 3.5 years has been observed manufacturing and using their own drinking tool. Nevertheless, all infants who remained at Bossou beyond that age were eventually seen to perform the behaviour. What stages do infants go through on their way to acquiring these skills?

3.2.1 Overview of the Stages of Learning

Inoue-Nakamura and Matsuzawa (1997) examined in detail the learning processes in young chimpanzees' acquisition of nut cracking. Three infants' progress was followed from 0.5 to 3.5 years of age over four consecutive field seasons. Fine-scale analysis of the infants' interactions with nuts and stones revealed that in the early stages of development such interactions were restricted to the manipulation of single objects on their own, such as holding a stone or rolling a nut. This stage was followed by multiple actions on multiple objects in combinations of increasing complexity and appropriateness in terms of the demands of the task. By the age of 1.5 years, infants had performed all the actions that constitute components of the nut-cracking sequence, albeit they had never combined them in the appropriate order. For example, they would pick up a nut and place it on an anvil, but then hit it with the hand, or pick up a hammer but use it to hit another stone or a nut not yet positioned on an anvil. It took 2 more years before the separate actions condensed into the correct temporal and spatial order and the infants were able to crack their first nut by themselves.

Drinking water with the aid of folded leaves differs from nut cracking in that it involves two distinct phases: a tool-making and a tool-using phase. Infant chimpanzees begin with the tool-using phase. At around the age of 2 years, they are first observed using leaf tools for drinking, but these tools are not yet the infants' own: they rely instead only on the discarded drinking tools of older individuals.[1] These they pick up off the ground, or occasionally take from their mother's hand, suck on, then place in the water and retrieve for drinking. Infants use such discarded leaf tools exclusively until the age of about 3.5 years, after which they begin to manufacture and use their own tools. In fact, tool manufacture begins slightly earlier than 3.5 years, but in those cases infants drop their own tools immediately after making them, picking up and using those discarded by adults instead. Infants' tools are generally smaller than those of adults (see next section), consisting of fewer leaves, or are made of species with leaves smaller than the *H. braunianum* favoured by adults. Tonooka (2001) also reports that younger chimpanzees are more likely to chew the leaves they place in their mouth, thus producing "sponges" rather than the much more common accordion-like folds.

3.2.2 Tool Efficiency

The skills of adult tool users appear almost stereotyped: in the case of nut cracking, for example, an individual chimpanzee's sitting posture, nut manipulation, hammer use, etc., all appear very similar from one cracking bout to the next.

[1]An earlier report (Tonooka 2001) estimated the age at which this tool use first appears in infant chimpanzees as 2.5 years. Our slightly earlier estimate may be due to methodological differences, such as the observation of a larger number of infants over several years, and our procedure to leave abandoned leaf tools in place, rather than remove them immediately after the chimpanzee party's departure, thus facilitating the behaviour in individuals not yet able to make their own tools.

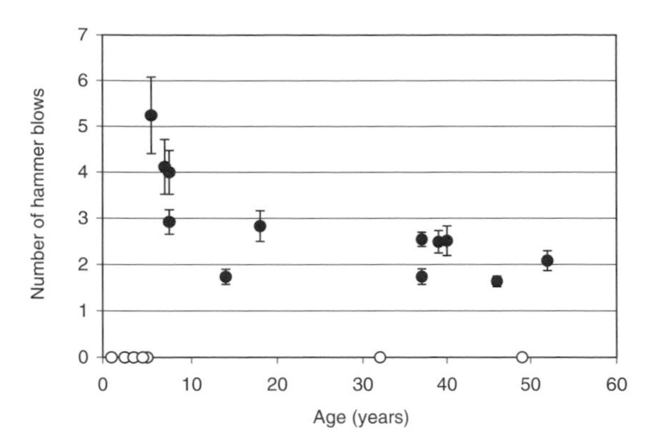

Fig. 9. Average number of hammer blows required to crack a single oil-palm nut as a function of age. Data were collected in 1999. Each *filled circle* represents the performance of a single individual who was able to crack nuts; *open circles* show age distribution of individuals who were never seen to succeed at cracking in 1999. Ages of oldest individuals are estimates. *Error bars* are standard errors of the mean

However, there are various ways in which the performance of younger chimpanzees changes as they hone their skills, such as settling on a particular hand to use for hammering after an initial ambidextrous phase. To assess whether more experience with a task leads to more efficient use of tools, we examined age-related differences in chimpanzees' performance of these tasks.

Figure 9 shows the number of hammer blows that individual chimpanzees used to open single oil palm nuts. The data show that, with increasing age, chimpanzees need progressively fewer blows to crack open nuts, reaching an asymptote around an average of only two blows per nut in the most experienced individuals. The increased effort that younger individuals seem to put into obtaining the same reward may be the result of lack of muscular development (weak blows), the choice of suboptimal tools for the task (such as hammers or anvils that are too small), or inferior technique (incorrectly aimed blows).

We also examined the efficiency of leaf tools as a function of age. In 2002, we collected tools discarded in the outdoor laboratory just after chimpanzees moved away following drinking bouts. These tools were immediately weighed and tested for their capacity to carry water by being dipped into a bucket of water and then squeezed over a measuring jug ten times consecutively. The amount of water thus carried by each tool was noted, and wherever possible assigned to the individual who manufactured and used the tool (as determined from onsite observation and video records). These data are plotted in Fig. 10. There was a significant correlation between tool weight and water-carrying capacity (Pearson's correlation; $n = 31$, $r = 0.83$, $P < 0.001$), which is unsurprising as larger tools would be expected to hold more water; however, the graph also shows that tools made by juveniles had much lower capacity than those of adults [one-way analysis of variance (ANOVA) on log-transformed data, $F_{1,30} = 23.58$, $P < 0.001$] because they were generally smaller ($F_{1,30} = 38.56$, $P < 0.001$).

3.2.3 Adjustments During Execution

Related to younger chimpanzees' generally lower efficiency in tool use is a lower fidelity to tools and techniques during the execution of the tasks. Juveniles in

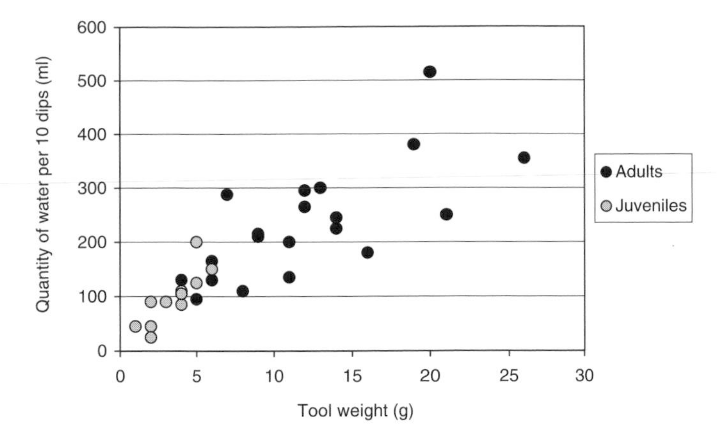

Fig. 10. Weight and efficiency of leaf tools made by individuals in different age classes. Each *circle* represents a single leaf tool; *solid circles* are tools made by adults and *shaded circles* are those made by juveniles. Data were collected in 2002

particular often change their hammers and anvils during bouts, sometimes selecting fresh stones, at other times taking those just abandoned by another individual. In addition, in their first year of successful cracking, young chimpanzees are often ambidextrous in hammer use and will switch hands in the course of a bout, something that is never observed in adults. During the use of leaf tools for drinking water, such hand changes occur even in adults, albeit at a much lower rate than in juveniles and infants (adults, 0.27 times per bout; juveniles, 1.3 times; infants, 0.83 times).

The switching of tools as well as hands during the execution of a tool-using task may represent a trial-and-error process in young individuals' learning. It may be that such incidents constitute a form of practice whereby chimpanzees develop their own favoured techniques.

3.3 Social Influences

Although both nut cracking and the use of leaves for drinking water are essentially solitary activities, members of chimpanzee parties travelling together often engage in these behaviours simultaneously. Furthermore, infants always travel with the mother and thus are exposed to her and other community members' tool-using activities long before they themselves begin to make any attempts. Similarly, juveniles who travel with the mother or other members of the community have access to the performance of skilled tool users. Thus, unless these tool-using skills rely entirely on genetically "preprogrammed" behaviour or individual trial-and-error learning (both of which are unlikely considering patterns of regional distribution; see Section 4), young chimpanzees are provided with a rich social environment that may well trigger and guide the acquisition process.

3.3.1 Scrounging and the Use of Abandoned Tools

Infants are allowed to scrounge freshly cracked nuts from their mothers (and also, to a lesser degree, from other individuals) and are allowed to interact with their stones during a nut-cracking bout. For example, infants reach out to touch their mother's anvil, even hold her hammering hand or arm as she is delivering blows, and take nuts and shells off her anvil. Such scrounged nuts constitute the infants' only tangible reinforcement in the nut-cracking context until the age of about 3.5 years when they begin to crack nuts by themselves. We have also observed that young infants, when held by their mothers engaged in leaf-tool use, sometimes reach into a tree hollow and dip their hand into the water within, which they can then lick off their fingers. Although not strictly speaking scrounging, this does parallel infants' access to oil palm kernels in an important way: through the mother's nut cracking or through the mother holding the infant up high enough to reach the water, infants may be "getting a taste" for what is to be gained from successful tool use well before they begin to attempt it themselves.

As mentioned earlier (Section 3.2.1), infants' tool use for drinking water begins at the age of around 2 years with the picking up of leaf tools abandoned by previous users. Moreover, infants have also been observed to take leaf tools from their mother's hand during a drinking bout and to continue to use the tool by themselves. A parallel in nut cracking is seen in infant and juvenile chimpanzees' propensity to use anvil–hammer sets freshly abandoned by adult users. Even if they are already engaged in cracking themselves, if an adult nearby walks off leaving behind his or her set, juveniles in particular will walk over and either crack in the adult's place or take one or both of the abandoned stones away with them.

Laboratory work confirms the importance of leftover tools in individuals' acquisition of tool-using skills. In a captive simulation of ant/termite fishing by wild chimpanzees, Hirata and Morimura (2000; see also Chapter 12 by Hirata in this volume) showed that adult chimpanzees naïve to the task (honey-fishing) were more successful in their attempts if they used those objects as tools that had just been abandoned by a previous user than if they made their own tool selection from the many different objects provided. When the experiment was repeated using mother–infant pairs of chimpanzees (Hirata and Celli 2003), the infants' tendency to scrounge from their mother performing the honey-fishing tasks was prominent: they would lick the honey off the mother's tool, or even attempt to steal the latter. Similarly for wild chimpanzees, such opportunities to scrounge, as well as the presence of leftover tools, may have much to contribute to individuals' acquisition of the task.

It is interesting to note that juveniles at Bossou, who also occasionally reuse discarded leaf-tools, are not granted the same liberties to scrounge as infants. Although the mother may show some degree of tolerance towards her juvenile offspring's attempts to interact with her objects during nut cracking, scrounging from others becomes impossible. Juveniles are often chased away when they approach an older individual engaged in nut cracking and are certainly not permitted to take nuts from them. This propensity may contribute to the end of the "critical period" for learning; once a juvenile, opportunities for direct interaction with older, skilled group members during the tool-using task disappear.

3.3.2 Conspecific Observation

While scrounging involves direct interaction with older tool users in the community, the social setting in which tool-using behaviour often takes place provides younger individuals not only with leftover tools but also with an opportunity to observe closely the actions of other group members. Such observation may be an important building block of socially transmitted behaviours.

We examined patterns in the observation of conspecifics carried out by individuals in the Bossou community (Biro et al. 2003; Sousa et al., in preparation). An episode of observation of one community member by another was said to take place when the latter approached the former to within about 1 metre and remained with gaze fixed upon the target individual's face or hands for at least 3s. Figure 11 shows the rates of occurrence of such observational episodes during periods in the outdoor laboratory when at least two individuals were present and at least one of them was engaged in nut cracking (including the handling of nuts and stones) or the use of leaves for drinking water. The data are strikingly similar for the two tool-using tasks, and three main conclusions can be drawn. (1) Adults are the most popular targets for observation by individuals in all three age classes. Juveniles are observed less often, while infants are almost never observed (the bar for infants in Fig. 11B corresponds to a single episode of observation of an infant by a juvenile). (2) Juveniles and infants are the most likely to observe, while adults are the least likely to act as observers.

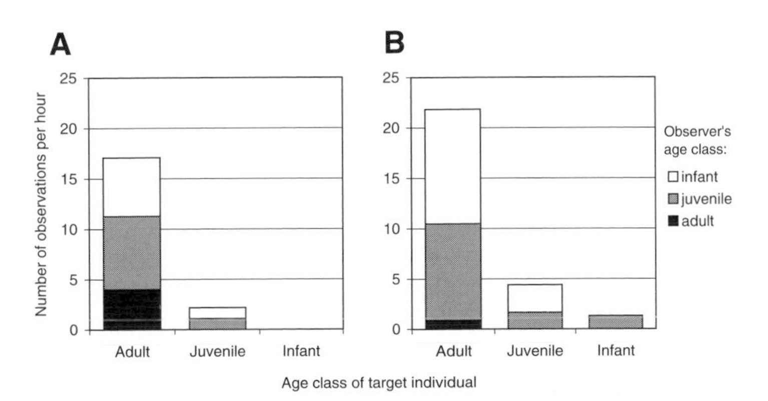

Fig. 11. Individuals in different age classes as targets of observation by conspecifics during (**A**) nut cracking and (**B**) using leaves for drinking water. *Bars of different colours* correspond to the observer's age class. Rates were calculated as number of episodes of observation divided by the amount of time individuals in the three different age classes spent in the outdoor laboratory engaged in the handling or cracking of nuts or in drinking water with the aid of leaf tools. Data were collected in 2000

(3) Individuals almost exclusively observe conspecifics in the same age group or older, but not younger, than themselves.

This pattern of conspecific observation has implications for any model of the social transmission of behaviour in chimpanzee communities. We return to this point in Section 4.2.

3.3.3 "Education by Master-Apprenticeship"

Previous work at Bossou has illuminated many aspects of the developmental changes that young chimpanzees go through as they learn to use tools (Inoue-Nakamura and Matsuzawa 1997; Tonooka 2001). Furthermore, elucidating the underlying mechanisms of learning and the role that skilled community members play in the acquisition process is crucial for understanding how community-specific behaviours are maintained among wild chimpanzees. Drawing together evidence from the individual as well as the social aspects of behavioural development, Matsuzawa et al. (2001) proposed a model to describe how tool-using skills may be propagated in wild chimpanzee communities, referred to as "education by master-apprenticeship." The central theme of this model is that infant chimpanzees, who remain in extremely close proximity to the mother for the first 4 to 5 years of life, are provided with an excellent setting in which learning aided by observation can take place, while the bulk of the work is then done at the individual level. The models whom the infants observe (primarily the mother) are highly tolerant: they will allow infants to scrounge, to observe from close range, and to interact with the objects involved in their actions, but they play no active role in the infant's learning. Infants in turn are driven by an intrinsic motivation to do as the others in the community [de Waal (2001) refers to this as "the desire to be like others"].

The underlying mechanism responsible for the social transmission of the skill from model to observer is a particularly intriguing, much-debated issue. Although imitative learning, where a model's behaviour is copied motor-pattern-by-motor-pattern by an observer, would in theory facilitate high-fidelity copies of behaviour, it does not seem to satisfactorily account for the years spent by infants gradually approximating the correct motor sequence necessary for the task. (Bear in mind that young chimpanzees receive no direct food reward from their interactions with nuts and stones for several years, yet they continue to handle these objects.) Facilitated by their tolerance, models draw attention to the tools and targets, as well as to the possible outcomes of a successful bout of tool-use. In that sense, infants' progress may be aided by a form of stimulus or local enhancement, or emulation learning (Tomasello 1996), where the details of the motor pattern have to be established through individual trial-and-error learning. This idea is supported also by our long-term record of laterality in tool use, which shows no consistent patterns in mother–infant congruence in handedness in either nut cracking or the use of leaves for drinking water. Near-perfect inter-sibling congruence in laterality, however, is a very striking feature of the data. We have tentatively suggested a way in which this pattern is consistent with the education by master-apprenticeship model (Biro et al. 2003): it may be that individual mothers provide specific learning environments for their offspring which will favour a particular hand for hammering across all siblings. For example, a mother who always places her offspring on her right during nut cracking may encourage the infant's right hand to be used for exploring objects within reach from an early age. The only individual who shows divergence from the pattern of inter-sibling consistency (Peley) was born to a non-nut-cracking mother (Pama), such that he—as well as his siblings—must have relied on other individuals as models for observation.

Various alternatives accounting for infants' learning can be discounted. In contrast with Boesch's (1991) observations of rare examples of active teaching by chimpanzee mothers at Taï, we have never encountered such cases at Bossou. Mothers do not mould the hands of their young, nor do they perform what appear to be deliberate slow-motion demonstrations of the appropriate techniques in front of them. At the other end of the scale, if models contributed nothing to learning and the behaviour arose independently in each individual in the community, then it would be difficult to account for regional variation in the presence or absence of specific tool-using skills in different communities. We turn our attention now to such inter-community differences and the processes that may contribute to their emergence, propagation, and maintenance in wild chimpanzee populations.

4 Regional Variation and Culture

In common with many other behavioural patterns (Whiten et al. 1999, 2001), nut cracking and the use of folded leaves for drinking water are not found in all wild chimpanzee communities across Africa. The reasons behind the patchy distribution of such behaviours and related issues of "cultural" variation are currently much debated. So far in the present chapter we have discussed possible mechanisms contributing to the maintenance of these behaviours within a community, but how can the absence of the same behaviours from other communities be explained? One simple hypothesis is that ecological factors, such as the absence of the target species from the chimpanzees' habitat, account for the absence of a particular behaviour; this is certainly true in some cases (Baldwin et al. 1991) but certainly not true in others. As mentioned at the start, for example, the cracking of hard-shelled nuts is restricted to West African communities (Boesch et al. 1994), yet the raw materials needed to perform the task are present in Central and East African communities' habitats. The use of leaves for drinking water is geographically more widespread (McGrew 1977; Nishida 1990; Quiatt and Kiwede 1994; Wrangham 1992; Boesch and Boesch 1990; Ghiglieri 1984), so much so that Whiten et al. (2001) putatively refer to it as a "chimpanzee universal." However, this universal is in fact the "leaf-sponge," a relatively rare variant at Bossou, whereas the folded leaves typical among Bossou chimpanzees have not been reported from any other site. Such community specificity in the characteristics of a tool is mirrored also in nut cracking, where although the goal of the behaviour is shared between Bossou and Taï, certain fine details are not: for example, only Taï chimpanzees are known to use wooden hammers and wooden anvils as well as stone, while those at Bossou use only stone.

In an attempt to examine questions related to such regional variation, we looked at the distribution of tool use (nut cracking, in particular) in detail at sites adjacent to Bossou and carried out a series of field experiments to explore possible mechanisms underlying cultural innovation and transmission.

4.1 Tool Use at Sites Adjacent to Bossou

Preliminary work has begun at three sites located at various distances from Bossou known to be inhabited by chimpanzees. The closest, Seringbara at the foot of the Nimba Mountains, is located about 6 km to the east of Bossou, and efforts are currently underway to habituate this community. From trace evidence (discarded tools and abandoned tool-using sites), it has already been possible to document the use of wands for ant dipping in this community, but not stone tools for nut cracking even though oil palms are available within the habitat (Humle and Matsuzawa 2001).

The second site, Yealé in the Ivory Coast, is 12 km southeast of Bossou (Matsuzawa and Yamakoshi 1996). At this site also, chimpanzees are known to ant dip, and furthermore traces of nut cracking have also been found. Yealé is home to two more species of nut besides the oil palm, neither of which occurs naturally at Bossou: the coula nut (*Coula edulis*) and the panda nut (*Panda oleosa*). Only coula and oil palm are cracked at Yealé; panda is not.

The third site, Diecke, 50 km to the west of Bossou, has been surveyed by Matsuzawa et al. (1999). Here, neither ant dipping nor oil-palm nut cracking has so far been found; however, chimpanzees do crack both coula and panda nuts.

Table 3 summarizes nut-cracking activity at these sites. Examining patterns in chimpanzees' utilization of different species of nut for cracking, it is clear that while in some cases the behaviour is ecologically impossible (coula and panda

Table 3. Species of nuts cracked by wild chimpanzees at Bossou, Seringbara, Yealé, and Diecke

		Species of nut		
Site	Distance from Bossou	Oil palm (*Elaeis guineensis*)	Coula (*Coula edulis*)	Panda (*Panda oleosa*)
Bossou		Yes	—	—
Seringbara	6 km	No	—	—
Yealé	12 km	Yes	Yes	No
Diecke	50 km	No	Yes	Yes

—, target nut species is not available at the site; No, no evidence of cracking by the chimpanzees has so far been found even though the nuts are available

are not available at Bossou and at Seringbara, and no oil palm at Diecke), in three cases such explanations are not sufficient. Oil-palm nuts at Seringbara and Diecke, and panda nuts at Yealé, are available but not cracked. Although Seringbara chimpanzees have not been found to nut crack at all, those at Diecke and Yealé are known to utilize some but not all of the nuts available in their habitat.

4.2 Nut-Cracking Field Experiment

At the heart of all models of culture lies an invention. Behaviours that show regional variation and are propagated through social learning must have originated with an inventor. Just as Imo, the Japanese macaque, has become famous as the first individual on Koshima island to perform sweet-potato washing (Kawai 1965; see also recent review by Hirata et al. 2001a), there must have lived many uncelebrated primates whose innovations have spread within their communities and often beyond. Such instances of invention are, however, extremely difficult to document in nature.

To investigate how novel tool-using behaviours (such as the cracking of previously neglected nut species) may emerge in wild chimpanzee communities, we carried out a field experiment involving the introduction of novel species of nuts in the Bossou community. We used the nuts available at neighbouring sites, but not at Bossou: coula and panda nuts (Fig. 12). Coula nuts were presented in five separate field seasons (January 1993, 1996, 2000, 2002, and 2005), whereas panda nuts were presented once (in January 2000). Initial presentation involved placing three of the unfamiliar nuts in the outdoor laboratory, along with the usual piles of oil-palm nuts, and replenishing them when they had been used up. In later years, coula were provided in two piles of about 30 nuts (see Matsuzawa 1994, Matsuzawa and Yamakoshi 1996, and Biro et al. 2003 for further detail). In each year, we continued presentation until all individuals present in the group had visited the outdoor laboratory at least four times with the novel nuts present (except in 2005, when one individual did not visit throughout the entire period of coula nut presentation).

Fig. 12. Three species of hard-shelled nut presented in the outdoor laboratory: oil-palm nuts (*left*), coula nuts (*centre*), and panda nuts (*right*). For each species, three stages are shown: fruit (**A**), hard-shelled contents of fruit before cracking (**B**), and after cracking, with edible kernels visible (**C**)

We examined Bossou chimpanzees' responses to these novel items. We classified behaviours displayed by individuals into three general categories: "ignore," "explore," and "crack." First, "ignore" described individuals who displayed no visible signs of interest toward the nuts: they neither approached them nor looked at or handled them in any way. Second, individuals were said to "explore" if they looked at closely, handled, sniffed, mouthed, or bit into nuts but did not attempt to crack them. Finally, "crack" included all attempts when a nut was placed on an anvil stone and when a hammer was used to deliver blows in the manner used for oil-palm nuts, whether or not the cracking efforts were eventually successful. In addition, individuals present at Bossou at the time of the experiments were also classified into three different age groups based on known aspects of life history (weaning and age of first parturition): infants from 0 to 4 years, juveniles from 5 to 8 years, and adults 9 years and above.

4.2.1 Initial Responses: Coula in 1993, Panda in 2000

Table 4 shows the proportion of individuals in the three different age groups who displayed each of the three responses during the very first presentation of coula nuts (in 1993) and panda nuts (in 2000). In both cases, a small proportion of the group attempted to crack the nuts: 3 of 17 individuals in the case of coula (one adult and two juveniles), and 4 of 20 for panda (two adults and two juveniles). This result was in sharp contrast with oil-palm nut cracking, which 10 of 17 individuals performed in 1993 and 13 of 20 in 2000.

Thus, the cracking of oil-palm nuts did not immediately generalize to novel species of nuts. Of the individuals who did attempt to crack, only in one case was cracking not preceded by some form of exploration: a single adult, Yo,

Table 4. Responses of chimpanzees in three different age classes to the initial presentation of novel species of nuts

Age group	Nut (year)	*n*	Crack	Explore	Ignore
Adult	Oil palm (2000)	10	8 (80%)	2 (20%)	0 (0%)
	Coula (1993)	9	1 (11%)	4 (44%)	4 (44%)
	Panda (2000)	10	2 (20%)	0 (0%)	8 (80%)
Juvenile	Oil palm (2000)	6	5 (83%)	1 (17%)	0 (0%)
	Coula (1993)	4	2 (50%)	2 (50%)	0 (0%)
	Panda (2000)	6	2 (33%)	2 (33%)	2 (33%)
Infant	Oil palm (2000)	4	0 (0%)	4 (100%)	0 (0%)
	Coula (1993)	4	0 (0%)	0 (0%)	4 (100%)
	Panda (2000)	4	0 (0%)	1 (25%)	3 (75%)

Data for oil-palm nuts are also shown for reference (data from 2000)
n, number of individuals within a particular age group in given year
Values show number of individuals in each age group who displayed the three different behaviours towards the nuts (see text for details)
Values in brackets are percentages of the total number of individuals in the respective age group

proceeded to crack coula nuts without any exploratory behaviours. All other individuals who cracked either type of nut (including Yo subsequently attempting panda nuts) did so after extensive sniffing and handling of the novel items. Yo also correctly selected ripe (dark) coula nuts over unripe (green) ones for cracking, even though neither showed obvious signs of either containing something edible inside or requiring the use of hammer and anvil. These observations have prompted the intriguing suggestion (Matsuzawa 1994; Matsuzawa and Yamakoshi 1996) that Yo may have been an immigrant, having transferred to Bossou some time before 1976 from a community where coula nut cracking was habitual. Her apparent lack of knowledge regarding panda nuts would suggest that of the nearby communities surveyed, she most likely hails from the region of Yealé, where both oil-palm and coula nuts are cracked but panda is not.

With both species of unfamiliar nuts, it appears that juveniles were more likely than adults to show interest in (explore or crack) the novel objects. Adults were relatively more conservative, being more likely to ignore (particularly in the case of panda) the unfamiliar nuts. The two adults who did crack panda nuts abandoned their attempts after a single successful bout, while juveniles continued to explore and try to crack over subsequent days of presentation.

4.2.2 Coula Cracking Through the Years 1993–2005

The repeated presentation of coula nuts revealed various trends across the years (Table 5). The proportion of individuals who cracked these nuts increased

Table 5. Responses of individuals in the three age groups to coula nuts as across the 5 separate years when these nuts were presented

Age group	Nut (year)	*n*	Crack	Explore	Ignore
Adult	Coula (1993)	9	1 (11%)	4 (44%)	4 (44%)
	Coula (1996)	9	3 (33%)	0 (0%)	6 (67%)
	Coula (2000)	10	4 (40%)	3 (30%)	3 (30%)
	Coula (2002)	9	6 (67%)	1 (11%)	2 (22%)
	Coula (2005)	7[a]	5 (72%)	1 (14%)	1 (14%)
Juvenile	Coula (1993)	4	2 (50%)	2 (50%)	0 (0%)
	Coula (1996)	5	3 (60%)	0 (0%)	2 (40%)
	Coula (2000)	6	4 (66%)	2 (33%)	0 (0%)
	Coula (2002)	4	3 (75%)	0 (0%)	1 (25%)
	Coula (2005)	3	2 (67%)	1 (33%)	0 (0%)
Infant	Coula (1993)	4	0 (0%)	0 (0%)	4 (100%)
	Coula (1996)	4	0 (0%)	0 (0%)	4 (100%)
	Coula (2000)	4	0 (0%)	4 (100%)	0 (0%)
	Coula (2002)	5	1 (20%)	2 (40%)	2 (40%)
	Coula (2005)	1	0 (0%)	1 (100%)	0 (0%)

[a]The 8th adult (Velu) present at Bossou in 2005 could not be tested as she did not visit the outdoor laboratory during the period of coula nut presentation; this individual had cracked coula nuts in previous years of presentation

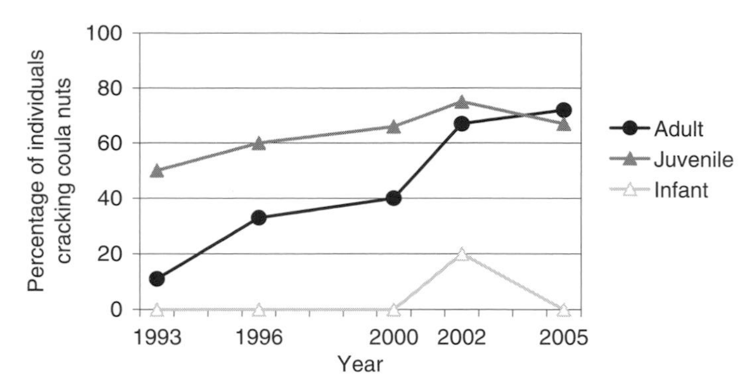

Fig. 13. Percentage of individuals in the three age classes who attempted to crack coula nuts on the five occasions when these nuts were presented at the outdoor laboratory

gradually among both adults and juveniles (Fig. 13), such that they are now roughly comparable to the proportion cracking oil-palm nuts. The rise among adults was partly the result of juvenile crackers from previous years reaching adulthood and partly adults who in previous years showed only exploratory behaviours eventually moving on to cracking.

Exploratory behaviours towards coula nuts also waned over the years. By 2002, seven of the ten individuals who cracked did so without any exploration on their first encounter of the year, as did five of the seven crackers in 2005. In addition, by 2005 only one individual (the adult female Fana) ignored coula nuts completely; this result again is comparable to oil-palm nuts (see Table 4), which all individuals of the group (except the youngest of infants) handle, scrounge on, or crack.

4.2.3 Conspecific Observation with Coula and Panda Nuts

As with oil-palm nuts, we recorded episodes of conspecific observation when coula and panda nuts were present. The overall patterns were similar to the data obtained from oil-palm nuts: adults were the most likely and infants the least likely to be observed, juveniles and infants performed most of the observing, and the targets of observation tended to be in the same age group or older than the observers themselves (Fig. 14). In addition, rates of observation were about twice as high when coula nuts were present as during oil-palm-only periods. Adults engaged in the cracking of coula nuts generated a great deal of interest in the rest of the group. Yo's cracking of coula nuts, in particular, often attracted several individuals (Fig. 15). In contrast, adults cracking panda nuts were observed at lower rates, partly because only relatively few interactions with the nuts occurred, and some of these took place with few other individuals present (for example, Yo was on her own in the outdoor laboratory when she first cracked panda nuts).

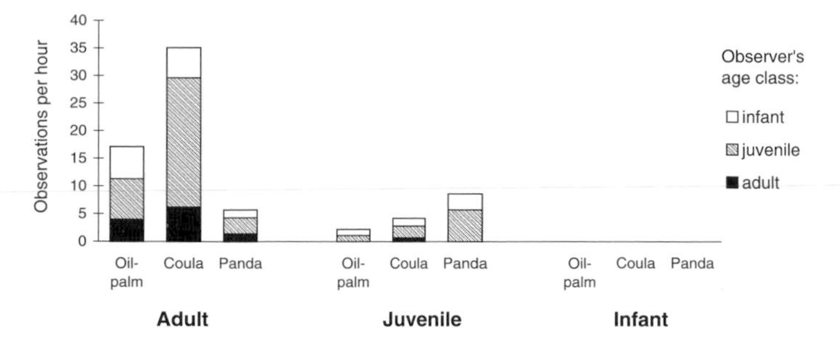

Fig. 14. Individuals in different age classes as targets of observation by conspecifics during the cracking of three different species of nuts. See legend to Fig. 11 for further detail. Data for all three nuts were collected in 2000

Fig. 15. An adult female (Yo) cracking coula nuts, while two juveniles observe her actions closely. A pile of coula nuts is visible on the right (*upward-pointing arrow*). A row of panda nuts that can be seen near the bottom-left corner (*left-pointing arrow*) was ignored by all three individuals present

4.2.4 Toward a Model for Cultural Innovation and Propagation

Together with our intensive observations of the development of nut-cracking skills at the individual level and patterns in the regional distribution of the behaviour, this experiment completes a three-way approach to tracing the emergence and propagation of cultural traditions in wild chimpanzee communities. As the regional survey has shown, ecological factors cannot fully account for the observed patterns in the utilization of different target species for cracking in communities adjacent to Bossou. In addition, intensive longitudinal study of developmental processes suggests that the social environment contributes significantly to the individual acquisition of the skill. There are two main ways in which such non-genetically based, socially transmitted behaviours can first appear in a community: through invention by one or more individuals independently, or through the arrival in the group of an immigrant possessing knowledge gained in her natal group.

We believe that our novel-nut experiments can contribute to our understanding of both these processes, as well as to the subsequent diffusion within the community that then serves to incorporate the behaviour into the group's repertoire. First, the contrast between the initial presentation of coula and panda nuts suggests that the actions of a knowledgeable immigrant (which Yo quite likely was) can have an effect on other group members' reactions to unfamiliar or neglected items in the environment. Yo's behaviour toward coula elicited high levels of interest among individuals in all age groups, such that most proceeded to at least investigate the novel nuts. Because our analysis shows that adult observers pay attention almost exclusively to the actions of other adults in the group, novel behaviours are likely to spread to them from adult performers only. That no individual at Bossou seemed to be familiar with panda nuts meant that adult chimpanzees—more conservative than juveniles or infants—showed only a transient interest in these nuts and abandoned them after a single successful attempt. Possibly, these adults gauged that the amount of effort that went into cracking panda nuts, the toughest of the three nuts by far (Boesch and Boesch 1983), was too high compared to the rewards gained (one of the adults was observed to strike the same panda nut 78 times consecutively before she was able to crack it, changing her hammer eight times and her anvil four times in the process!). Meanwhile, the more persistent interest shown by juveniles towards panda nuts, as well as the generally higher levels of interest in coula nuts from the start, might suggest that newly invented behaviours may be most likely to originate in this group. Bearing in mind patterns of conspecific observation, such new inventions would then be likely to spread horizontally to other juveniles or downwards to infants, but not upwards to adults.

Over the five different years of presentation, coula-nut cracking seems to have been assimilated by members of the Bossou community, even though encounters with these nuts were brief and occurred as much as 3 to 4 years apart. Most individuals now crack coula nuts without any form of prior exploratory behaviour. It is difficult to predict what the fate of panda-nut cracking would be if these nuts were reintroduced at Bossou. Without a knowledgeable adult as a model for observation but given sustained interest among juveniles, it may be that this nut would also come to be accepted over time, albeit less quickly, through the maturation of younger innovators and their contemporaries.

It may be reasonable to suggest that the rates at which migration, innovation, and within-community transmission take place will ultimately influence how quickly novel cultural traditions are assimilated by wild chimpanzee communities. The finding that communities that share migrants do not necessarily possess identical behavioural repertoires confirms that transmission within communities plays an indispensable role in the maintenance of such traditions. Exploring the channels through which information travels from knowledgeable individuals within and across generations will illuminate how community-specific behaviours come to be established. The emergence of "cultural zones" (Matsuzawa et al. 2001; Biro et al. 2003) where sets of neighbouring communities come to develop similar but not necessarily identical behavioural traditions may be the result of such interplay between inter-community migration and within-community propagation.

5 Future Perspectives

This chapter has provided an overview of results obtained through the long-term study of two chimpanzee tool-using behaviours at Bossou. Observations and field experiments continue at this site year after year, and neighbouring communities continue to be explored in efforts to build a comprehensive picture of the cultural life of chimpanzees in this corner of Africa (Humle and Matsuzawa 2001, 2002, 2004). Genetic analyses of these populations are also helping to illuminate local migration patterns (Shimada et al. 2004) and thus possible channels for the flow of knowledge between communities.

Current research at the outdoor laboratory is examining further the developmental and social aspects of skill acquisition in both nut cracking and the use of leaves for drinking water. One of the focuses of the work concerns object manipulation during the performance of these tasks (M. Hayashi et al., in preparation). For example, chimpanzees occasionally adjust the position of their anvil, rotating or rolling the stone before placing the next nut on the upper surface. Whether these adjustments are based on an understanding of the properties of the objects involved (such as a slanted upper surface not being able to support a round object on top) are being examined by presenting chimpanzees with ready-made anvil–hammer sets in which the anvil is positioned incorrectly. Fine-scale analysis of the actions performed by individuals when encountering such situations will shed light on chimpanzees' perception of the physical world.

Meanwhile, at the Primate Research Institute of Kyoto University, researchers are investigating tool use by chimpanzees in a captive setting, including experiments on nut cracking (Hayashi et al. 2005), the use of leaves for drinking (Tonooka et al. 1997; Celli et al. 2004), and the transmission of such behaviours from mother to infant (Hirata and Celli 2003). In addition, a new series of experiments that began with the birth of three chimpanzee infants in 2000 are examining various issues related to cognitive development (Matsuzawa 2003; see also other chapters in this book). Many of these are intricately related to the developmental aspects discussed in the present chapter. For example, Hayashi and Matsuzawa (2003) have shown that in infant chimpanzees' free manipulation of objects, the appearance of "inserting" actions precedes that of "stacking". Infants from the age of about 1 year begin to insert one object into another, but it takes another year before they begin to spontaneously stack objects on top of one another. The significance of this finding becomes clear when data from the wild are considered in parallel. The use of leaves for drinking water is essentially an inserting action whereas nut cracking requires precise stacking of objects: much as the captive data would predict, in infants at Bossou the former emerges earlier than the latter.

Field and laboratory work can thus go hand-in-hand to shed light on various aspects of wild chimpanzee cognition and behaviour. The combination of such parallel efforts is likely to emerge as a whole greater than the sum of its parts.

Acknowledgements

We thank the Direction National de la Recherche Scientifique et Technique, République de Guinée, for permission to conduct field work at Bossou. The research was supported by Grants-in-Aid for scientific research from the Ministry of Education, Science, Sports, and Culture of Japan (grants 07102010, 12002009, 10CE2005, and the 21COE program). We would like to express our gratitude to the following people who have been involved in research at Bossou over the years and thus contributed to the data reported here: Gen Yamakoshi, Rikako Tonooka, Noriko Inoue-Nakamura, Tatyana Humle, Hiroyuki Takemoto, Satoshi Hirata, Gaku Ohashi, Makoto Shimada, Takao Fushimi, Osamu Sakura, Masako Myowa-Yamakoshi, Maura Celli, Tomomi Ochiai, and Misato Hayashi. We are also grateful to Yukimaru Sugiyama, who began the study of wild chimpanzees at Bossou, and to Guanou Goumy, Tino Camara, Paquillé Cherif, Pascal Goumy, Marcel Doré, Bonifas Zogbila, Jiles Doré, and Henry Gbelegbe for assistance in the field.

References

Baldwin JP, Sabater PIJ, McGrew WC, Tutin CEG (1981) Comparisons of nests made by different populations of chimpanzees (*Pan troglodytes*). Primates 22:474–486

Biro D, Inoue-Nakamura N, Tonooka R, Yamakoshi G, Sousa C, Matsuzawa T (2003) Cultural innovation and transmission of tool use in wild chimpanzees: evidence from field experiments. Anim Cogn 6:213–223

Boesch C (1991). Teaching among wild chimpanzees. Int J Primatol 41:530–532

Boesch C, Boesch H (1983) Optimisation of nut cracking with natural hammers by wild chimpanzees. Behaviour 83:265–286

Boesch C, Boesch H (1984) Mental map in wild chimpanzees: an analysis of hammer transports for nut cracking. Primates 25:160–170

Boesch C, Boesch H (1990) Tool use and tool making in wild chimpanzees. Folia Primatol 54:86–99

Boesch C, Marchesi P, Marchesi N, Fruth B, Joulian F (1994) Is nut cracking in wild chimpanzees a cultural behavior? J Hum Evol 26:325–338

Celli M, Hirata S, Tomonaga M (2004) Socioecological influences on tool use in captive chimpanzees. Int J Primatol 25:1267–1281

de Waal FBM (2001) The ape and the sushi master. Basic Books, New York

Ghiglieri MP (1984) The chimpanzees of Kibale Forest. Columbia University Press, New York

Goodall J (1964) Tool using and aimed throwing in a community of free-living chimpanzees. Nature (Lond) 201:1264–1266

Hayashi M, Matsuzawa T (2003) Cognitive development in object manipulation by infant chimpanzees. Anim Cogn 6:225–233

Hayashi M, Mizuno Y, Matsuzawa T (2005) How does stone-tool use emerge? Introduction of stones and nuts to naïve chimpanzees in captivity. Primates 46:91–102

Hirata S, Celli ML (2003) Role of mothers in the acquisition of tool use behaviours by captive infant chimpanzees. Anim Cogn 6:235–244

Hirata S, Morimura N (2000) Naïve chimpanzees' (*Pan troglodytes*) observation of experienced conspecifics in a tool-using task. J Comp Psychol 114:291–296

Hirata S, Myowa M, Matsuzawa T (1998) Use of leaves as cushions to sit on wet ground by chimpanzees. Am J Primatol 44:215–220

Hirata S, Watanabe K, Kawai M (2001a) "Sweet-potato washing" revisited. In: Matsuzawa T (ed) Primate origins of human cognition and behaviour. Springer, Tokyo, pp 487–508

Hirata S, Yamakoshi G, Fujita S, Ohashi G, Matsuzawa T (2001b) Capturing and toying with hyraxes (*Dendrohyrax dorsalis*) by wild chimpanzees (*Pan troglodytes*) at Bossou, Guinea. Am J Primatol 53:93–97.

Humle T, Matsuzawa T (2001) Behavioural diversity among the wild chimpanzee populations of Bossou and neighbouring areas, Guinea and Cote d'Ivoire, West Africa. Folia Primatol 72:57–68

Humle T, Matsuzawa T (2002) Ant-dipping among the chimpanzees of Bossou, Guinea, and some comparisons with other sites. Am J Primatol 58:133–148

Humle T, Matsuzawa T (2004) Oil palm use by adjacent communities of chimpanzees at Bossou and Nimba Mountains, West Africa. Int J Primatol 25:551–581

Inoue-Nakamura N, Matsuzawa T (1997) Development of stone tool use by wild chimpanzees (*Pan troglodytes*). J Comp Psychol 111:159–173

Kawai M (1965) Newly acquired pre-cultural behaviour of the natural troop of Japanese monkeys on Koshima islet. Primates 6:1–30

Lonsdorf EV, Hopkins WD (2005) Wild chimpanzees show population-level handedness for tool use. Proc Natl Acad Sci U S A 102:12634–12638

Matsuzawa T (1991) Chimpanzee mind (in Japanese). Iwanami, Tokyo

Matsuzawa T (1994) Field experiments on use of stone tools by chimpanzees in the wild. In: Wrangham R, McGrew W, de Waal F, Heltne P (eds) Chimpanzee cultures. Harvard University Press, Cambridge, pp 351–370

Matsuzawa T (1996) Chimpanzee intelligence in nature and captivity: isomorphism of symbol use and tool use. In: McGrew WC, Marchant LF, Nishida T (eds) Great ape societies. Cambridge University Press, Cambridge, pp 196–209

Matsuzawa T (1997) The death of an infant chimpanzee at Bossou, Guinea. Pan Africa News 4:4–6

Matsuzawa T (1999) Communication and tool use in wild chimpanzees: cultural and social contexts. In: Hauser M, Konishi M (eds) The design of animal communication. MIT Press, Cambridge, pp 645–671

Matsuzawa T (2003) The Ai project: historical and ecological contexts. Anim Cogn 6:199–211

Matsuzawa T, Yamakoshi G (1996) Comparison of chimpanzee material culture between Bossou and Nimba, West Africa. In: Russon AE, Bard K, Taylor Parker S (eds) Reaching into thought: the minds of the great apes. Cambridge University Press, Cambridge, pp 211–232

Matsuzawa T, Takemoto H, Hayakawa S, Shimada M (1999) Diecke forest in Guinea. Pan Africa News 6:10–11

Matsuzawa T, Biro D, Humle T, Inoue-Nakamura N, Tonooka R, Yamakoshi G (2001) Emergence of culture in wild chimpanzees: education by master-apprenticeship. In: Matsuzawa T (ed) Primate origins of human cognition and behaviour. Springer, Tokyo, pp 557–574

Matsuzawa T, Humle T, Koops K, Biro D, Hayashi M, Sousa C, Mizuno Y, Kato A, Yamakoshi G, Ohashi G, Sugiyama Y, Kourouma M (2005) Wild chimpanzees at Bossou-Nimba: deaths through a flu-like epidemic in 2003 and the Green Corridor Project (in Japanese with English summary). Primate Res 20:45–55

McGrew WC (1977) Socialization and object manipulation of wild chimpanzees. In: Chevalier-Skolnikoff S, Poirier FE (eds) Primate biosocial development. Garland, New York, pp 304–309

McGrew WC (1992) Chimpanzee material culture: implications for human evolution. Cambridge University Press, Cambridge

McGrew WC (2004) The cultured chimpanzee: reflections in cultural primatology. Cambridge University Press, Cambridge

McGrew WC, Tutin CEG (1978) Evidence for a social custom in wild chimpanzees. Man 13:234–251

Nishida T (1990) A quarter century of research in the Mahale Mountains: an overview. In: Nishida T (ed) The chimpanzees of the Mahale Mountains: sexual and life history. University of Tokyo Press, Tokyo, pp 3–35
Quiatt D, Kiwede ZT (1994) Leaf sponge drinking by the Budongo forest chimpanzees. Am J Primatol 33:236 (abstract)
Sakura O, Matsuzawa T (1991) Flexibility of wild chimpanzee nut cracking behavior using stone hammers and anvils: an experimental analysis. Ethology 87:237–248
Shimada MK, Hayakawa S, Humle T, Fujita S, Hirata S, Sugiyama Y, Saitou N (2004) Mitochondrial DNA genealogy of chimpanzees in the Nimba Mountains and Bossou, West Africa. Am J Primatol 64:261–275
Sugiyama Y (1995) Drinking tools of wild chimpanzees at Bossou. Am J Primatol 37:263–269
Sugiyama Y (1998) Local variation of tool-using repertoire in wild chimpanzees. In: Nishida T (ed) Comparative study of the behaviour of the genus *Pan* by compiling video ethogram. Nissindo, Kyoto, pp 82–91
Sugiyama Y (1999) Socioecological factors of male chimpanzee migration at Bossou, Guinea. Primates 40:61–68
Sugiyama Y, Koman J (1979) Tool-using and making behavior in wild chimpanzees at Bossou, Guinea. Primates 20:513–524
Tomasello M (1996) Do apes ape? In: Heyes CM, Galef BG (eds) Social learning in animals: the roots of culture. Academic Press, San Diego, pp 319–346
Tonooka R (2001) Leaf-folding behaviour for drinking water by wild chimpanzees (*Pan troglodytes verus*) at Bossou, Guinea. Anim Cogn 4:325–334
Tonooka R, Tomonaga M, Matsuzawa T (1997) Acquisition and transmission of tool use and making for drinking juice in a group of captive chimpanzees (*Pan troglodytes*). Jpn Psychol Res 39:253–265
Whiten A, Goodall J, McGrew WC, Nishida T, Reynolds V, Sugiyama Y, Tutin CEG, Wrangham RW, Boesch C (1999) Cultures in chimpanzees. Nature (Lond) 399:682–685
Whiten A, Goodall J, McGrew WC, Nishida T, Reynolds V, Sugiyama Y, Tutin CEG, Wrangham RW, Boesch C (2001) Charting cultural variation in chimpanzees. Behaviour 138:1481–1516
Wrangham RW (1992) Living naturally: aspects of wild chimpanzee management. In: Erwin J, Landon JC (eds) Chimpanzee observation and public health. Diagnon, Rockville, pp 71–81

Subject Index

A

D

F

G

H

I

J

K

L

M

N

O

P

Q

R

S

T

U

V

W